Sources, Sinks and Sustainability

Source–sink theories provide a simple yet powerful framework for understanding how the patterns, processes and dynamics of ecological systems vary and interact over space and time.

Integrating multiple research fields, including population biology and landscape ecology, this book presents the latest advances in source–sink theories, methods and applications in the conservation and management of natural resources and biodiversity. The interdisciplinary team of authors uses detailed case studies, innovative field experiments and modeling, and syntheses to incorporate source–sink ideas into research and management, and explores how sustainability can be achieved in today's increasingly fragile human-dominated ecosystems.

Providing a comprehensive picture of source–sink research as well as tangible applications to real-world conservation issues, this book is ideal for graduate students, researchers, natural-resource managers and policy makers.

JIANGUO (JACK) LIU is Rachel Carson Chair in Sustainability and University Distinguished Professor as well as Director of the Center for Systems Integration and Sustainability (CSIS) at Michigan State University. His major research interests include landscape ecology, conservation ecology, and the integration of ecology with social sciences and policy for understanding and achieving sustainability.

VANESSA HULL is a PhD student of Jianguo (Jack) Liu at the Center for Systems Integration and Sustainability (CSIS) at Michigan State University. Her research interests include animal behavior and ecology, landscape ecology and conservation biology.

ANITA T. MORZILLO is an Assistant Professor, Senior Research, in the Department of Forest Ecosystems and Society at Oregon State University. Her major research interests include wildlife ecology and management, human dimensions of natural resources, landscape ecology, systems ecology, urban ecology, and integrating ecology and social science for natural resource management.

JOHN A. WIENS is Chief Conservation Science Officer at PRBO Conservation Science, former Lead and Chief Scientist at The Nature Conservancy, and former Distinguished Professor at Colorado State University. His broad interests include landscape ecology and the ecology of birds and insects in arid environments. His current scientific work focuses on critical issues of conservation in a rapidly changing environment resulting from climate change, economic globalization, land use change and human demands on natural ecosystems.

Cambridge Studies in Landscape Ecology

Cambridge Studies in Landscape Ecology presents synthetic and comprehensive examinations of topics that reflect the breadth of the discipline of landscape ecology. Landscape ecology deals with the development and changes in the spatial structure of landscapes and their ecological consequences. Because humans are so tightly tied to landscapes, the science explicitly includes human actions as both causes and consequences of landscape patterns. The focus is on spatial relationships at a variety of scales, in both natural and highly modified landscapes, on the factors that create landscape patterns, and on the influences of landscape structure on the functioning of ecological systems and their management. Some books in the series develop theoretical or methodological approaches to studying landscapes, while others deal more directly with the effects of landscape spatial patterns on population dynamics, community structure, or ecosystem processes. Still others examine the interplay between landscapes and human societies and cultures.

The series is aimed at advanced undergraduates, graduate students, researchers and teachers, resource and land use managers, and practitioners in other sciences that deal with landscapes.

The series is published in collaboration with the International Association for Landscape Ecology (IALE), which has Chapters in over 50 countries. IALE aims to develop landscape ecology as the scientific basis for the analysis, planning and management of landscapes throughout the world. The organization advances international cooperation and interdisciplinary synthesis through scientific, scholarly, educational and communication activities.

Other titles in series

Globalisation and Agricultural Landscapes
Edited by Jørgen Primdahl, Simon Swaffield
978-0-521-51789-8 (hardback)
978-0-521-73666-4 (paperback)

Key Topics in Landscape Ecology
Edited by Jianguo Wu, Richard J. Hobbs
978-0-521-85094-0 (hardback)
978-0-521-61644-7 (paperback)

Issues and Perspectives in Landscape Ecology
Edited by John A. Wiens, Michael R. Moss
978-0-521-83053-9 (hardback)
978-0-521-53754-4 (paperback)

Ecological Networks and Greenways
Edited by Rob H. G. Jongman, Gloria Pungetti
978-0-521-82776-8 (hardback)
978-0-521-53502-1 (paperback)

Transport Processes in Nature
William A. Reiners, Kenneth L. Driese
978-0-521-80049-5 (hardback)
978-0-521-80484-4 (paperback)

Integrating Landscape Ecology into Natural Resource Management
Edited by Jianguo Liu, William W. Taylor
978-0-521-78015-5 (hardback)
978-0-521-78433-7 (paperback)

EDITED BY
JIANGUO LIU
MICHIGAN STATE UNIVERSITY
VANESSA HULL
MICHIGAN STATE UNIVERSITY
ANITA T. MORZILLO
OREGON STATE UNIVERSITY
JOHN A. WIENS
PRBO CONSERVATION SCIENCE AND UNIVERSITY OF WESTERN AUSTRALIA

Sources, Sinks and Sustainability

CAMBRIDGE
UNIVERSITY PRESS

CAMBRIDGE UNIVERSITY PRESS
Cambridge, New York, Melbourne, Madrid, Cape Town,
Singapore, São Paulo, Delhi, Tokyo, Mexico City

Cambridge University Press
The Edinburgh Building, Cambridge CB2 8RU, UK

Published in the United States of America by Cambridge University Press, New York

www.cambridge.org
Information on this title: www.cambridge.org/9780521199476

First published 2011

Printed in the United Kingdom at the University Press, Cambridge

A catalogue record for this publication is available from the British Library

Library of Congress Cataloguing in Publication data
Sources, sinks, and sustainability / [edited by] Jianguo Liu ... [et al.].
p. cm. – (Cambridge studies in landscape ecology)
Includes bibliographical references and index.
ISBN 978-0-521-19947-6 (hardback) – ISBN 978-0-521-14596-1 (paperback)
1. Animal populations–Research. 2. Habitat selection. 3. Animals–Dispersal. 4. Ecological heterogeneity. 5. Ecosystem management.
I. Liu, Jianguo, 1963–
QL752.S677 2011
577.8′8–dc23
2011011504

ISBN 978-0-521-19947-6 Hardback
ISBN 978-0-521-14596-1 Paperback

Additional resources for this publication at www.cambridge.org/9780521199476

Contents

Contributors

Mark C. Andersen
Department of Fish Wildlife and Conservation Ecology, New Mexico State University, Las Cruces, NM 88003, USA

Craig W. Benkman
Department of Zoology and Physiology, University of Wyoming, Laramie, WY 82071–3166, USA

Eric J. Berry
Biology Department, Saint Anselm College, Manchester, NH 03102, USA

Ellen I. Damschen
Department of Zoology, University of Wisconsin, Madison, WI 53706, USA

Brent J. Danielson
Department of Ecology, Evolution, and Organismal Biology, Iowa State University, 253 Bessey Hall, Ames, IA 50011, USA

Ben Davies
University of Aberdeen, Aberdeen Centre for Environmental Sustainability, (ACES) Tillydrone Avenue, Aberdeen AB24 2TZ, UK

Jeffrey M. Diez
School of Natural Resources and Environment, Dana Building, 440 Church Street, Ann Arbor, MI 48109–1041, USA

John M. Drake
Odum School of Ecology, University of Georgia, 151 Ecology Building, Athens, GA 30602, USA

John B. Dunning Jr.
Department of Forestry and Natural Resources, Purdue University, 195 Marsteller Street, West Lafayette, IN 47907–2033, USA

Bryan A. Endress
Center for Conservation and Research for Endangered Species, Zoological Society of San Diego, Escondido, CA 92027, USA

Matthew A. Etterson
US Environmental Protection Agency, Mid-Continent Ecology Division, 6201 Congdon Boulevard, Duluth, MN 55804, USA

Matthew R. Falcy
Department of Ecology, Evolution, and Organismal Biology, Iowa State University, 253 Bessey Hall, Ames, IA 50011, USA

Jennifer M. Fraterrigo
Department of Natural Resources and Environmental Sciences, University of Illinois, 1102 South Goodwin Avenue, Urbana, IL 61801, USA

Itamar Giladi
Department of Life Sciences, Ben-Gurion University of the Negev, 84105 Beer-Sheva, Israel

Alessandro Gimona
The James Hutton Institute, Craigiebuckler, Aberdeen AB15 8QH, UK

David L. Gorchov
Miami University, Department of Botany, 336 Pearson Hall, Oxford, OH 45056, USA

Russell Greenberg
Smithsonian Migratory Bird Center, National Zoological Park, PO Box 37012–MRC 5503, Washington, DC 20013, USA

Nick M. Haddad
Department of Biology, Box 7617, North Carolina State University, Raleigh, NC 27695–7617, USA

Andrew Hansen
Department of Ecology, College of Letters and Science, Montana State University – Bozeman, PO Box 173460, Bozeman, MT 59717–3460, USA

Robert D. Holt
Department of Biology, University of Florida, 111 Bartram, PO Box 118525, Gainesville, FL 32611–8525, USA

Jeffrey P. Hoover
Illinois Natural History Survey, Institute of Natural Resource Sustainability, University of Illinois at Urbana-Champaign, Champaign, IL 61820, USA

Alexandra Houston
4758 Soria Drive, San Diego, CA, 92115, USA

Brian Hudgens
Institute for Wildlife Studies, PO Box 1104, Arcata, CA 95518, USA

Vanessa Hull
Center for Systems Integration and Sustainability, Michigan State University, 115 Manly Miles Building, East Lansing, MI 48823, USA

Ferenc Jordán
The Microsoft Research – University of Trento, Centre for Computational and Systems Biology, Piazza Manci 17, Trento 38123, Italy

Douglas J. Levey
Department of Biology, University of Florida, Gainesville, FL 32611–8525, USA

Romuald N. Lipcius
Virginia Institute of Marine Science, The College of William and Mary, 1208 Greate Road, Gloucester Point, VA 23062, USA

Jianguo Liu
Center for Systems Integration and Sustainability, 1405 S. Harrison Road, Suite 115 Manly Miles Building, Michigan State University, East Lansing, MI 48823, USA

Stephen F. Matter
Department of Biological Sciences, 1402 Crosley Tower, University of Cincinnati, Cincinnati, OH 45221–0006, USA

Douglas W. Morris
Department of Biology, Lakehead University, 955 Oliver Road, Thunder Bay, ON, Canada P7B 5E1

Anita T. Morzillo
Oregon State University, Department of Forest Ecosystems and Society, 321 Richardson Hall, Corvallis, OR 97331, USA

Brian J. Olsen
5751 Murray Hall, School of Biology and Ecology, University of Maine, Orono, ME 04469, USA

John L. Orrock
Department of Zoology, University of Wisconsin, Madison, WI 53706

Gilberto Pasinelli
Swiss Ornithological Institute, CH-6204 Sempach, Switzerland, and, Institute of Evolutionary Biology and Environmental Studies, University of Zurich, CH-8057 Zurich, Switzerland

Scott M. Pearson
Department of Natural Sciences, Mars Hill College, Mars Hill, NC 28754, USA

David K. Person
Alaska Department of Fish and Game, Division of Wildlife Conservation, Ketchikan, AK 99901, USA

J. Gary Polhill
The James Hutton Institute, Craigiebuckler, Aberdeen AB15 8QH, UK

H. Ronald Pulliam
Odum School of Ecology, University of Georgia, Athens, GA 30605, USA

Juliet R. C. Pulliam
Department of Biology, PO Box 118525, University of Florida, Gainesville, FL 32611-8525, USA
Formerly of:
Fogarty International Center, National Institutes of Health, Bethesda, MD 20892, USA

Sanjay Pyare
Program in Environmental Sciences, University of Alaska Southeast, Juneau, AK 99801, USA

Gina M. Ralph
Virginia Institute of Marine Science, The College of William and Mary, 1208 Greate Road, Gloucester Point, VA 23062, USA

Scott K. Robinson
Florida Museum of Natural History, University of Florida, PO Box 117800, Gainesville, FL 32611, USA

Jens Roland
Department of Biological Sciences, University of Alberta, Edmonton, Alberta, Canada T6G 2E9

Jonathan P. Runge
Colorado Division of Wildlife, 317 West Prospect Road, Fort Collins, CO 80526, USA

Daniel M. Scheiman
Audubon Arkansas, 4500 Springer Boulevard, Little Rock, AR 72206, USA

Karin Schiegg
Psychiatric University Clinic, Seinaustrasse 9, CH-8001 Zurich, Switzerland, and, Institute of Evolutionary Biology and Environmental Studies, University of Zurich, Winterthurerstrasse 190, CH-8057 Zurich, Switzerland

W. Gregory Shriver
257 Townsend Hall, Department of Entomology and Wildlife Ecology, University of Delaware, Newark, Delaware 19716–2160, USA

Adam M. Siepielski
Department of Biological Sciences, Dartmouth College, 7 Lucent Drive, Centerra Biolabs, Lebanon, NH 03766, USA

Winston P. Smith
USDA Forest Service, Pacific Northwest Research Station, Forestry Sciences Laboratory, 3625 93rd Avenue, SW, Olympia, WA 98512, USA

Joshua J. Tewksbury
Department of Biology, University of Washington, Box 351800, 24 Kincaid Hall, Seattle, WA 98195–1800, USA

Beatrice Van Horne
Pacific Northwest Research Station, USDA Forest Service, Corvallis, OR 97331, USA

Aimee J. Weldon
Potomac Conservancy, 8601 Georgia Avenue, Suite 612, Silver Spring, MD 20910, USA

John A. Wiens
PRBO Conservation Science, 3820 Cypress #11, Petaluma, CA 94954, USA

Stephen R. Wing
Department of Marine Science, University of Otago, 310 Castle Street, Dunedin, Aotearoa, New Zealand

Preface

Organisms and populations are discontinuously distributed in space and change over time. As a result, conserving and managing ecological systems requires an understanding of how these systems and their patterns, processes and dynamics vary and interact in space and time. More than two decades ago, H. Ronald Pulliam developed a conceptual framework of spatial population dynamics to address this need. In his 1988 paper ("Sources, sinks, and population regulation," *American Naturalist* 132: 652–661), Pulliam created a framework that envisioned that populations in "sink" (poor) habitats would rely on inputs from "source" (good) habitats in order to persist. The dynamics of population segments across heterogeneous landscapes were linked. This simple yet powerful framework has inspired numerous studies and has provided the foundation for rapid advances in ecological theory and practice.

To reflect upon and synthesize the development of thinking and research inspired by Pulliam's framework, a symposium on "Sources, Sinks, and Sustainability across Landscapes" was held at the 2008 annual conference of the US Regional Association of the International Association for Landscape Ecology (US-IALE) in Wisconsin, USA. The symposium, organized in honor of Pulliam's retirement, amply illustrated his many contributions to ecology, animal behavior, evolution, and other fields, through his former roles as Regents Professor and Director of University of Georgia's Institute of Ecology, Director of the National Biological Service, Science Advisor to the Secretary of the Interior, and the President of the Ecological Society of America. The 30 presenters from around the world included Pulliam's former students and postdoctoral associates, as well as other leading scholars who have been influenced by Pulliam's work.

This book draws on the presentations at the symposium and the excitement they generated to integrate source–sink ideas into research and management. It discusses how sustainability can be achieved in today's increasingly fragile and human-dominated ecosystems, and consists of five interrelated sections. The first section contains an introductory chapter that highlights the impact of Pulliam's 1988 paper on the scientific and management communities. The chapter provides an overview of trends in the large volume of citations of Pulliam (1988) during the first 20 years since the paper's publication (1988–2008), and discusses the major contributions of the paper to ecological theory and natural resource management, as well as extensions of the source–sink concept to other disciplines.

The book then proceeds with three major sections, each with an overview and seven chapters, that address advances in source–sink theory, progress in source–sink methodology, and improvement in source–sink management. The section on source–sink theory presents recent advances in the theoretical framework originally put forth to characterize sources and sinks, namely with regard to novel implications for evolutionary theory and extensions of the source–sink concept to characterizing the ever-increasing human impacts on ecosystems. The section on source–sink methodology explores new approaches for estimating demographic parameters for source–sink models, emerging modeling frameworks that capture source–sink dynamics across heterogeneous space, and original experiments that test fundamental aspects of source–sink theory. The section on source–sink management addresses applications of source–sink concepts to a number of important topics in natural resource management, including reserve design, marine and estuarine ecosystem protection, and impacts of habitat fragmentation and overharvesting of resources on populations and habitats.

The book concludes with a chapter synthesizing the preceding sections. Drawing upon insights from the literature and other chapters in this volume, the synthesis chapter describes four "realities" that may influence the utility of source–sink theory for conservation and resource management. These realities include the scientific and management need for detailed information that is difficult to obtain, the embedding of sources and sinks in heterogeneous landscapes, the dynamic nature of sources and sinks, and the importance of choosing appropriate scales when considering source–sink systems. The chapter also discusses the implications of these realities for making source–sink concepts operational in natural resource management.

The unifying theme of this book is a shared commitment to integrating multiple research fields, particularly population biology and landscape ecology. Key concepts in population biology such as survival, habitat selection, evolution, competition, and niche theory are explicitly linked to central issues in landscape ecology, including landscape structure, pattern, process, function, scale, and spatial and temporal dynamics. These complex interactions between sources and sinks drive the dynamics of populations across landscapes and have significant implications for understanding and managing both natural systems and coupled human–natural systems.

The 54 contributors to this book have conducted research throughout the world in a wide variety of landscapes, including agricultural systems, grasslands, forests, wetlands and marine systems. The study organisms include birds, fish, insects, mammals, trees and other plants. The topics are equally diverse: biodiversity, climate change, ecosystem services, invasive species, land cover and land use change, water availability, natural disasters, natural resource management, and sustainability. The advanced tools and methods used include remote sensing, geographic information systems, computer simulations, system modeling, and spatial statistics.

We have designed this book to inform and meet the needs of ecologists, population biologists, conservation biologists, natural-resource managers, policy makers, sustainability scholars, graduate students, and advanced undergraduate students. Source–sink concepts, however, extend far beyond species and their habitats. Related ideas are also found in physiology, carbon emissions and sequestration, air pollution, and the trading of goods and products between different locations around the globe. Thus, many ideas in this book will also be helpful to scientists and students in other disciplines.

There has been astounding progress in source–sink and related concepts in the past two decades. Future applications of these concepts will continue to develop. It is our hope that the information and novel approaches presented in this book can advance our understanding of source–sink dynamics and improve the management and conservation for the sustainability of ecological systems in a human-dominated world.

Jianguo Liu
Vanessa Hull
Anita T. Morzillo
John A. Wiens

Acknowledgments

First of all, we thank the contributors to this book for their time, effort, enthusiasm and cooperation in making this edited volume possible. The Organizing Committee of the 2008 annual conference of the US Regional Association of the International Association for Landscape Ecology (US-IALE) graciously included our symposium "Sources, Sinks and Sustainability across Landscapes: A Symposium in Honor of H. Ronald Pulliam" in the conference program. This symposium was the inspiration for this book. We appreciate the organizational support from Sarah Goslee (program chair of the 2008 US-IALE conference) as well as Monica Turner and Phil Townsend (local hosts of the conference). We are also grateful to the presenters and other participants at the symposium for their excellent presentations and lively discussion.

We gratefully acknowledge the following individuals who served as reviewers for chapters included in this volume: Niels Anten (Utrecht University, The Netherlands), Robert Askins (Connecticut College), Michael Barfield (University of Florida), Linda Beaumont (Macquarie University, Australia), Steven Beissinger (University of California at Berkeley), Matthew Betts (Oregon State University), Louis Botsford (University of California at Davis), David Boughton (NOAA Fisheries Service), Paul-Marie Boulanger (Institut pour un Développement Durable, Belgium), François Bousquet (CIRAD, France), Judith Bronstein (University of Arizona), Loren Burger (Mississippi State University), Kevin Crooks (Colorado State University), Diane Debinski (Iowa State University), Miguel Delibes (Spanish Council for Scientific Research (CSIC), Spain), John DiBari (Sonoran Institute), Jay Diffendorfer (USGS), Martin Drechsler (Helmholtz Centre for Environmental Research, Germany), Sam Droege (US Geological Survey), Curtis Flather (USDA Forest Service), Kathryn Flinn (McGill University, Canada), Mark Gibbs (Commonwealth Scientific and Industrial Research Organisation (CSIRO), Australia), Jacob Goheen (University of Wyoming), Richard Gomulkiewicz (Washington State

University), Andrew Gonzalez (McGill University, Canada), Antoine Guisan (University of Lausanne, Switzerland), Ilkka Hanski (University of Helsinki, Finland), Selina Heppell (Oregon State University), James Herkert (The Nature Conservancy), Robert Hilderbrand (University of Maryland), Robert Holt (University of Florida), Niclas Jonzén (Lunds University, Sweden), Ronen Kadmon (Hebrew University, Israel), Tadeusz Kawecki (University of Lausanne, Switzerland), Michael Kearney (University of Melbourne, Australia), William Kristan III (California State University, San Marcos), Mikko Kuussaari (Finnish Environment Institute, Finland), Joshua Lawler (University of Washington), Shawn Leroux (McGill University, Canada), Susan Loeb (USDA Forest Service, Clemson University), Todd Lookingbill (University of Richmond), Brian Maurer (Michigan State University), Nancy McIntyre (Texas Tech University), Emily Minor (University of Illinois at Chicago), William Newmark (Utah Museum of Natural History), Reed Noss (University of Central Florida), Craig Pease (Vermont Law School), Julien Pottier (University of Lausanne, Switzerland), Larkin Powell (University of Nebraska–Lincoln), Ronald Pulliam (University of Georgia), Seth Riley (National Park Service), Jeanne Robertson (University of Idaho), Manojit Roy (University of Florida), Santiago Saura Martínez de Toda (University of Lleida, Spain), Robert Scheller (Portland State University), Robert Schooley (University of Illinois), Vesa Selonen (University of Turku, Finland), Nicholas Shears (University of California at Santa Barbara), Jonathan Silvertown (The Open University, UK), Susan Skagen (US Geological Survey), Tamara Ticktin (University of Hawaii at Manoa), Dean Urban (Duke University), Beatrice Van Horne (USDA Forest Service), Karl Vernes (University of New England, Australia), Jeffrey Walters (Virginia Polytechnic Institute and State University), Robert Warren II (Yale University), Michael Wilberg (University of Maryland), Kimberly With (Kansas State University), George Wittemyer (Colorado State University), and Douglas Yu (Kunming Institute of Zoology, Chinese Academy of Sciences). Their efforts to provide constructive comments for authors, sometimes multiple times, greatly helped improve the quality of this book.

We are indebted to Alan Crowden at the British Ecological Society, Lynette Talbot, Zewdi Tsegai, and others at Cambridge University Press for their tireless efforts in preparing this book for publication. We also thank Sue Faivor, Michael Harris, Blake House, Michael Hoxsey, Shuxin Li, and Danielle Truesdell from the Center for Systems Integration and Sustainability (CSIS) at Michigan State University for providing administrative and technical assistance for this book.

We also thank the National Science Foundation, National Aeronautics and Space Administration, John Simon Guggenheim Memorial Foundation, and Michigan Agricultural Experiment Station for financial support.

Part I

Introduction

1

Impact of a classic paper by H. Ronald Pulliam: the first 20 years

Summary

The central message of Pulliam's classic paper, "Sources, sinks, and population regulation" (1988), was that population dynamics change across heterogeneous landscapes, and the persistence of populations in "sink" habitats relies on inputs from "source" habitats. Pulliam's paper has gained widespread attention from the scientific and natural resource management communities. Here, we first provide the context in which the paper was developed and illustrate the paper's overall impact during the past two decades. We then outline the contributions of Pulliam's paper to the theories underlying niche concept, population dynamics and distribution, and community structure. Furthermore, we briefly discuss how Pulliam's message has spread to other disciplines such as microbiology, economics, and public health. We also provide examples to demonstrate the paper's influence on sustainable natural resource management in issues such as control of invasive species, design of protected areas, and harvesting of resources. Considering the growing impact of Pulliam's work during the past 20 years, it is likely that this influential paper will continue to inspire scientific discovery and applications in the future.

Development of the paper and model structure

Twenty years after its publication, the highly cited paper "Sources, sinks, and population regulation" (Pulliam 1988) still resonates with scientists and managers in the field of ecology and beyond. This paper presented the first

Sources, Sinks and Sustainability, ed. Jianguo Liu, Vanessa Hull, Anita T. Morzillo and John A. Wiens. Published by Cambridge University Press.

comprehensive source–sink model of population dynamics across heterogeneous landscapes. In order to analyze the impact of this work, it is necessary to describe related concepts that had emerged in the field of ecology before Pulliam's (1988) publication.

Prior to the 1960s, ecological studies of population dynamics were largely non-spatial, such that populations were often treated as abstract entities independent of their surrounding environments. Incorporation of spatial context began with the emergence of new ideas that considered species–location relationships, such as island biogeography theory (MacArthur and Wilson 1963), which sought to understand the relationships between species richness and the degree of insularity of habitats. Shortly thereafter, the idea of a "metapopulation" was put forth, which encouraged ecologists to view species not as single entities, but as collections of local populations that differed from one another in their demographic characteristics (Levins 1969; Hanski and Gilpin 1991).

Also published during the same time period was a seminal work by Fretwell and Lucas (1970), which was the first comprehensive theoretical framework constructed to explain the relationship between territoriality and habitat distribution in animals. The authors adopted a novel approach to explore theory about habitat selection by explicitly accounting for multiple different habitat types as they related to successful animal reproduction. Ecologists then began to link the concepts of heterogeneity in landscapes and variability in demographic processes, and began to ask questions about the consequences of this linkage as related to population dynamics and community structure (Levin 1974; Wiens 1976; Holt 1987). Soon came a deeper understanding of species dispersal and how it allows populations to recover from local extinctions (Levin 1974; Fahrig and Merriam 1985).

The source–sink terminology adopted by Pulliam and others to characterize population dynamics across heterogeneous space was first suggested by Lidicker (1975). Lidicker described the phenomenon of a "dispersal sink," defined as a relatively unoccupied area where dispersed individuals congregate after being excluded from areas of relatively higher-quality habitat. Van Horne (1983) elaborated on implications of the "dispersal sink" phenomenon by explaining how individuals may settle in lower-quality sink habitats at higher densities than expected when nearby high-quality source habitats are fully occupied. Holt (1985) integrated the concepts explored by Lidicker and Van Horne into a two-patch predator–prey model that linked a source habitat and a sink habitat by dispersal of predators in search of prey. Using the two-patch predator–prey framework, Holt was the first to model distinct spatial and temporal patterns of population dynamics that resulted from dispersal within a landscape of varying habitat quality.

Pulliam (1988) expanded the conceptual frameworks introduced by Lidicker, Van Horne and Holt in order to model habitat selection by individuals across sources and sinks explicitly. The model presented in Pulliam (1988) was a conceptually simple difference equation model, initially developed by Cohen (1969), which assumed that a population in equilibrium would comply with the following structure:

$$b_j + i_j - d_j - e_j = (bide)_j = 0$$

where j is a population of interest and b, i, d, and e correspond to birth, immigration, death, and emigration, respectively. Within this conceptual framework, sources were defined as areas where birth exceeded death and emigration exceeded immigration at equilibrium. In contrast, sinks were defined as areas where death exceeded birth and immigration exceeded emigration at equilibrium.

Pulliam presented a simple example by quantifying the probability of adults and juveniles surviving a winter non-breeding season. The number of individuals alive at the end of winter was expressed as:

$$n_1(t+1) = P_A n_1(t) + P_J \beta_1 n_1(t) = \lambda_1 n_1$$

where P_A is probability of adult survival, P_J is probability of juvenile survival, β is the number of juveniles alive at the end of the previous breeding season, and λ is the finite rate of increase for the population. For instances that consider more than one habitat, the subscript 1 can be changed to reflect j habitats. In such cases, a source habitat would be characterized by $\lambda > 1$ and a sink habitat would have $\lambda < 1$.

Pulliam demonstrates how the model assumes that a population in a source habitat would increase at a rate of $\lambda_1 = P_A + P_J\beta_1$ until all sites are occupied, in which case individuals would emigrate to sinks. Because a sink is defined by supporting a population with a value of λ of <1, the model assumes that a local population in a sink cannot persist without immigration from a source. Pulliam suggests a useful application of this model for estimating equilibrium population size, which is estimated by:

$$n^*_2 = \hat{n}(\lambda_1 - 1)/(\lambda_2 - 1)$$

where $\hat{n}$ is the population size in the source habitat, λ_1 is the finite rate of increase of the population in the source habitat, and λ_2 is the finite rate of increase of the population in the sink habitat.

Overall, Pulliam's (1988) model provides a simple and useful framework for understanding population heterogeneity and sustainability across space, including mechanisms and consequences of dispersal driven by varying habitat quality. These concepts had previously emerged in various forms in the ecological

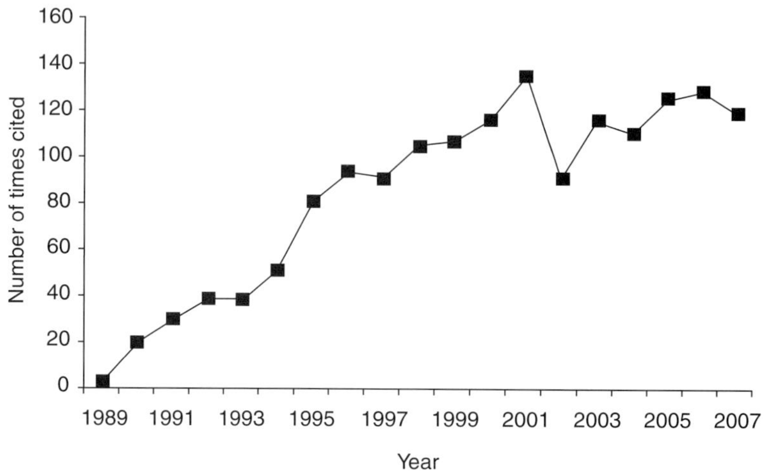

FIGURE 1.1.
Number of citations of Pulliam (1988) in scientific journals by year from 1989 to 2007.

literature (Levin 1974; Wiens 1976; Holt 1985). Pulliam helped to bring many ideas related to population dynamics processes together and framed them in a cohesive, yet conceptually simple model that resonates well with scientists and natural resource managers.

Overview of the impact

Following publication of his 1988 paper, Pulliam was quoted, in a "Research News" article published in *Science* entitled "Sources and sinks complicate ecology," as saying, "I hope that my model, like any new model, will stimulate new research and lead to the reexamination of existing datasets" (Lewin 1989: 478). It appears that the research and applications inspired by Pulliam's paper during the past 20 years have exceeded his expectations (H. R. Pulliam, personal communication).

Although it is impossible to evaluate the total impact of Pulliam (1988), part of the paper's contribution can be assessed by citations of it by other researchers. As of December 31, 2007, Pulliam (1988) had been cited 1,600 times in journals indexed by the Institute for Scientific Information (ISI). Of these citations, 457 explicitly mentioned sources and/or sinks in the abstract, title or keywords. The papers citing Pulliam (1988) ranged across 53 subject areas and 271 individual scientific journals, and were written by 3,109 individual authors from 1,039 different institutions in 76 countries around the world. The number of new citations per year has increased over time, with more than 100 citations each year for the most recent 5-year period of analysis (2003–2007; see Fig. 1.1).

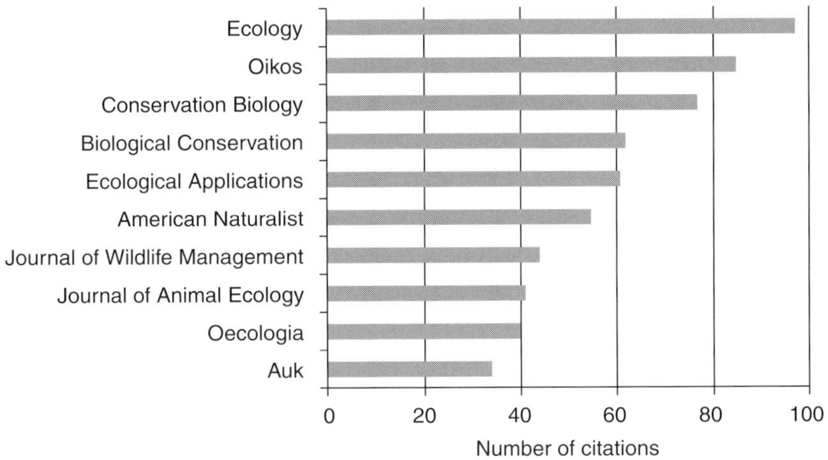

FIGURE 1.2.
Journals containing the greatest number of citations of Pulliam (1988) in the period 1989–2007.

Papers containing many of the 1,600 citations were published in the top peer-reviewed journals in the field of ecology (e.g., *Ecology*; see Fig. 1.2). Beyond ecology, the number of subject areas identified in the citations by ISI increased from 3 in 1989 to 53 in 2007. For example, citations rapidly widened in scope to other disciplines such as mathematics, law, medicine and international relations. In addition, a search in Google Books for the terms "Pulliam" and "1988" yielded 302 books containing citations of the work, of which 95 were published since the year 2000. Many of these were textbooks spanning disciplines including landscape ecology, marine biology, animal behavior, natural resource policy, and toxicology.

Impact on the theoretical development of ecology

Pulliam's paper has had a strong influence on the theoretical development of ecology. In the 1970s and 1980s, many researchers (e.g., Levin 1974; Wiens 1976) observed that there was a tendency in the ecological literature for scientists to analyze population demographic data from individual isolated locations. Lewin (1989) noted that adopting Pulliam's source–sink modeling framework meant that ecologists had to acknowledge that a high degree of complexity existed within ecological systems. As Pulliam stated, researchers were "looking at only half of the picture" (Lewin 1989: 477) and were perplexed by the fact that populations persisted in sink habitats even when the local birth rates were less than death rates (Lewin 1989). The ability of Pulliam's model to capture such demographic phenomena so succinctly helped to contribute to the growing literature on mathematically characterizing heterogeneous population demographics across habitats of varying quality.

Pulliam (1988) discussed three main implications of his work. The first was a clarification of the "niche concept," which has been discussed and debated for decades (Hutchinson 1958; Colwell and Futuyma 1971; Holt 1984) and is still a popular topic of research today. Pulliam drew upon the definition of the ecological niche put forth by Grinnell (1917) as "the range of values of environmental factors that are necessary and sufficient to allow a species to carry out its life history" (James *et al.* 1984: 18). The Grinnellian definition was further developed by Hutchinson (1958), who introduced the realized niche, which was smaller than the previously defined fundamental ecological niche. Pulliam (1988) noted that it is possible that the realized niche is larger than predicted because individuals occupy sink habitats that may not provide sufficient resources for a species to persist without receiving inputs from source habitats. Pulliam stated that "in such cases, it can be said that the fundamental niche is smaller than the realized niche" (Pulliam 1988: 659). Pulliam (1988) encouraged others to think more deeply about the meaning of the "niche" and how it may be a misleading indicator of suitable habitat for species, as originally defined. This realization, along with ideas from other scholars emerging around the same time (e.g., Holt and Gaines 1992), called for a more nuanced definition which recognized that the niche can vary over space and evolutionary time (see also Holt, Chapter 2, this volume).

The second main implication of the Pulliam (1988) paper was species conservation, which is dependent upon processes relating to population dynamics and distributions. Pulliam stated: "Given that a species may commonly occur and successfully breed in sink habitats, an investigator could easily be misled about the habitat requirements of a species" (Pulliam 1988: 659). In fact, in some cases, a large proportion of a population could exist in sink habitat. Pulliam's source–sink model, along with concepts explored in other literature emerging around that time (Fahrig and Merriam 1985; Hanski and Gilpin 1991), helped explain what were once perplexing patterns in population dynamics. For example, researchers came to appreciate that immigration allowed a local population to remain stable despite the fact that the overall population was declining (e.g., Brawn and Robinson 1996). Harrison (1991) highlighted how Pulliam's (1988) model added to the growing literature on metapopulation dynamics by illustrating how classifying local populations as sources or sinks could change predictions about extinction probability if sources are assumed to be extinction-resistant. In addition, habitat characteristics are dynamic, so that habitats considered sources at one point in time may change to become sinks at another time (and vice versa). Therefore, temporal variability in source–sink dynamics has come to be appreciated as an important factor determining sustainability of populations in such heterogeneous environments (Johnson 2004).

The third implication of Pulliam (1988) dealt with understanding and modeling community structure. Pulliam noted that "in extreme cases, the local assemblages of species may be an artifact of the type and proximity of neighboring habitats and have little to do with the resources and conditions at the study site" (Pulliam 1988: 660). He later realized that this phenomenon was not isolated to extreme instances, but may be true for most cases in nature and thus warrants further study (H. R. Pulliam, personal communication). This revelation, within the context of other related literature (Levin 1974; Holt 1984), inspired researchers to appreciate that the factors driving community dynamics often exist at a broader level than the local habitat. For instance, Danielson (1991) highlights how Pulliam's (1988) model provided a useful framework for linking competitive interactions between species to underlying variation in habitat quality when community dynamics are dependent on landscape structure.

Like all models, Pulliam's (1988) model is not without its limitations. One limitation is that the original model did not consider that individuals in sources could exclude individuals in sinks. Inspired in part by Pulliam (1988), Pulliam and Danielson (1991) proposed an alternative "ideal preemptive distribution" theory, such that individuals choose the best (source) habitat available that is not otherwise occupied and, in turn, preempt others from occupying it. They also found that the inability of individuals to find source habitat in times of abundant sink habitat could result in a threat to the entire population.

Another limitation of Pulliam's (1988) model is that it did not consider emigration from sinks to sources. Morris (1991) considered this possibility and questioned the model's assumption of dispersal as an evolutionarily stable strategy. He postulated that sink-to-source migration could make source-to-sink migration an evolutionarily stable strategy (see also Morris, Chapter 3, this volume), but otherwise may be a non-adaptive response to overpopulation in sources. Doncaster *et al.* (1997) proposed an alternative to the form of dispersal predicted by Pulliam's source–sink model by suggesting that, rather than being dominated by immigration from sources to sinks, dispersal could consist of balanced amounts of emigration and immigration among neighboring patches. A number of studies ensued that tested theoretical and empirical evidence for each of the two dispersal theories (Diffendorfer 1998; Tattersall *et al.* 2004). As illustrated above, many individuals inspired by Pulliam's (1988) idea were able to adapt it or expand upon the original form for application to other cases.

Applications to other scientific disciplines

The ideas put forward by Pulliam have influenced many other scientific disciplines. Beyond the ecology literature, Pulliam (1988) has been cited more often than other papers dealing with the source–sink concept published around the same time. Perhaps this is because Pulliam presented the idea in a conceptually simple and straightforward manner that made it accessible to individuals working in other fields, who did not have previous exposure to theoretical ecology.

The expansion of citations of Pulliam (1988) began within the natural sciences in the 1990s in genetics (e.g., Blumler *et al.* 1991), paleogeography (e.g., Holterhoff 1996), oceanography (e.g., Botsford *et al.* 1994) and, most notably, microbiology (Sokurenko *et al.* 2006). References to Pulliam's work are frequent in the recent microbiology literature focusing on virulence evolution in microorganisms (Sokurenko *et al.* 2006). Source–sink modeling has enabled microbiologists to understand how *E. coli* adapts to source and sink microhabitats within an individual host organism (Chattopadhyay *et al.* 2007), and how viruses that adapt to sink hosts become extinct when these hosts are depleted (Dennehy *et al.* 2006).

The ideas put forward in Pulliam (1988) have also been integrated into the literature of the social sciences, such as economics. For instance, Sanchirico and Wilen (1999) drew upon Pulliam (1988) in their research related to renewable resource exploitation. They created a spatially integrated bioeconomic model that incorporated landscape ecology concepts to explain economic behavior across heterogeneous space. This economic model was linked to a source–sink model similar to Pulliam's. Together, the two models helped to evaluate distribution of effort, a function of opportunity cost to run a fishery and available fish biomass. As illustrated by the models, the authors noted that characterizing sources or sinks requires consideration not only of the biological factors but also the economic factors within a system. Recognizing that dispersal of harvestable biomass within a system changes economic behavior represented a novel application of source–sink theory.

The ideas from Pulliam (1988) have also extended to the field of public health. Some examples include the management of human cases of plague in Madagascar (Duplantier *et al.* 2005) and polio in India (Grassly *et al.* 2006), which targeted source and sink foci of disease based on the prevalence and nature of spread through the human population. Thus, for disease surveillance, recognizing the heterogeneity of the landscape helped explain how disease, although believed to be eradicated, may seem to "hide" in pathogenic sinks. Additionally, Perron *et al.* (2007) examined how antibiotic resistance may be promoted by mutations induced by the dispersal of a pathogenic bacteria from sources to sinks. The extension of the concepts of sources and sinks to fields such as public health demonstrates the broad extent to which the message in Pulliam's paper has been applied to disciplines outside of ecology.

Applications to natural resource management

The essential message of Pulliam (1988) has been well received by the natural resource management community. Pulliam's paper has been more frequently cited in the management-related literature than other papers focusing on source–sink theory, again probably because of its accessibility to non-theoretical ecologists and its ease of application to real-world management issues.

In an automated search of the abstracts, keywords and titles of the 1,600 citations of Pulliam (1988), we found the words "policy," "management," and/or "conservation" used in 47% of the citations. Among the works that cited Pulliam (1988), some prominent themes included management of protected areas (258 articles), resource consumption (harvest; 110 articles), and invasive species (85 articles). We also found that applications have widened in scope over time to include broad management concerns such as climate change, ecosystem services, toxic pollution, and radioactive contamination.

Applications of Pulliam (1988) to protected areas management have included both terrestrial and marine systems. McCoy *et al.* (1999) presented an assessment of the Conservation Reserve Program for the management of grassland bird populations. They found that identifying reserves as source or sink habitats was useful in quantifying the impact of protection on the viability of the bird populations, and concluded that the reserves were sources for some species but sinks for others. Spatially explicit models designed by Crowder *et al.* (2000) suggested that placing no-take marine reserves in source habitats resulted in an increase in the fish population. In contrast, placement of no-take reserves in either randomly selected habitats or sink habitats resulted in no significant increase in the population compared with having no reserves. Thus, explicitly incorporating source–sink information could improve efforts to identify and establish reserves and sustain populations (see also Hansen, Chapter 16, this volume).

Efforts to manage resource consumption (harvest) have been informed by Pulliam (1988). Lundberg and Jonzen (1999) presented two-patch habitat models outlining multiple scenarios for harvesting a theoretical animal population under a maximum sustained-yield strategy across sources and sinks. The models highlighted that harvesting in either sources or sinks destabilized the population (Lundberg and Jonzen 1999). The highest yield was obtained by harvesting at a higher rate in the sink habitat than the source habitat (Lundberg and Jonzen 1999). Novaro *et al.* (2005) suggested that Pulliam's

(1988) source–sink model can provide managers with reliable tools for harvest quotas because the model explicitly captures the essence of hunting-induced "sinks" distributed across heterogeneous space that may affect the population demographics of game species.

To our knowledge, the first citation of Pulliam (1988) relating to invasive species was by Horvath *et al.* (1996), who characterized the source–sink dynamics of zebra mussel populations. The authors suggested that some streams were at greater risk of invasion by zebra mussels than previously believed because stream connectivity to nearby lakes, which functioned as source populations, made particular streams susceptible to becoming sink habitats. As another example, the results of a study about invasion of house mice in southeastern Australia (Singleton *et al.* 2007) suggested that delineation of sources in farm buildings and linkages to nearby sinks in sensitive woodland habitats would be helpful for developing strategies to manage these destructive outbreaks.

Other applications of Pulliam (1988) have addressed climate change and ecosystem services. Davis *et al.* (1998) discussed methods for predicting species responses to global climate change and refuted the "climate envelope" approach to climate trend prediction, which relies on overlaying current population distributions with climate data. Specifically, Davis *et al.* (1998) suggested that climate change researchers need to account for sinks when projecting species occurrence, which may not reflect habitat requirements of the species completely and are not sustainable without sources. Such evaluations are timely, considering the recent focus on niche modeling for evaluating the effects of global climate change on ecological systems (Araujo *et al.* 2005; Luoto *et al.* 2005), as these studies often fail to address source–sink dynamics. With regard to ecosystem services, Loreau *et al.* (2003) incorporated source–sink theory into a meta-ecosystem framework to quantify ecosystem services across heterogeneous space. Dispersal was used as a key component for modeling the flow of ecosystem services across what Loreau *et al.* (2003) termed "global source–sink" systems, or ecosystems linked by resource exchange.

The fields of toxic pollution and radioactive contamination research provide other examples of recent citations of Pulliam (1988). A study by Rowe *et al.* (2001) on source–sink dynamics of southern toads found that habitats contaminated with trace elements became sinks for the declining toad population. Delineating local habitats as sources and sinks helped to target affected areas for future toad population management. Møller *et al.* (2006) modeled the impact of the Chernobyl radioactive contamination site on barn swallow population dynamics. The authors documented a new method of identifying sources and sinks according to patterns in stable isotopes in the feathers of barn swallows. The stable-isotope signatures were different across source and sink landscapes because the isotopes related to the content and location of food obtained by the birds (Møller *et al.* 2006). In addition, the authors argued that sink populations have a greater variance in the isotope ratio because sink habitats rely on the dispersal of individuals from other habitats in order to persist.

Although we have focused on the successful application of the ideas expressed in Pulliam (1988) to natural resource management in this section, it is important to note that barriers to implementation exist as well. One frequently discussed barrier is the difficulty in identifying sources and sinks in real landscapes, where it is costly and impractical to obtain extensive demographic data over large spatial and temporal scales (Dias 1996; Roberts 1998). Such data limitations are common in ecological research and can limit the utility of theories for on-the-ground management (Turner *et al.* 1995; DeAngelis and Mooij 2005).

Sources and sinks can also be misidentified, leading to ineffective natural resource management. Watkinson and Sutherland (1995) recognized that some habitat patches could be misidentified as sinks without sufficient demographic data. Although deaths could exceed births and immigration could exceed emigration, a particular habitat may not be a true sink if it is able to support local populations without inputs from sources when population levels are maintained below carrying capacity. In the same study, the authors postulated a new definition of "pseudo-sink" to reflect the condition in which high immigration rates raise the population above a patch's carrying capacity and thus increase mortality. In other words, a pseudo-sink can support a smaller population below carrying capacity in the absence of immigration, whereas a true sink cannot support any population without immigration. Dias (1996) highlighted examples of researchers misclassifying sources and sinks in short-term studies, where there could be unusually high or low reproductive rates or rates of dispersal when compared with longer time scales.

Runge *et al.* (2006) critiqued the common practice of delineating sources and sinks based on apparent survival (e.g., as obtained by mark–recapture techniques) rather than emigration. Thus, one could potentially mistake emigration for mortality and be unable to identify how an emigration-dominated sink habitat contributes to the overall population (Runge *et al.* 2006). Pulliam *et al.* (Chapter 9 in this volume) address the implications of wrongly identifying sources and sinks as a result of inadequate dispersal data. They also provide a potential solution in the form of a maximum likelihood model of demographic parameters that minimizes estimation bias. Ultimately, commitment to long-term management of population dynamics, collaboration across management boundaries that bisect sources and sinks in real landscapes, and appreciation of the benefits of monitoring populations with respect to habitat-specific demographic parameters and heterogeneous reproductive rates and dispersal will be necessary in order to overcome such barriers (for further discussion, see Wiens and Van Horne, Chapter 23, this volume).

The examples highlighted in this section have demonstrated the breadth of applications of Pulliam (1988) to natural resource management. As challenges to sustainable resource management continue, the ideas presented in Pulliam (1988) may function as a practical tool to assist in a better understanding of the dynamics of vulnerable populations across changing landscapes.

Conclusions

Pulliam's (1988) paper has made significant contributions to science during the past two decades. It had implications for the theoretical development of ecology with respect to three main areas: the niche concept, population dynamics and distribution, and community structure. The applications of Pulliam's work have extended to many scientific disciplines ranging from economics to microbiology to public health. Pulliam's work has also informed natural resource management. The model is a useful tool that has helped predict population dynamics and distribution in a habitat-specific manner, rather than with simple population size measures alone, and has helped to integrate the landscape perspective with assessments of population vulnerability and resource sustainability. Considering the growing impacts of Pulliam's work during the past 20 years, there is little doubt that its impact will continue to grow in the future.

Acknowledgments

We acknowledge the contribution of John DiBari, William McConnell, H. R. Pulliam, John Wiens, and two anonymous reviewers for their helpful comments on earlier drafts that greatly improved this chapter. We also thank the National Science Foundation, the National Aeronautics and Space Administration, and the United States Department of Agriculture for providing funding. The information in this document has been funded in part by the US Environmental Protection Agency. It has been subjected to review by the National Health and Environmental Effects Research Laboratory's Western Ecology Division and approved for publication. Approval does not signify that the contents reflect the views of the Agency, nor does mention of trade names or commercial products constitute endorsement or recommendation for use. This is contribution number WED-09-022 of the Western Ecology Division.

References

Araújo, M. B., R. G. Pearson, W. Thuiller and M. Erhard (2005). Validation of species–climate impact models under climate change. *Global Change Biology* **11**(9): 1504–1513.

Blumler, M. A., R. Byrne, *et al.* (1991). The ecological genetics of domestication and the origins of agriculture (and comments and reply). *Current Anthropology* **32**(1): 23–54.

Botsford, L. W., C. L. Moloney, A. Hastings, J. L. Largier, T. M. Powell, K. Higgins and J. F. Quinn (1994). The influence of spatially and temporally varying oceanographic conditions on meroplanktonic metapopulations. *Deep Sea Research – Topical Studies in Oceanography* **41**(1): 107–145.

Brawn, J. D. and S. K. Robinson (1996). Source–sink population dynamics may complicate the interpretation of long-term census data. *Ecology* **77**(1): 3–12.

Chattopadhyay, S., M. Feldgarden, S. J. Weissman, D. E. Dykhuizen, G. van Belle and E. V. Sokurenko (2007). Haplotype diversity in "source-sink" dynamics of *Escherichia coli* urovirulence. *Journal of Molecular Evolution* **64**(2): 204–214.

Cohen, J. E. (1969). Natural primate troops and a stochastic population model. *American Naturalist* **103**(933): 455–477.

Colwell, R. K. and D. J. Futuyma (1971). On the measurement of niche breadth and overlap. *Ecology* **52**(4): 567–576.

Crowder, L. B., S. J. Lyman, W. F. Figueira and J. Priddy (2000). Source–sink population dynamics and the problem of siting marine reserves. *Bulletin of Marine Science* **66**(3): 799–820.

Danielson, B. J. (1991). Communities in a landscape: the influence of habitat heterogeneity on the interactions between species. *American Naturalist* **138**(5): 1105–1120.

Davis, A. J., L. S. Jenkinson, J. H. Lawton, B. Shorrocks and S. Wood (1998). Making mistakes when predicting shifts in species range in response to global warming. *Nature* **391**(6669): 783–786.

DeAngelis, D. L. and W. M. Mooij (2005). Individual-based modeling of ecological and evolutionary processes. *Annual Review of Ecology, Evolution, and Systematics* **36**(1): 147–168.

Dennehy, J. J., N. A. Friedenberg, R. D. Holt and P. E. Turner (2006). Viral ecology and the maintenance of novel host use. *American Naturalist* **167**(3): 429–439.

Dias, P. C. (1996). Sources and sinks in population biology. *Trends in Ecology and Evolution* **11**(8): 326–330.

Diffendorfer, J. E. (1998). Testing models of source–sink dynamics and balanced dispersal. *Oikos* **81**(3): 417–433.

Doncaster, C. P., J. Clobert, B. Doligez, L. Gustafsson and E. Danchin (1997). Balanced dispersal between spatially varying local populations: an alternative to the source–sink model. *American Naturalist* **150**(4): 425–445.

Duplantier, J. M., J. B. Duchemin, S. Chanteau and E. Carniel (2005). From the recent lessons of the Malagasy foci towards a global understanding of the factors involved in plague reemergence. *Veterinary Research* **36**(3): 437–453.

Fahrig, L. and G. Merriam (1985). Habitat patch connectivity and population survival. *Ecology* **66**(6): 1762–1768.

Fretwell, S. D. and H. L. Lucas (1970). On territorial behaviour and other factors influencing habitat distribution in birds. I. Theoretical development. *Acta Biotheoretica* **19**: 16–36.

Grassly, N. C., C. Fraser, J. Wenger, J. M. Deshpande, R. W. Sutter, D. L. Heymann and R. B. Aylward (2006). New strategies for the elimination of polio from India. *Science* **314**(5802): 1150–1153.

Grinnell, J. (1917). The niche-relationships of the California thrasher. *Auk* **34**(4): 427–433.

Hanski, I. and M. Gilpin (1991). Metapopulation dynamics: brief history and conceptual domain. *Biological Journal of the Linnean Society* **42**(1–2): 3–16.

Harrison, S. (1991). Local extinction in a metapopulation context: an empirical evaluation. *Biological Journal of the Linnean Society* **42**(1–2): 73–88.

Holt, R. D. (1984). Spatial heterogeneity, indirect interactions, and the coexistence of prey species. *American Naturalist* **124**(3): 377–406.

Holt, R. D. (1985). Population dynamics in two-patch environments: some anomalous consequences of an optimal habitat distribution. *Theoretical Population Biology* **28**: 181–208.

Holt, R. D. (1987). Prey communities in patchy environments. *Oikos* **50**(3): 276–290.

Holt, R. D. and M. S. Gaines (1992). Analysis of adaptation in heterogeneous landscapes: implications for the evolution of fundamental niches. *Evolutionary Ecology* **6**(5): 433–447.
Holterhoff, P. F. (1996). Crinoid biofacies in Upper Carboniferous cyclothems, midcontinent North America: faunal tracking and the role of regional processes in biofacies recurrence. *Palaeogeography Palaeoclimatology Palaeoecology* **127**(1–4): 47–81.
Horvath, T. G., G. A. Lamberti, D. M. Lodge and W. L. Perry (1996). Zebra mussel dispersal in lake–stream systems: source–sink dynamics? *Journal of the North American Benthological Society* **15**(4): 564–575.
Hutchinson, G. E. (1958). Concluding remarks. *Cold Spring Harbor Symposia on Quantitative Biology* **22:** 415–427.
James, F. C., R. F. Johnston, N. O. Wamer, G. J. Niemi and W. J. Boecklen (1984). The Grinnellian niche of the wood thrush. *American Naturalist* **124**(1): 17–47.
Johnson, D. M. (2004). Source–sink dynamics in a temporally heterogeneous environment. *Ecology* **85**(7): 2037–2045.
Levin, S. A. (1974). Dispersion and population interactions. *American Naturalist* **108**(960): 207–228.
Levins, R. (1969). Some demographic and genetic consequences of environmental heterogeneity for biological control. *Bulletin of the Entomological Society of America* **15:** 237–240.
Lewin, R. (1989). Sources and sinks complicate ecology. *Science* **243**(4890): 477–478.
Lidicker, W. Z. (1975). The role of dispersal in the demography of small mammals. In *Small Mammals: Their Productivity and Population Dynamics* (F. B. Golley, K. Petrusewicz and L. Ryszkowski, eds.). Cambridge University Press, New York: 103–128.
Loreau, M., N. Mouquet and R. D. Holt (2003). Meta-ecosystems: a theoretical framework for a spatial ecosystem ecology. *Ecology Letters* **6**(8): 673–679.
Lundberg, P. and N. Jonzen (1999). Optimal population harvesting in a source–sink environment. *Evolutionary Ecology Research* **1**(6): 719–729.
Luoto, M., J. Poyry, R. K. Heikkinen and K. Saarinen (2005). Uncertainty of bioclimate envelope models based on the geographical distribution of species. *Global Ecology and Biogeography* **14**(6): 575–584.
MacArthur, R. H. and E. O. Wilson (1963). An equilibrium theory of insular zoogeography. *Evolution* **17**(4): 373–387.
McCoy, T. D., M. R. Ryan and L. W. Burger Jr. (1999). Conservation Reserve Program: source or sink habitat for grassland birds in Missouri? *Journal of Wildlife Management* **63**(2): 530–538.
Møller, A. P., K. A. Hobson, T. A. Mousseau and A. M. Peklo (2006). Chernobyl as a population sink for barn swallows: tracking dispersal using stable-isotope profiles. *Ecological Applications* **16**(5): 1696–1705.
Morris, D. W. (1991). On the evolutionary stability of dispersal to sink habitats. *American Naturalist* **137**(6): 907–911.
Novaro, A. J., M. C. Funes and R. S. Walker (2005). An empirical test of source–sink dynamics induced by hunting. *Journal of Applied Ecology* **42**(5): 910–920.
Perron, G. G., A. Gonzalez and A. Buckling (2007). Source–sink dynamics shape the evolution of antibiotic resistance and its pleiotropic fitness cost. *Proceedings of the Royal Society B – Biological Sciences* **274**(1623): 2351–2356.
Pulliam, H. R. (1988). Sources, sinks, and population regulation. *American Naturalist* **132**(5): 652–661.
Pulliam, H. R. and B. J. Danielson (1991). Sources, sinks, and habitat selection: a landscape perspective on population dynamics. *American Naturalist* **137**(Suppl.): S50–S66.
Roberts, C. M. (1998). Sources, sinks, and the design of marine reserve networks. *Fisheries* **23**(7): 16–19.
Rowe, C. L., W. A. Hopkins and V. R. Coffman (2001). Failed recruitment of southern toads (*Bufo terrestris*) in a trace element-contaminated breeding habitat: direct and indirect effects

that may lead to a local population sink. *Archives of Environmental Contamination and Toxicology* **40**(3): 399–405.
Runge, J. P., M. C. Runge and J. D. Nichols (2006). The role of local populations within a landscape context: defining and classifying sources and sinks. *American Naturalist* **167**(6): 925–938.
Sanchirico, J. N. and J. E. Wilen (1999). Bioeconomics of spatial exploitation in a patchy environment. *Journal of Environmental Economics and Management* **37**(2): 129–150.
Singleton, G. R., C. R. Tann and C. J. Krebs (2007). Landscape ecology of house mouse outbreaks in south-eastern Australia. *Journal of Applied Ecology* **44**(3): 644–652.
Sokurenko, E. V., R. Gomulkiewicz and D. E. Dykhuizen (2006). Opinion: source–sink dynamics of virulence evolution. *Nature Reviews Microbiology* **4**(7): 548–555.
Tattersall, F. H., D. W. Macdonald, B. J. Hart and W. Manley (2004). Balanced dispersal or source–sink: do both models describe wood mice in farmed landscapes? *Oikos* **106**(3): 536–550.
Turner, M. G., G. J. Arthaud, R. T. Engstrom, S. J. Hejl, J. Liu, S. Loeb and K. McKelvey (1995). Usefulness of spatially explicit population models in land management. *Ecological Applications* **5**(1): 12–16.
Van Horne, B. (1983). Density as a misleading indicator of habitat quality. *Journal of Wildlife Management* **47**(4): 893–901.
Watkinson, A. R. and W. J. Sutherland (1995). Sources, sinks and pseudo-sinks. *Journal of Animal Ecology* **64**(1): 126–130.
Wiens, J. A. (1976). Population responses to patchy environments. *Annual Review of Ecology and Systematics* **7**(1): 81–120.

Part II

Advances in source–sink theory

Since the theory underpinning sources and sinks was proposed, it has advanced considerably as a useful means of conceptualizing and modeling the spatial–temporal dynamics of animal and plant populations across heterogeneous landscapes. The initial theory was a relatively simple representation of organisms in habitats with varying quality that contribute to the persistence or sustainability of subpopulations and the entire population. Over time, the theory has evolved to be more comprehensive and to better reflect the complexity that exists in real-world populations.

This section presents some of the latest developments in source–sink theory and their implications for understanding population dynamics and species persistence. The first three chapters expand source–sink theory to evaluate how populations evolve in complex landscapes. In the first chapter of this section (Chapter 2), Holt describes ways in which a species' niche is maintained as a result of niche conservatism or is altered through adaptive evolution. His work portrays how interactions between populations and their spatially heterogeneous environment relate to species persistence. To make his abstract ideas concrete and explicit, he uses a real source–sink system (the sea rocket *Cakile edentula* in coastal sand dunes of Nova Scotia) that illustrates a number of typical characteristics of many source–sink systems. In Chapter 3, Morris sheds new light on habitat selection and dispersal by exploring alternatives (evolutionary attractors) to the ideal free distribution theory. The two alternatives include an inclusive fitness strategy that leads to maximum population growth and a cooperative strategy whereby unrelated organisms build coalitions and drive the emigration of unaligned organisms. Furthermore, he discovers that sinks can dampen otherwise unstable population dynamics and enhance the probability of population persistence even when serving as ecological traps. In Chapter 4, Benkman and Siepielski investigate interactions between different species and the persistence of multiple species within a community context over evolutionary time. They conduct their study in a complex system that is defined by a seed-dispersal mutualism between Clark's nutcrackers (*Nucifraga columbiana*) and limber pine (*Pinus flexilis*), which is in turn influenced by a seed predator, the red squirrel (*Tamiasciurus hudsonicus*). Analyses of complex interactions between these species provide new insights on the evolution of species mutualism and the impacts of antagonists on source and sink habitat for mutualistic species.

The original source–sink theory was developed in natural systems where human impacts were assumed to be minimal or were not explicitly considered. However, direct and indirect human disturbances have been increasing around the world and there are virtually no pristine systems left on Earth. Instead, coupled human and natural systems are increasingly common. Both natural and human disturbances (e.g., land use, climate change) can alter the demographic makeup of populations and the underlying habitat, and can vary with respect to type (e.g., different causes), frequency, and magnitude. Singular and combined effects of natural and human disturbances on source–sink systems can be extreme to the point that sources can become sinks and sinks can become sources. The next four chapters of this section (Chapters 5–8) focus on theoretical models that explicitly address the effects of disturbances on source–sink dynamics.

Although the likely impacts of climate change on the geographic distributions of species have been extensively researched, the associated demographic effects are less well studied. To simulate the effects of climate change on the demography of a set of interconnected stream fish populations in different "numerical experiments," Andersen (Chapter 5) uses metapopulation models and stochastic stage-structured multiregional models. His results suggest that climate change destabilizes populations by altering dispersal rates and spatial population structure, making populations more vulnerable to additional natural or human disturbances.

Although spatial heterogeneity, niche breadth, and environmental stochasticity are widely identified as important determinants of population structure, their combined effects have rarely been examined. In Chapter 6, Pearson and Fraterrigo explore the combined effects using simulated populations with life-history traits similar to perennial forest herbaceous plants in the southern Appalachian Mountains of the USA. They find that the effects of habitat fragmentation, spatial variation in habitat quality, and niche breadth lead to varying demographic outcomes among habitat patches of similar size and shape. Furthermore, these effects outweigh the impacts of climatic stochasticity on population persistence. The results suggest that forest perennials may be more sensitive to habitat fragmentation and heterogeneity of habitat quality than to interannual climatic variation and frequencies of extreme events.

A cycle of habitat disturbance and subsequent recovery can cause a patch to alternate between a population source and a population sink, but the impacts of such a disturbance–recovery cycle and subsequent source–sink dynamics are underappreciated. In Chapter 7, Falcy and Danielson develop a model of source–sink dynamics that examines the impacts of habitat disturbance, recovery in a putative source patch, and population decline in the sink on metapopulation persistence. They find that decreasing the rate of population decline in the sink can have a much larger impact on metapopulation persistence than increasing the rate of habitat recovery in the source. The impact is amplified with increased disturbance frequency. The results suggest that sinks can become crucial for population recovery and persistence because putative sources affected by a disturbance may rely on individuals from sinks for recolonization.

Although the roles of sinks in natural systems have been widely acknowledged, the effects of sinks and associated human socio-economic factors (e.g., policy and market conditions) in coupled human and natural systems are largely unexplored. For example, little is known about the effects of conservation incentives on environmental changes such as the creation of habitats. To elucidate how present and future environmental variation might affect source–sink dynamics in managed landscapes and determine what roles sinks might play in species persistence, it is essential to understand the effects of socio-economic factors on the dynamics of habitat availability and quality. The final chapter in this section (Chapter 8) assesses the effects of sinks on species persistence and richness in different socio-economic and policy contexts. By running simulations using a spatially explicit, agent-based model of land use decision making and a spatially explicit metacommunity model, Gimona and colleagues find that not accounting for population sinks can have serious impacts on a species' persistence. Furthermore, sinks have more influence on species that are associated with habitat of moderate profitability from land use under free-market conditions.

Collectively, the chapters in this section develop new twists on traditional source–sink theory, providing fertile ground for more detailed modeling and field studies to explore the ideas and conclusions presented here. But they also point toward the next-generation development of a "meta" source–sink theory that integrates these ideas into a more unified and cohesive body of theory.

ROBERT D. HOLT

2

Evolution in source–sink environments: implications for niche conservatism

Summary

Demographic sources and sinks arise from the interplay of spatial variations in birth and death rates, and movement between habitats. One way to view sources and sinks is that, in the former, individuals are well adapted to the local environment, whereas in the latter, individuals are poorly adapted. This raises the question of how adaptive evolution might influence the evolutionary stability of source–sink population structures. When can a species' niche evolve, so that a habitat – now a sink – becomes a source? This chapter provides an overview of theoretical investigations into this question. The scenarios considered include the fate of single favorable mutants that improve adaptedness to a sink environment, quantitative genetic variation for single traits determining local fitness, and the influence of reciprocal dispersal from sinks to sources. The overall conclusion across models is that the harsher the sink (as assessed in terms of absolute fitness), the harder it may be for adaptive evolution to sculpt adaptation sufficiently to permit population persistence. Theoretical studies show that the rate of immigration can have a variety of impacts upon evolution in sinks, depending upon many details of genetics, life history, and demography. Such theoretical exercises are not merely academic exercises, because source–sink dynamics naturally arise in a wide range of applied evolutionary contexts (such as the control of agricultural pests, and in disease emergence across host species) where the management aim is to prevent evolution in focal species in particular habitats.

Sources, Sinks and Sustainability, ed. Jianguo Liu, Vanessa Hull, Anita T. Morzillo and John A. Wiens. Published by Cambridge University Press.

Introduction

The world is spatially heterogeneous at almost any scale one cares to examine. Understanding how species respond to this heterogeneity is a core concern of both ecology and evolution. Reflecting demographic responses to spatial heterogeneity, local populations across a species' range may differ in many ways – including differences in abundance, local fitness, temporal variance in fitness, rates of immigration or emigration, and expected times to extinction. These demographic responses, in turn, can influence the direction of adaptive evolution in local populations, or across a species as a whole.

Source–sink dynamics emerge from such spatial heterogeneities. The celebrated paper by Ron Pulliam in *The American Naturalist* in 1988 helped stimulate an explosion of interest in source–sink population dynamics. The clarity of thinking and prose in that short paper set a high standard that we should all try to emulate. Ron argued that a particularly important aspect of spatial heterogeneity is that many individuals in a species may live in habitats (sinks) where local recruitment does not replace local losses due to mortality, yet the population in these habitats may persist because of recurrent immigration from external sources. Source–sink dynamics have clear and obvious implications for many issues in basic and applied ecology, such as resource management, pest control, and conservation. It may not be so obvious at first glance, but source–sink dynamics also play a central role in key issues in evolutionary biology. I take this opportunity to reflect on some of the important evolutionary questions that are illuminated by a focus on environments with sources and sinks. Tad Kawecki (2004, 2008) has provided thoughtful, detailed reviews of both theoretical and empirical studies of adaptation in sinks, so I will not attempt to be comprehensive, but instead give a somewhat personal perspective on this topic. I will make a few simple conceptual points that are often obscured in the published literature, will then sketch out some of the principal conclusions of theoretical studies of adaptive evolution in source–sink systems, and will end by pointing out the implications for interspecific interactions.

Understanding evolution in source–sink systems is at the heart of the fundamental problem of understanding the evolution and conservatism of species' niches. The term "*niche*" – as I use it here – denotes the suite of abiotic conditions, resources, and abundances of interacting species that collectively permit a population of a given species to persist (deterministically) in a particular habitat, without replenishment by immigration (Hutchinson 1957; Holt and Gaines 1992; Holt 2009). In adaptive radiations, species' niches can change, sometimes very substantially (Price 2007). Likewise, in applied ecology, whenever pest species adapt to control measures, or pathogens such as HIV adapt to spread in novel hosts, this involves niche evolution.

Yet the history of life is also replete with examples of niche conservatism, where a lineage (a species or entire clade) maintains much the same niche over substantial swaths of its evolutionary history (Bradshaw 1991; Wiens and Graham 2005; Futuyma 2010; Wiens *et al.* 2010). Adaptation in marginal environments can at times be constrained by the sheer absence of relevant genetic variation (Bradshaw 1991; Blows and Hoffmann 2005). But even with appropriate genetic variation, demographic constraints can slow, and at times prevent, adaptive evolution (Holt and Gaines 1992; Kawecki 1995; Holt 1996a, 1996b) and lead to (for instance) evolutionary limits on species' ranges (Antonovics *et al.* 2001; Polechova *et al.* 2009; Sexton *et al.* 2009; Bridle *et al.* 2010; Turner and Wong 2010). An explicit concern with the demographic interplay of sources and sinks is integral to the emerging theoretical framework for understanding niche conservatism and evolution.

An exemplary real-world source–sink system: the sea rocket

To make these abstract ideas a bit more concrete, it is useful to consider a real source–sink system as we explore them. A lovely example of a natural sink population is provided by the elegant study by Paul Keddy of a flowering plant, the sea rocket (*Cakile edentula*), in coastal sand dunes of Nova Scotia (Keddy 1981, 1982; see Fig. 2.1). This example shows several features that characterize many source–sink systems.

1. What one sees as one walks along the beach is that the plant is most abundant in the center of the dunes, and sparser near the sea and on the landward side. This does not mean that environmental conditions for individual sea rockets are particularly favorable inside the dune. Keddy measured per capita birth and death rates and found that births were less than deaths in the dune, for all measured densities. So conditions in the dune center are outside the sea rocket's niche. Many (though not all) sink populations are outside their species' niches.
2. The reason this sink population persists is that recurrent wind from the ocean strips seeds from the favorable habitat on the sea verge (lowering density there), and deposits them in the dune interior. Emigration can thus lower population density (near the sea margin in this example) and immigration can raise it (from zero to high values in the dune interior).
3. The size of a sink population reflects both movement rates and its rate of population decline. The dune center is closer to the source than is the landward side, and so doubtless receives more immigrant seeds. Demographic studies (Keddy 1981) show that the landward side also has higher death and lower birth rates. The dune interior is a "mild" sink, with a substantial

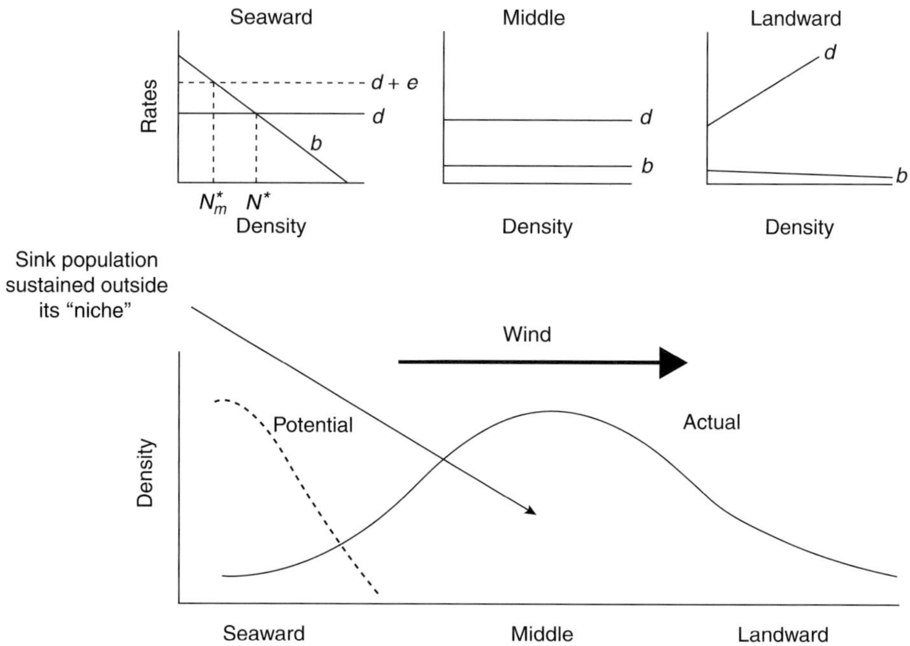

FIGURE 2.1.
Source–sink dynamics in the sea rocket, *Cakile edentula*, on Nova Scotian sand dunes (adapted from Keddy 1981, 1982). See Eq. (2.1) for definitions of symbols.

population maintained by immigration; the landward side of the dune, by contrast, is a "harsh" sink, with few individuals present.

4. When considering two habitats coupled by dispersal, if at demographic equilibrium there is asymmetry in the flows of individuals between them, a source–sink population structure will emerge (Pulliam 1988). Passive dispersal in a heterogeneous landscape is a potent mechanism for generating sink populations (Holt 1985). Sea rocket seeds passively waft on the wind. Given the persistent directionality of wind from the ocean, this physical transport process across the dunes is strongly asymmetric, and maintains sink populations. Other dispersal mechanisms can also produce sinks. For instance, the model used by Ron Pulliam in his 1988 paper had asymmetric movement arising from interference competition (e.g., for territories in birds), forcing individuals (e.g., fledgling birds) out of high- into low-quality habitats. Pulliam's assumption should hold for many birds and mammals that contend for territories and actively disperse from their natal habitats because of interference or resource limitation. An assumption of passive dispersal is more reasonable for plants and invertebrates transported by wind or currents away from their natal habitats. Even with no directional individual movements,

given variation in carrying capacity (K), more individuals tend to leave high-K areas than return, creating asymmetric flows and thus source–sink dynamics (Holt 1985) by pushing low-K populations above equilibrial density ("pseudo-sinks"; see Watkinson and Sutherland 1995) or by sustaining populations at sites with no positive K at all.

5. Given recurrent immigration, a sink population can persist even though its local demographic parameters are density independent, as shown by the sea rocket in the center of the dune. But a sink population can also experience very strong density dependence, as in the sea rocket on the landward dune edge.

Keddy focused on the ecology of the sea rocket. But if this species has genetic variation in traits that influence fitness (a fair assumption for most species), the sea rocket might eventually adapt to the dune sink environment. In effect, evolution could then transform these sink populations into future potential sources inside the sea rocket's niche. All of the ecological and demographic features of the sea rocket source–sink system noted above can, in principle, strongly influence the likelihood of such evolution.

General questions about adaptive evolution in sinks

We can pose a series of abstract questions about adaptive evolution in source–sink systems, and then develop models tailored to address those questions. To make this exercise concrete, I again use the sea rocket as a hypothetical example.

How is adaptive evolution in a sink influenced by

- **the severity of the sink environment?** Keddy showed that the absolute fitness of the sea rocket (as measured by per capita birth minus death rates) is higher in the center of the dune than on the landward side. How does spatial variance in fitness influence the likelihood of adaptive evolution in each location?
- **the number of immigrants, per unit time?** An increase in productivity in the seaward population should increase the number of seeds deposited by the wind into the dunes. Would a boost in immigration rate facilitate, or hamper, adaptive evolution in the sink?
- **the constancy or variability of this immigration rate?** The wind on the coast is notoriously fickle: calm one day, a gale the next. On some coastlines, gentle winds may not effectively disperse seeds into the dune interior, but occasional storms can come along and generate sharp dispersal pulses. Sink populations may become established in sporadic colonizing episodes, with little subsequent immigration. After the

colonizing pulse, the sink population is on its own, and its dynamics should resemble a closed population faced with an abrupt, unfavorable environmental change (Gomulkiewicz and Holt 1995; Orr and Unckless 2008). By contrast, were there a regular, strong steady wind, fresh immigrants would arrive each and every generation. Recurrent dispersal leads to an admixture of immigrants and residents, and gene flow can occur. Is adaptation most likely if immigration is sporadic, or instead continual?

- **the rate of emigration?** Winds blowing across the dune can move seeds out of the sink habitat in a landward direction. How does emigration impact niche evolution in a sink?
- **the directionality of dispersal?** The wind usually blows from the ocean, and so what happens in the sink populations of the sea rocket should not strongly influence the dynamics of the seaward sources. But one could imagine other scenarios in which the wind constantly changes direction, so that dispersal occurs equitably in both directions. In this case, there can be a dispersal loop across generations from the source, to the sink, back to the source. How does this influence the likelihood of adaptation to the sink, and is there a pattern of dispersal that is optimal for evolution in the sink?
- **the constancy or temporal variability of the sink environment itself?** Sand dunes dry rapidly, so temporal variation in rainfall could lead to temporal variation in habitat quality and thus fitness. Does environmental variation make adaptive evolution more difficult, or instead could it facilitate adaptation in the sink?
- **tradeoffs with conditions in the source?** If the wind does occasionally blow seeds from the dune back onto the seaward edge, and these seeds germinate and produce seeds that in turn are blown back into the interior of the dune, one has to consider how evolution weights conditions across the two habitats in determining the overall direction of evolution, and the potential for local adaptation to the sink. Adaptation to the sink conditions of the dune interior might be constrained because it is too costly in terms of reduced fitness back in the source habitat.
- **density dependence in the sink?** If resources or germination sites are scarce on the inland dune margin, that may explain the strong density dependence Keddy observed there. If numbers are low, in some species there may be positive density dependence as well (e.g., because pollinators are attracted to dense populations). What is the evolutionary impact of negative and positive density dependence upon adaptation to sink conditions?
- **interspecific interactions in the sink?** On the landward side of the dune, it is likely that there is strong competition inflicted on the sea

rocket by other plant species, and possibly impacts from generalist herbivores as well. What are the consequences of interspecific interactions for adaptation by the sea rocket to sink conditions? More broadly, how do source–sink dynamics color coevolutionary processes?

We will return to the example of the sea rocket after I have walked through the ideas presented below.

The demographic context of sink evolution

A consideration of source–sink dynamics helps highlight the demographic context in which niche evolution necessarily plays out and indeed may be key to understanding niche conservatism (as argued in Holt and Gaines 1992; for more recent references, see Holt 2009; Wiens *et al.* 2010). It is useful to go back to basics.

The dynamics of a local population in a given area without significant age or stage structure (where the population is assumed to be common enough to ignore demographic stochasticity) with continuous, overlapping generations can be represented as:

$$\frac{dN_j}{dt} = N_j(b_j - d_j) + I_j - e_j N_j \tag{2.1}$$

where N_j is population density at locality j, b_j, d_j and e_j are per capita birth, death and emigration rates, respectively, and I_j is the rate of immigration into this location. (Below, when I come to explicit genetic models, as is customary in population genetics I will use a discrete-generation formulation.)

In contrast to the other components of demography, the term for immigration in Eq. (2.1) is more usefully expressed as an absolute rate, rather than as a per capita rate. The reason is that the rate of immigration is driven more strongly by external forces than by local factors, and expressing this rate on a per capita basis (using local density) may make little sense. For instance, if the starting condition is zero population density but positive immigration, the initial "per capita" immigration rate (using local density for the "per capita") is infinite – a quantity which seems artifactual.

What counts as a "local population" of course depends upon spatial scale – at sufficiently large scales there will be no immigration from outside. All the demographic rates in Eq. (2.1) can, in principle, depend on local population density, spatially varying environmental variables such as temperature or resource abundances, interspecific interactions, and so forth. Immigration and emigration rates may be density independent or density dependent, and will themselves vary with environmental conditions.

What is the precise relationship between source–sink dynamics and the ecological niche of a species? We focus on a single local population *j* in a constant environment. Consider first a limiting case where there is only a minimal rate of immigration from external sources (so $I \approx 0$), and any individuals present also emigrate at a similarly low rate (so $e \approx 0$). Therefore the population is essentially closed. We want to distinguish habitats where a population will persist from those where it becomes extinct. If $b_j < d_j$ for all population sizes N_j, then locality *j*, by definition, has conditions (temperature, resource availability, etc.) outside the fundamental niche of the species (Hutchinson 1957, 1978) and, from Eq. (2.1), the population faces inevitable extinction. This habitat is what Kawecki (2004) calls an *absolute sink*, "where the intrinsic growth rate is negative and so births can never compensate for deaths." (The intrinsic growth rate is defined as the difference between birth and death rates at very low density.)

By contrast, if at a very low value of N_j, $b_j > d_j$ (i.e., the intrinsic growth rate r_j is positive), then if a colonizing propagule shows up and is not depleted by emigration, it (deterministically) can increase (until limited by resource availability or other factors). In principle that habitat is not an absolute sink, as it has conditions that meet the species' niche requirements.

There is an ambiguity in Hutchinson's initial formalization of the niche, which he defined in terms of the "indefinite persistence" of a species (Holt 2009). This ambiguity arises because of Allee effects: positive density dependence in demographic rates at some (typically low) densities, so that at very low values of *Nj*, $b_j < d_j$, but at higher *Nj*, $b_j > d_j$. There is an increasing recognition of the importance of Allee effects in population ecology (Courchamp *et al.* 2008). Mechanisms leading to positive density dependence are diverse, ranging from the need to find mates in sexual species, to predator satiation leading to indirect positive density dependence in their prey (Holt *et al.* 2004b), to many kinds of positive impacts a species might have upon ecosystem processes (Wilson and Agnew 1992). The upshot is that species can sometimes persist indefinitely in environments where they have a negative intrinsic growth rate (Holt 2009). So habitats may be absolute sinks in the narrow sense that, when a species is rare, its births do not match its deaths, but not in a broader sense, because there is some density above which the population can potentially persist.

As an example, Figure 2.2 contrasts the relationship between per capita growth rate and density for populations with logistic-like growth (Fig. 2.2A), to that for populations with Allee effects (Fig. 2.2B). Growth curves are shown for three habitats, with 1 being the best, and 3 the worst. Solid dots indicate equilibria in the absence of dispersal. If a population has a logistic-like pattern of growth, it achieves its maximal growth at low density. Without immigration, either this population cannot persist at all, or it has a positive carrying capacity.

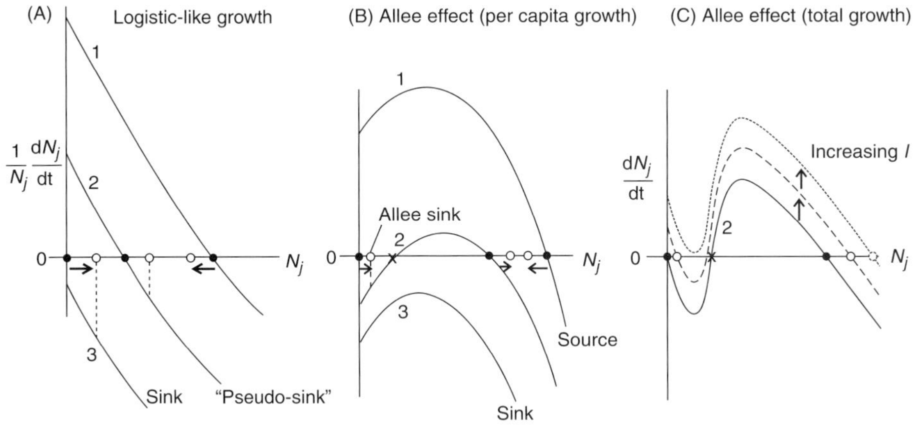

FIGURE 2.2.
Density dependence and source–sink dynamics. A: Three habitats with logistic-like growth (negative density dependence at all densities); the intrinsic growth rate and the carrying capacity increase from bottom to top (habitat 3 to 1), and per capita growth rate declines with local density, N_j. Black dots are stable equilibrial densities, with no movement. An asymmetric flow of individuals from the high-*K* habitat (habitat 1) into the other habitats alters densities, as shown by the white dots. For instance, immigration can sustain a population in an absolute sink (habitat 3), but individual fitness there is depressed by immigration. Habitat 2 is not an absolute sink, but is turned into a sink population (a "Pseudosink") because immigration pushes numbers above local carrying capacity. Again, immigration increases population size, depressing local fitness. B: Three habitats with Allee effects at low densities. A sink population can again be sustained by immigration (habitat 3, indicated by the thin dashed line), but now fitnesses there are elevated by immigration. In habitat 2, with immigration from habitat 1, the habitat may hold a pseudo-sink population that can persist at lower density were immigration to be cut off, or instead a sink population at low density that cannot persist without immigration (see text). **C:** Total growth rate versus density with an Allee effect. The model is $dN/dt = Ng(N) + I$, where N is density and g is per capita growth. The per capita growth rate shown for habitat 2 in (B) implies a total growth comparable to the solid line. Adding constant immigration elevates the curve by a fixed amount. With low immigration, two alternative stable states are present (a pseudo-sink at high numbers, and an "Allee sink" at low numbers). At yet higher immigration, the lower equilibrium disappears.

With a strong Allee effect, by contrast, alternative equilibria (extinction or persistence) may occur in a given environment (as in habitat 2 in Fig. 2.2B), and so a species' "establishment niche" (where it can increase when rare) may differ sharply from its "population persistence niche" (where it can persist, once established in sufficient numbers) (Holt 2009). For our purposes, we will largely set aside this ambiguity in the meaning of the term "niche."

Now, we allow ongoing dispersal, so immigration and emigration are nonzero. What is the relationship between a species' niche and source–sink dynamics? There are only three possible ways that a population in habitat *j* can be in equilibrium at a non-zero density N_j^*, where the asterisk denotes equilibrium (expanding on a passage in Pulliam 1988):

$$\text{I.}\quad b_j = d_j \quad \text{and} \quad I_j = e_j N_j^*. \tag{2.2}$$

If immigration equals emigration, then births must also just match deaths. Even though there is ongoing dispersal, the population is at its local carrying capacity (Holt 1984). A pattern of dispersal that does not alter local abundance (because input equals output) is called "balanced dispersal" (Doncaster *et al.* 1997; Diffendorfer 1998; Morris and Diffendorfer 2004). Balanced dispersal emerges when individuals move so as to approximate an "*ideal free distribution*" (Fretwell 1972), where fitnesses are equilibrated over space, so no occupied habitat is a sink. If condition (2.2) is met, and the equilibrium is locally stable, it is fair to assume that the habitat has conditions inside the species' niche; if movement were to be cut off, the population could persist because local births match local deaths.

$$\text{II.}\quad b_j > d_j \quad \text{and} \quad I_j < e_j N_j^*. \tag{2.3}$$

Now there is a surfeit of births over deaths, so the habitat is clearly within the niche. The population stays in balance because there is a net export of individuals to the external environment. This population could be deemed a "*source population*." In Figure 2.2A, if habitat 1 is coupled to habitats 2 and 3 by dispersal, and per capita dispersal rates are constant and moderate in magnitude, more individuals should leave habitat 1 than return from the other habitats. Therefore, the population in habitat 1 will equilibrate at a density lower than the local carrying capacity, so births exceed deaths (a potential equilibrium, given net emigration from habitat 1, is indicated by an open dot).

$$\text{III.}\quad b_j < d_j \quad \text{and} \quad I_j > e_j N_j^*. \tag{2.4}$$

In this final case, the local population is intrinsically in decline, as measured by its local demographic rates, and is kept in balance only because more individuals enter than leave. This could be called a "*sink population*" (these definitions of "source" and "sink" match those proposed by Pulliam 1988, as noted above). In Figure 2.2A, habitats 2 and 3 when coupled to habitat 1 are sink populations maintained at higher densities by immigrants from habitat 1 (the open circles).

In the above expressions, all the "parameters" (b_j, d_j, e_j, and I_j) could actually be functions of local density, as well as of the densities of interacting species (competitors, predators, etc.) and the values of environmental factors (temperature, etc.). Therefore, whether or not a habitat satisfying Eq. (2.4) is within the niche cannot be determined from simply inspecting these demographic relationships, because of density dependence. If births and/or deaths are negatively density dependent, so that the net per capita growth rate declines with increasing density (e.g., as in a logistic growth model), cutting off immigration could lead to compensatory increases in local growth rates as numbers decline. If a positive carrying capacity exists the population could then equilibrate at some positive local carrying capacity $K < N_j^*$. This is the "*pseudo-sink*" of Watkinson and Sutherland (1995), modeled in Holt (1983, 1985). A pseudo-sink is a demographic sink, in that locally births are less than deaths and there are more immigrants than emigrants, but because the population can persist in the absence of immigration, albeit at lower numbers, conditions there are within the niche. So habitat 2 in Figure 2.2A is a pseudo-sink – a demographic sink (when receiving immigrants), but within the niche.

The other equilibrium that needs to be examined is zero density, $N_j = 0$. According to Eq. (2.1), $N_j = 0$ is not an equilibrium if there is any immigration from external sources. If there is emigration, but no immigration, zero density is a stable equilibrium if $b_j < d_j + e_j$. Thus, a species may go extinct from habitat patches where its niche requirements are in fact being met (i.e., $b_j > d_j$), because there is too great a rate of loss to the external environment due to emigration. This, in essence, is the process driving extinction in deterministic minimum patch size models for passively dispersing organisms (e.g., the KISS model for phytoplankton; Kierstead and Slobokin 1953), where losses across the patch edge into an unfavorable matrix overwhelm the reproductive capacity of the local population.

Thus, each of the three possible ways in which a local population can be in demographic equilibrium (as expressed by cases I through III above) are all consistent with the local habitat having conditions within the species' niche. As emphasized by Pulliam (2000), when there is dispersal, one may not be able to make strong inferences about whether or not local environmental conditions are within a species' niche from static distributional data. The above thoughts suggest that the same holds true even if one knows how birth rates and death rates vary across space. One also needs to know something about patterns and rates of dispersal and about how density dependence operates, in order to make sound inferences about niches from static data, particularly at fine spatial scales.

We can further categorize sink populations. If there is no emigration, then individuals enter, but do not leave. This is the "*black-hole sink*" that my colleagues and I have examined in several places (e.g., Holt and Gomulkiewicz 1997a, 1997b; Gomulkiewicz *et al.* 1999). If $b_j < d_j$, the population equilibrates at

$$N_j^* = \frac{I_j}{d_j - b_j}. \tag{2.5}$$

If we further assume that births and deaths are density independent in a black-hole sink, then without immigration the population declines to extinction, so the habitat is clearly an absolute sink, with conditions outside the species' niche.

If there are Allee effects (with the general form as shown for habitat 2 in Fig. 2.2B) and immigration but no emigration, a given habitat can have either two equilibria or one equilibrium, depending on the magnitude of the immigration rate. Figure 2.2C shows total growth rate in these habitats at different rates of immigation. At zero immigration (solid line), a population will go to extinction if it starts at low density, but can increase to its carrying capacity if sufficiently abundant. At a low but constant rate of immigration (the dashed line), stable low- and high-density sink populations can be alternative equilibria in the same habitat. The higher-density equilibrium is a pseudo-sink, because the population is pushed above its carrying capacity. The low-density equilibrium exists because there is a negative intrinsic growth rate and losses are replenished by immigration. But calling the habitat an "absolute sink" does not seem quite right, since the habitat can in fact potentially sustain a population. There is no accepted terminology for such cases, but we might call such a sink an "*Allee sink*," or a "*conditional sink habitat*." At yet higher immigration rates (the dashed line in Figure 2.2 (c)), the low-density equilibrium disappears entirely, and one will only see a pseudo-sink.

If there is emigration, growth rates are depressed over all densities. With an Allee effect, a gradual increase in emigration rates can cause a population to suddenly collapse from a high carrying capacity to zero density.

Important subtleties arise in defining sources and sinks when multiple habitats are linked by dispersal (Figueira and Crowder 2006; Runge *et al.* 2006). In general, the long-term contribution of an individual in a given habitat patch to the overall population, taking into account not just its own survival and reproduction, but that of its offspring, and their offspring in turn, and on into the future (its reproductive value), cannot be assessed simply by assessing the input–output status of the habitat in which that individual lives (Rousset 1999), but instead requires an accounting of the entire network of dispersal between patches, weighted appropriately by patch-specific fitnesses (Figueira and Crowder 2006; Runge *et al.* 2006). I briefly touch on some of these complexities below.

Evolution of local adaptation in a sink population: the fate of single favorable mutants

To summarize the above points: a source–sink population structure reflects how birth, death and dispersal rates depend on local habitat conditions. These rates also depend upon the phenotypic traits of organisms. Given genetic variation, evolution can occur, which in turn can change the spatial pattern of demographic rates, so that populations in absolute sinks are transformed into potential source populations. When this process occurs, it amounts to evolution in the niche of the species – which can now persist in habitats where it previously faced extinction without recurrent immigration. Conversely, a species initially well adapted to one habitat within its niche, but distributed over a range of habitat types, may over the course of time shift its pattern of utilization of other habitats, and even lose the capacity to persist in its ancestral habitat (Holt *et al.* 2003). Niche expansions, shifts and contractions are all potential outcomes in the evolution of species' ecological niches.

To develop evolutionary models within the demographic scaffolding of source–sink dynamics, we have to make assumptions about the nature of genetic variation influencing fitness. For simplicity, let us start by assuming that the population described by Eq. (2.1) has haploid or clonal genetic variation (the conditions for an allele to increase when rare also carry over to one-locus sexual models), and that the population is initially genetically homogeneous. Assume that an allele arises by mutation in this sink population and improves fitness there. Fitness of an allele can be defined as its expected instantaneous growth rate, which is the difference between the local birth and death rates of individuals with the allele, or $F = b - d$. If the allele improves the fitness of individuals who carry it in the local environment (as measured by its intrinsic growth rate when rare) by an amount $\delta > 0$, to a fitness of F' (say by an increase in birth rate from b to $b' = b + \delta$), the allele obviously has a higher relative fitness in the local environment, regardless of whether or not the habitat is a source or sink. But will it be able to increase deterministically in frequency?

In a spatially closed population in demographic equilibrium, the answer is "yes." The resident type must have a growth rate of 0 at equilibrium, so the mutant type has a growth rate of δ, and hence it will spread. But in a spatially open population, we need to pay attention to dispersal as well as local fitnesses. It is useful to consider first a limiting case, where we focus on evolution in the sink and ignore evolutionary dynamics outside it. If emigrants get mixed thinly into a very large and spatially widespread external population (what Levin 1976 calls a "bath"), we can assume to a first approximation that no individuals who leave will have descendants represented in the immigrant stream. Alternatively, there can be completely asymmetrical flows along a chain of habitats (e.g., imagine passive dispersers living in a mountain stream, with periodic waterfalls along its course dividing it into discrete habitats), with immigrants arriving from "upstream" habitats, and emigrants leaving to enter "downstream" habitats, but with no backflow. In effect, our allele is what Slatkin (1985) calls a "private allele" of the focal population.

In an open population, the net growth rate of the new allele has to account for losses by emigration as well as births and deaths (by assumption, there is no immigration of this allele). Its growth rate is $\frac{dN'}{dt} = (F' - e)N'$, where N' is the abundance of the new allele, F' its local absolute fitness, and e is the per capita rate of emigration (assumed equal for the resident type and the new allele). The resident clone has a net total growth rate (expressed per capita) of $\frac{1}{N}\frac{dN}{dt} = (F - e) + I / N$, so the realized growth rate of the resident at equilibrium, taking out immigration, is $F - e = -I/N < 0$ (the equilibrial density of the resident is $N^* = I/(e - F)$). For the growth rate of the new mutant to be positive, since δ is the increase in fitness enjoyed by the mutant (so that $F' = F + \delta$), it must be the case that $F' - e > 0$, or

$$\delta > e - F > 0. \tag{2.6}$$

This simple inequality defines a threshold effect on fitness required for this novel mutation to be retained by selection, when it is rare – see Gomulkiewicz *et al.* (1999); Kawecki (2000: 1317) derived the same result for a specific model. If mutants occur rarely (so that they can be considered one at a time), and most mutations have a very small effect upon fitness (the classical Darwinian assumption), inequality (2.6) implies that there is a constraint on the evolution of local adaptation in *any* population that receives immigrants and sends emigrants on a one-way trip to the external world, regardless of whether that population is an ideal free population, a demographic source, or a demographic sink. The inequality leads to two immediate conclusions.

First, the harsher the sink, the lower is F, so the greater the threshold that must be surmounted for a fitter allele to be retained in the sink environment. If most alleles that arise via mutation have small effects upon fitness (relative to the immigrant type), then most will go extinct, and adaptation to the sink will be slow, and not observed at all over reasonable time scales.

Second, the greater the rate of emigration e, the greater the fitness threshold that must be surpassed before an allele can be captured by selection. Increases in emigration rate thus hamper local adaptation.

Intriguingly, to reach these conclusions, we did not actually rely upon the assumption that the local population is a sink. The local F could be zero (our "ideal free" case), or even positive (i.e., a source), and (2.6) would still hold. The Appendix to this chapter provides a worked example for a consumer–resource interaction. So there is a kind of intriguing constraint on the evolution of local adaptation in *any* population that sends emigrants on a one-way trip to the external world, if that population is open and receives immigrants as well.

That is not to say that source–sink distinctions between habitats do not matter. If we look at the limiting case of low movement rates, then in both ideal free, source and pseudo-sink populations, we expect $F \approx 0$ (in the case of the pseudo-sink, the population will be only slightly perturbed above its local carrying capacity by a trickle of immigrants). In this case, even mutants with a very small effect on fitness have a chance of being captured by natural selection, and adaptation to the local environment can be honed. By contrast, in an absolute sink environment, $F < 0$ even at negligible immigration rates, so mutants with a small effect upon fitness cannot be captured by selection, and emigration just makes things worse.

If in a sink environment one can ignore density dependence, then in our asexual model, changes in the rate of immigration do not affect the fate of a new mutation, but changes in emigration rates assuredly do. The reason is that the realized growth rate in the sink of the new mutant is its intrinsic growth rate, minus losses to emigration. So increasing the rate of export to the external world depresses the realized local growth rate of a new allele and therefore makes it less likely to persist.

If there is density dependence in the sink, changes in immigration rates can either hamper or facilitate adaptive evolution, depending on the nature of density dependence in the sink (Gomulkiewicz *et al.* 1999; Holt *et al.* 2004b). When there is no density dependence, fitness in the sink is unchanged by altering immigration, and so the criterion given by (2.6) for selective retention of a novel allele is independent of the rate of immigration. In a genetically fixed population with continuous growth and direct density dependence, an increase in the rate of immigration increases equilibrial population size (Holt 1983). But with negative density dependence in the sink (e.g., because immigrants use up resources, as in the example shown in the Appendix), this increased abundance lowers the baseline fitness, F, of the population. This in turn increases the magnitude of δ needed for a favorable mutant to be retained by selection, and could prevent alleles with a moderate effect upon fitness from increasing in frequency. Immigration can thus hamper selection for ecological reasons (Holt 1997; Kawecki and Holt 2002).

By contrast, if there is an Allee effect, increases in immigration can enhance fitness and thus reduce this threshold value of δ. In this case, a moderate amount of immigration can indirectly facilitate adaptive evolution, by reducing the fitness benefit required for a novel allele to increase when rare. Thus, increased immigration can at times foster adaptation to the sink and thus expansion of the niche (Holt *et al.* 2004b).

Gomulkiewicz *et al.* (1999) analyze, in some detail, a discrete-generation model for selection at a diploid locus in a black-hole sink. They show that there is an absolute fitness criterion for selection to retain a locally favorable allele. Figure 2.3 (adapted from Holt and Gomulkiewicz 1997a) schematically explains why this is the case, for selection at a diploid locus with alleles A and

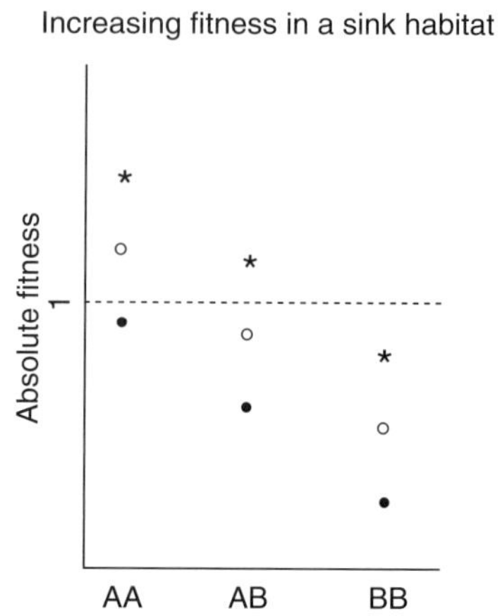

FIGURE 2.3.
Genotypic fitness in a sink for a one-locus diploid model (adapted from Holt and Gomulkiewicz 1997a). The immigrant genotype is assumed fixed for the locally less-fit allele B, and allele A has arisen by local mutation. In all cases, the fitness of the heterozygote, and the homozygote for allele A, is greater than the fitness of the homozygote for B. The three symbols denote sinks differing in harshness, with increasing harshness from asterisks, to open circles, to solid circles. With random mating, rare allele A will be found mainly in heterozygotes. As explained in the main text, the criterion for an increase in A is that the absolute fitness of the heterozygote must exceed unity. In the absence of density dependence, this absolute fitness criterion is independent of the rate of immigration or the fitness of the less-fit homozygote. Thus in the harsh sink, allele A is lost. In the mild sink, allele A can increase when rare. In the intermediate sink, allele A cannot increase when rare. However, the homozygote for A has absolute fitness greater than one, and there is a threshold gene frequency above which it can then increase rather than decline.

B. The source population is fixed for allele B, and so all immigrants are BB. In all cases, each copy of allele A carried by an individual has an additive effect on fitness, increasing fitness in the heterozygote by a fixed amount, with an equal increase when homozygous. The black dots indicate fitnesses in a relatively harsh sink. The white dots and asterisks indicate sinks which are progressively less harsh. In the harsh sink, even though allele A has a larger relative fitness than allele B, we know that it cannot increase in frequency because each individual carrying it is not replacing itself. So its numbers will decline, even as the numbers of allele B are maintained by immigration. In the relatively benign sink indicated by the asterisks, both AB and AA have fitnesses exceeding one, so each copy of allele A more than replaces itself, and it is expected to be retained and to spread in the sink population. In the intermediate sink, if the allele is rare, with random mating it will be expected to be found only in heterozygotes, and so it should decline in frequency because each such allele is not replacing itself. But if sufficiently frequent, enough homozygotes AA may be present for the allele to increase.

Formally, Gomulkiewicz *et al.* (1999) include density-dependent fitnesses and show that when the favorable allele is rare in a stable sink population, the gene frequency recursion over one generation is described by

$$p_{t+1} \approx p_t W_{AB}(N_{BB}^*),$$

where p_t is the frequency of allele A at generation t, and W_{AB} (N_{BB}^*) is the absolute fitness of the heterozygote, when the immigrant homozygote is at its equilibrial density. Negative density dependence depresses fitness, and if it pushes heterozygote fitness below a value of 1, the favored allele will disappear from the population, when initially rare. Immigration tends to increase the abundance of the resident genotype (Holt 1983), and so immigration can indirectly affect selection via its effect on absolute fitness.

If the environment is variable, or the population is unstable, the gene frequency recursion, iterated over t generations, is $p_t \approx p_0 W_g^t$, where $W_g = \left[\Pi_{i=1}^{t} W_t(N_{BB}(t), t)\right]^{1/t}$ is the geometric mean absolute fitness of the heterozygote over this time-span (LoFaro and Gomulkiewicz 1999). Since geometric means are dominated by low values, Gomulkiewicz *et al.* (1999) suggest that temporal variation in the environment, affecting fitness either directly or indirectly via changes in density, could hamper selection. Clearly, any year of zero fitness for the allele will expunge it from the population, so extreme temporal variability in sink fitness will usually hamper adaptive evolution.

The above models are deterministic and ignore mutation and genetic drift. A full accounting of evolution in sink environments must consider the origin and maintenance of genetic variation, not only selection. Mutations can occur either in the source or sink, and favorable alleles can be lost due to drift. Gomulkiewicz *et al.* (1999) develop a branching process approach to this problem which leads to the following approximation for the overall rate of establishment of alleles permitting local persistence (and hence niche evolution): $2\varepsilon(N^*)(N^* \upsilon + I p_{source})$, where N^* is the equilibrial population abundance maintained by I immigrants per generation, υ is the local mutation rate, p_{source} is the gene frequency of the favorable mutation found in the source, and ε is the probability of persistence of the descendants of a given copy of a favorable allele after it appears in the sink at some given future time, expressed as a function of abundance. The term in the second set of parentheses expresses the rate at which novel favorable alleles are expected to arise either by immigration or by mutation in the sink, and this term increases with immigration rate (both directly and indirectly via the effect of immigration on N^*). The probability of establishment for a given allele declines with decreasing fitness of the heterozygote, and if fitness declines strongly with increasing immigration (via density dependence), evolution could be constrained by high immigration rates.

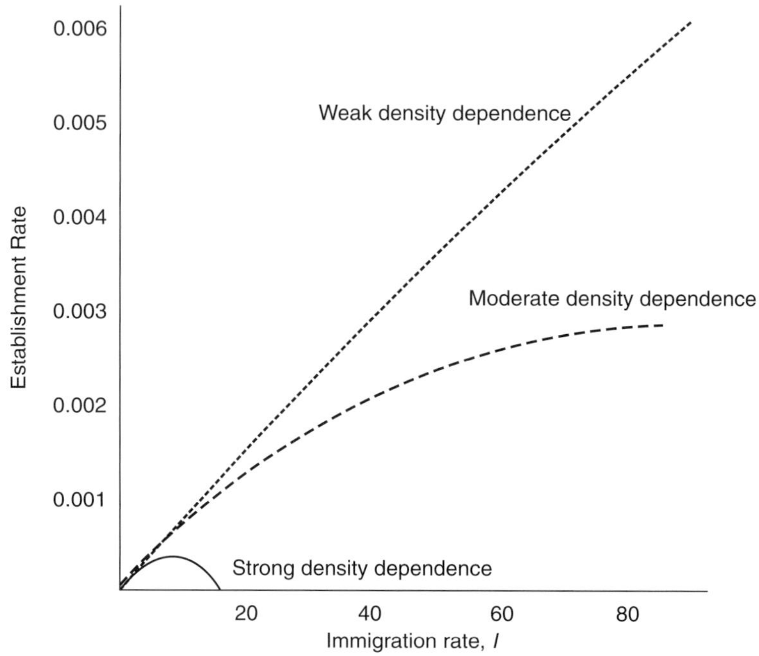

FIGURE 2.4.
The probability of establishment of a favorable allele in a sink (adapted from Gomulkiewicz *et al.* 1999). When established, this allele will transform the sink population into a persistent population that does not need immigration to survive. The progression from short dashes, to long dashes, to the solid line corresponds to increasingly strong negative density dependence in the sink (see text and Gomulkiewicz *et al.* 1999 for more details).

Putting these two effects together leads to the prediction that there will be an intermediate rate of immigration that is most favorable for evolution; this rate will be quite low if density dependence in the sink is strong, and may be very high if density dependence is weak. Figure 2.4 shows examples of the per generation rate of establishment predicted by this model.

These results do not explicitly account for emigration, which can easily be included by multiplying local fitness by $1 - \mu$, where μ is the proportion of individuals that emigrate each generation. If one works back through the analyses of Gomulkiewicz *et al.* (1999), it can be readily seen that when fitnesses are density independent, emigration always makes it more difficult for a local allele to spread when rare.

Perron *et al.* (2007) have recently created experimental sinks in a laboratory microcosm, using the evolution of resistance to antibiotics in *Pseudomonas aeruginosa* as a model system. The number of immigrants was low, relative to the potential carrying capacity, so density dependence was likely to be negligible. Their "mild" sink for this clonal organism contained a single antibiotic; their "harsh" sink contained two antibiotics simultaneously. The rate of adaptation was slower in the harsh sink environment. In each environment, adaptation permitting persistence occurred more rapidly with higher immigration rates. Some replicates in the harsh sink never adapted during the time scale of the experiment. These experimental results are qualitatively consistent with the theoretical prediction that local adaptation is easier in mild than in harsh sink environments, and that immigration can facilitate local adaptation.

Evolution of local adaptation in sink environments: quantitative genetic approaches

The above models assume that adaptation is determined by a single major gene locus (or even clonal variation), and so we focused on the fate of a single allele at this locus. Many traits of ecological relevance are instead influenced by multiple loci, each of small effect, with many alleles segregating at each locus. In sexual species, the traits of offspring will be a combination of parental traits, and mating between immigrants and residents in sinks can have an important impact upon the likelihood of adaptation there. Assuming that traits undergoing selection in a sink are influenced by a quantitative trait leads to some conclusions that parallel those presented above, and others that differ. It would take too much space to lay out the full models here, so I instead focus on reviewing the results.

Holt *et al.* (2004a) developed a deterministic quantitative genetic model for evolution in a sink, assuming fixed heritability, and a parallel individual-based simulation model, which allows heritability to change because of mutation, drift and selection, in order to explore the impact of temporal variation on sink evolution. In the sink, selection is on survival to adulthood. The relationship between survival of an individual and its phenotype z is assumed to be given by a Gaussian function $W(z)=\exp[-(z-\theta)^2/(2\omega^2)]$, where θ is the optimum phenotype in the sink, and $1/\omega^2$ is the strength of selection. Individuals in the source have an average phenotype of 0, so θ is a measure of the degree to which immigrants are maladapted in the sink. Immigration occurs after selection, and there is random mating between surviving residents and immigrants. The effective rate of immigration (which determines the strength of gene flow) is $m_t = I/(N_t + I)$, which varies if the population size itself varies over time (N_t is the sink population size and I the number of immigrants per generation). In Holt *et al.* (2004a), it is assumed that the immigration rate into the sink is constant across generations.

Figure 2.5 shows examples of the equilibria predicted from the deterministic model, with no density dependence in the sink. At low degrees of maladaptation (low θ) there is only one equilibrium, corresponding to the upper heavy curve, which describes an adapted population with a mean genotype near the

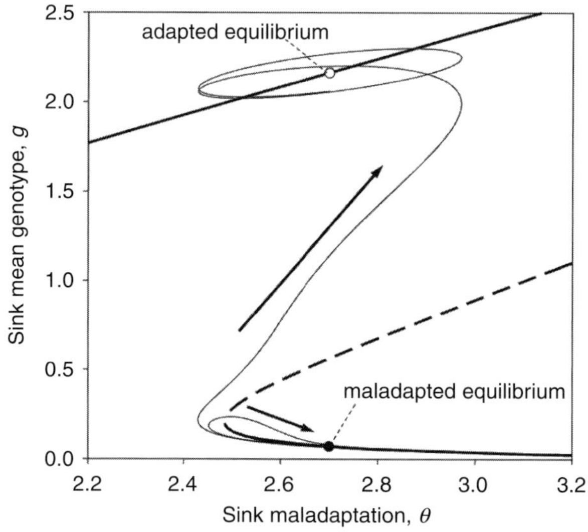

FIGURE 2.5.
Adaptation to a sink for a quantitative genetic model. The model, which assumes polygenic inheritance for a single trait experiencing selection, density-independent growth below a ceiling, and recurrent immigration, is described in detail in Holt *et al.* (2004a). The heavy solid lines describe stable equilibria of the model. At low maladaptation in the sink, the population will adapt (and can then persist without recurrent immigration). At higher maladaptation, the population may have two alternative evolutionary equilibria, in one of which it is severely maladapted to the local environment. This maladaptation is maintained by recurrent gene flow from the source. Moderate temporal variation in the environment (i.e., in the locally optimal phenotype) can permit a species to "escape" the maladapted state (see text and Holt *et al.* 2004a for more details).

sink optimum (somewhat lower due to recurrent gene flow). If a population is initially maladapted, but at low levels, it becomes adapted. But at higher degrees of maladaptation, there are two locally stable equilibria, one that is relatively well adapted (upper curve) and one that stays maladapted (lower heavy solid curve, with mean genotype near 0). The latter exists because random mating between maladapted immigrants and better-adapted residents depresses the reproductive success of the latter, thus weakening the response to selection.

The fate of a new mutant may be determined as much by its genetic environment as by the external environment. In a sexual, outcrossing species, once adaptation in the sink gets off the ground, further immigration from the source will tend to hamper adaptation, because relatively well-adapted residents can suffer a reduction in fitness because of mating with relatively maladapted immigrants. One way to understand why alternative stable states emerge is to consider the indirect influence of selection on the strength of gene flow. If selection is sufficiently strong to increase absolute fitness over a generation, this increases population size, and this in turn reduces m_t for the following generation. This sets up a positive feedback, where selection becomes yet more effective in shifting the population toward its new optimum. Conversely, if absolute fitness is initially low, the population will be largely composed of immigrants, and mating with them will reduce any advantage held by residents who survived the last round of selection. Again, this is a positive feedback, as initially maladapted populations tend to become yet more maladapted (due to recurrent gene flow) as their numbers decline (Ronce and Kirkpatrick 2001; Tufto 2001). Comparable results emerge in individual-based models, except that at moderate to intermediate levels of maladaptation, where the only long-term equilibrium is adaptation to the sink, there can be long periods of maladaptation observed before a rapid transition to the adapted state (Holt *et al.* 2003).

Sustained maladaptation is thus more likely if immigrants are initially strongly maladapted to the sink environment. If selection is on survival, reduced fecundity can further constrain selection (Holt and Gomulkiewicz 2004; Boulding 2008). This result qualitatively matches the conclusion discussed above, that adaptation should be more difficult in harsher sinks where the adaptive threshold required to capture a locally favored allele is higher. The detailed causal mechanism, however, is different, since it involves recurrent gene flow directly hampering adaptation. Moreover, stable maladaptation in a sink is more likely when selection acts on characters with low heritability (Holt and Gomulkiewicz 1997b; Boulding and Hay 2001; Holt *et al.* 2004a). Initially adapted local populations with recurrent immigration of maladapted individuals can also risk losing adaptation if numbers are perturbed toward low density. Ronce and Kirkpatrick (2001) coined the pungent phrase "migrational meltdown" to describe how a species could collapse from habitat generalization (adapted to two habitats) to habitat specialization (adapted to only one) because of gene flow overwhelming local selection in one of the habitats. These authors considered two patches with equal reciprocal dispersal rates, but comparable processes are at work in a black-hole sink maintained by recurrent immigration.

In this model, negative density dependence tends to further constrain adaptation in the sink. Indeed, if density dependence is strong, at high degrees of sink maladaptation, the only equilibrium that may exist for the sink is the maladapted one. Again, this result parallels the conclusions we reached using the haploid and diploid models above. Negative density dependence in sinks tends to make adaptation to conditions there more difficult, and can aggravate the effects of migrational load (see also Kawecki 2000).

The one-locus model sketched above suggests that temporal variation should tend to hamper adaptation to sink environments. Different results can emerge with multilocus variation. Figure 2.5 provides one example, showing that sometimes temporal variation in the environment can facilitate adaptive evolution in a sink. We start with a constant environment (indicated by the black dot), where the sink population is at its maladapted equilibrium. The environment then begins varying cyclically, with increasing amplitude (while maintaining the same mean). Initially, this leads to only modest fluctuations in the population state (thin line in Fig. 2.5, marked with arrows), but eventually the sink maladaptation is sufficiently weak that only the adapted equilibrium exists, and the sink population begins to grow as it adapts. As noted above, this sets up a positive feedback, because burgeoning local numbers weaken gene flow and permit selection to be more effective. During a transient phase of improved conditions, the population can escape the trough of maladaptation and become sufficiently adapted and numerous that it can remain reasonably well adapted, tracking the moving environment. When the environment then returns to its initial constant state, the population has moved to a new, adapted local equilibrium (the white dot). Similar results emerge if the variation is not cyclical, but stochastic with a positive autocorrelation, and also show up in the individual-based model. If there is very large amplitude variation in the optimum, however, adaptation can be lost. The bottom line is that a moderate amount of temporal variation in the environment can foster adaptation in a sink.

This seems to contradict the claim made above that environmental variation inhibits adaptation, which was based on geometric mean fitness in the diploid model. That claim rested on the assumption that gene frequency stays low enough that nearly all copies of the favorable allele are found in heterozygotes. At higher gene frequencies, mating between heterozygotes can produce homozygotes for the favorable allele. Moderate temporal variation can increase the scope for this to happen, provided that initial gene frequency is not too low, and therefore enhance the scope for adaptive evolution.

One difference that arises between these models is in the effect of immigration. We noted above, for the one-locus model, that in the absence of density dependence, increased immigration tends to facilitate adaptive evolution by providing genetic variation. In the individual-based quantitative genetic models explored in the research summarized above, this facilitative effect of immigration on the evolution of local adaptation is largely canceled out by disruption of adaptation by recurrent gene flow due to adult dispersal, followed by mating between immigrants and residents (Holt *et al.* 2005).

If instead of having a recurrent flow of immigrants, we consider single bouts of colonization, with long time periods in any given sink between episodes of dispersal from a source, then increasing the number of individuals found in a colonizing propagule substantially enhances the probability of successful colonization "outside the niche," even for adult dispersal (Holt *et al.* 2005;

Holt and Barfield 2011). One should also keep in mind that evolution in sink environments is likely to reflect the complex interplay of multiple evolutionary processes that go well beyond the models reviewed above. For instance, the accumulation of deleterious mutations in sinks may hamper adaptation (Kawecki *et al.* 1997), as could inbreeding depression because of mating between kin in low-density populations (Willi *et al.* 2006).

Coupled source–sink evolution

I have summarized up to now theoretical studies of evolution in black-hole sinks, where there is a one-way flow from source to sink. The absence of such evolution is tantamount to niche conservatism. There are some natural situations which are likely to fit (to a reasonable approximation) the black-hole sink scenario, with no back-dispersal to the source. But more often, one might expect that some individuals from the sink (or their descendants) will find their way back to the source. This leads to a number of interesting complications, which are still not fully understood.

In considering this scenario, Tad Kawecki (1995) and I (Holt 1996a, 1996b); see also Holt and Gaines 1992) first took an evolutionary ecology or adaptive dynamics approach to the problem, where we assumed that a genetically homogeneous population is at demographic equilibrium in two habitats coupled by dispersal, and then alleles of very small effect are introduced. To determine the fate of these alleles, we used a sensitivity analysis, following standard protocols (Caswell 1989; Rousset 1999; Kawecki 2004). For novel alleles of small effect upon fitness, their distribution across the two habitats is governed by the pattern of distribution of the resident type. So, if dispersal is infrequent between the source and sink, few individuals should be found in the sink, and they will be of low reproductive value. The strength of selection favoring an allele of small effect should thus be weak. By contrast, if dispersal is frequent in both directions, this increases the number of individuals exposed to sink conditions, and equalizes their reproductive value (because many of their descendants are found back in the source). In evolutionary analyses of coupled source–sink systems, it has to be remembered that the reproductive value of an individual in a sink (or source) is not equivalent to the growth rate just of that habitat, but instead reflects the long-term contribution of an individual in that habitat to the entire population (Rousset 1999), and so reflects the entire network of movement between habitats which governs the distribution of offspring, grand-offspring, and so on across space.

Moreover, as emphasized by Tad Kawecki in several places (Kawecki 2000, 2004; Kawecki and Holt 2002), results based on sensitivity analyses really only pertain to alleles that have small effects upon fitness, relative to dispersal rates, and do not capture the impact of movement upon genetic differentiation between populations. When fitness effects are large, frequency differences emerge between habitats, which then implies that a new allele experiences a different "weighting" of the habitats than does the ancestral resident. As Kawecki (2000: 1317) notes, even for quite simple haploid models "the relationship between the dispersal rate and the conditions for the invasion of a mutant allele is complex and differs qualitatively between alleles with small and large effects." If an allele arises by mutation that can permit persistence in the sink in one fell swoop, but is lethal in the source (i.e., its fitness effects are very large), then the optimal movement pattern for this allele is clearly to have no movement from the sink back to the source and, if movement can occur before selection, to move at maximal rate from the source to the sink. The details of genetic architecture (e.g., the average effects of alleles upon fitness, and the number of loci influencing fitness) can thus have a strong impact on the likelihood of adaptation in a sink, and on the conditions which favor such adaptation. Lower dispersal rates also permit the build-up of locally differentiated gene pools when alleles have large effects upon fitness, but this may be more difficult if multiple loci with small allelic effects upon fitness are involved (Kawecki 2008). This implies that the pattern of movement that is optimal for adaptation to the sink depends upon the magnitude of allelic effects upon fitness, as well as upon the strength of the tradeoffs in fitness across habitats.

Theory has only just begun to address the implications for niche conservatism and evolution of the rich diversity found among organisms in the structure of life histories, mating systems, behavior, genetic architecture, and environmental context. For instance, the order of events in the life history can play a crucial role in adaptation to a sink. In the models reviewed above, it was assumed that adults immigrate, followed by random mating and then selection of the offspring. An increase in immigration rate then increases the reproductive "load" experienced by relatively adapted residents due to mating with maladapted immigrants and can lead to permanent maladaptation. If instead, juveniles immigrate, and selection occurs before mating, many maladapted individuals will be weeded out from the local mating pool, reducing this reproductive load (Ronce and Kirkpatrick 2001; Holt and Barfield 2011). Also, in individual-based models of sink populations, if sink abundance is low, genetic variation can be limiting. Increased immigration into the sink can then facilitate adaptation, because of the infusion of genetic variation from the source (Barton 2001; Holt and Barfield 2011). As another example, Kawecki (2003) found that strongly female-biased dispersal (the norm in birds, but not in mammals) could facilitate adaptation to marginal habitats. In plants, pollen dispersal provides a different conduit for gene flow than does the movement of seeds (Antonovics 1976). In the sea rocket, pollen flow from the adapted seaward edge could help constrain adaptation to the dune interior. Behavior can likewise have large effects upon niche evolution. Sexual selection can assist the evolution of adaptation to a sink, if females choose males with locally appropriate traits (Proulx 2002). Habitat selection can either hamper or facilitate niche evolution, depending on how sensitive individuals are to their own genotypes in making decisions to move between habitats (Holt and Barfield 2008). In general, one expects a kind of coevolution between movement strategies and the ability to utilize local environments (Holt 1997, 2003; Cohen 2006) that should determine the evolutionary stability or transience of sink habitats within a species' geographical range.

Analyses of evolution along smooth gradients (e.g., Kirkpatrick and Barton 1997) can lead to somewhat different conclusions than models of discrete sources and sinks (Kawecki 2008). Further work is needed to determine whether these differences reflect subtle effects of the ecological assumptions (e.g., the juxtaposition of distinct habitats versus smooth transitions along gradients) or other assumptions about the genetic architecture of the traits built into the models.

Concluding thoughts

To conclude, I return to the series of questions posed initially about adaptive evolution in sinks. The theoretical studies sketched above do not provide complete answers to any of these questions, but do hint at the range of potential outcomes.

So, what does current theory say about the effect of each of the following on adaptive evolution in sinks, leading potentially to niche evolution?

The severity of the sink environment

One generalization that transcends many differences between models of evolution in source–sink systems is that the harsher the sink environment (in an absolute sink, *sensu* Kawecki 2008), the less effective selection may be for improving adaptation there. For instance, with unidirectional flow in the clonal model above, there is an absolute fitness constraint which is more difficult to surmount, as it requires alleles of large positive effect upon fitness. For the sea rocket, adaptive evolution may be more likely in the center of the dune than on the landward side.

The rate of immigration

In some cases, immigration provides a potent source of genetic variation for selection, and increased immigration fosters adaptation to the sink. But when maladapted immigrants mate with better-adapted residents, this cuts in the other direction, leading to a migrational load and effects such as migrational meltdown, arising because gene flow swamps selection (Ronce and Kirkpatrick 2001; Lenormand 2002). These disparate impacts of dispersal can lead to many differences among species in their likelihood of adaptation to sink environments (Garant *et al.* 2007). Which effect of immigration dominates depends on many factors, such as the timing of dispersal in the life history of a species and the genetic architecture governing traits determining local adaptation. For the sea rocket, demographic immigration is by seed dispersal. If selection acts on juvenile survival, then maladapted individuals brought in by the wind are not likely to make it to where they can reproduce with residents. The main effect of immigration may then be to facilitate adaptation, by adding an additional source of genetic variation. The prediction one could make for the sea rocket might not apply to other species in its community.

Temporal variability in immigration rate

If immigration occurs in sporadic colonizing pulses, there is no recurrent confounding of selection by gene flow. An increase in "propagule pressure" (the product of the number of individuals per colonizing episode and the number of such episodes) will in general hasten adaptation to a sink (Holt *et al.* 2005). The whole issue of how temporal variation in dispersal influences adaptive evolution has been largely neglected in the literature. Along the coastline, different patterns of variation in the wind could lead to different patterns of immigration into sea rocket sink habitats, which might then experience emergent spatial differences in the likelihood of local adaptation in the sinks.

The rate of emigration

If emigrants leave, and their descendants do not return, emigration is equivalent to increased mortality in the sink. So one-way emigration should make adaptation to the sink more difficult. In the sea rocket, any emigration from the center of the dune to the dune edge makes adaptation to the center even more difficult.

The directionality of dispersal, and tradeoffs

When there are reciprocal movements between sources and sinks, one has to consider both the fitness consequences of a given allele in each habitat, and the patterns of movement in both directions. There are many subtleties that can arise. For instance, there are two complementary ways in which one can think about bidirectional flows. At the population perspective, when emigrants from the sink can enter the source, they can potentially influence evolution there, with feedback effects on the sink. From the perspective of the gene lineage stemming from a new mutation, descendants will be found in both habitats, so the overall fitness describing the rate of growth of the lineage will in a sense be a weighted average over both habitats, where the "weights" usually involve a nonlinear function incorporating movement rates and local fitnesses. Usually these weights will be biased toward habitats already within the niche, which – if movement is limited – will be where most individuals occur (Holt and Gaines 1992; Kawecki 1995, 2000; Cohen 2006); selection in averaging across the two habitats tends to discount conditions in the harsh sink. But using the fraction of individuals found in a given habitat can be a poor indication of the direction of evolution if habitats are not absolute sinks, but are instead pseudo-sinks (Kawecki and Holt 2002). With bidirectional dispersal and tradeoffs in fitness between the source and sink, selection can actively weed out alleles that might improve adaptation to the sink if they are too costly in the source; this is more likely to occur if the sink is harsh to start with. In the sea rocket, it seems likely that it faces adaptive tradeoffs in morphology, physiology, and other traits between the sea margin and the dune interior, and that specialization to the sea margin could preclude adaptation to the dune.

So niche conservatism might be expected in source–sink systems when there are sharp differences in fitness, adaptive optima, and population size between sources and sinks. But exceptions can occur, for instance due to strong asymmetries in movement (Holt 1996a; Kawecki and Holt 2002).

Temporal variability in the sink environment

A geometric mean fitness argument for a single-locus model suggests that temporal variation in selection should make adaptation to a sink more difficult. But a quantitative genetic model showed that temporal variation could sometimes facilitate adaptive evolution, if moderate in magnitude and with some positive autocorrelation. The reason is that a few years of moderate conditions may permit adaptation, which then leads to increases in population size and so weakens the effect of gene flow. As with immigration, different patterns may emerge depending upon the details of genetic architecture for traits controlling fitness. In the sea rocket, temporal variability in the quality of the dune sink habitat could potentially facilitate its adaptive expansion there.

Interspecific interactions

The above discussion focused entirely on a single species. But the *reason* why a habitat is a sink in the first place may be because of strong negative interspecific interactions, or the absence of positive interactions (see Benkman and Siepielski, Chapter 4, this volume). For instance, generalist predators or other natural enemies may inflict high mortality, creating sink conditions. In the case of sea rocket, it could be that rabbits are present and abundant on the landward side of the dune, and find sea rocket seedlings to be tasty morsels. Or the sea rocket might be a poor competitor for light in the thick vegetation behind the dune. Or an essential mutualist might be scarce or absent, such as specialized mycorrhizae needed for germination. Interspecific interactions could magnify and sharpen source–sink conditions that exist for other reasons, magnifying the decline in demographic performance of the sea rocket from the center of the dune to the landward side. Conversely, source–sink dynamics can have implications for the evolutionary dimension of local interspecific interactions.

Assume, for a moment, that a prey species is immigrating into a black-hole sink where a generalist predator is present, sustained by alternative food sources. The question we are interested in is how evolution will alter anti-predator adaptations in the immigrant prey. As noted above, theoretical studies suggest that the harsher the demographic conditions of the sink are, the harder it may be for adaptive evolution to occur, particularly if available genetic variants have a small effect upon local fitness. Thus, the more effective the resident predator is at capturing the focal prey species, or the more abundant that predator is in the sink, the *less* effective natural selection will be in the prey for sculpting anti-predator morphological or behavioral defenses. My friend Bill Kunin once quipped that the ecological folk wisdom is that "evolution works hardest where the shoe pinches worst." In sink environments, this is exactly the opposite: the harsher the sink, the harder it may be for adaptive evolution to transform the sink into a source.

The effect of changing immigration rates into the sink on adaptation to the predator may depend upon the population dynamics of the predator itself. If the predator has fixed abundance, and a functional response that can be saturated, increasing rates of immigration of the prey species should reduce the per capita mortality rate experienced by the prey. This is an Allee effect, and so increasing immigration (up to a point) implies that the sink is less severe, and so according to the arguments presented above, the prey should also be better able to adapt to the resident predator. Conversely, if the predator has a pronounced numerical response to the immigrant prey, or switches its attention to this prey as it gets more common, increasing the prey immigration rate boosts predation, and so makes the sink harsher. In this case, the prey may be less able to adapt to the predator in the sink.

Finally, if the habitat is a harsh "intrinsic sink" (a sink in the absence of the predator), for instance due to unfavorable abiotic conditions or scant resources, predation just makes things worse. Evolution is then impotent at sculpting anti-predator adaptations in the sink, at least in a black-hole sink. Imagine that an allele comes along permitting the prey to completely escape predation. In an intrinsic source habitat, such an allele would sweep through the population. But if the habitat is an intrinsic sink, individuals carrying this allele, even if they escape predation completely, still have an absolute fitness of less than one, and so these alleles will disappear from the sink population. Sink environments can thus impose a kind of constraint on coevolutionary responses by one species to another.

These seemingly abstract observations could have important applied implications, for instance to the evolutionary stability of biological control of agricultural pests, and to the evolution of host–pathogen interactions in heterogeneous host populations and communities. Our understanding of source–sink dynamics in heterogeneous landscapes has been greatly advanced over the past 20 years, stimulated in large measure by the clarity and eloquence of Ron Pulliam's 1988 exposition of this theme. I believe that a clear analysis of source–sink dynamics is also of fundamental importance for many topics in evolutionary biology, such as niche evolution and conservatism, and the evolutionary dimension of interspecific interactions, and that the time is ripe for a deepened theoretical and empirical understanding of this theme.

Acknowledgments

I have had the good fortune to have discussed and written papers on the themes explored in this chapter with many friends and collaborators over the years, in particular Mike Barfield, Doug Futuyma, Richard Gomulkiewicz, John Thompson, Samantha Forde, Paul Turner, Mark McPeek, Michael Hochberg, Andy Gonzalez, and Tad Kawecki. I thank you all. I also thank the University of Florida Foundation for its support, as well as NSF and NIH, and Jack Liu and his associates for their invitation to participate in the Pulliam Festschrift.

References

Antonovics, J. (1976). The nature of limits to natural selection. *Annals of the Missouri Botanical Garden* **63**: 224–247.

Antonovics, J., T. J. Newman and B. J. Best (2001). Spatially explicit studies on the ecology and genetics of population margins. In *Integrating Ecology and Evolution in a Spatial Context* (J. Silvertown and J. Antonovics, eds.). Blackwell Scientific, Oxford, UK: 91–116.

Arditi, R., N. Perrin and H. Saiah (1991). Functional responses and heterogeneities: an experimental test with cladocerans. *Oikos* **60**: 69–75.

Barton, N. (2001). Adaptation at the edge of a species' range. In *Integrating Ecology and Evolution in a Spatial Context* (J. Silvertown and J. Antonovics, eds.). Blackwell Scientific, Oxford, UK: 365–392.

Blows M. W. and A. A. Hoffmann (2005). A reassessment of genetic limits to evolutionary change. *Ecology* **86**: 1371–1384.

Boulding, E. G. (2008). Genetic diversity, adaptive potential, and population viability in changing environments. In *Conservation Biology: Evolution in Action* (S. P. Carroll and C. W. Fox, eds.). Oxford University Press, Oxford, UK: 199–219.

Boulding, E. G. and T. K. Hay (2001). Genetic and demographic parameters determining population persistence after a discrete change in the environment. *Heredity* **8**: 313–324.

Bradshaw, A. D. (1991). Genostasis and the limits to evolution. *Philosophical Transactions of the Royal Society of London (B)* **333**: 289–305.

Bridle, J. R., J. Polechova, M. Kawata and R. K. Butlin (2010). Why is adaptation prevented at ecological margins? New insights from individual-based simulations. *Ecology Letters* **13**: 485–494.

Caswell, H. (1989). *Matrix Population Models*. Sinauer Press, Sunderland, MA.

Cohen, D. (2006). Modeling the evolutionary and ecological consequences of selection and adaptation in heterogeneous environments. *Israel Journal of Ecology and Evolution* **52**: 467–485.

Courchamp F., L. Berec and J. Gascoigne (2008). *Allee Effects in Ecology and Conservation*. Oxford University Press, Oxford, UK.

Diffendorfer, J. E. (1998). Testing models of source–sink dynamics and balanced dispersal. *Oikos* **81**: 417–433.

Doncaster, C. P., J. Clobert, B. Doligez, L. Gustafsson and E. Danchin (1997). Balanced dispersal between spatially varying local populations: an alternative to the source–sink model. *American Naturalist* **150**: 425–445.

Figueira, W. F. and L. B. Crowder (2006). Defining patch contribution in source–sink metapopulations: the importance of including dispersal and its relevance to marine systems. *Population Ecology* **48**: 215–224.

Fretwell, S. D. (1972). *Populations in a Seasonal Environment.* Princeton University Press, Princeton, NJ.

Futuyma, D. J. (2010). Evolutionary constraint and ecological consequences. *Evolution* **64**: 1865–1884.

Garant, D., S. E. Forde and A. P. Hendry (2007). The multifarious effects of dispersal and gene flow on contemporary adaptation. *Functional Ecology* **21**: 434–443.

Gomulkiewicz, R. and R. D. Holt (1995). When does evolution by natural selection prevent extinction? *Evolution* **49**: 201–207.

Gomulkiewicz, R., R. D. Holt and M. Barfield (1999). The effects of density dependence and immigration on local adaptation and niche evolution in a black-hole sink environment. *Theoretical Population Biology* **55**: 283–296.

Holt, R. D. (1983). Immigration and the dynamics of peripheral populations. In *Advances in Herpetology and Evolutionary Biology* (K. Miyata and A. Rhodin, eds.). Museum of Comparative Zoology, Harvard University, Cambridge, MA: 680–694.

Holt, R. D. (1984). Spatial heterogeneity, indirect interactions, and the coexistence of prey species. *American Naturalist* **124**: 377–406.

Holt R. D. (1985). Population dynamics in two-patch environments: some anomalous consequences of an optimal habitat distribution. *Theoretical Population Biology* **28**: 181–208.

Holt, R. D. (1996a). Adaptive evolution in source–sink environments: direct and indirect effects of density-dependence on niche evolution. *Oikos* **75**: 182–192.

Holt, R. D. (1996b). Demographic constraints in evolution: towards unifying the evolutionary theories of senescence and niche conservatism. *Evolutionary Ecology* **10**: 1–11.

Holt, R. D. (1997). On the evolutionary stability of sink populations. *Evolutionary Ecology* **11**: 723–731.

Holt, R. D. (2003). On the evolutionary ecology of species ranges. *Evolutionary Ecology Research* **5**: 159–178.

Holt, R. D. (2009). Bringing the Hutchinsonian niche into the 21st century: ecological and evolutionary perspectives. *Proceedings of the National Academy of Sciences of the USA* **106**: 19659–19665.

Holt, R. D. and M. Barfield (2008). Habitat selection and niche conservatism. *Israel Journal of Ecology and Evolution* **54:** 295–309.
Holt, R. D. and M. Barfield (2011). Theoretical perspectives on the statics and dynamics of species' ranges. *American Naturalist* **177:** in press.
Holt R. D. and M. S. Gaines (1992). Analysis of adaptation in heterogeneous landscapes: implications for the evolution of fundamental niches. *Evolutionary Ecology* **6:** 433–447.
Holt R. D. and R. Gomulkiewicz (1997a). How does immigration influence local adaptation? A reexamination of a familiar paradigm. *American Naturalist* **149:** 563–572.
Holt, R. D. and R. Gomulkiewicz (1997b). The evolution of species' niches: a population dynamic perspective. In *Case Studies in Mathematical Modelling: Ecology, Physiology, and Cell Biology* (H. Othmer, F. Adler, M. Lewis and J. Dallon, eds.). Prentice-Hall, Englewood, NJ: 25–50.
Holt, R. D. and R. Gomulkiewicz (2004). Conservation implications of niche conservatism and evolution in heterogeneous environments. In *Evolutionary Conservation Biology* (R. Ferrière, U. Dieckmann and D. Couvet, eds.). Cambridge University Press, Cambridge, UK: 244–264.
Holt R. D., R. Gomulkiewicz and M. Barfield (2003). The phenomenology of niche evolution via quantitative traits in a "black-hole" sink. *Proceedings of the Royal Society of London (B)* **270:** 215–224.
Holt, R. D., R. Gomulkiewicz and M. Barfield (2004a). Temporal variation can facilitate niche evolution in harsh sink environments. *American Naturalist* **164:** 187–200.
Holt, R. D., T. M. Knight and M. Barfield (2004b). Allee effects, immigration, and the evolution of species' niches. *American Naturalist* **163:** 253–262.
Holt, R. D., M. Barfield and R. Gomulkiewicz (2005). Theories of niche conservatism and evolution: could exotic species be potential tests? In *Species Invasions: Insights into Ecology, Evolution, and Biogeography* (D. Sax, J. Stachowicz and S. D. Gaines, eds.). Sinauer Associates, Sunderland, MA: 259–290.
Hutchinson, G. E. (1957). Concluding remarks. *Cold Spring Harbor Symposia on Quantitative Biology* **22:** 415–427.
Hutchinson, G. E. (1978). *An Introduction to Population Ecology*. Yale University Press, New Haven, CT.
Kawecki, T. J. (1995). Demography of source–sink populations and the evolution of ecological niches. *Evolutionary Ecology* **9:** 38–44.
Kawecki, T. J. (2000). Adaptation to marginal habitats: contrasting influence of dispersal on the fate of rare alleles with small and large effects. *Proceedings of the Royal Society of London (B)* **267:** 1315–1320.
Kawecki, T. J. (2003). Sex-biased dispersal and adaptation to marginal habitats. *American Naturalist* **162:** 415–426.
Kawecki, T. J. (2004). Ecological and evolutionary consequences of source–sink population dynamics. In *Ecology, Genetics, and Evolution of Metapopulations* (I. Hanski and O. E. Gaggiotti, eds.). Elsevier Academic Press, Burlington, MA: 387–414.
Kawecki, T. J. (2008). Adaptation to marginal habitats. *Annual Review of Ecology, Evolution, and Systematics* **39:** 321–342.
Kawecki, T. J. and R. D. Holt (2002). Evolutionary consequences of asymmetric dispersal rates. *American Naturalist* **160:** 333–347.
Kawecki, T. J., N. H. Barton and J. D. Fry (1997). Mutational collapse of fitness in marginal habitats and the evolution of ecological specialization. *Journal of Evolutionary Biology* **10:** 407–429.
Keddy, P. A. (1981). Experimental demography of the sand-dune annual, *Cakile edentula*, growing along an environmental gradient in Nova Scotia. *Journal of Ecology* **69:** 615–630.
Keddy, P. A. (1982). Population ecology on an environmental gradient: *Cakile edentula* on a sand dune. *Oecologia* **52:** 348–355.
Kierstead, H. and L. B. Slobodkin (1953). The size of water masses containing plankton blooms. *Journal of Marine Research* **12:** 141–147.
Kirkpatrick, M. and N. H. Barton (1997). Evolution of a species' range. *American Naturalist* **150:** 1–23.

Lenormand, T. (2002). Gene flow and the limits to natural selection. *Trends in Ecology and Evolution* **17**: 183–189.
Levin, S. (1976). Spatial patterning and the structure of ecological communities. *Lectures in Mathematics in the Life Sciences* **8**: 1–35.
LoFaro, T. and R. Gomulkiewicz (1999). Adaptation versus migration in demographically unstable populations. *Journal of Mathematical Biology* **38**: 571–584.
Morris, D. W. and J. E. Diffendorfer (2004). Reciprocating dispersal by habitat-selecting white-footed mice. *Oikos* **107**: 549–558.
Orr, H. A. and R. L. Unckless (2008). Population extinction and the genetics of adaptation. *American Naturalist* **172**: 160–169.
Perron, G. G., A. Gonzalez and A. Buckling (2007). Source–sink dynamics shape the evolution of antibiotic resistance and its pleiotropic fitness cost. *Proceedings of the Royal Society of London (B)* **274**: 2351–2356.
Polechova, J., N. Barton and G. Marion (2009). Species range: adaptation in space and time. *American Naturalist* **174**: E186–E204.
Price, T. (2007). *Speciation in Birds*. Roberts & Co., Publishers, Greenwood Village, CO.
Proulx, S. R. (2002). Niche shifts and expansion due to sexual selection. *Evolutionary Ecology Research* **4**: 351–369.
Pulliam, H. R. (1988). Sources, sinks, and population regulation. *American Naturalist* **132**: 652–661.
Pulliam, H. R. (2000). On the relationship between niche and distribution. *Ecology Letters* **3**: 349–361.
Ronce, O. and M. Kirkpatrick (2001). When sources become sinks: migrational meltdown in heterogeneous habitats. *Evolution* **55**: 1520–1531.
Rousset, F. (1999). Reproductive value vs. sources and sinks. *Oikos* **86**: 591–596.
Runge, J. P., M. C. Runge and J. D. Nichols (2006). The role of local populations within a landscape context: defining and classifying sources and sinks. *American Naturalist* **167**: 925–938.
Sexton, J. P., P. J. McIntyre, A. L. Angert and K. J. Rice (2009). Evolution and ecology of species range limits. *Annual Review of Ecology, Evolution, and Systematics* **40**: 415–436.
Slatkin, M. (1985). Rare alleles as indicators of gene flow. *Evolution* **39**: 53–65.
Tufto, J. (2001). Effects of releasing maladapted individuals: a demographic-evolutionary model. *American Naturalist* **158**: 331–340.
Turner, J. R. G. and H. Y. Wong (2010). Why do species have a skin? Investigating mutational constraint with a fundamental population model. *Biological Journal of the Linnean Society* **101**: 213–227.
Watkinson, A. R. and W. J. Sutherland (1995). Sources, sinks and pseudo-sinks. *Journal of Animal Ecology* **64**: 126–130.
Wiens, J. J. and C. H. Graham (2005). Niche conservatism: integrating evolution, ecology, and conservation biology. *Annual Review of Ecology, Evolution, and Systematics* **36**: 519–539.
Wiens, J. J., D. D. Ackerly, A. P. Allen, B. L. Anacker, L. B. Buckley, H. V. Cornell, E. I. Damschen, T. J. Davies, J. A. Grytnes, S. P. Harrison, B. A. Hawkins, R. D. Holt, C. M. McCain and P. R. Stephens (2010). Niche conservatism as an emerging principle in ecology and conservation biology. *Ecology Letters* **13**: 1310–1324.
Willi, Y., J. Van Buskirk and A. A. Hoffmann (2006). Limits to the adaptive potential of small populations. *Annual Review of Ecology, Evolution, and Systematics* **37**: 433–458.
Wilson J. B. and A. D. Agnew (1992). Positive-feedback switches in plant communities. *Advances in Ecological Research* **23**: 263–336.

Appendix: Local adaptation in a one-way flow environment

Consider a chain of habitats coupled by unidirectional movement of a species. For instance, Arditi *et al.* (1991) carried out an interesting lab experiment with cladocerans growing in beakers, arranging the beakers in a chain of serially arranged compartments, with unidirectional flow along the chain of compartments. With some ingenious tinkering, this experiment was set up so that each compartment had its own food base, but the cladocerans moved only in one direction (downstream) between them. Such an experiment might mimic the life of a non-volant aquatic organism in a mountain stream on a steep gradient, for instance, where the ancestor lived in a river coursing across a gentle plain which then experienced tectonic uplift, leading to a descendant population connected unidirectionally in a chain of local populations.

If we look at a single compartment along this gradient, the following simple model could describe the interaction between a resident cladoceran clone (of density N_1), moving unidirectionally between these compartments, a mutant clone (of density N_2) that has arisen in a particular focal compartment (and so is not contained in the immigrant stream), and the algal food resource (of density R) that they share:

$$\frac{dN_1}{dt} = I + a_1 N_1 R - (m + e_1) N_1$$
$$\frac{dN_2}{dt} = a_2 N_2 R - (m + e_2) N_2$$
$$\frac{dR}{dt} = rR(1 - R/K) - R(a_1 N_1 + a_2 N_2). \tag{2.A1}$$

These equations assume that the biotic resource in the compartment grows logistically, with r and K, respectively, being its intrinsic growth rate and carrying capacity. The resident consumer is initially found both upstream and in the focal compartment, and I is the rate of inflow of individuals from upstream. Consumers die at a rate m, which we assume is the same for both clones. Births are determined by a linear functional response, defined for each consumer clone by a fixed attack rate a_i (we assume there is a constant conversion of consumption into births, and scale consumer densities so as not to have to deal with it). Thus, the local growth of clone i (our measure of fitness) is $F_i = b_i - d_i = a_i R - m$. We assume that both clones passively emigrate downstream at a per capita rate e_i.

In the absence of any flows (in or out), the resident consumer, if it can persist, equilibrates at a resource–consumer equilibrium:

$$N_1^* = \frac{r}{a_1}\left(1 - \frac{m}{a_1 K}\right),\ R_{(1)}^* = \frac{m}{a_1}. \tag{2.A2}$$

The notation "(i)" indicates that resource levels are being evaluated when consumer i alone is present, and the asterisk denotes equilibrium. If $m > a_1 K$, then the prey base is insufficient to sustain the consumer, and the habitat is an absolute sink (Kawecki 2004). Adding emigration can make persistence impossible, even if the habitat is not intrinsically a sink. Adding recurrent immigration, by contrast, allows a population to be present, even if the habitat is a sink because of insufficient resources or high mortality.

If the habitat is not an absolute sink ($m < a_1K$) and there is no emigration, immigration will increase population size above the carrying capacity (N_1^*) indicated above, and the habitat will become a "pseudo-sink" (Watkinson and Sutherland 1995). However, with emigration, this outcome is not so clear. If we assume for simplicity that consumption keeps resource levels down to levels at which intrinsic density regulation is weak, then we can set $K = \infty$ and avoid some messy algebra. With this assumption,

$$N_1^* = \frac{r}{a_1}, \; R_{(1)}^{*I} = \left(\frac{m+e_1}{a_1}\right) - \frac{I}{r}. \tag{2.A3}$$

Note that now the expression for the resource includes the effects of consumer movement; resource levels are higher at higher emigration rates, and lower at higher immigration rates; the I in the superscript indicates that resource levels are evaluated when there is consumer immigration and emigration. The equilibrium exists if the expression for resource abundance is positive, which is likely if attack rates are low; consumer death rates or emigration are high; consumer immigration is low; and resource growth rates are high. For a given influx of consumers, more consumers and resources will be sustained if the resource has higher productivity (as measured by r).

According to Pulliam's (1988) definition, a habitat is a sink if immigration exceeds emigration. Conversely, it is a source if emigration exceeds immigration.

For our compartment to be a source thus requires that

$$I < e_1N^* = \frac{e_1r}{a_1}, \text{ or } \frac{I}{r} < \frac{e_1}{a_1}. \tag{2.A4}$$

If this holds, then from Eq. (2.A3) it is clear that the equilibrium exists. If it does not hold, there is a range of values for I/r where the equilibrium in Eq. (2.A3) still exists, and the compartment is a demographic sink. When the equilibrium in Eq. (2.A3) does not exist, it is because the immigration rate is sufficiently great that the resource is driven to extinction, and the consumer ends up at an abundance of $I/(m + e_1)$, immigrating into a sink and dying, or emigrating but not reproducing there. Inspection of the isoclines shows that both equilibria are stable (details not shown).

So, using Pulliam's definition, the focal habitat along a gradient can be a source or a sink, depending on the relative magnitudes not just of immigration rate and per capita emigration rates but of local attack rates and resource renewal rates as well.

With this machinery in hand, we now can examine the fate of a novel mutant type with higher relative fitness because of its higher rate of resource uptake. When rare, it experiences resource levels at the rate set by the resident (and immigrant) type, given in Eq. (2.A3). The per capita growth rate of the novel clone is

$$\frac{1}{N_2}\frac{dN_2}{dt} = a_2 R^{*I}_{(1)} - (m+e_2) = a_2\left[\left(\frac{m+e_1}{a_1}\right) - \frac{I}{r}\right] - (m+e_2)$$

which is greater than 0 when

$$\left[R^{*I}_{(1)} - R^{*I}_{(2)}\right] > \frac{I}{r} \tag{2.A5}$$

where $R^{*I}_{(i)}$ is the equilibrial level of the resource, when clone i is present alone with both immigration and emigration (Eq. (2.A3), with the subscript 1 replaced by 2 to denote species 2). The left side of Eq. (2.A5) is a measure of the competitive superiority of clone 2, as assessed by the level to which it can potentially depress limiting resource levels in this open system. Clearly, for the new type to spread, it must be superior in resource competition. In a closed community, this would suffice for it to invade (and eventually outcompete and supplant the resident type). In this open community, however, there is a threshold value of competitive superiority that must be achieved for the new type to increase in frequency. Immigration of the resident consumer lowers resource levels, and so reduces the initial fitness of the new type. Moreover, the left-hand side of Eq. (2.A5), if it is positive, decreases with e_2. This implies that increasing emigration makes it harder for a given novel clone to be retained by selection. Finally, increases in resource productivity (r) reduce the threshold, and so make it easier for a superior competitor to invade. The conclusions of this specific model thus buttress the claims made in the main text. After this allele becomes established, then after some algebraic manipulation it can be shown that the original sink has now become a source, with no backflow of the novel allele from the external environment. Moreover, if we cut off immigration, the local population persists, so the conditions for adaptation to the sink are tantamount to niche evolution. The sink is more likely to be evolutionarily stable (*qua* sink) if immigration is high or the habitat is unproductive, in both cases because of the indirect negative density dependence resulting from resource competition. Note that because of our assumption of a one-way flow for dispersal, we were able to ignore evolution "downstream" from the focal habitat. If we allow reciprocal dispersal, then we would need to account for selection averaged across a number of habitats coupled by asymmetric dispersal (Kawecki and Holt 2002).

DOUGLAS W. MORRIS

3

Source–sink dynamics emerging from unstable ideal free habitat selection

Summary

Adaptive theories of source–sink regulation assume that dispersal maximizes individual fitness. Fitness in these models is improved through the long-term benefits of habitat-dependent dispersal rates, optimized habitat choices, pulsed dispersal, advantages to relatives, or natural selection of dispersal unrelated to habitat use. But source–sink dynamics may often simply represent ideal free habitat selection in unstable populations. The prevalence of source–sink systems suggests that there may be other evolutionary attractors. I explore two candidates: an inclusive fitness strategy that maximizes population growth, and a cooperative strategy whereby unrelated individuals form coalitions whose combined aggression forces the emigration of unaligned individuals. Although these forms of source–sink dynamics can displace ideal free habitat selection, they create high-fitness patches available for counter-invasion. Regardless of whether there are cycling evolutionary attractors for different forms of source–sink systems, computer simulations reveal crucial roles for sinks in damping otherwise unstable dynamics. Sinks, even when acting as ecological traps, increase the probabilities of persistence and, if the sink is not too severe, can create the illusion that they are unimportant in stabilizing source populations. Alteration or removal of these critical sinks can doom source populations to wildly fluctuating dynamics and extinction. The lesson for conservation in mosaic landscapes is clear: all habitat is critical.

Sources, Sinks and Sustainability, ed. Jianguo Liu, Vanessa Hull, Anita T. Morzillo and John A. Wiens. Published by Cambridge University Press.

Introduction

Source–sink dynamics are rooted firmly in theories of habitat selection and dispersal. Early models demonstrated that source–sink dynamics emerge in temporally varying environments through passive dispersal that equalizes population density (Holt 1985), or with adaptive responses of "subordinate" individuals. Dispersal by subordinates can be mediated either through direct interference from dominant individuals (Fretwell and Lucas 1969) or through the indirect effects of breeding-site preemption (Pulliam 1988; Pulliam and Danielson 1991). Other theories of habitat selection demonstrate that adaptive source–sink regulation can evolve when individuals maximize their inclusive fitness (Morris *et al.* 2001), and in fluctuating populations with reciprocating pulses of adaptive dispersal (Morris *et al.* 2004a, 2004b). But it is also clear that source–sink dynamics can be non-adaptive to actively emigrating individuals. Non-adaptive active dispersal is most common when organisms are caught in an ecological trap (Dwernychuk and Boag 1972; Schlaepfer *et al.* 2002; Kristan 2003) whereby formerly reliable cues of habitat quality have been altered by disturbance (Shochat *et al.* 2005). Non-adaptive habitat choice may also occur when dominant individuals capitalize on strength and size asymmetries to drive subordinates (including offspring) away from otherwise favorable habitats, and when dispersal has evolved for reasons not associated with habitat quality (Morris 1991).

These few examples suggest that we should more fully explore theories of habitat selection to search for additional ecological and evolutionary mechanisms that can create source–sink dynamics. And, while we do that, we must also evaluate the implications of source–sink systems to our understanding of population dynamics, and our ability to manage and conserve species in mosaic landscapes. So I explore the creation of source–sink dynamics under ideal free habitat selection with the aim of assessing the degree to which sink habitats can buffer otherwise unstable dynamics. I then investigate other forms of adaptive, and potentially non-adaptive, emigration by asking, in a general sense:

1. whether an ideal free distribution yields higher fitness than other strategies;
2. what mechanisms of emigration might evolve if habitat selection has a different type of evolutionary attractor?

I begin by briefly reviewing density-dependent habitat selection to show how source–sink dynamics can emerge from ideal free habitat selection when emigrants maximize their inclusive fitness. I then generalize the inclusive-fitness model to explore the conditions under which adaptive emigration that maximizes individual fitness may be trumped by forced emigration maximizing population growth rate. This approach suggests that unrelated individuals may frequently cooperate to exclude others from preferred habitats. I illustrate some of the consequences of ideal free versus inclusive-fitness habitat selection in source–sink systems, then conclude by reconsidering the role of density-dependent dispersal in population regulation, and by suggesting experiments to test the theory.

Ideal free habitat selection and source–sink regulation

Ideal free habitat selection emerges when individuals choose habitats to maximize fitness and are free to occupy the habitats they choose (Fretwell and Lucas 1969; Fretwell 1972). Such individuals can occupy sink habitats whenever their populations in the source habitat exceed carrying capacity (negative population growth; e.g., Holt 1997). Whether these individuals actually enter the sink will depend on its basic quality, on the number of "excess" individuals living in the source, and on density-dependent feedback on source fitness. Once the sink is occupied, the number of individuals living in each habitat, the duration of their occupancy in the sink, and the effects on overall population dynamics will also depend on the relationship between fitness and density in each habitat. Contrary to classical models, where dynamic patterns are determined only by maximum population growth rates, dynamics with habitat selection depend critically on the density and frequency dependence of habitat selection.

We can make these generalities explicit by simulating ideal free choice between source and sink habitats. In order to keep the models tractable, imagine an asexual species undergoing discrete pulses of reproduction (Fig. 3.1). Each period of reproduction is followed by ideal free dispersal that equalizes mean fitness in the two habitats. Only the source habitat is occupied when population size is less than or equal to the source carrying capacity. Both habitats are occupied at higher population sizes when population growth is negative. Population growth in each habitat occurs via the Ricker equation:

$$N_{i(t+1)} = N_{i(t)} e^{r\left(1 - \frac{N_{i(t)}}{K_i}\right)} \tag{3.1}$$

where N is the number of individuals living in habitat i at times t and $t+1$, r is the intrinsic rate of population growth, and K is the habitat's carrying capacity. This model produces unstable population dynamics in single habitats whenever $r > 2$ (see, e.g., May and Oster 1976; Holt 1997).

The ideal free habitat selection strategy can be revealed by calculating the per capita population growth rate in habitats 1 and 2 as estimates of fitness (divide both sides of Eq. (3.1) by N_t; use natural logarithms to eliminate the exponent), then setting those estimates equal to one another:

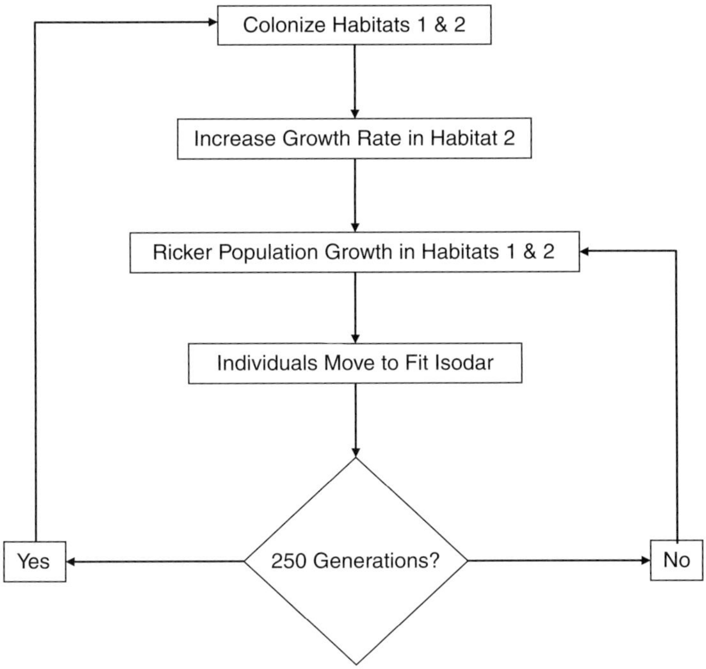

FIGURE 3.1.
A flow chart summarizing computer simulations of population growth and habitat selection. In the simulations used here, habitat 2 corresponds to the source habitat.

$$r_2 - \frac{r_2}{k_2} N_2 = r_1 - \frac{r_1}{K_1} N_1. \tag{3.2}$$

Solving for N_2

$$N_2 = \frac{(r_2 - r_1)}{r_2} K_2 + \frac{r_1 K_2}{r_2 K_1} N_1 \tag{3.3}$$

yields the system's habitat isodar, the set of densities in each habitat such that mean fitness is equal at all population sizes where both habitats are occupied. The isodar intercept represents the density where individuals shift from specialists occupying one habitat to generalists occupying both. Note that in Eqs. (3.1) and (3.2) the ratio $\frac{r_i}{K_i}$ is the slope of population growth rate against population density in each habitat. Thus, when applying the model to sink habitats (e.g., Eq. 3.14), I use only the slope (i.e., K_{sink} is imaginary).

Computer simulations (see below) reveal that ideal free habitat selection yields source–sink dynamics when populations are unstable, that sink habitat for ideal free habitat selectors can buffer populations from extinction, and that the quality of the sink determines the stability of the population's dynamics. But it would be wise, before we embark on a research program aimed at testing these predictions, to evaluate whether or not the ideal free distribution is an appropriate model for habitat selection.

Evolutionary stability of ideal free habitat selection

There is a rather simple way to assess whether an ideal free distribution (IFD) yields higher fitness than other habitat-selection strategies. Imagine that we can draw the adaptive landscape of habitat selection between two habitats. We do this by first assuming that we know the relationship between fitness and density in each habitat. Then, for each value of total population size that we wish to assess, we calculate the mean fitness for every possible distribution of individuals in the two habitats. We then plot mean fitness against both total population size and the proportional occupation of one of the two habitats. The graph yields a fitness landscape (Wright 1931) that explicitly captures the density and frequency dependence of habitat selection. The best strategy of density-dependent habitat selection corresponds to the set of proportions of habitat occupancy that maximize mean fitness at each population size. So to assess whether an alternative strategy can outperform ideal free habitat selection, we simply superimpose the IFD solution onto the fitness landscape. Apaloo *et al.* (2009) provide explicit directions on the use of fitness landscapes to assess evolutionarily stable strategies.

Figure 3.2 illustrates that ideal free habitat selection does indeed maximize fitness at all population sizes for two habitats with equivalent fitness at low density but different linear rates of decline in fitness with increasing density. The proportional occupation of habitat that maximizes fitness is the same at all population sizes (i.e., there is a single strategy of habitat selection).

Figure 3.3 illustrates one of many alternatives where the optimal strategy of habitat selection varies with density. Here, innate fitness at low density, as well as the density-dependent decline in fitness, differs between the two habitats. The convergent population regulation (Morris 1988) illustrated here could occur, for example, if the "preferred" habitat yielding high fitness at low density is also subject to higher rates of predation. The ideal free distribution yields lower mean fitness at many population sizes than does the frequency-dependent strategy that maximizes fitness (Eqs. (3.6) and (3.7) explain why). It is thus important to explore which other reasonable strategies come closer to maximizing fitness than does the ideal free distribution.

Inclusive fitness and the MAXN strategy of habitat selection

Morris *et al.* (2001) developed an inclusive fitness model to assess the conditions under which emigrants should sacrifice their own individual

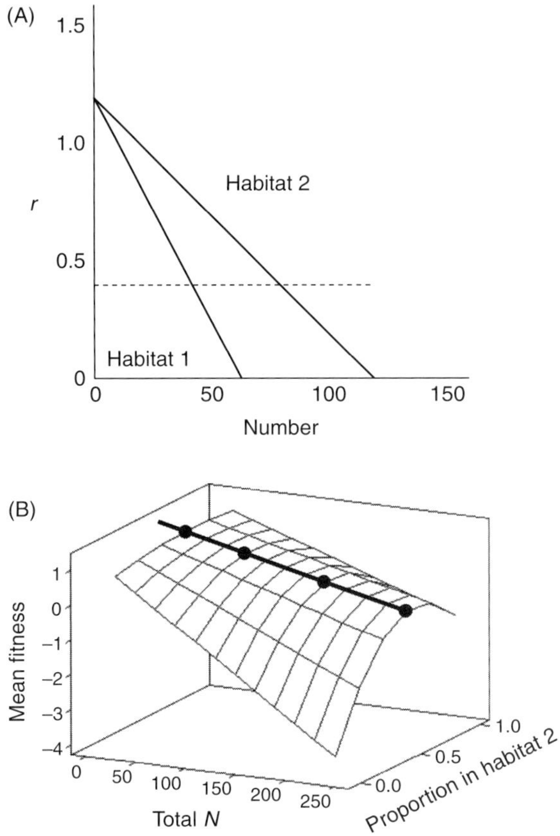

FIGURE 3.2.
The fitness landscape (B) that emerges when maximum fitness in two habitats is identical at low population size, but where the linear relationship between fitness and density differs between them (A). The landscape has a single invariant "ridge" of maximum density (solid line) that corresponds to an invariant proportional occupation of the landscape by ideal and free habitat selectors (dashed line in A, dots in B, the habitat-matching rule of Pulliam and Caraco 1984).

fitness for the benefit of relatives. According to that general model, individuals should achieve a stable distribution between habitats only when

$$g(N_1, N_2) = R(N_2 f'_2 - N_1 f'_1) \tag{3.4}$$

where $g(N_1, N_2)$ describes the decision function associated with migration, R is the coefficient of relatedness, and $N_i f'_i$ is the change in fitness in habitat i with changes in population density when the system otherwise lies at an ideal free distribution (Morris *et al.* 2001). Individuals maximizing their inclusive fitness will migrate from habitat 1 to 2 when $g(N_1, N_2)$ is positive, from 2 to 1 when the function is negative, and will stay at the ideal free distribution when the function is zero. If, for the sake of the argument, we assume that fitness declines linearly with increasing density such that

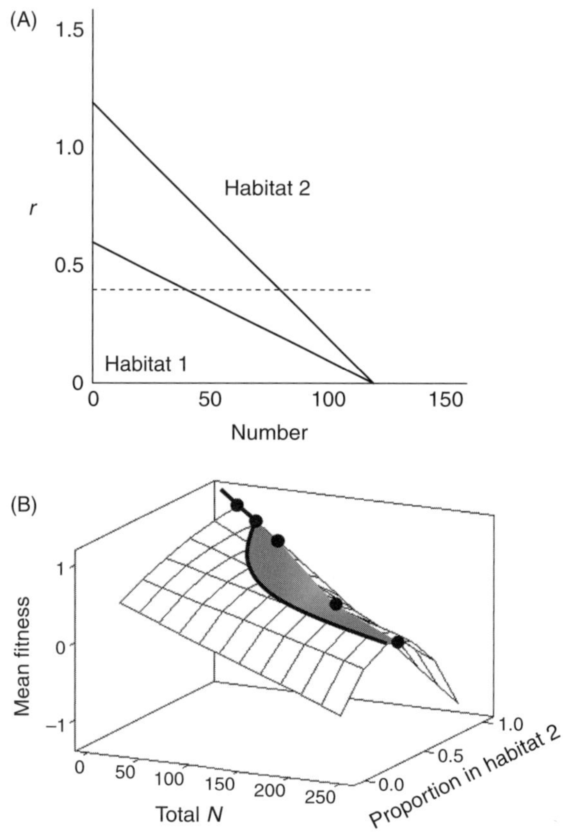

FIGURE 3.3.
An example of a fitness landscape (B) emerging when maximum fitness, as well as the relationship between fitness and density, differs between two habitats (A). The strategy maximizing fitness (solid line, MAXN) yields different proportions of individuals in each habitat at different population sizes (density and frequency dependent). The ideal free strategy (dots), even though it is also density and frequency dependent, does not maximize fitness at most population sizes (shading represents the zone where the MAXN strategy yields higher mean fitness).

$$\frac{1}{N_i}\frac{dN_i}{dt} = r_{\max i} - b_i N_i \tag{3.5}$$

where $r_{\max}$ is the maximum fitness at low density and b is the rate of decline in fitness with increasing density, then Morris *et al.* (2001) showed that adaptive dispersal between two habitats will cease for related individuals only when

$$N_2 = \left[\frac{(r_{\max 2} - r_{\max 1})}{(b_2)(1+R)}\right] + \left[\frac{b_1}{b_2}\right] N_1. \tag{3.6}$$

If individuals are perfectly related to one another (for instance, as in the parts of modular organisms), then this strategy also maximizes total population growth rate (Morris *et al.* 2001) because individuals overexploit the poor habitat, where each has a small effect on fitness, to maximize mean fitness from the more efficient use of the rich habitat by others.

Let us compare this inclusive fitness "MAXN strategy" with the ideal free distribution for unrelated individuals. Substituting $b_1 = \frac{r_1}{K_1}$ and $b_2 = \frac{r_2}{K_2}$ into Eq. (3.3) and solving for N_2 yields the ideal free isodar: the set of densities where individual fitness is identical in both habitats,

$$N_2 = \left[\frac{(r_{max2} - r_{max1}}{b_2}\right] + \left[\frac{b_1}{b_2}\right] N_1. \tag{3.7}$$

Thus ideal free and inclusive fitness isodars (Eqs. (3.7) and (3.6), respectively) are identical only when individuals are unrelated. Otherwise, mean fitness and mean population growth rate will always be greater for the inclusive-fitness MAXN strategy than it is for ideal free habitat selection (Morris *et al.* 2001). When habitat selection maximizes inclusive fitness, some individuals will sacrifice their own individual fitness by dispersing to the habitat with the lowest density-dependent decline in fitness so that their relatives can gain even higher fitness in the better habitat (undermatching of resources and source–sink regulation). Like worker bees in a hive, some individuals sacrifice their direct fitness in order to maximize inclusive fitness.

Note that Eqs. (3.6) and (3.7) can be readily applied to source–sink dynamics. Assuming that habitat 1 is the sink, then r_{max1} is negative, and the isodar intercept is increased relative to a system with only source habitats of differing quality. But the increase is less under the MAXN strategy (Eq. 3.6) than it is with ideal free habitat selection (Eq. 3.7) because related individuals "cooperate" to undermatch resources for their mutual benefit. The two strategies differ in the proportional use of habitat.

Knowing that source–sink dynamics emerge in populations of related individuals, we now ask whether or not cooperation can also replace ideal free habitat selection when all individuals are unrelated. To make the answer as clear as possible, we also imagine that there is only one habitat. All emigrants would thus be forced into a "black-hole" sink such that neither they, nor their offspring, can contribute to the source population. We will first evaluate what density of individuals should remain in the habitat if all individuals are identically related to one another. If we assume that population growth obeys the Ricker equation (Eq. 3.1), then the expected fitness of an individual that remains in the population is given by

$$W_{(stay)} = \left[(N-1)\left(r-\frac{rN}{K}\right)\right]+\left[r-\frac{rN}{K}\right] \tag{3.8}$$

where the term in the first square bracket yields the total fitness accrued by all $N-1$ relatives and the term enclosed by the second square bracket is the fitness of the target individual. Differentiating Eq. (3.8) yields the well-known solution that population growth rate will be maximized at $N=\frac{K}{2}$.

We now contrast Eq. (3.8) with the fitness of an individual leaving the population for the black-hole sink:

$$W_{(leave)} = \left[(N-1)\left(r-\frac{rN}{K}\right)\right]-\left[r-\frac{rN}{K}\right]. \tag{3.9}$$

The term in the second square bracket is the value of fitness that the individual loses by leaving the "source" population (note that this formulation does not include the expectation of fitness that the individual might attain should it survive and reproduce successfully elsewhere). Equation (3.9) maximizes fitness at $N=\frac{K}{2}+1$.

So the decision on whether an individual should stay or leave the population is without conflict. If $N\le\frac{K}{2}$, then stay; if $N\ge\frac{K}{2}+1$, then leave. Populations of perfectly related individuals should maintain a density that maximizes the population growth rate. The decision on whether or not an individual should leave (Eq. 3.9) is easily generalized to include differing degrees of relatedness by

$$W_{(leave)} = \left[R(N-1)\left(r-\frac{rN}{K}\right)\right]-\left[r-\frac{rN}{K}\right].$$

Thus, an individual should willingly leave the population whenever

$$N\ge\frac{1}{2}\left(K+\frac{R+1}{R}\right). \tag{3.10}$$

Thus, as R becomes small, no individual in a population occupying a single habitat will sacrifice its own fitness by migrating for the benefit of others. If, however, two or more habitats are occupied, then individuals will distribute themselves among the habitats according to the MAXN strategy (Eq. 3.6) in order to maximize their inclusive fitness.

Now we imagine that individuals can cooperate to force another to disperse away from the habitat. We further imagine that the collective behavior to banish an individual has negligible direct cost to the cooperating individuals. Cooperation in this game involves the risk that the individual will itself be the target for expulsion. An individual should cooperate as long as its expectation of fitness gain by banishing another individual exceeds the expected fitness loss if it is itself targeted for dispersal. Fitness of the cooperative strategy is given by

$$W=\left[r-\frac{r(N-1)}{K}\right]-\left[\left(r-\frac{rN}{K}\right)\frac{1}{N}\right] \tag{3.11}$$

where the term in the left-hand square bracket represents the fitness of a cooperating individual that succeeds in banishing an intraspecific competitor without incurring any cost. The term in the right-hand bracket is the expected loss in fitness that would occur if the individual is itself expelled (fitness at density N multiplied by the probability that it is the one individual banished from the population). Equation (3.11) yields a maximum fitness at

$$N=\sqrt{K}. \tag{3.12}$$

Thus we see that even if dispersal is equivalent to death (because the emigrants' genes never return to the source population), individuals should still cooperate to force dispersal. Source–sink dynamics will then dominate habitat selection if the cooperation cost is not excessive. And when these source–sink dynamics occur, the proportion of individuals forced to disperse declines with increasing carrying capacity. The model becomes more complicated if the banished individual can find refuge in an alternative habitat. The emigrant's fitness will depend not only on its probability of being banished, but also on its expectation of fitness in the second habitat. At equilibrium, the density of individuals in the second habitat will obey its own version of Eq. (3.12). Thus, for only two habitats, the fitness of a cooperating individual in habitat 2, that cannot return if banished to habitat 1, is given by

$$W_2=\left\{r_2\frac{r_2(N_2-1)}{K_2}\right\}-\frac{1}{N_2}\left\{\left[r_2-\frac{r_2N_2}{K_2}\right]-\left[\left(r_1-\frac{r_1N_1}{K_1}\right)-\frac{1}{N_1}\left(r_1-\frac{r_1(N_1+1)}{K_1}\right)\right]\right\} \tag{3.13}$$

which clearly becomes much more complex as additional habitats are occupied. Equation (3.13) reveals an interesting habitat hierarchy. Positive fitness in secondary habitats reduces the cost of dispersal, and a greater proportion of individuals would be forced to migrate. Negative fitness in secondary habitats (sinks) increases the cost of dispersal and a smaller proportion would migrate. In each case the differences in proportional occupation cascade downward as habitats of lower and lower quality are occupied. Nevertheless, for any set of habitat-specific values for r, each habitat would attain a constant density at equilibrium. Habitat occupation promises to be far less predictable for the ideal free and MAXN strategies of habitat selection, and it is thus interesting to explore their respective dynamics with computer simulations.

Simulating ideal free versus MAXN strategies of source–sink habitat selection

I contrasted the population dynamics emerging from ideal free and MAXN strategies of habitat selection by assuming a life history with discrete intervals of reproduction (Eq. 3.6) followed by pulses of dispersal between two habitats (Fig. 3.1). Source habitats were those that maintained a positive growth rate at low density. Sink habitats had negative population growth at all densities and could thus be maintained only by immigration from the source. Since the simulations allowed back-migration from the sink to source, both habitats could contribute at different times to the metapopulation, and would qualify as sources using the C' statistic proposed by Runge *et al.* (2006). Yet over the period of any one simulation, only one habitat (source) had a net positive contribution to the total population.

Each set of simulations used identical time steps, initial population densities, and population growth parameters to contrast the dynamics between source and sink habitats corresponding to ideal free and MAXN strategies. Dispersal did not incur a fitness cost. After populations had grown (Eq. 3.1) for a single generation, the simulations moved individuals from one habitat to another to fit the expectations of each habitat-selection strategy. I calculated the slope of fitness with density then rearranged Eqs. (3.3) and (3.6) in terms of total population size. The rearrangement was necessary because the original equations can otherwise be solved only by iteration (the equilibrium densities, N_i^*, in each habitat are unknown, whereas total population size before dispersal is known). Thus, for the ideal free simulation, the model placed the appropriate number of individuals satisfying

$$N_2^* = \frac{\left[(N_1+N_2)+\left(\frac{r_2-r_1}{b_1}\right)\right]}{\left(\frac{b_2}{b_1}\right)+1} \qquad (3.14)$$

and $N_1^* = (N_1 + N_2) - N_2^*$ into each habitat. The MAXN strategy did the same except that the slope was adjusted for the degree of relatedness (Eq. 3.6).

The simulation proceeded as follows (Fig. 3.1). The sink habitat was parameterized with a negative r and a density-dependent decline in fitness. The source habitat was designated with a positive r and $K = 200$. A small (but constant) number of individuals colonized each habitat. The source habitat's r was increased by a small increment, while K was held constant. The quality of, and density dependence in, the sink also remained constant. The population then alternated growth with dispersal for 250 generations. Partial individuals produced by the population-growth equation were rounded to integers. The final 200 generations of density, dispersal and fitness data were saved, the habitats were recolonized, and then the model was repeated. Stable dynamics or limit cycles were attained, at moderate values of r, after 50 generations in all simulations.

The biological reasoning behind the simulations is that managers faced with a source–sink system may attempt to improve the quality of the source habitat. Improvements could include removal or control of predators and pathogens. Each of these treatments will likely increase r, but would have no direct effect on habitat productivity (K can be considered constant).

Many species of management concern possess relatively low population growth rates and, even with the removal of predators or pathogens, may never exhibit intrinsically unstable dynamics. The success of numerous hatcheries and nurseries demonstrates, however, that many other species have sufficiently high fecundity that can yield high growth rates when juvenile mortality is reduced. Numerous pest species do possess high rates of population growth that can cause population irruptions from favorable source habitat into sinks (e.g., from field margins into cropland). In each instance it will be valuable to know the general conditions that can damp population instability in these source–sink landscapes.

The simulations contrasted three scenarios.

1. *Severe sinks* (e.g., unfavorable and small areas where crowding has a major effect on population growth rate) had negative growth rates and a steep decline in growth rate with increasing population size.
2. *Mild sinks* had barely negative growth rates at low density and a minor decline in population growth rate with density (as could occur if much of the landscape was composed of sink habitat).
3. *Ecological traps* were selected even at low density because they appeared to yield positive fitness, but were actually severe sinks.

Figure 3.4 illustrates the source-habitat population dynamics across a range of r values in the absence of habitat selection. The population is stable at low r, branches into two and then multipoint cycles at $r > 2$, and becomes extinct at $r \geq 3.5$. This population represents the control against which we can assess differences caused by source–sink simulations.

Figures 3.5 and 3.6 illustrate some of the dynamics that can emerge under ideal free habitat selection when sources are connected to nearby sinks. If the sink is severe (Fig. 3.5), then the dynamics in the source habitat are similar to those that occur in the absence of habitat selection (Fig. 3.4). There are two exceptions:

1. Population fluctuations are less extreme in the source–sink system.
2. The source–sink population persists across a wider range of r.

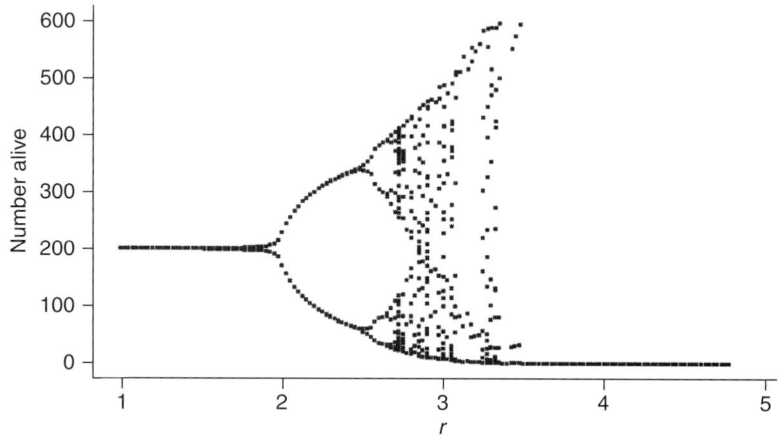

FIGURE 3.4.
A bifurcation diagram illustrating the population dynamics of a single population growing according to the Ricker equation (Eq. 3.1). The abscissa represents iterated values of *r* in the source habitat. Simulations were constrained to operate on integers (partial individuals were rounded to the nearest whole number). The first 50 generations of each 250-generation simulation were excluded prior to analysis. Parameter values were as follows: $N_0 = 100$; $K = 200$; *r* incremented in units of 0.025.

Meanwhile, population density in the sink can exceed that in the source even though dynamics in the sink always retain potential for local extinction. These local extinctions occur whenever the source population size is less than the isodar intercept (Eq. 3.3).

Simulations using mild sink habitats with weak density dependence stabilized population dynamics in both habitats, and across a vast range of *r* (Fig. 3.6). Dynamics stabilized because the sink absorbed a large number of immigrating source individuals with little change in fitness, while emigrants from the source habitat had a large effect on the fitness of the individuals that remained behind. To visualize the relevance of this effect imagine, following population growth in a population (or a habitat) with a large *r*, that the source population has grown well above *K*. In the absence of habitat selection toward the sink, the source population would collapse in the next generation. But high emigration to the sink reduces the excess in the source and buffers population decline.

The rate of population decline is low in the sink. So, following population "growth" in the subsequent generation, individuals from the sink can immigrate into the source and increase source density beyond *K* (but less than in the previous growth phase). Both populations will again decline, but at a lower rate than in the preceding generation. This repeating process damps population oscillations until each habitat attains a constant density. At relatively low values of *r*, the source population stabilizes at *K* and the sink is unoccupied. At higher values of *r*, however, reproductive potential is so great that the movement of a

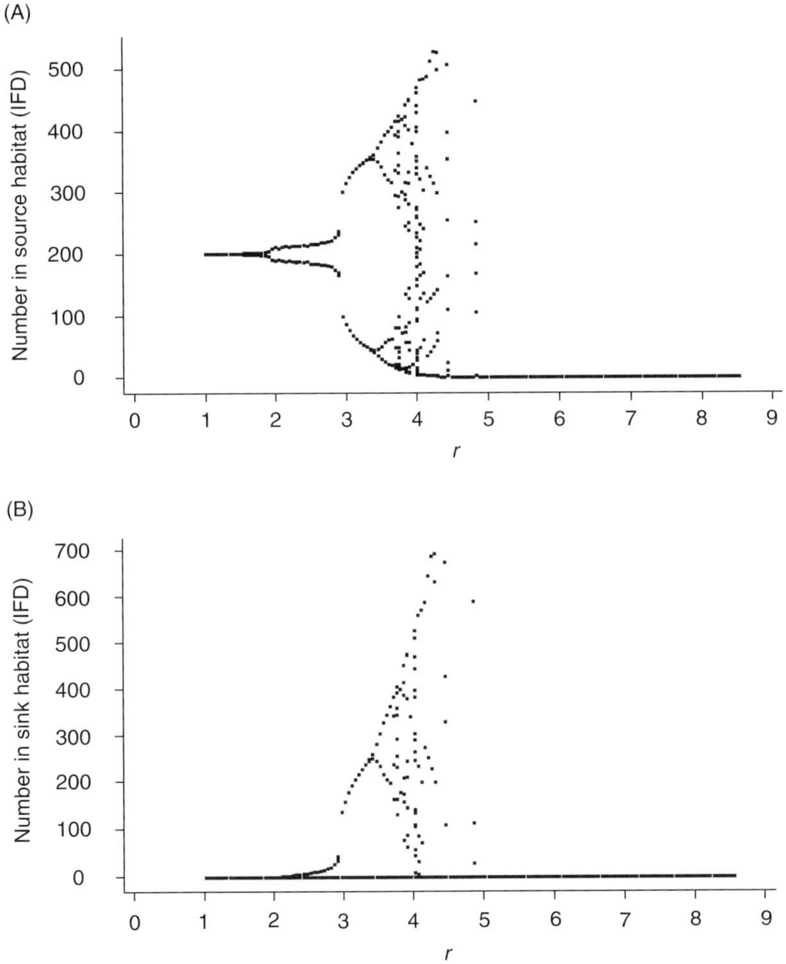

FIGURE 3.5.
Ideal free population dynamics in source (A) and sink (B) habitats when the sink is relatively severe with strong density-dependent feedback on fitness. The first 50 generations of each 250-generation simulation were excluded prior to analysis. Please note the differences in scale on the ordinate. The abscissa represents iterated values of r in the source habitat. Parameter values as follows: initial N (source) = 100, (sink) = 10; K (source) = 200; r (sink) = −0.1; slope of fitness with density (sink) = −0.01; r (source) incremented in units of 0.05.

single individual from source to sink is sufficient to maintain a small population of individuals in the sink habitat (Fig 3.6B). This rather peculiar situation arises, however, as an artifact of restricting the model to whole numbers.

The implications of Figure 3.6 are extraordinary and profound. An ecologist confronted with a source and high-quality sink habitat might observe a highly stable population exploiting only the source. But imagine that the "insignificant" sink habitat is altered to yield lower fitness or increased density

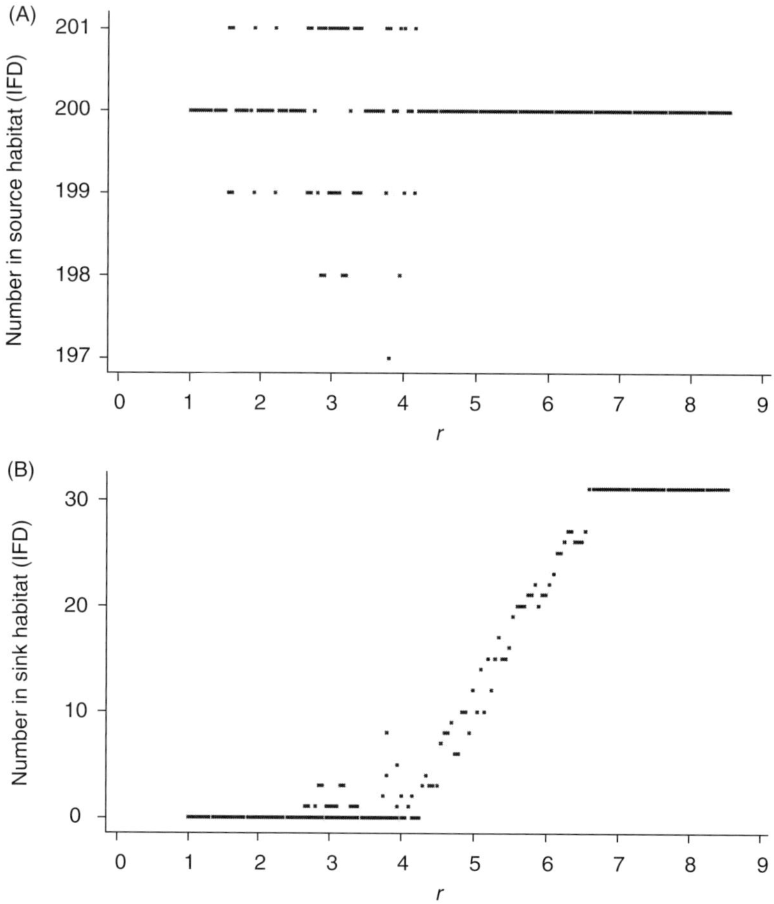

FIGURE 3.6.
Ideal free population dynamics in source (A) and sink (B) habitats when the sink is mild with weak density-dependent feedback on fitness. Population growth in the source habitat is zero across most values of *r*. Occupation of sink habitat with 31 individuals at high *r* is an artifact of rounding to whole numbers (negative population growth in the sink is so small that all individuals survive indefinitely). In real systems these individuals would die, after which the sink would be unoccupied. The first 50 generations of each 250-generation simulation were excluded prior to analysis. Please note the differences in scale on the ordinate. The abscissa represents iterated values of *r* in the source habitat. Parameter values as follows: initial *N* (source) = 100, (sink) = 10; *K* (source) = 200; *r* (sink) = −0.01; slope of fitness with density (sink) = −0.0002; *r* (source) incremented in units of 0.05.

dependence. Then any stochastic variation in the source, particularly for a species with high *r*, could suddenly convert the population into one with massive fluctuations in abundance (e.g., Fig. 3.5), or cause rapid extinction.

The MAXN strategy revealed similar dynamics but with noticeable differences (Figs. 3.7 and 3.8). First, the source population is more stable for species with low *r*. Second, the number of individuals in the sink habitat increases

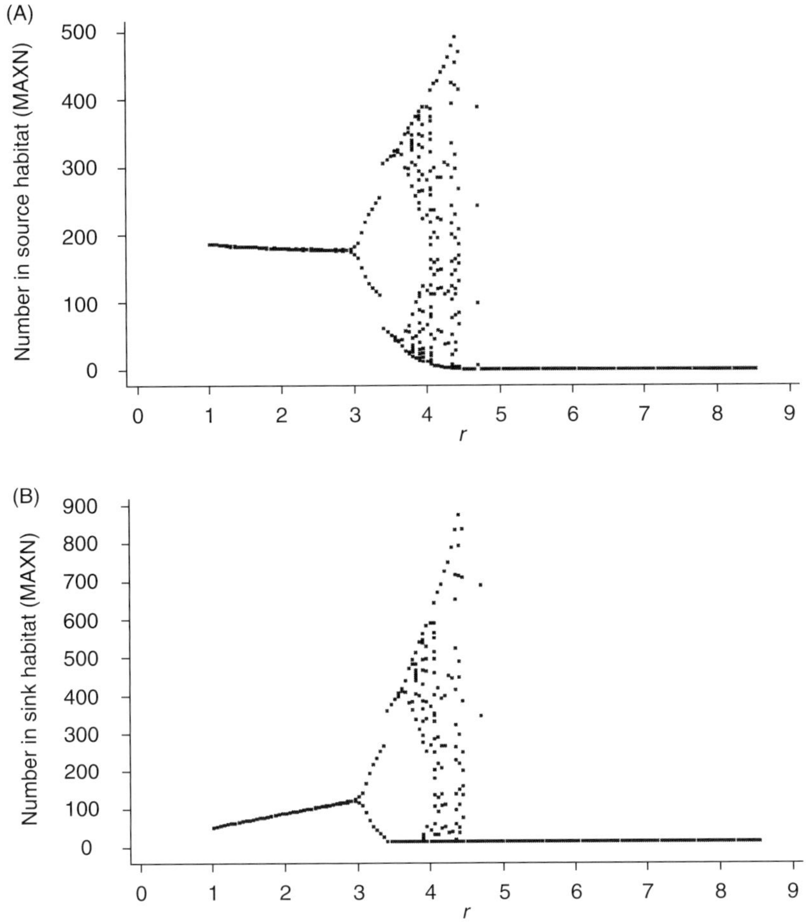

FIGURE 3.7.
Population dynamics in source (A) and sink (B) habitats when the sink is of relatively low quality with strong density-dependent feedback on fitness (MAXN strategy). The first 50 generations of each 250-generation simulation were excluded prior to analysis. Please note the differences in scale on the ordinate. The abscissa represents iterated values of r in the source habitat. $R = 1$; all other parameter values as in Figure 3.5.

across low values of r and attains a higher population size than under ideal free habitat selection. Third, sink habitat can always be unoccupied under ideal free habitat selection, but this occurs at only relatively large values of r in the MAXN strategy ($r \geq 3.40$ in Fig. 3.7).

My final simulation evaluated the effect of source–sink dynamics when the sink habitat represents an ecological trap (an attractive sink; Figs. 3.9 and 3.10). The simulation assumes that the cues used by individuals to assess habitat quality indicate a constant fitness advantage, at low density, for the sink. As density increases, the perceived quality of the trap deteriorates in direct proportion to the density-dependent decline of fitness in this attractive sink.

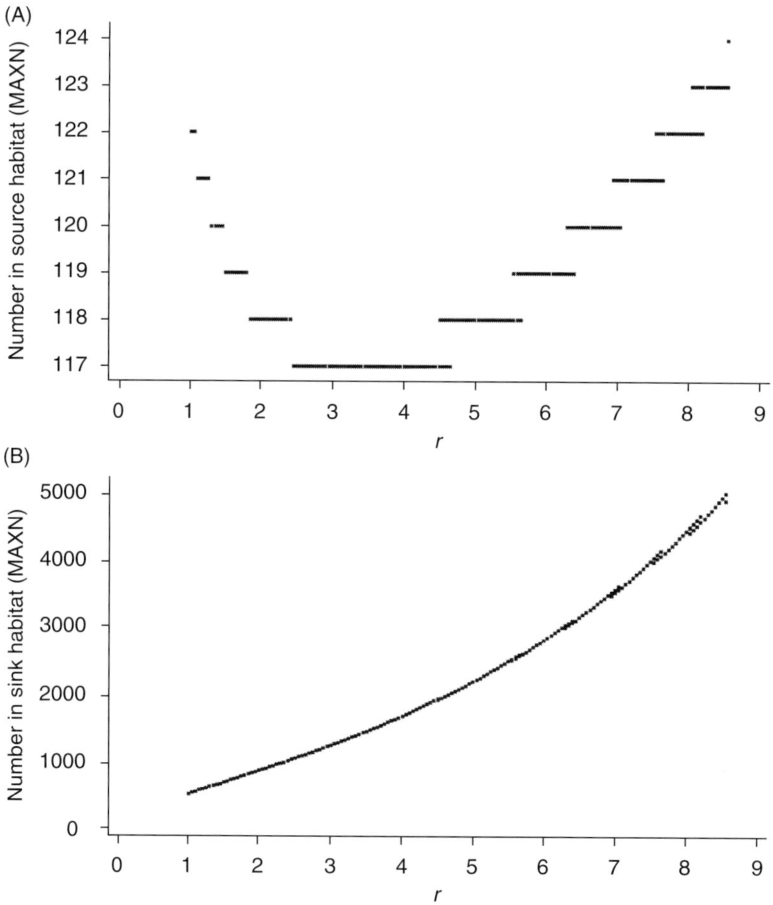

FIGURE 3.8.
Population dynamics in source (A) and sink (B) habitats when the sink is of relatively high quality with weak density-dependent feedback on fitness (MAXN strategy). The first 50 generations of each 250-generation simulation were excluded prior to analysis. Please note the differences in scale on the ordinate. The abscissa represents iterated values of *r* in the source habitat. $R = 1$; all other parameter values as in Figure 3.6.

Each successive iteration of the model increased maximum fitness in both the source and sink, but the slope of population growth with density was altered only in the source (because *K* was constant).

Although the pattern in the dynamics is similar to the control, there are again two exceptions (Figs. 3.9 and 3.10). Fluctuations in source habitat are less than those occurring without habitat selection, and the value of *r* that precipitates extinction is also increased. These important, and perhaps unexpected, results contrast sharply with the usual interpretation that populations trapped by unreliable cues are more prone to extinction than are populations using reliable cues of habitat quality. This is not to suggest that individuals

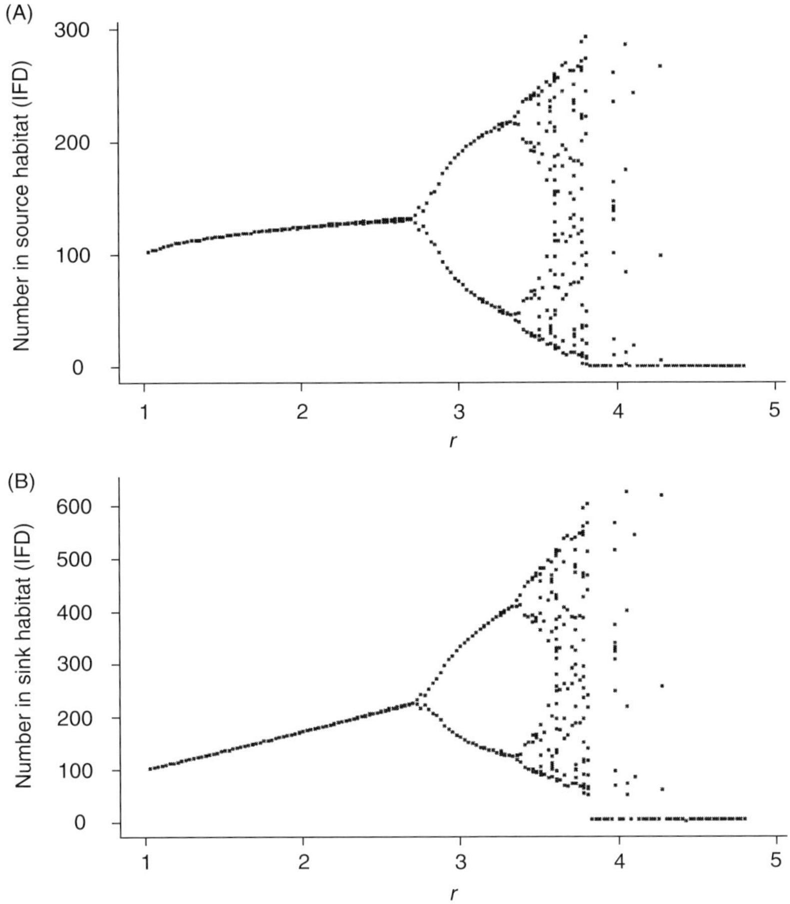

FIGURE 3.9.
Ideal free population dynamics in source (A) and sink (B) habitats when the sink habitat is perceived to have higher quality than the source (an ecological trap). The first 50 generations of each 250-generation simulation were excluded prior to analysis. Please note the differences in scale on the ordinate. The abscissa represents iterated values of r in the source habitat. Parameter values as follows: initial N (source) = 100, (sink) = 10; K (source) = 200; r (sink) = −0.01; slope of fitness with density (sink) = −0.01; perceived r (sink) = r (source) +0.5; r (source) incremented in units of 0.05.

never settle in attractive traps with very low recruitment. Grassland birds nesting in hayfields may fail to raise any surviving offspring. The point is that whenever the trap is not too severe, it can act to stabilize metapopulation dynamics.

Populations obeying the MAXN strategy had similar dynamics to those undergoing ideal free habitat selection. Densities in the MAXN source habitat were more stable at low values of r than were IFD populations. MAXN populations also maintained a higher density in the sink (trap) habitat.

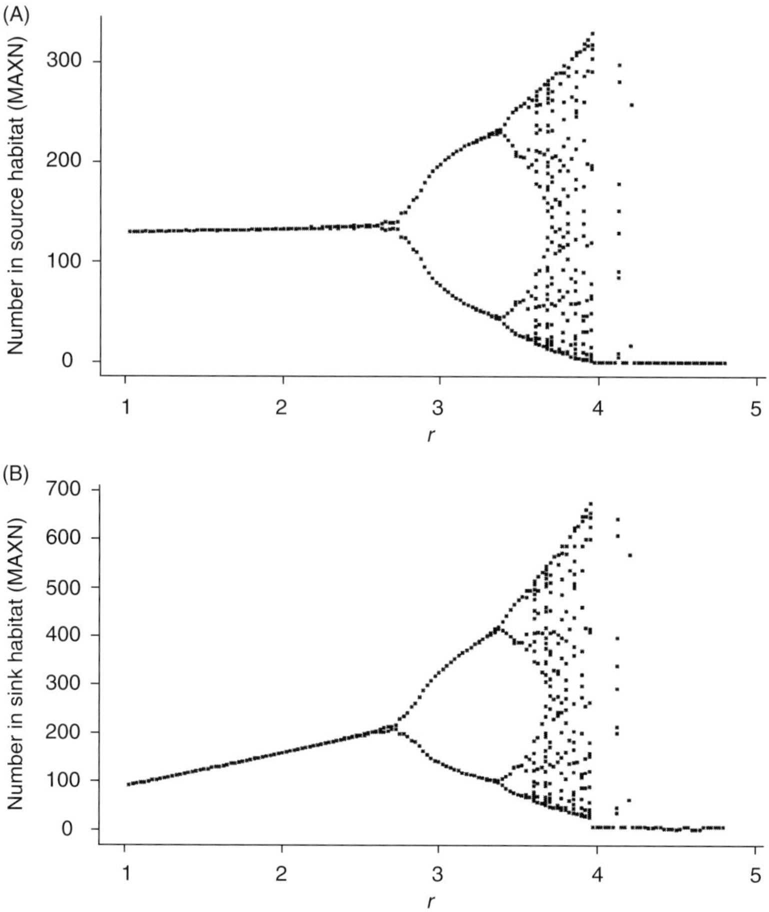

FIGURE 3.10.
Population dynamics in source (A) and sink (B) habitats for a population following the MAXN habitat-selection strategy when the sink habitat is perceived to have higher quality than the source (an ecological trap). The first 50 generations of each 250-generation simulation were excluded prior to analysis. Please note the differences in scale on the ordinate. The abscissa represents iterated values of r in the source habitat. $R = 1$; all other parameter values as in Figure 3.9.

Discussion and implications

Ecologists studying source–sink dynamics, and especially those using the source–sink framework as a guide for management and conservation, should proceed cautiously. As noted by Holt (1997), source–sink dynamics are a necessary byproduct of density-dependent habitat selection in populations with unstable dynamics. And although I simulated those dynamics by varying intrinsic population growth rates, source–sink regulation can emerge whenever populations fluctuate above and below a source habitat's carrying capacity. Thus, to a rough approximation, we can imagine that any population with either intrinsic or externally forced time lags (as through predator–prey interactions) will have the potential for source–sink regulation.

Paradoxically, if the quality of the sink habitat is relatively high, and fitness declines slowly with increasing density (as might often occur if sink area is large relative to source), then the characteristic unstable dynamics in the source habitat that force occupation of the sink may disappear completely. The sink habitat may be unoccupied. Thus, a conservation or resource manager might easily perceive that the sink habitat has no value in regulating the population and approve decisions that reduce its quality or area. The sink would have less capacity to absorb immigrants and thereby buffer population change. Any stochastic change in source density could then ramify quickly as oscillating source–sink dynamics. If the change to the sink habitat is large, or if the species has a high capacity for reproduction, the population could easily be pushed to extinction.

"Invisible" source–sink regulation, where sinks are unoccupied but essential to generate stability in the source, may be commonplace. And if it is, there are serious implications for conservation strategies requiring the identification and preservation of "critical habitat" (as demonstrated in Gimona *et al.*, Chapter 8, this volume). All habitat in a source–sink system may be critical. Research programs aimed at conservation and management must be ever vigilant for the role that habitat selection plays in population regulation, and especially so for species at risk.

It is thus important for ecologists to find ways to identify sources, sinks and ecological traps (e.g., Runge *et al.* 2006). The task will not be easy because traps that are also sinks can yield quite stable dynamics at low population growth rates, and unstable dynamics typical of source habitat at higher values of r. Although estimates of fitness may help to differentiate source from sink and trap habitats (Shochat *et al.* 2005), fitness alone is likely to be insufficient because ideal free habitat selectors should equalize fitness between source and sink. The situation is complicated because managers will often be motivated to search for sink habitats when faced with declining populations. So negative fitness alone cannot, for the IFD, differentiate source from sink. Dispersal, too, may be unreliable because large sink populations may nevertheless export individuals to the source habitat.

The "contribution metric" of Runge *et al.* (2006) that incorporates migration and fitness can also be ambiguous because it evaluates source versus sink over a single time step. Net contribution over longer time intervals would thus be much more reliable, but may not be capable of distinguishing intrinsic instability caused by high growth rates from extrinsic stochasticity. None of these "solutions" would help a manager trying to distinguish an unoccupied mild sink that stabilized source density (e.g., Fig. 3.6) from much more severe unsuitable habitat.

Typical removal or resource-addition experiments designed to assess habitat quality may also fail if the source–sink system is left intact. Individuals will neutralize the treatment by moving to equalize fitness. And they will do so regardless of which habitat we manipulate. It may thus be necessary to remove habitat selection by using enclosures before culling the enclosed populations, to reveal the underlying relationship between fitness, or at least population growth rate, and density in each habitat.

Alternative habitat-selection strategies that do not equalize fitness augur hope as well as potential failure. MAXN-type strategies, such as produced by despotic, preemptive, inclusive-fitness and cooperative habitat selection, suggest that population growth should always be positive in the source habitat. But there are at least two problems with this perspective.

1. Inclusive-fitness distributions decay toward the ideal free alternative as the degree of genetic relatedness declines.
2. MAXN strategies, even though they maximize per capita population growth rates, are themselves invasible to unrelated individuals living in sink or otherwise low-quality habitats (Morris *et al.* 2001).

It is thus convenient to think of MAXN as a precursor to despotism, and to think of despotism and territorial defense as mechanisms to maintain the fitness advantage that originated among cooperating habitat-selecting relatives.

Be that as it may, any weakness in group or territorial defense will open the door to unrelated intruders who may precipitate "cycles" of habitat-selection strategies alternating between MAXN and ideal free alternatives. Further theory will be required to assess whether such invasion and counter-invasion cycles of "pure" strategies actually occur, or whether some form of stable MAXN–free mixed strategy can evolve.

Previous theories of habitat selection have not included the possibility that multiple unrelated individuals might cooperate to increase their individual fitness. The simple models developed here suggest that such strategies are possible, but that they may be difficult to disentangle from other alternatives in heterogeneous complexes of multiple habitats. And they too predict stable dynamics in systems where we might otherwise expect to find rather large differences in population size through time or space. If all habitats are of similar size, the model predicts directional dispersal from high- to low-quality habitat. But again all bets are off if low-quality habitat is extensive. Even though each individual may produce relatively fewer recruits, the combined production of offspring could cause a reversal in the normal flow of emigrants from high- to low-quality habitat.

Source–sink dynamics are a recurring theme in population ecology (Anderson 1970; MacArthur 1972; Lidicker 1985) motivated through a broader research program on the role of dispersal in population regulation. Early work by Krebs *et al.* (1969), on enclosed populations, clearly implicated dispersal as a key determinant of rodent population regulation, and led Lidicker (1975) to classify emigrants into pre-saturation and saturation dispersal classes. Although there are both semantic and conceptual difficulties with this classification (Anderson 1989), populations following the MAXN strategy provide a reasonably comprehensive mechanism to explain dispersal across a range of population densities. Emigrants leave high-quality habitats at relatively low densities to settle in lower-quality sinks. But if the population has high reproductive potential and lives in a landscape including sinks of very low quality, or if sinks are misinterpreted as high quality (traps), then dynamics are unlikely to stabilize and can yield alternating cycles of "pre-saturation" and "saturation" dispersal.

At present, therefore, it seems unlikely that we will be able to devise failproof methods that use existing patterns of density, and perhaps even movement or fitness, to distinguish between source, sink and trap habitats. Rather, we may frequently need to design manipulative experiments that force populations away from existing modes of regulation in order to infer the relative qualities of their constituent habitats. Such experiments will often need to eliminate, even if just temporarily, the dispersal causing spatial regulation.

Source–sink dynamics have a more subtle and potentially very significant evolutionary effect. If dispersing individuals can match habitat choice with phenotypic variance, then populations initially maladapted to sink habitat can evolve not only to exploit it, but to do so with positive population growth (Holt and Barfield 2008). Holt and Barfield's sink pre-adaptation model demonstrates that adaptation to sink habitats is most likely when there is weak density dependence and when the fitness penalty of using the sink is small. These are the same conditions that lead to stable source dynamics. Stochasticity enhances dispersal (Morris 2003) as well as the use of sink habitats (D. W. Morris, unpublished simulations), and creates the opportunity for sink adaptation. But if the population has evolved toward high population growth potential, adaptation that connects sink to source habitat may convert otherwise stable dynamics to erratic shifts that increase the probability of extinction. Populations occupying high-quality sinks, with no other habitats available with lower fitness expectations, may evolve themselves out of existence.

Acknowledgments

I thank J. Liu, V. Hull, A. Morzillo and J. Wiens for inviting me to participate in, and the US Regional Association of the International Association for Landscape Ecology for hosting, an excellent symposium that celebrated one of Ron Pulliam's many influential contributions to ecology and evolutionary biology. We, and those who follow us, are deeply indebted to Ron's example and intellect. R. Holt kindly pointed me toward relevant literature on source–sink dynamics in unstable populations and cogent anonymous reviews helped me to improve this contribution. I thank Canada's Natural Sciences and Engineering Research Council for its continuing support of my research in the broad fields of evolutionary and landscape ecology.

References

Anderson, P. K. (1970). Ecological structure and gene flow in small mammals. *Symposium of the Zoological Society of London* **26:** 90–92.

Anderson, P. K. (1989). *Dispersal in Rodents: A Resident Fitness Hypothesis*. Special Publication No. 9, The American Society of Mammalogists.

Apaloo, J., J. S. Brown and T. L. Vincent (2009). Evolutionary game theory: ESS, convergence stability, and NIS. *Evolutionary Ecology Research* **11:** 489–515.

Dwernychuk, L. W. and D. A. Boag (1972). Ducks nesting in association with gulls: an ecological trap? *Canadian Journal of Zoology* **50:** 559–563.

Fretwell, S. D. (1972). *Populations in a Seasonal Environment*. Princeton University Press, Princeton, NJ.

Fretwell, S. D. and H. L. Lucas Jr. (1969). On territorial behavior and other factors influencing habitat distribution in birds. *Acta Biotheoretica* **14:** 16–36.

Holt, R. D. (1985). Population dynamics in two-patch environments: some anomalous consequences of an optimal habitat distribution. *Theoretical Population Biology* **28:** 181–208.

Holt, R. D. (1997). On the evolutionary stability of sink populations. *Evolutionary Ecology* **11:** 723–731.

Holt, R. D. and M. Barfield (2008). Habitat selection and niche conservatism. *Israel Journal of Ecology and Evolution* **54:** 295–310.

Krebs, C. J., B. L. Keller and R. H. Tamarin (1969). *Microtus* population biology: demographic changes in fluctuating populations of *M. ochrogaster* and *M. pennsylvanicus* in southern Indiana. *Ecology* **50:** 587–607.

Kristan, W. B. III (2003). The role of habitat selection behavior in population dynamics: source–sink systems and ecological traps. *Oikos* **103:** 457–468.

Lidicker, W. Z. Jr. (1975). The role of dispersal in the demography of small mammals. In *Small Mammals: Their Productivity and Population Dynamics* (F. B. Golley, K. Petrusewicz and L. Ryszkowski, eds.). Cambridge University Press, Cambridge, UK: 103–128.

Lidicker, W. Z. Jr. (1985). Population structuring as a factor in understanding microtine cycles. *Acta Zoologica Fennica* **173:** 23–27.

MacArthur, R. H. (1972). *Geographical Ecology*. Harper and Row, New York.

May, R. M. and G. F. Oster (1976). Bifurcations and dynamic complexity in simple ecological models. *American Naturalist* **110:** 573–599.

Morris, D. W. (1988). Habitat-dependent population regulation and community structure. *Evolutionary Ecology* **2:** 253–269.

Morris, D. W. (1991). On the evolutionary stability of dispersal to sink habitats. *American Naturalist* **138:** 702–716.

Morris, D. W. (2003). Shadows of predation: habitat-selecting consumers eclipse competition between coexisting prey. *Evolutionary Ecology* **17:** 393–422.

Morris, D. W. and J. E. Diffendorfer (2004a). Reciprocating dispersal by habitat selecting white-footed mice. *Oikos* **107**: 549–558.
Morris, D. W., J. E. Diffendorfer and P. Lundberg (2004b). Dispersal among habitats varying in fitness: reciprocating migration through ideal habitat selection. *Oikos* **107**: 559–575.
Morris, D. W., P. Lundberg and J. Ripa (2001). Hamilton's rule confronts ideal-free habitat selection. *Proceedings of the Royal Society of London B* **268**: 291–294.
Pulliam, H. R. (1988). Sources, sinks, and population regulation. *American Naturalist* **132**: 652–661.
Pulliam, H. R. and T. Caraco (1984). Living in groups: is there an optimal group size? In *Behavioural Ecology: An Evolutionary Approach*, 2nd edition (J. R. Krebs and N. B. Davies, eds.). Sinauer Associates, Sunderland, MA: 122–147.
Pulliam, H. R. and B. J. Danielson (1991). Sources, sinks, and habitat selection: a landscape perspective on population dynamics. *American Naturalist* **137**(Suppl.): S50–S66.
Runge, J. P., M. C. Runge and J. D. Nichols (2006). The role of local populations within a landscape context: defining and classifying sources and sinks. *American Naturalist* **167**: 925–938.
Schlaepfer, M. A., M. C. Runge and P. W. Sherman (2002). Ecological and evolutionary traps. *Trends in Ecology and Evolution* **17**: 474–480.
Shochat, E., M. A. Pattern, D. W. Morris, D. L. Reinking, D. H. Wolfe and S. K. Sherrod (2005). Ecological traps in isodars: effects of tallgrass prairie management on bird nest success. *Oikos* **111**: 159–169.
Wright, S. (1931). Evolution in Mendelian populations. *Genetics* **16**: 97–159.

CRAIG W. BENKMAN AND ADAM M. SIEPIELSKI

4

Sources and sinks in the evolution and persistence of mutualisms

Summary

Pulliam's (1988) model of sources and sinks demonstrated the importance of considering spatial variation in demographic rates for understanding population persistence. One of the factors contributing to such spatial variation is variation in the occurrence of other species, including prey, predators and mutualists. Here we consider how such variation in community context affects what could be termed sources and sinks in the evolution of species interactions. We focus on the seed dispersal mutualism between Clark's nutcrackers (*Nucifraga columbiana*) and limber pine (*Pinus flexilis*), and how the presence and absence of a seed predator, the red squirrel (*Tamiasciurus hudsonicus*), likely causes the mutualists to experience demographic sinks and sources, respectively. Although sink populations of limber pine mostly represent the later stages in forest succession, when limber pine trees are older, species interactions within the source and sink populations will affect the evolution and maintenance of the seed dispersal mutualism. In general, the persistence of mutualisms is probably dependent on the amount of habitat that lacks a competitively superior antagonist (i.e., a "source" habitat) and on whether selection exerted by antagonists conflicts with selection exerted by mutualists. Because most mutualisms are vulnerable to exploitation by antagonists, and the distributions of antagonists are unlikely to overlap completely with mutualists, we believe that such a source–sink perspective will be useful for examining the evolution and persistence of mutualisms.

Sources, Sinks and Sustainability, ed. Jianguo Liu, Vanessa Hull, Anita T. Morzillo and John A. Wiens.
Published by Cambridge University Press.

Introduction

Few species occur in completely homogeneous landscapes. Consequently, the survival and reproductive success of individuals in a population vary spatially. Pulliam (1988), in particular, demonstrated that accounting for variation in demographic rates between different habitats is critical for understanding the persistence of populations and thus for providing key information for land managers. Habitats vary in quality for many reasons, including spatial variations in abiotic factors and in the distribution of enemies such as predators or parasites. Such spatial variation presumably underlies much of the variation in demographic rates between sources and sinks (Pulliam 1988). For example, sinks have often been attributed to high abundances of predators or parasites (e.g., Lloyd *et al.* 2005). This variation in the occurrence of other species (i.e., the community context) across a species' range also underlies much of the geographic variation found in the form and evolution of species interactions (Thompson 2005). For instance, variation in the distribution of enemies not only affects the demographic rates of the victim species but can also have profound consequences for their evolution in response to interactions with these and other species.

How an antagonist can affect the persistence and evolution of a mutualism has been the focus of recent theoretical and empirical studies, in large part because most mutualisms are exploited by antagonistic species (Bronstein 2001; Yu 2001). Especially important in theoretical studies is the competitive strength of antagonistic species relative to the mutualistic species. When the mutualist is competitively superior to the antagonist, then persistence is possible under a range of conditions (Wilson *et al.* 2003). This competitive asymmetry is assumed in some theoretical studies (e.g., Ferrière *et al.* 2007). In contrast, mutualisms in which an antagonistic species is competitively superior to one of the mutualistic species may require some form of refuge from the antagonist for the mutualism to persist, especially when resources provided by the mutualism do not solely limit the antagonist. Such refuges can arise when mutualists have better colonization abilities than antagonists in metacommunities (Yu 2001). Refuges can also arise because antagonistic species avoid certain habitats. Indeed, superior competitors often have more restricted distributions along environmental gradients than inferior competitors (Colwell and Fuentes 1975). This is the type of example we will consider further. In particular, we focus on the seed dispersal mutualism between Clark's nutcrackers (Aves: *Nucifraga columbiana*) and limber pine (*Pinus flexilis*), which is constrained by a competitively superior seed predator and antagonist of the mutualism, the red squirrel (*Tamiasciurus hudsonicus*). We provide background on these species and frame the evolution of their interactions using a simple graphical model. Although the nutcracker–pine system does not fit the classic source–sink model in which dispersal of surplus individuals from source habitats maintains populations in sink habitats (Pulliam 1988), the distribution of antagonists does potentially affect the distribution of source and sink habitats for mutualists (as is also suggested in Holt, Chapter 2, this volume). Thus, we evaluate how the competitive and evolutionary effects of antagonists in sink habitats affect the persistence and evolution of mutualisms.

Natural history of the study system

Limber pine seeds are large (>90 mg) and wingless, with little potential to be carried by the wind. Thus, they are reliant on animals, especially Clark's nutcrackers, for seed dispersal (Lanner 1996). Nutcrackers, in turn, rely throughout the winter and into spring on the thousands of pine seeds that they cache in the fall in small clusters of 2–5 or more seeds in generally suitable germination sites several centimeters underground (Lanner 1996). Red squirrels are important seed predators of many conifers, including limber pine, in North America (Smith and Balda 1979), rapidly harvesting whole cones and burying them, usually at the base of a large tree near the center of their territory (Benkman *et al.* 1984). A few of the seeds buried in their middens escape predation and germinate (Vander Wall 1990). Those few seeds that fall out of cones that open in the midden and then germinate will likely be killed in subsequent years because red squirrels re-use their middens year after year, turning over both midden material and any seedlings that emerge. Limber pine occurs predominantly in mountainous areas from the east side of the Sierra Nevada, east across much of the Great Basin, and into the Rocky Mountains. Clark's nutcrackers occur throughout the range of limber pine, whereas red squirrels are absent from the Great Basin and are also absent or uncommon in open stands of limber pine. The red squirrel's congener, the Douglas squirrel (*T. douglasii*), occurs in the Sierra Nevada (hereafter we refer only to red squirrels).

Nutcrackers and limber pine have adaptations that enhance their mutualism. Nutcrackers have long pointed bills for shredding and reaching between cone scales to extract the underlying seeds. Nutcrackers have also evolved a sublingual pouch that holds 30 or more grams of seeds (over 20% of the nutcracker's body mass) and can fly up to 22 km to cache them (Vander Wall and Balda 1981). These traits, plus strong flight capabilities, allow nutcrackers to cache thousands of seeds, while their exceptional spatial memory enables them to find and recover these buried seeds. Caches are widely dispersed and are not defendable against ground-foraging seed predators such as rodents, and not all caches can be remembered (Balda and Kamil 1992). This presumably explains why nutcrackers cache many more seeds than they will need in any given year; nutcrackers have been estimated to cache two to three times the number of pine seeds required (reviewed in Vander Wall 1990; Lanner 1996). Because remaining seeds are cached in sites favorable for germination, and a large fraction of the seeds may not be retrieved, nutcrackers act as a mutualist to the pine. However, the benefit to nutcrackers of caching additional seeds should be expected to decelerate (Fig. 4.1A), especially as the number of seeds cached exceeds expected demand. The extent to which nutcrackers cache additional seeds will depend on both the benefits and costs of caching to themselves, because caching will only occur when and where the benefits of caching exceed the costs (Fig. 4.1A). When seeds are abundant, the costs of harvesting and caching seeds will simply increase in proportion to the number of seeds cached. However, searching effort per seed will increase as seeds are depleted from the cones (Benkman *et al.* 1984; Vander Wall 1988), causing the costs of acquiring and caching additional seeds to accelerate (Fig. 4.1A). The optimal number of seeds cached is where the difference between the benefits and costs is maximized (Fig. 4.1A). The abovementioned adaptations of nutcrackers for harvesting and caching seeds reduce the costs, while those related to recovery increase the benefits, and together these effects act to shift the optimum number of seeds cached to larger values (Fig. 4.1A; shifts in benefits not shown). Benefits to the pine are a byproduct of the benefit to nutcrackers of caching seeds in excess of need and do not come with fitness costs to nutcrackers (i.e., a byproduct mutualism; Connor, 1995; Sachs and Simms 2006).

Limber pine increasingly benefits as nutcrackers cache seeds beyond their needs (Fig. 4.1B), because this increases the proportion of cached seeds that will not be retrieved and will thus potentially germinate. This favors the evolution of cone and seed traits that reduce the costs of harvesting and caching seeds, and shifts the optimum to a greater number of seeds cached by moving the cost curve for nutcrackers down and to the right (from c_1 to c_2 in Fig. 4.1A). Traits that facilitate the harvest and caching of seeds by nutcrackers, and result in the preferential harvest of seeds by nutcrackers, include greater numbers of seeds per cone and thinner cone scales and seed coats (Siepielski and Benkman 2007a). Decreases in seed coat thickness reduce seed mass and volume and thus increase the number of seeds that can be carried during caching flights (Benkman 1995a; Siepielski and Benkman 2007a). Such cone and seed traits characterize bird-dispersed pines (Lanner 1996; Siepielski and Benkman 2007b). The preference by nutcrackers for trees that exhibit this set of traits is a form of partner choice that provides greater benefits to these individuals and favors the evolution of the mutualism (Foster and Wenseleers 2006).

In the absence of selection by antagonists such as pre-dispersal seed predators, limber pine may continue to evolve traits that facilitate the harvest and caching of their seeds by nutcrackers. However, as limber pine loses its seed defenses, seeds also become increasingly vulnerable to seed predators that

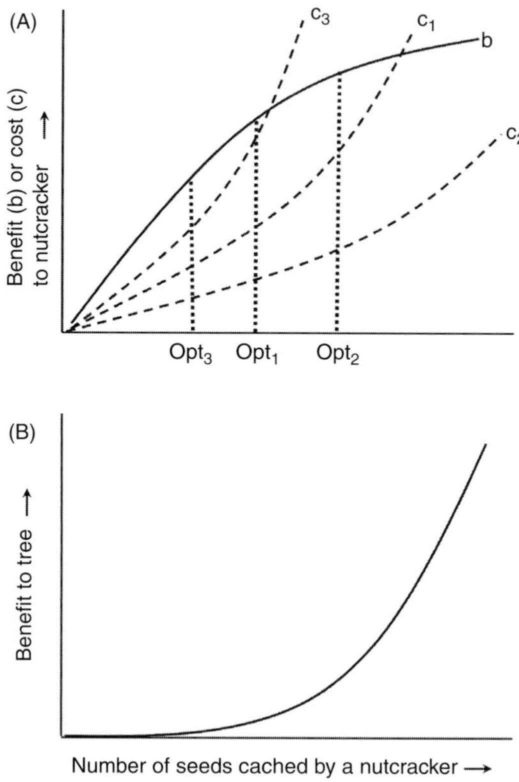

FIGURE 4.1.
(A) The optimal number of seeds that Clark's nutcrackers should cache (i.e., natural selection favors) resides where the difference between the benefits (b; solid curve) and costs (c_1, c_2, c_3; dashed curves) of caching seeds is maximized (dotted vertical lines; Opt_1, Opt_2, Opt_3). The benefits from caching seeds increase with increases in the number of seeds cached, but these benefits then decelerate, especially as the number of seeds cached begins to exceed the number of seeds that can be recovered. Adaptations that increase the likelihood of seed recovery such as improved memory or caching seeds in areas where seeds are unlikely to be pilfered by others will shift the benefit curve upward (not shown) and thereby shift the optimum number of seeds cached to a larger value. Costs of caching accelerate as seeds are depleted, while adaptations that facilitate seed harvesting and caching such as nutcracker foraging and seed transporting adaptations, or reductions in seed defenses will shift the cost curve downward (e.g., from c_1 to c_2) and favor an increase in the number of seeds cached (from Opt_1 to Opt_2). Phenotypic selection and preemptive competition by a seed predator such as the red squirrel can shift upward the seed-caching cost curve (from c_1 or c_2 to c_3) for nutcrackers, favoring a reduction in the number of seeds cached (from Opt_1 or Opt_2 to Opt_3). (B) The benefit to trees remains initially negligible as the number of seeds cached increases, because nutcrackers will recover all or nearly all of the seeds. As the number of seeds cached by a given nutcracker increases beyond what is likely to be harvested, the benefits will accelerate because the nutcrackers recover a decreasing proportion of these additional cached seeds.

normally avoid seeds protected in closed conifer cones (e.g., *Spermophilus lateralis*; C. W. Benkman, personal observation). This increased susceptibility to seed predators will potentially act to set a limit on the extent to which limber pine evolves to become accessible to nutcrackers. Seed predators may even act to reverse the course of cone evolution. Red squirrels have a competitive advantage over nutcrackers because red squirrels rapidly cut and cache closed cones full of seeds and thereby remove and bury most of the seed crop before nutcrackers have a chance to cache many seeds (Benkman *et al.* 1984). This preemptive competition depresses seed availability for nutcrackers, which in turn will increase the costs of harvesting seeds (Benkman *et al.* 1984; Vander Wall 1988) and thereby shift the cost curve upward, favoring a reduction in the number of seeds cached by nutcrackers (Fig. 4.1A). A decline in the number of seeds cached will, in turn, reduce the benefit to the tree (Fig. 4.1B). The cost curve is shifted further upward (c_3 in Fig. 4.1A) because red squirrels also exert selection pressure on cone and seed traits that opposes the selection pressure exerted by nutcrackers (Siepielski and Benkman 2007a, 2007b). This selection by red squirrels causes the evolution of cone and seed traits that slow the seed harvesting and caching rates of nutcrackers (Benkman 1995a; Siepielski and Benkman 2007a). These shifts in cone and seed traits between areas with and without red squirrels are striking (Fig. 4.2) and are replicated among conifers dispersed by corvids (Siepielski and Benkman 2007a, 2007b). The shifts in limber pine cone and seed traits in regions with red squirrels reduce by about 70% the number of seeds potentially dispersed by nutcrackers in comparison with regions without red squirrels. Such a reduction occurs because, with increasing seed defenses, about 55% fewer seeds are produced per unit of reproductive allocation (assuming that energy is limiting for reproduction; Benkman 1995a; Siepielski and Benkman 2007a), and the probability of a seed being harvested by nutcrackers decreases by 34% (Siepielski and Benkman 2008a). These evolutionary effects of red squirrels, in combination with their competitive effects, greatly reduce the potential for seed dispersal by nutcrackers in regions with red squirrels.

Red squirrels, and other tree squirrels (i.e., *T. douglasii*, *Sciurus* spp.), alter the evolutionary trajectory of seed dispersal mutualisms between corvids and pines and, when common throughout the range of a pine, appear to prevent this mutualism from evolving and persisting (Benkman 1995b; Siepielski and Benkman 2007b). This effect of red squirrels is consistent with the theory that mutualisms are less likely to persist when an antagonistic species (i.e., red squirrels) has a competitive advantage over the mutualist (i.e., nutcrackers) (Ferrière *et al.* 2007). A strong effect is especially likely when the antagonist is not an obligate specialist on the mutualism and therefore is less limited by the resources provided by the mutualists. This interpretation is

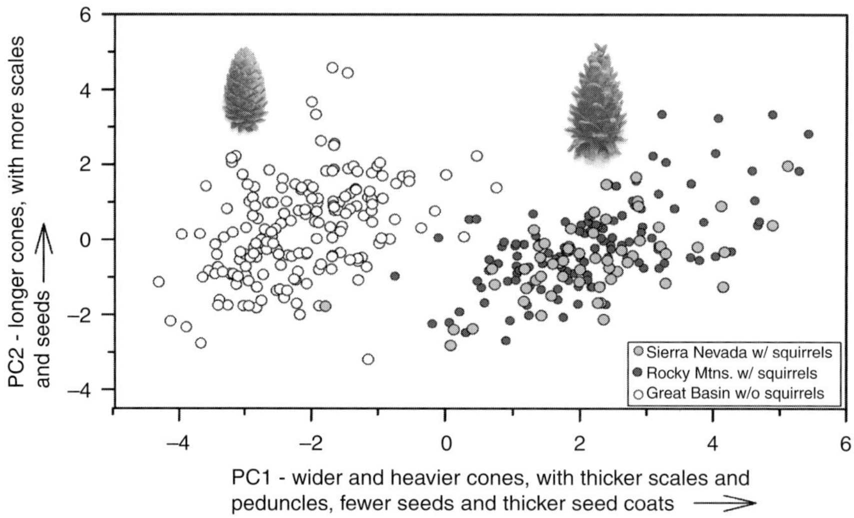

FIGURE 4.2.
The first two principal components (PC1, PC2) of seven limber pine cone and seed traits and representative cones from areas with pine squirrels (*Tamiasciurus* spp.; on right) and without pine squirrels (on left) (from Siepielski and Benkman 2007a). The Rocky Mountains have red squirrels (*T. hudsonicus*) while the Sierra Nevada has the ecologically equivalent Douglas squirrel (*T. douglasii*).

also supported by the observation that large-seeded pines that occur consistently in forested habitats with tree squirrels (*Tamiasciurus* or *Sciurus*) do not rely on nutcrackers or other seed-caching corvids for seed dispersal. In contrast, large-seeded pines that occur in more open habitats where tree squirrels are consistently scarce evolve adaptations that facilitate the mutualism with seed-caching corvids (Benkman 1995b; Siepielski and Benkman 2007b). Limber pine is an intriguing species because in much of its geographic range it occurs in areas where red squirrels are common, at least locally. In regions with red squirrels (e.g., Rocky Mountains), limber pine cones have enhanced defenses that are effective against red squirrels (Benkman 1995a; Siepielski and Benkman 2007a) and appear to rely more heavily on ground-foraging scatter-hoarding rodents (Siepielski and Benkman 2008a) that are also important for the dispersal of other large-seeded pines in forested areas (Vander Wall 2003). Presumably, nutcrackers need to cache some minimum number of seeds for them to act as mutualists to pines (Fig. 4.1B) and for the mutualism to evolve.

Sources and sinks in the mutualism

Pulliam (1988) used a simple habitat-specific demographic model to show that the equilibrium proportion of the population residing in the source

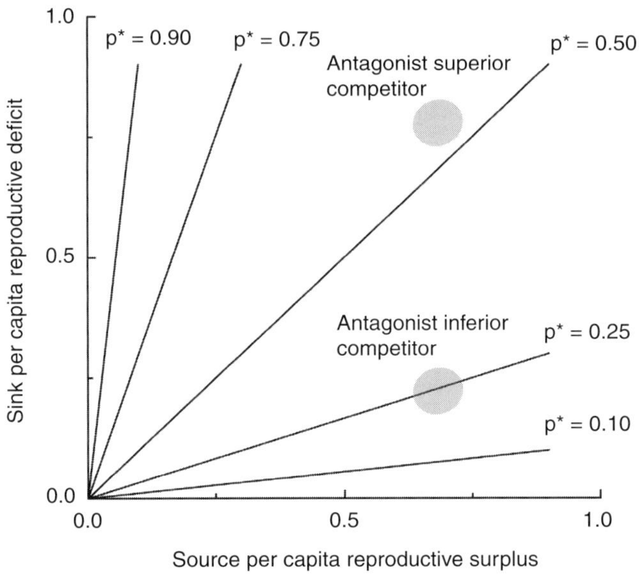

FIGURE 4.3.
The equilibrium proportion of the population (p*) residing in source habitat depends on the per capita reproductive surplus in the source habitats ($\lambda_1 - 1$, where λ_1 is the finite rate of increase in the source habitat) and deficit in sink habitats ($1 - \lambda_2$, where λ_2 is the finite rate of increase in the sink habitat) (modified from Pulliam 1988). Shaded areas represent how the equilibrium conditions might vary with variation in the competitive ability of the antagonistic species.

habitat varies inversely with the ratio of the per capita reproductive surplus (i.e., young produced beyond those needed for replacement) in the source habitat, to the per capita reproductive deficit (mortality exceeds reproduction) in the sink habitat (Fig. 4.3). Thus, at equilibrium, a very small proportion of the population can occur in the source habitat (e.g., 0.1), given a large reproductive surplus in the source habitat and a small reproductive deficit in the sink habitat (Fig. 4.3). Conversely, most of the population needs to be in source habitat (e.g., 0.9) in order to persist if the reproductive surplus in the source habitat is relatively small compared with the reproductive deficit in the sink habitat (Fig. 4.3).

Several lines of evidence indicate that areas with red squirrels act as sink (or at least "pseudo-sink") habitats for limber pine and nutcrackers, whereas areas without red squirrels act as source habitats for limber pine and perhaps nutcrackers. First, the fate of seeds in cones shifts from a high probability of predation by red squirrels to a high probability of being harvested and cached by nutcrackers in the absence of squirrels. For example, red squirrels harvested about 80% of the limber pine seed crop in a forested habitat in northern Arizona (Benkman *et al.* 1984; see Hutchins and Lanner 1982 for similar evidence for whitebark pine, *P. albicaulis*, in habitat occupied by red squirrels), whereas nutcrackers harvested an estimated 70–98% of the limber pine seed crop in a range in the Great Basin lacking red squirrels (Lanner and Vander Wall 1980). Moreover, most of the limber pine seeds harvested by nutcrackers in the latter study area were cached (Vander Wall 1988).

Second, and as predicted if competition and selection exerted by red squirrels limit seed dispersal by nutcrackers, stand densities of limber pine are uniformly low in regions with red squirrels (Rocky Mountains), but in regions where the squirrels are absent (Great Basin), stand densities are on average about two times greater (Siepielski and Benkman 2008b). This pattern is consistent with red squirrels limiting recruitment and stand densities; while, in the absence of red squirrels, recruitment increases and stand densities increase until they apparently become limited by annual precipitation (Siepielski and Benkman 2008b).

Although the above comparisons are based on large-scale geographic comparisons between regions with and without red squirrels, we believe it is reasonable to infer that local variation in the occurrence of red squirrels among habitats may act similarly to determine the local distribution of sources and sinks. Within the Rocky Mountains, red squirrels become less abundant as tree density declines, and many patches of more open woodland lack red squirrels because they rely on trees for food, cover, and to escape predation (Smith 1968). Likewise, because red squirrels avoid crossing large openings between forests, red squirrels have not crossed the large expanses of sagebrush-steppe to colonize the forests atop the mountains in the Great Basin. The more open patches of forest – whether formed because of rocky substrates or because the site is early in succession after a disturbance – are likely to represent source habitats for limber pine.

Even though more densely forested habitats (including other tree species in addition to limber pine) are likely to represent sink habitats for limber pine in regions with red squirrels, and more open habitats represent sources, sink habitats are unlikely to be maintained by emigration from source habitats as envisioned by the source–sink model (Pulliam 1988). Because limber pine is an early successional species that colonizes open habitat such as after a fire, and in many areas dense forests develop, habitats begin as source habitats and over time shift to become sink habitats. Thus, the proportion of the population that occupies sink habitat depends on conditions that affect the probability of open woodland becoming dense forest (e.g., soil moisture, substrate). Consequently, we will not focus on the source–sink dynamics of limber pine and nutcrackers. Instead, our aim is to consider how the proportion of source and sink habitats, and the evolution of the interaction, affect the persistence of the mutualism.

Sources, sinks, and the evolution and persistence of mutualisms

We assume that the presence of an antagonistic species (e.g., red squirrels) causes what would otherwise be a source habitat to become a sink habitat, and that the size of the per capita reproductive deficit increases in proportion to the competitive effect of the antagonistic species. If the antagonistic species is an inferior competitor relative to the mutualist, then the difference between habitats in per capita growth rates will be small and the reproductive deficit, if there is a sink habitat, will also be small (Fig. 4.3). On the other hand, if the antagonistic species is a superior competitor relative to the mutualist, as are red squirrels relative to nutcrackers (Benkman *et al.* 1984), then the reproductive deficit in the sink habitat will be large (Fig. 4.3). With an increasing competitive impact from the antagonist (this could arise from an increase in either the competitive ability or the density of the antagonist), a smaller proportion of the population of mutualists will occur in sink habitats at equilibrium (Fig. 4.3). At the extreme, strong antagonists could prevent mutualists from co-occurring with them. This presumably explains why bird-dispersed pines tend to occur where tree squirrels are uncommon or absent (Benkman 1995b; Siepielski and Benkman 2007b).

We can also incorporate into the model an evolutionary response to phenotypic selection exerted by the antagonist. The form of selection experienced by the pine depends on how incremental changes in seed defenses affect the seed harvesting abilities of both nutcrackers and red squirrels. Evolution in response to selection in the sink is most likely if the increase in fitness in the sink (i.e., a decrease in per capita reproductive deficit) is greater than the decrease in fitness in the source (Holt 1996). In the extreme case, fitness increases in the sinks but is unaltered in the source (vertical arrow in Fig. 4.4). This would occur, for example, if red squirrels exerted selection on traits independent of those involved in the mutualism. In the case of phenotypic selection exerted on limber pine by red squirrels, the response to selection results in a decrease in the availability of seeds to both red squirrels and nutcrackers because traits that are favored by selection exerted by red squirrels make seeds less accessible to nutcrackers (Fig. 4.1A; Siepielski and Benkman 2007a). In Figure 4.4 we show such an evolutionary response by decreasing both the per capita reproductive deficit in the sink and the per capita reproductive surplus in the source (arrow angled down and to the left in Fig. 4.4). The per capita reproductive surplus in the source declines because – as described earlier – with increasing seed defenses, fewer seeds are produced per unit of reproductive allocation (Benkman 1995a; Siepielski and Benkman 2007a) and the probability of a seed being harvested by nutcrackers decreases (Siepielski and Benkman 2008a). The per capita reproductive deficit in the sink also declines because red squirrels

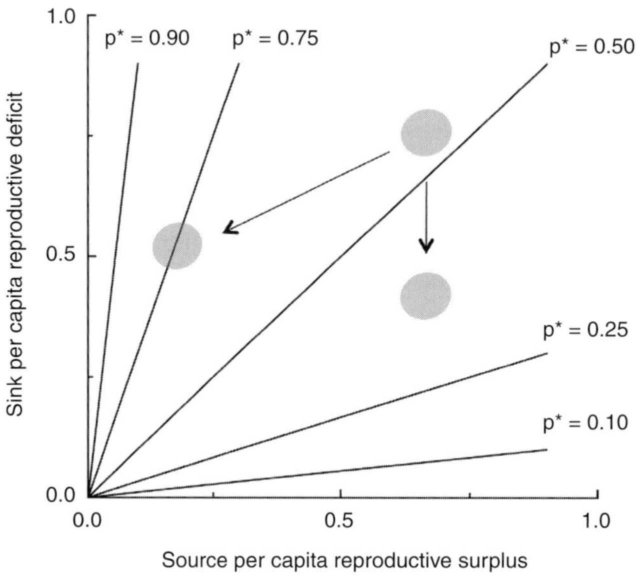

FIGURE 4.4.
The equilibrium proportion of the population (p*) residing in source habitat will vary depending on whether selection exerted by the antagonist conflicts with selection exerted by the mutualist. The diagonal arrow directed down and to the left from the upper shaded area represents the evolutionary response when selection exerted by the antagonist conflicts with that exerted by the mutualist. The descending vertical arrow represents the evolutionary response when selection exerted by the antagonist is independent of that exerted by the mutualist.

avoid cones that are well defended (Siepielski and Benkman 2007a). However, nutcrackers would also be impeded, reducing the extent of the decline in the reproductive deficit. Thus, we show that the decrease in the reproductive deficit is less than the decrease in the reproductive surplus (Fig. 4.4). The result is that at equilibrium, a smaller proportion of the mutualist populations will occupy sink habitat (and less total habitat) when the antagonistic species is both a superior competitor and exerts a strong selection that conflicts with the selection exerted by the mutualist.

Although our model predicts that evolution in the pine in response to selection exerted by red squirrels causes a proportionately greater reduction in fitness in sources than increases in fitness in sinks, such a selective impact is only likely when red squirrels occupy a large fraction of the habitat (Kawecki 1995; Holt 1996). As the proportion of the squirrel population in sink habitat increases, the selective impact of these antagonists on the pine increases because they will exert selection on an increasingly larger fraction of the population of mutualists (Kawecki 1995; Holt 1996). Evolution in response to selection in the sink is also most likely when fitness in sinks is not too low, because with declines in fitness in the sink few, if any, individuals in sinks will

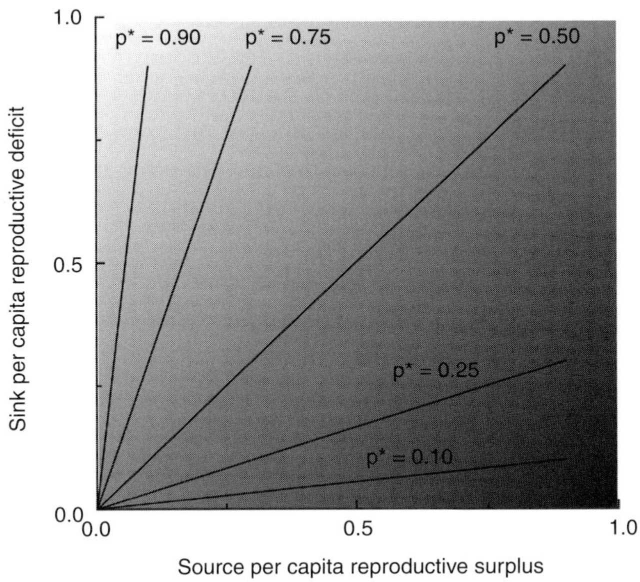

FIGURE 4.5.
The strength of selection by an antagonist increases with increases in the proportion of habitat it occupies (i.e., sink habitat) and when the per capita reproductive deficit in the sink is small. Darker shading represents greater potential selective impact by the antagonist.

contribute to subsequent generations (Holt 1996). Thus, we expect that the selective impact of antagonists should increase as we move from the upper left down to the lower right of Figure 4.5.

This raises a paradox. We have argued that red squirrels greatly depress the fitness of pines (i.e., fitness is very low in sink habitats and therefore few, if any, individuals contribute to future generations), yet pines nevertheless evolve in response to selection exerted by red squirrels. This paradox arises because we have assumed that there is just one dispersal agent, namely nutcrackers, that remove seeds from cones. However, an alternative dispersal agent is scatter-hoarding, ground-foraging rodents, which disperse seeds that have fallen to the ground (Vander Wall 2003). Moreover, selection exerted by red squirrels favors secondary dispersal by scatter-hoarding rodents and results in an increase in their importance as seed dispersers (Siepielski and Benkman 2008a). Thus, the evolutionary response to selection exerted by red squirrels causes a decrease in pine fitness via reduced seed dispersal by nutcrackers, but results in an increase in fitness via increased seed dispersal by scatter-hoarding rodents. When we include scatter-hoarding rodents, the fitness of pines in the presence (and absence) of red squirrels increases (the per capita reproductive deficit is no longer very large). Consequently, the evolutionary responses to selection exerted by red squirrels are likely to cause a greater decrease in the per capita reproductive deficit in the sink than the concurrent decrease in the per capita reproductive surplus in the source. Both of these conditions favor an evolutionary response to selection exerted by red squirrels (Holt 1996). We anticipate that seed predators that are superior competitors and exert conflicting selection pressures to those exerted by seed dispersers will often favor the evolution of alternative seed dispersal mutualisms for the plant.

Although divergent selection between local areas with and without red squirrels may cause limber pine to diverge between habitats, as we have found between regions with and without red squirrels (Fig. 4.2), we have not considered adaptive divergence between subpopulations in source and sink habitats. This is justified for two reasons. First, nutcrackers disperse pine seeds long distances that would potentially span numerous source and sink habitats (Vander Wall and Balda 1981), which reduces the chance of local divergence. Second, limber pine is often an early successional species. Over the course of succession, a given site will initially lack red squirrels and act as a source habitat. Then, as the forest fills in, red squirrels will colonize, causing the habitat to shift to become a sink. In systems where gene flow is more limited, and source and sink habitats are more discrete, we would anticipate local divergence between areas with and without the antagonist. Interestingly, we found a bimodal distribution of limber pine cone traits in one region of the Rocky Mountains where there are local areas with and without red squirrels; elsewhere, including the Great Basin where red squirrels are absent, we found unimodal distributions of cone traits (Siepielski and Benkman 2010). Perhaps in this region of the Rocky Mountains, the more open forests remain open and nutcrackers consistently disperse seeds into more open habitat, whereas limber pine in the more closed forests may rely on local recruitment from seeds dispersed by rodents. With local divergence, the shift in the equilibrium conditions would be similar to what is predicted to occur when selection exerted by red squirrels and nutcrackers is uncorrelated (Fig. 4.4). This would enable the mutualists to increase in sink and perhaps in source habitats, and would thus stabilize the mutualism in the presence of an antagonist.

The evolution and the dissolution of sinks

Whether antagonists will act to create sinks for mutualists will depend on the extent to which mutualists can evolve to lessen the impact of antagonists without incurring too many costs (in terms of compromising the benefits to mutualists and resources allocated). Whether plants have evolved to lessen the impact of antagonists while attracting mutualists has been recognized as an important problem and has been elucidated in a few systems. For example, plants that are dispersed by animals need to differentially attract seed dispersers over seed predators and pathogens (Herrera 1982; Cipollini and Levey 1997) and plants need to attract pollinators while deterring nectar robbers

(Galen and Cuba 2001; Irwin *et al.* 2004) and herbivores (Adler and Bronstein 2004).

In some cases, mutualists and antagonists have similar preferences and thus exert conflicting selection, like that exerted by nutcrackers and red squirrels on limber pine. For example, ants that steal nectar from and often sever the style of *Polemonium viscosum* prefer the same corolla shapes that bumblebee pollinators do (Galen and Cuba 2001). Because this causes conflicting selection pressures, it may have led to local variation in corolla shape depending on the abundance of ants (Galen and Cuba 2001). In some examples where mutualists and antagonists have similar preferences, plants provide rewards that differentially favor mutualists over antagonists. For example, when *Ipomopsis aggregata* produces more dilute nectar it can deter nectar robbers and thereby increase the attractiveness of flowers to pollinators (Irwin *et al.* 2004). By increasing the ratio of elaiosome size to seed size, certain plants can increase seed removal by high-quality seed dispersers relative to seed predators (Hughes and Westoby 1992). Likewise, increasing nutrients in high-quality fruits appears to differentially favor fruit removal by seed dispersers over pathogens (Cazetta *et al.* 2008). In these last two studies, the seed dispersers are superior competitors to the antagonists and the plants have apparently evolved to exploit this asymmetry. Presumably, red squirrels are such effective preemptive competitors for pine seeds because they cache whole cones, whereas seed predators on fruits cannot cache fruits because they rot (Janzen 1977) and their competitive effect is limited to short-term consumption.

In other cases, plants can directly deter antagonists without deterring their mutualists. That is, the evolutionary response to selection exerted by the antagonist is independent of that exerted by the mutualist (see, e.g., Fig. 4.4). Examples include chilies (*Capsicum* spp.) that produce capsaicin, which deters seed predators while not deterring seed dispersers (Levey *et al.* 2006); ant-dispersed plants that shed their seeds during the day when ants, but not granivorous rodents, are active (Ness and Bressmer 2005); and plants that produce nectar mostly during periods when pollinators, but not nectar robbers, are active (Carpenter 1979). These examples show that the impact of antagonists can be reduced without deterring mutualists, and suggest situations when a source–sink perspective may be less relevant. Nevertheless, antagonists are not always easy to deter and can be superior competitors for resources provided by one of the mutualists. Such examples include antagonists that dominate and overwhelm mutualisms between hummingbirds that pollinate plants (McDade and Kinsman 1980) and between ants that defend plants (Yu *et al.* 2001) and disperse their seeds (Fedriani *et al.* 2004). Similar to our studies with red squirrels, the antagonists in the first two studies were mostly confined to dense patches of plants so that smaller low-density patches of plants acted as source habitats to the mutualists. These studies indicate that antagonists have played an important role in the ecology and evolution of many mutualisms and that a source–sink perspective can provide valuable insight into the evolutionary ecology of mutualisms.

Finally, we are biased toward studying mutualisms that have evolved and persisted in the face of antagonists, and thus may underestimate the impact of antagonists. Comparative phylogenetic analyses may prove very helpful in evaluating the extent to which antagonists have altered or prevented the evolution of certain mutualisms. For example, if antagonists alter the evolution of mutualisms, it would then be worthwhile to examine the gains and losses of fleshy fruits in plant clades not only from the perspective of the benefits of endozoochorous seed dispersal in relation to vegetation structure and dynamics (Bolmgren and Eriksson 2005), but also in relation to the occurrence of antagonists.

Conclusions

The abundance of many antagonists varies spatially and temporally (e.g., Irwin and Maloof 2002; Fedriani *et al.* 2004). Especially relevant to a source–sink approach are examples in which the antagonist is a superior competitor for resources provided by one of the mutualists. However, whether the reproductive surpluses in areas without the antagonist maintain the mutualistic populations in areas with the antagonist is unknown (but would be a very interesting hypothesis to test). Moreover, although we suspect that the distribution of antagonists, especially those that are superior competitors, rarely overlap completely with mutualists, most studies focus on a single or only a few locations (J. L. Bronstein, personal communication; J. N. Thompson, personal communication). Consequently, the extent to which the theoretical framework provided by Pulliam (1988) is applicable to the ecology of mutualisms also remains unknown. Nevertheless, the evolution and persistence of many mutualisms will likely depend on the amount of habitat that lacks a competitively superior antagonist (i.e., source habitat) and on whether selection exerted by antagonists conflicts with selection exerted by mutualists. Although an increasing number of studies have examined how mutualistic populations have evolved in the context of antagonists, our understanding of the evolution and persistence of mutualisms will be enhanced by evaluating to what extent antagonists co-occur with mutualists and exert selection that conflicts with that exerted by the mutualists. Such studies, combined with phylogenetic analyses, may allow us to evaluate the extent to which antagonists alter the evolution of mutualisms.

Acknowledgments

We thank two anonymous reviewers, the editors, and especially Judie Bronstein, for comments and encouragement. Our research was support by National Science Foundation grants (DEB-0455705 and DEB-0515735).

References

Adler, L. S. and J. L. Bronstein (2004). Attracting antagonists: does floral nectar increase leaf herbivory? *Ecology* **85**: 1519–1526.
Balda, R. P. and A. C. Kamil (1992). Long-term spatial memory in Clark's nutcracker, *Nucifraga columbiana*. *Animal Behaviour* **44**: 761–769.
Benkman, C. W. (1995a). The impact of tree squirrels (*Tamiasciurus*) on limber pine seed dispersal adaptations. *Evolution* **49**: 585–592.
Benkman, C. W. (1995b). Wind dispersal capacity of pine seeds and the evolution of different seed dispersal modes in pines. *Oikos* **73**: 221–224.
Benkman, C. W., R. P. Balda and C. C. Smith (1984). Adaptations for seed dispersal and the compromises due to seed predation in limber pine. *Ecology* **65**: 632–642.
Bolmgren, K. and O. Eriksson (2005). Fleshy fruits: origins, niche shifts, and diversification. *Oikos* 109: 255–272.
Bronstein, J. L. (2001). The exploitation of mutualisms. *Ecology Letters* **4**: 277–287.
Carpenter, F. L. (1979). Competition between hummingbirds and insects for nectar. *American Zoologist* **19**: 1105–1114.
Cazetta, E., H. M. Schaefer and M. Galetti (2008). Does attraction of frugivores or defense against pathogens shape fruit pulp composition? *Oecologia* **155**: 277–286.
Cipollini, M. L. and D. J. Levey (1997). Secondary metabolites of fleshy vertebrate-dispersed fruits: adaptive hypotheses and implications for seed dispersal. *American Naturalist* **150**: 346–372.
Colwell, R. K. and E. R. Fuentes (1975). Experimental studies of the niche. *Annual Review of Ecology and Systematics* **6**: 281–310.
Connor, R. C. (1995). The benefits of mutualism: a conceptual framework. *Biological Reviews* **70**: 427–457.
Fedriani, J. M., P. J. Rey, J. L. Garrido, J. Guitián, C. M. Herrera, M. Medrano, A. M. Sánchez-Lafuente and X. Cerdá (2004). Geographic variation in the potential of mice to constrain an ant–seed dispersal mutualism. *Oikos* **105**: 181–191.
Ferrière, R., M. Gauduchon and J. L. Bronstein (2007). Evolution and persistence of obligate mutualists and exploiters: competition for partners and evolutionary immunization. *Ecology Letters* **10**: 115–126.
Foster, K. R. and T. Wenseleers (2006). A general model for the evolution of mutualisms. *Journal of Evolutionary Biology* **19**: 1283–1293.
Galen, C. and J. Cuba (2001). Down the tube: pollinators, predators, and the evolution of flower shape in the alpine skypilot, *Polemonium viscosum*. *Evolution* **55**: 1963–1971.
Herrera, C. M. (1982). Defense of ripe fruit from pests: its significance in relation to plant–disperser interactions. *American Naturalist* **120**: 218–241.
Holt, R. D. (1996). Adaptive evolution in source–sink environments: direct and indirect effects of density-dependence on niche evolution. *Oikos* **75**: 182–192.
Hughes, L. and M. Westoby (1992). Effect of diaspore characteristics on removal by seeds adapted for dispersal by ants. *Ecology* **73**: 1300–1312.
Hutchins, H. E. and R. M. Lanner (1982). The central role of Clark's nutcracker in the dispersal and establishment of whitebark pine. *Oecologia* **55**: 192–201.
Irwin, R. E. and J. E. Maloof (2002). Variation in nectar robbing over time, space, and species. *Oecologia* **133**: 525–533.
Irwin, R. E., L. S. Adler and A. K. Brody (2004). The dual role of floral traits: pollinator attraction and plant defense. *Ecology* **85**: 1503–1511.
Janzen, D. H. (1977). Why fruits rot, seeds mold, and meat spoils. *American Naturalist* **111**: 691–713.
Kawecki, T. J. (1995). Demography of source–sink populations and the evolution of ecological niches. *Evolutionary Ecology* **9**: 38–44.
Lanner, R. M. (1996). *Made for Each Other: A Symbiosis of Birds and Pines*. Oxford University Press, New York.

Lanner, R. M. and S. B. Vander Wall (1980). Dispersal of limber pine seed by Clark's nutcrackers. *Journal of Forestry* **78**: 637–639.

Levey, D. J., J. J. Tewksbury, M. L. Cipollini and T. A. Carlo (2006). A field test of the directed deterrence hypothesis in two species of wild chili. *Oecologia* **150**: 61–69.

Lloyd, P., T. E. Martin, R. L. Redmond, U. Langner and M. M. Hart (2005). Linking demographic effects of habitat fragmentation across landscapes to continental source–sink dynamics. *Ecological Applications* **15**: 1504–1514.

McDade, L. A. and S. Kinsman (1980). The impact of floral parasitism in two neotropical hummingbird-pollinated plant species. *Evolution* **34**: 944–958.

Ness, J. H. and K. Bressmer (2005). Abiotic influences on the behaviour of rodents, ants, and plants affect an ant–seed mutualism. *Ecoscience* **12**: 76–81.

Pulliam, H. R. (1988). Sources, sinks, and population regulation. *American Naturalist* **132**: 652–661.

Sachs, J. L. and E. L. Simms (2006). Pathways to mutualism breakdown. *Trends in Ecology and Evolution* **21**: 585–592.

Siepielski, A. M. and C. W. Benkman (2007a). Convergent patterns in the selection mosaic for two North American bird-dispersed pines. *Ecological Monographs* **77**: 203–220.

Siepielski, A. M. and C. W. Benkman (2007b). Selection by a pre-dispersal seed predator constrains the evolution of avian seed dispersal in pines. *Functional Ecology* **21**: 611–618.

Siepielski, A. M. and C. W. Benkman (2008a). A seed predator drives the evolution of a seed dispersal mutualism. *Proceedings of the Royal Society of London Series B* **275**: 1917–1925.

Siepielski, A. M. and C. W. Benkman (2008b). Seed predation and selection exerted by a seed predator influence tree densities in sub-alpine communities. *Ecology* **89**: 2960–2966.

Siepielski, A. M. and C. W. Benkman (2010). Conflicting selection from an antagonist and a mutualist enhances phenotypic variation in a plant. *Evolution* **64**: 1120–1128.

Smith, C. C. (1968). The adaptive nature of social organization in the genus of tree squirrels *Tamiasciurus*. *Ecological Monographs* **38**: 31–63.

Smith, C. C. and R. P. Balda (1979). Competition among insects, birds and mammals for conifer seeds. *American Zoologist* **19**: 1065–1083.

Thompson, J. N. (2005). *The Geographic Mosaic of Coevolution*. University of Chicago Press, Chicago, IL.

Vander Wall, S. B. (1988). Foraging of Clark's nutcrackers on rapidly changing pine seed resources. *Condor* **90**: 621–631.

Vander Wall, S. B. (1990). *Food Hoarding in Animals*. University of Chicago Press, Chicago, IL.

Vander Wall, S. B. (2003). Effects of seed size of wind-dispersed pines (*Pinus*) on secondary seed dispersal and the caching behavior of rodents. *Oikos* **100**: 25–34.

Vander Wall, S. B. and R. P. Balda (1981). Ecology and evolution of food-storage behavior in conifer-seed-caching corvids. *Zeitschrift für Tierpsychologie* **56**: 217–242.

Wilson, W. G., W. F. Morris and J. L. Bronstein (2003). Coexistence of mutualists and exploiters on spatial landscapes. *Ecological Monographs* **73**: 397–413.

Yu, D. W. (2001). Parasites of mutualisms. *Biological Journal of the Linnean Society* **72**: 529–546.

Yu, D. W., H. B. Wilson and N. E. Pierce (2001). An empirical model of species coexistence in a spatially structured environment. *Ecology* **82**: 1761–1771.

MARK C. ANDERSEN

5

Effects of climate change on dynamics and stability of multiregional populations

Summary

Climate change is one of the greatest long-term potential threats to the functional integrity of the biosphere. Although the likely effects of climate change on ecosystem function and the geographic distributions of organisms have been extensively studied, their demographic effects are less well understood. In order to examine the effects of climate change on populations in their landscape context, I integrate results from two different modeling approaches to examine the effects of climate change on the demography and dispersal of organisms. I use simple two-patch metapopulation models and more complex stochastic stage-structured multiregional models of stream fish populations. Plausible effects of climate change on dispersal rates, and on spatial population structure, may destabilize metapopulations and make them susceptible to further anthropogenic or natural perturbations. The findings suggest several hypotheses to be tested empirically, and imply that future biodiversity conservation strategies will need to account for the landscape-level effects of climate change and attendant changes in land use.

Background

Climate change over the next century is likely to lead to substantial economic and environmental costs (IPCC 2007a, 2007b). It will also most likely result in significant shifts in land use as human populations respond

Sources, Sinks and Sustainability, ed. Jianguo Liu, Vanessa Hull, Anita T. Morzillo and John A. Wiens. Published by Cambridge University Press.

to climatic shifts (Schroter *et al.* 2005; Jetz *et al.* 2007; Williams *et al.* 2007). Confronting these costs and impacts will require adaptation to new conditions and mitigation of the impacts and their causes, both in human social and economic structures and in our approach to conservation and management of biodiversity (Andersen 2007). In terms of biodiversity conservation, although climate change can be considered a specific instance of the broad category of human impacts, climate change is qualitatively different from other human impacts on biodiversity in both its extent and its severity.

Research on the effects of climate change on biodiversity has focused on three major areas:

1. Experimental studies of treatments mimicking global warming have been shown to influence the properties of terrestrial (Mitchell *et al.* 2003; Zavaleta and Hulvey 2004; De Boeck *et al.* 2007) and aquatic systems (Vinebrooke *et al.* 2004; Christensen *et al.* 2006). The evidence for interactions among multiple climate change-related stressors is equivocal (Mitchell *et al.* 2003; Zavaleta *et al.* 2003; Christensen *et al.* 2006). However, when they do occur, such interactions may cause the impacts of climate change to be even more severe than anticipated (Korner 2003; Jenssen 2006).
2. Detailed bioclimatic models of range shifts predict species' future geographic ranges based on global circulation models of future climate (Thuiller 2003; Hijmans and Graham 2006; Beaumont *et al.* 2007). Limited environmental tolerances or dispersal ability, or the presence of dispersal barriers, may lead to extinction for many species whose predicted ranges have little or no overlap with their current geographic ranges (Midgley *et al.* 2002, 2006; Broennimann *et al.* 2006). In addition, future range shifts may completely reshuffle ecological communities, creating novel and unprecedented ecological interactions between species (Thuiller *et al.* 2006).
3. Some research teams are performing GIS-based integrated analyses of potential interactions between climate change and other regional variables such as land use, and their predicted impacts on other regional variables such as hydrology, forest products production, and agricultural production, as well as biodiversity (Metzger *et al.* 2005; Schroter *et al.* 2005). The impacts of global warming will be heterogeneous and context-dependent (Metzger *et al.* 2006; Rounsevell *et al.* 2006; Jetz *et al.* 2007), and may provide opportunities for biodiversity conservation (e.g., through farm abandonment) as well as threats to biodiversity (Holman *et al.* 2005; Rounsevell *et al.* 2006).

However, not all previous research has focused on the potential impacts of climate change. For example, direct effects of climate change have been documented for development, survival, geographic range, and abundance of insect herbivores (Bale *et al.* 2002), altitudinal range and community composition of vascular plants (le Roux and McGeoch 2008), distribution and abundance of wintering wading birds (Maclean *et al.* 2008), and arrival dates of migrating North American passerines (Miller-Rushing *et al.* 2008). The broad range of documented examples of climate change impacts on biodiversity emphasizes the need for conservation action as well as research.

Management of biodiversity is typically targeted at populations, either through direct management interventions or indirectly through management of habitats (Sinclair *et al.* 2006; Mills 2007). However, the broader landscape context of populations also influences their dynamics (Pulliam 1988; Pulliam *et al.* 1992; Donovan and Thompson 2001). The effects of climate change on habitats will occur as effects on landscape composition and dynamics. Thus effective management of populations in the face of climate change will require a landscape perspective.

The potential landscape-level effects of climate change are varied and diverse. These effects may include changes in habitat quality, connectivity, and heterogeneity (Hilderbrand *et al.* 2007; Rowe 2007; Semlitsch 2008), as well as on dynamic processes influencing vegetation structure and composition (Rupp *et al.* 2000; Schumacher *et al.* 2006; Yao *et al.* 2006). These effects may influence population dynamics through effects on habitat selection and foraging behaviors, disruption of dispersal and migration, and reshuffled interactions with predators, competitors, pathogens, and mutualists.

The broad objectives of the research reported here are

1. to demonstrate how relatively simple models can illuminate the population-level effects of landscape changes, particularly those that might be caused by climate change;
2. to suggest some landscape patterns and processes as possible targets for management and mitigation of the effects of climate change on populations of concern.

Specifically, I use two-patch models to contrast the effects on population dynamics of changes in average rates of population growth and dispersal with the effects of changes in the asymmetry of population growth and dispersal. I also use multiregional models specifically formulated to describe hypothetical stream fish metapopulations in order to explore the effects of patch deletions on metapopulation properties. Taken together, the findings imply that the landscape consequences of climate change may destabilize metapopulations and make them more susceptible to further disturbance.

Research methods

Two-patch models

Analyses of the simplest possible metapopulation models, consisting of only two patches linked by dispersal, have contributed much to our understanding of theoretical population dynamics (Hastings 1993; Lloyd 1995; Kendall and Fox 1998) and are being applied to analysis of experimental data as well (Serra *et al.* 2007).

Assuming Ricker density dependence in the two populations (where the dynamics depends on local density rather than global population size) and assuming constant dispersal at rate D_{ij} from patch i to patch j, the dynamics of the system are given by

$$N_{1,t+1} = \lambda_1 N_{1,t} \exp(-\beta_1 N_{1,t}) - D_{12} N_{1,t} + D_{21} N_{2,t} \tag{5.1a}$$

$$N_{2,t+1} = \lambda_2 N_{2,t} \exp(-\beta_2 N_{2,t}) - D_{21} N_{2,t} + D_{12} N_{1,t} \tag{5.1b}$$

where $N_{i,t}$ is the size of the population in patch i at time t, λ_i is the growth rate of population i in the absence of density dependence, and β_i is the attenuation of population growth rate with increasing population density for population i.

For a similar system, it is possible to derive an upper bound on the Lyapunov exponent that depends only on the dispersal rate (Sole and Gamarra 1998), thus establishing an explicit link between chaotic dynamics and the correlations introduced by dispersal in determining population synchrony. It has also been shown that dispersal in systems such as this can stabilize chaotic local dynamics (Hastings 1993; Lloyd 1995). The simple fact that there are two populations in the system does not have as much impact on the dynamics of the system as differences between the habitats in which the local populations live (Kendall and Fox 1998). These differences, which may also be influenced by the form of density dependence, in particular by the presence of an Allee effect (Amarasekare 1998), are at the heart of the concept of source–sink dynamics (Pulliam 1988). Models such as this have been directly applied to the study of laboratory microcosms, with varying degrees of success (Donahue *et al.* 2003; Serra *et al.* 2007).

My approach to this model was to pick bifurcation parameters of ecological interest and examine their effects. In particular, I looked at the effects on system dynamics of changes in both the average values and the difference between patches in the value of the rates of population growth and dispersal. I performed four "numerical experiments" using the model described above.

1. The difference between the growth rates in the two patches was held constant while varying the average growth rate.
2. The average growth rate in the two patches was held constant while varying the difference between the growth rates. For both experiment 1 and experiment 2, the dispersal rates were held constant at 0.1.
3. The two dispersal rates were increased while keeping them equal to each other.
4. The two dispersal rates were allowed to be different, and the difference between them was increased while holding their average value constant.

Experiments 1 and 2 examine the effects of changing population growth while holding dispersal constant. Experiments 3 and 4 examine the effects of changing dispersal rates while holding population growth rates constant. Experiments 1 and 3 examine the effects of changes in average values of population model parameters, while experiments 2 and 4 examine the effects of changes in asymmetry in model parameters between the two populations. As climate change proceeds, overall regional habitat quality may change, leading to changes in population growth across an entire metapopulation (experiment 1); however, climate change may also lead to greater habitat heterogeneity, leading to more variation in growth rates of local populations across the landscape (experiment 2). In addition, climate change may affect the characteristics of the habitat matrix in which local populations are embedded and through which dispersing individuals move. Two possible effects of changes in the matrix habitat are changes in overall rates of dispersal (experiment 3) and changes in heterogeneity in dispersal rates across different local populations (experiment 4). The latter, in particular, may strongly influence the source or sink status of local populations.

Multiregional models

Multiregional models are a class of vector-state metapopulation models (Hastings 1991) with age- or stage-structured vital rates and movement rates (Lebreton 1996; Caswell 2001). Consider a metapopulation consisting of, say, three local populations, each following a projection matrix model. Then the dynamics of the overall metapopulation may be written as

$$N_t^{+1} = \boldsymbol{P}_t N_t \quad (5.2)$$

where

$$N = \begin{bmatrix} N_{1,t} \\ N_{2,t} \\ N_{3,t} \end{bmatrix} \tag{5.3}$$

and

$$P_t = \begin{bmatrix} P_{1,t} & D_{1\to 2} & D_{1\to 3} \\ D_{2\to 1} & P_{2,t} & D_{2\to 3} \\ D_{3\to 1} & D_{3\to 2} & P_{3,t} \end{bmatrix}. \tag{5.4}$$

Here the $N_{i,t}$ terms are the population structure vectors in local population i at time t, the $\boldsymbol{P}_{i,t}$ terms are the population projection matrices in local population i at time t, and the $\boldsymbol{D}$ matrices contain age-specific rates of movement of individuals between local populations. For a stream fish metapopulation, the local populations will inhabit specific stream reaches. Juveniles will tend to move from upstream or headwater populations into downstream populations by drift, while typically only adult fish will be capable of movement from downstream to upstream reaches. The $\boldsymbol{D}$ matrices will reflect these differences in age-specific movement rates and directions. For some fishes, reproduction may be limited in the warmer downstream reaches. Mortality may be higher in the warmer downstream reaches as well, due to thermal tolerances, due to competition with (possibly introduced) fish species, or due to predation by (again possibly non-native) piscivorous fishes. The $\boldsymbol{P}_{i,t}$ matrices will reflect these differences in age-specific vital rates. Note that climate change may also alter the projection matrices for local populations and, in extreme cases, result in the loss of local populations if local conditions move outside of the species' niche boundaries.

For the results reported here, we examined four different arrangements of local populations, as shown in Figure 5.1. These scenarios consisted of: one upstream population and one downstream population (scenario A); one upstream, one midstream, and one downstream population (scenario B); two headwater populations and one downstream population (scenario C); and two headwater populations, one upstream population, and one downstream population (scenario D). Note that we can compare these model scenarios to examine the effects of patch deletions. For example, by comparing results for scenario B with those for scenario A we can study the effect of extirpation of the furthest-downstream reach (perhaps due to climate change or to the construction of a diversion). Computer programs to simulate the various model scenarios were written in MATLAB® (The MathWorks, Inc.).

Simulated populations included three stage classes. Mean values of the vital rates were chosen so that upstream reaches had slightly higher expected

TABLE 5.1. *Growth rates (in the absence of density dependence) of populations in the various scenarios shown in Figure 5.1. Population numbers (left-hand column) correspond to the labels of numbered populations in Figure 5.1; scenario letters A–D in this table correspond to the scenarios in Figure 5.1.*

Scenario	A	B	C	D
Population				
1	1.2268	1.2268	1.2268	1.2268
2	0.9597	0.9597	1.2268	1.2268
3	–	0.9597	0.9597	0.9597
4	–	–	–	0.9597

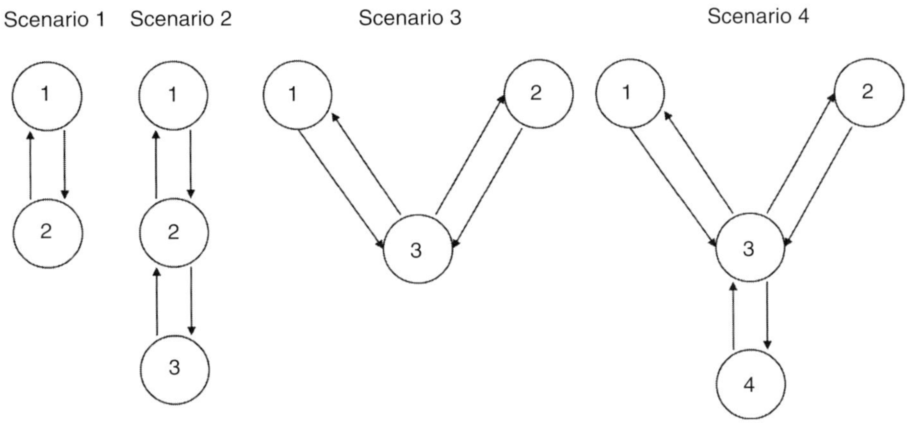

FIGURE 5.1.
Metapopulation scenarios considered in the multiregional models. Local populations are indicated by numbered circles; arrows indicate dispersal connections between them. Upstream populations are at the top of the figure, downstream populations at the bottom.

growth rates than downstream reaches (Table 5.1). Fecundities were log-normally distributed, while dispersal rates and survival rates were beta-distributed (Table 5.2). Production of juveniles was subject to Ricker-model density dependence. This was the only type of density dependence in the model; thus there is no carrying capacity built into the model as such, although the local populations will not grow without limit. Stochasticity was introduced into the simulations as either independent identically distributed (iid) or Markov fluctuations in the population vital rates; Markov fluctuations used a 10-state Markov process with the states randomly generated for each replicate simulation run. In addition, all populations were subject to a 0.05 probability per year of catastrophic reproductive failure in one of two ways: in local reproductive failure, local populations fail independently; while in regional reproductive failure, all local

TABLE 5.2. *Summary of model parameters and default values. See text for explanation of computation of survival rates from s and t parameters. See Table 5.1 for population growth rates (in the absence of density dependence) resulting from default parameter values. Where two default values are listed, the first value applies to local populations shown as having a growth rate of 1.2268 in Table 5.1, while the second applies to all other local populations. The coefficient of variation for all fecundity, stage survival and transition, and dispersal rates was 0.1.*

Parameter	Explanation	Default values
f2xbar	Mean fecundity of stage 2	1.25
f3xbar	Mean fecundity of stage 3	1.10
Beta	Ricker beta parameter (attenuation of fecundity)	0.01
p33xbar	Mean survival of stage 3	0.6
s1xbar	Mean overall stage 1 survival	0.7, 0.5
t1xbar	Mean stage 1 maturation rate	0.7, 0.5
s2xbar	Mean overall stage 2 survival	0.7, 0.5
t2xbar	Mean stage 2 maturation rate	0.7, 0.5
d1upijxbar	Mean upstream dispersal rate from population *i* to population *j*, stage 1	0.01
d1downijxbar	Mean downstream dispersal rate from population *i* to population *j*, stage 1	0.5
d2upijxbar	Mean upstream dispersal rate from population *i* to population *j*, stage 2	0.05
d2downijxbar	Mean downstream dispersal rate from population *i* to population *j*, stage 2	0.05
d3upijxbar	Mean upstream dispersal rate from population *i* to population *j*, stage 3	0.5
d3downijxbar	Mean downstream dispersal rate from population *i* to population *j*, stage 3	0.01

populations fail at once. For each scenario, and for each type of stochasticity (iid or Markov) and each type of reproductive failure (local or regional), 100 replicate simulations were run, each consisting of 250 time steps.

Results are presented for selected pairwise comparisons of the scenarios shown in Figure 5.1; each of these comparisons represents a potential "before–after" pair of stream reach configurations in which one stream reach in the "before" scenario is absent in the "after" scenario. For example if one of the two headwater populations in scenario C becomes too warm to be habitable (due to climate change), scenario A will be the result. As another example, if the furthest-downstream reach in scenario D becomes too warm to be occupied (again due to climate change, although this could just as well be due to some other factor such as damming or diversion), scenario C will be the result.

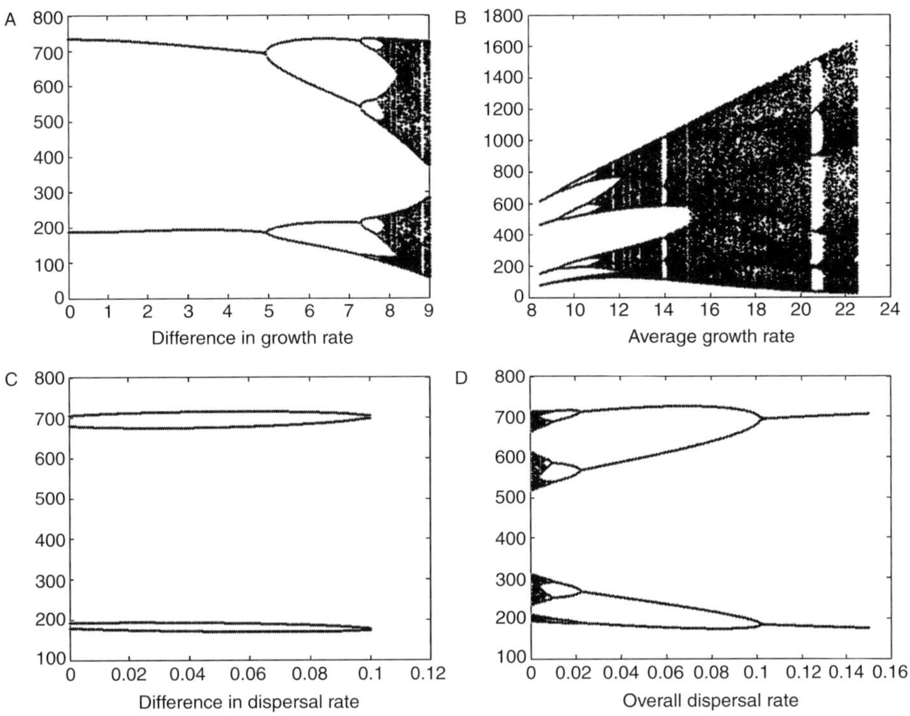

FIGURE 5.2.
Bifurcation diagrams for the four "numerical experiments" using the two-patch model described in the text. The bifurcation parameter is plotted on the horizontal axis and the values taken on by the total population (size of population 1 plus size of population 2) are plotted on the vertical axis. The bifurcation parameters are as follows: (A) Increasing difference in population growth rate between the two patches while holding average population rate constant at 10. (B) Increasing average growth rate over the two patches while holding the difference between the growth rates of the two patches constant at 15. (C) Increasing difference in dispersal rate between the two patches from a minimum difference of zero to a maximum difference of 0.1, while holding the average of the two dispersal rates constant at 0.1. (D) Increasing the dispersal rate between the two patches from 0.001 to 0.15, with no difference between the two patches.

Results

I can present here only a small subset of the potential results from analyses of the two-patch and multiregional models described above. The multiregional models, in particular, although relatively simple, are very rich in the range of ecological questions to which they might provide insight. Rather than attempting a more complete exploration of these models, I present a selection of results with particular relevance to the potential impacts of climate change (and other human impacts) on populations in their landscape context.

Figure 5.2 shows bifurcation diagrams from the four numerical experiments with the two-patch model of Eqs. (5.1a) and (5.1b), with the steady-state values of total population size for the two-patch system plotted as functions of

TABLE 5.3. *Effects of patch loss on variances of local population size in the multiregional stream fish metapopulation model for all combinations of iid and Markov stochasticity and local and regional catastrophic reproductive failure. Scenario letters refer to the population scenarios in Figure 5.1; note that all scenario transitions involve the loss of one local population.*

Scenario transition	IID/local	IID/regional	Markov/local	Markov/regional
C to A	Lower	Lower	Lower	Higher
B to A	Lower	Lower	Lower	Lower
D to B	Upstream higher, downstream lower	Lower	Lower	Lower
D to C	Upstream lower, downstream higher	Upstream lower, downstream higher	Lower	Lower

the value of a changing parameter. Figure 5.2A shows the effect of increasing differences in population growth rate between the two patches on population size, while holding the average growth rate constant. Figure 5.2B shows the effect of increasing the average growth rate in the two patches for a constant difference in growth rates; like Figure 5.2A, this increase leads down the well-known path of period-doubling bifurcations and chaos (Flake 1998: Ch. 10; Sprott 2003: Ch. 2). In both of these figures, the two populations in the system behave as though they are relatively loosely coupled; changes in the growth rates drive strong changes in the dynamics. The effects of changes in dispersal rates are not as pronounced. Figure 5.2C shows that increasing the asymmetry of dispersal between the two patches has little effect on population sizes, other than possibly inducing small oscillations over a limited range. Figure 5.2D shows the effect of changes in the average dispersal rate on population size; note that a decrease in average dispersal can destabilize the system, leading to bifurcations and chaos. This occurs because a decrease in average dispersal leads to a decoupling of the two populations, allowing their dynamics to be more strongly driven by local conditions.

As mentioned above, the multiregional models provide a rich source of simulation results that could be presented and analyzed in a number of different ways. Here I present only results for variances of population sizes and for correlations between the sizes of local populations in the various stream reaches. Table 5.3 shows the effects of four different patch loss scenarios on local population variances of stream fish metapopulations simulated by the multiregional stochastic model described above. Note that patch deletions tend to result in lower population variances, possibly by leading to tighter coupling of local population dynamics throughout the metapopulation. I also found that, for all scenarios, local population variances tend to be lower in downstream reaches as well. This may be because downstream reaches, with possibly high rates of immigration from upstream, average across population fluctuations throughout the river system.

Results for correlations between local population sizes are not tabulated, because they were consistent across all patch-deletion scenarios. For all scenarios, correlations between local population sizes decrease with increasing separation between the two stream reaches being compared, as one might expect simply because of the decay in spatial autocorrelation with increasing distance. In addition, patch deletions invariably led to higher correlations between local populations in different stream reaches. As for variances, this may reflect a tighter coupling of the dynamics of local population across the entire metapopulation following patch loss.

Conclusions

It might be argued that the simple two-patch model and the relatively complex multiregional model are in fact both quite simple relative to the complexity of system-specific spatially explicit models. However, even for such simple models as those presented here, it is likely that sufficient data may not be available to permit application to specific situations. Conversely, the available data will (or at least should) constrain the complexity of models that one might wish to formulate for specific systems and locations (see Wiens and Van Horne, Chapter 23, this volume). The multiregional models discussed here could be considered templates for such applications. However, additional factors might need to be considered, including other environmental stressors. Also, models intended for applications must be constructed so that they can be used to test scientific hypotheses (Hilborn and Mangel 1997).

The bifurcation diagrams of Figure 5.2B and 5.2D show that steeper gradients in habitat quality can destabilize metapopulations, and that decreasing average dispersal rates can also destabilize metapopulations. Climate change is predicted to lead to the compression of many habitat gradients (McRae *et al.* 2008). Climate change may also induce changes in landscape composition that lead to changes in the suitability of matrix habitat for dispersal (Trivedi *et al.* 2008). Thus both the effects identified in my analysis of the two-patch model are plausible consequences of landscape-level effects of climate change. Results from the multiregional model show that patch deletions can make stream fish metapopulations less variable, but also more highly correlated. These effects may combine to make these populations more susceptible to further disturbance. Similarly, populations already impacted by anthropogenic disturbance may prove to be more susceptible to climate change impacts. Thus my findings fulfill the first objective presented above, to demonstrate how the models can illuminate the population-level effects of climate change-induced landscape change.

My findings do not as clearly or directly address my second objective, to suggest possible targets for management and mitigation interventions. However, some potential avenues for exploration are evident. The results for the two-patch model suggest that areas in which habitat zones or gradients may be compressed, such as high latitudes and high elevations, may be particularly susceptible to the detrimental effects of climate change on biodiversity. These findings are in agreement with those from other approaches (Guisan and Thuiller 2005; Midgley *et al.* 2006). Results for the two-patch model also show that species occupying patchy habitats may be susceptible not just to impacts of climate change on the habitats in which they live, but also on the habitats through which they disperse (as shown also by McRae *et al.* 2008). The results for the multiregional model show that both the number and spatial arrangement of local fish populations in different reaches of a river or stream may influence the coupling between local populations, thus influencing their susceptibility to perturbation. Other model results show that small increases in local carrying capacities of stream fish metapopulations may greatly increase system persistence, and that higher correlations between local populations may decrease persistence (Hilderbrand 2003). Empirical studies have shown that effects of climate change on temperature and flow regime (via effects on precipitation) can interact with land use to drive stream fish populations into decline (Peterson and Kwak 1999; Stranko *et al.* 2008). Stream channel and watershed restoration may prove essential to mitigating these effects (Peterson and Kwak 1999).

Since plausible effects of climate change may destabilize metapopulations, and may make them susceptible to further anthropogenic or natural perturbations, future conservation planning efforts on behalf of biodiversity need to account for the landscape-level effects of climate change and attendant changes in land use (Hannah *et al.* 2002). As species' geographic ranges shift, population parameters such as vital rates and dispersal rates will also shift, in possibly predictable ways (see also, Etterson *et al.*, Chapter 13, this volume). My findings should be interpreted as hypotheses to be tested in the context of the types of large-scale mechanistic studies that others have advocated (Cumming 2007; Peters *et al.* 2008), rather than as firm conclusions. In particular, the results presented here may lead to studies that help to demonstrate the importance of changes in the strength of connections between local populations.

Acknowledgments

I would like to thank Jack Liu, Anita Morzillo, Vanessa Hull, and John Wiens for their hard work on the symposium and on this volume, and Ron Pulliam for his many contributions to our field. I also thank Dave Cowley for numerous discussions of stream fish population structure and dynamics. Comments from three anonymous reviewers led to significant improvements in the manuscript. This research was supported in part by the New Mexico State University Agricultural Experiment Station.

References

Amarasekare, P. (1998). Interactions between local dynamics and dispersal: insights from single species models. *Theoretical Population Biology* **53**: 44–59.

Andersen, M. A. (2007). Global climate change and biodiversity conservation: the role of ecological risk assessment. *Forum on Public Policy* **3**: 267–276.

Bale, J. S., G. J. Masters, I. D. Hodkinson, C. Awmack, T. M. Bezemer, V. K. Brown, J. Butterfield, A. Buse, J. C. Coulson, J. Farrar, J. E. G. Good, R. Harrington, S. Hartley, T. H. Jones, R. L. Lindroth, M. C. Press, I. Symrnioudis, A. D. Watt and J. B. Whittaker (2002). Herbivory in global climate change research: direct effects of rising temperature on insect herbivores. *Global Change Biology* **8**: 1–16.

Beaumont, L. J., A. J. Pitman, M. Poulsen and L. Hughes (2007). Where will species go? Incorporating new advances in climate modelling into projections of species distributions. *Global Change Biology* **13**: 1368–1385.

Broennimann, O., W. Thuiller, G. Hughes, G. Midgley, J. M. R. Alkemade and A. Guisan (2006). Do geographic distribution, niche property and life form explain plants' vulnerability to global change? *Global Change Biology* **12**: 1079–1093.

Caswell, H. (2001). *Matrix Population Models: Construction, Analysis, and Interpretation*, 2nd edition. Sinauer Associates, Sunderland, MA.

Christensen, M. R., M. D. Graham, R. D. Vinebrooke, D. L. Findlay, M. J. Paterson and M. A. Turner (2006). Multiple anthropogenic stressors cause ecological surprises in boreal lakes. *Global Change Biology* **12**: 2316–2322.

Cumming, G. S. (2007). Global biodiversity scenarios and landscape ecology. *Landscape Ecology* **22**: 671–685.

De Boeck, H. J., C. M. H. M. Lemmens, B. Gielen, H. Bossuyt, S. Malchair, M. Carnol, R. Merckx, R. Ceulemans and I. Nijs (2007). Combined effects of climate warming and plant diversity loss on above- and below-ground grassland productivity. *Environmental and Experimental Botany* **60**: 95–104.

Donahue, M. J., M. Holyoak and C. Feng (2003). Patterns of dispersal and dynamics among habitat patches varying in quality. *American Naturalist* **162**: 302–317.

Donovan, T. M. and F. R. Thompson (2001). Modeling the ecological trap hypothesis: a habitat and demographic analysis for migrant songbirds. *Ecological Applications* **11**: 871–882.

Flake, G. W. (1998). *The Computational Beauty of Nature*. The MIT Press, Cambridge, MA.

Guisan, A. and W. Thuiller (2005). Predicting species distribution: offering more than simple habitat models. *Ecology Letters* **8**: 993–1009.

Hannah, L., G. Midgley and D. Millar (2002). Climate change-integrated conservation strategies. *Global Ecology and Biogeography* **11**: 485–495.

Hastings, A. (1991). Structured models of metapopulation dynamics. *Biological Journal of the Linnean Society* **42**: 57–71.
Hastings, A. (1993). Complex interactions between dispersal and dynamics: lessons from coupled logistic equations. *Ecology* **74**: 1362–1372.
Hijmans, R. J. and C. H. Graham (2006). The ability of climate envelope models to predict the effect of climate change on species distributions. *Global Change Biology* **12**: 2272–2281.
Hilborn, R. and M. Mangel (1997). *The Ecological Detective: Confronting Models with Data*. Princeton University Press, Princeton, NJ.
Hilderbrand, R. H. (2003). The roles of carrying capacity, immigration, and population synchrony on persistence of stream-resident cutthroat trout. *Biological Conservation* **110**: 257–266.
Hilderbrand, R. H., R. H. Gardner, M. J. Ratnaswamy and C. E. Keller (2007). Evaluating population persistence of Delmarva fox squirrels and potential impacts of climate change. *Biological Conservation* **137**: 70–77.
Holman, I. P., R. J. Nicholls, P. M. Berry, P. A. Harrison, E. Audsley, S. Shackley and M. D. A. Rounsevell (2005). A regional, multi-sectoral and integrated assessment of the impacts of climate and socio-economic change in the UK. Part II. Results. *Climatic Change* **71**: 43–73.
IPCC (2007a). *Climate Change 2007: Impacts, Adaptation, and Vulnerability*. Contribution of Working Group II to the Fourth Assessment Report of the Intergovernmental Panel on Climate Change. Cambridge University Press, Cambridge, UK.
IPCC (2007b). *Climate Change 2007: The Physical Science Basis*. Contribution of Working Group I to the Fourth Assessment Report of the Intergovernmental Panel on Climate Change. Cambridge University Press, Cambridge, UK.
Jenssen, B. J. (2006). Endocrine-disrupting chemicals and climate change: a worst-case combination for Arctic marine mammals and seabirds? *Environmental Health Perspectives* **114**: 76–80.
Jetz, W., D. S. Wilcove and A. P. Dobson (2007). Projected impacts of climate and land-use change on the global diversity of birds. *PLoS Biology* **5**: 1211–1219.
Kendall, B. E. and G. A. Fox (1998). Spatial structure, environmental heterogeneity, and population dynamics: analysis of the coupled logistic map. *Theoretical Population Biology* **54**: 11–37.
Korner, C. (2003). Ecological impacts of atmospheric CO_2 enrichment on terrestrial ecosystems. *Philosophical Transactions of the Royal Society of London, Series A – Mathematical Physical and Engineering Sciences* **361**: 2023–2041.
le Roux, P. C. and M. A. McGeoch (2008). Rapid range expansion and community reorganization in response to warming. *Global Change Biology* **14**: 2950–2962.
Lebreton, J. D. (1996). Demographic models for subdivided populations: the renewal equation approach. *Theoretical Population Biology* **49**: 291–313.
Lloyd, A. L. (1995). The coupled logistic map: a simple model for the effects of spatial heterogeneity on population dynamics. *Journal of Theoretical Biology* **173**: 217–230.
Maclean, I. M. D., G. E. Austin, M. M. Rehfisch, J. Blew, O. Crowe, S. Delany, K. Devos, B. Deceuninck, K. Gunther, K. Laursen, M. Van Roomen and J. Wahl (2008). Climate change causes rapid changes in the distribution and site abundance of birds in winter. *Global Change Biology* **14**: 2489–2500.
McRae, B. H., N. H. Schumaker, R. B. McKane, R. T. Busing, A. M. Solomon and C. A. Burdick (2008). A multi-model framework for simulating wildlife population response to land-use and climate change. *Ecological Modelling* **219**: 77–91.
Metzger, M. J., R. Leemans and D. Schroter (2005). A multidisciplinary multi-scale framework for assessing vulnerabilities to global change. *International Journal of Applied Earth Observation and Geoinformation* **7**: 253–267.
Metzger, M. J., M. D. A. Rounsevell, L. Acosta-Michlik, R. Leemans and D. Schroter (2006). The vulnerability of ecosystem services to land use change. *Agriculture Ecosystems and Environment* **114**: 69–85.

Midgley, G., L. Hannah, D. Millar, M. C. Rutherford and L. W. Powrie (2002). Assessing the vulnerability of species richness to anthropogenic change in a biodiversity hotspot. *Global Ecology and Biogeography* **11**: 445–451.
Midgley, G., G. O. Hughes, W. Thuiller and A. G. Rebelo (2006). Migration rate limitations on climate change-induced range shifts in Cape Proteaceae. *Diversity and Distributions* **12**: 555–562.
Miller-Rushing, A. J., T. L. Lloyd-Evans, R. B. Primack and P. Satzinger (2008). Bird migration times, climate change, and changing population sizes. *Global Change Biology* **14**: 1959–1972.
Mills, L. S. (2007). *Conservation of Wildlife Populations: Demography, Genetics, and Management*. Blackwell Publishing, Malden, MA.
Mitchell, C. E., P. E. Reich, D. Tilman and J. V. Groth (2003). Effects of elevated CO_2, nitrogen deposition, and decreased species diversity on foliar fungal plant disease. *Global Change Biology* **9**: 438–451.
Peters, D. P. C., P. M. Groffmann, K. J. Nadelhoffer, N. B. Grimm, S. L. Collins, W. K. Michener and M. A. Huston (2008). Living in an increasingly connected world: a framework for continental-scale environmental science. *Frontiers in Ecology and the Environment* **6**: 229–237.
Peterson, J. T. and T. J. Kwak (1999). Modeling the effects of land use and climate change on riverine smallmouth bass. *Ecological Applications* **9**: 1391–1404.
Pulliam, H. R. (1988). Sources, sinks, and population regulation. *American Naturalist* **132**: 652–661.
Pulliam, H. R., J. B. Dunning Jr. and J. Liu (1992). Population dynamics in complex landscapes: a case study. *Ecological Applications* **2**: 165–177.
Rounsevell, M. D. A., P. M. Berry and P. A. Harrison (2006). Future environmental change impacts on rural land use and biodiversity: a synthesis of the ACCELERATES project. *Environmental Science and Policy* **9**: 93–100.
Rowe, R. J. (2007). Legacies of land use and recent climatic change: the small mammal fauna of the mountains of Utah. *American Naturalist* **170**: 242–257.
Rupp, T. S., A. M. Starfield and F. S. Chapin (2000). A frame-based spatially explicit model of subarctic vegetation response to climatic change: comparison with a point model. *Landscape Ecology* **15**: 383–400.
Schroter, D., W. Cramer, R. Leemans, I. C. Prentice, M. B. Araujo, N. W. Arnell, A. Bondeau, H. Bugmann, T. R. Carter, C. A. Garcia, A. C. de la Vega-Leinert, M. Erhard, F. Ewert, M. Glendening, J. I. House, S. Kankaanpaa, R. J. T. Klein, S. Lavorel, M. Lindner, M. J. Metzger, J. Meyer, T. D. Mitchell, I. Reginster, M. D. A. Rounsevell, S. Sabate, S. Sitch, B. Smith, J. Smith, P. Smith, M. T. Sykes, K. Thonicke, W. Thuiller, G. Tuck, S. Zaehle and B. Zierl (2005). Ecosystem service supply and vulnerability to global change in Europe. *Science* **310**: 1333–1337.
Schumacher, S., B. Reineking, J. Sibold and H. Bugmann (2006). Modeling the impact of climate and vegetation on fire regimes in mountain landscapes. *Landscape Ecology* **21**: 539–554.
Semlitsch, R. D. (2008). Differentiating migration and dispersal processes for pond-breeding amphibians. *Wildlife Management* **72**: 260–267.
Serra, H., I. C. Rodrigues da Silva, P. F. Mancera, L. Faria, C. J. Von Zuben, F. J. Von Zuben, S. F. dos Reis and W. A. C. Godoy (2007). Stochastic dynamics in exotic and native blowflies: an analysis combining laboratory experiments and a two-patch metapopulation model. *Ecological Research* **22**: 686–695.
Sinclair, A. R. E., J. M. Fryxell and G. Caughley (2006). *Wildlife Ecology, Conservation, and Management*, 2nd edition. Blackwell Publishing, Malden, MA.
Sole, R. V. and J. G. P. Gamarra (1998). Chaos, dispersal and extinction in coupled ecosystems. *Journal of Theoretical Biology* **193**: 539–541.
Sprott, J. C. (2003). *Chaos and Time-Series Analysis*. Oxford University Press, Oxford, UK.
Stranko, S. A., R. H. Hilderbrand, R. P. Morgan, M. W. Staley, A. J. Becker, A. Roseberry-Lincoln, E. S. Perry and P. T. Jacobson (2008). Brook trout declines with land cover and temperature changes in Maryland. *North American Journal of Fisheries Management* **28**: 1223–1232.

Thuiller, W. (2003). BIOMOD: optimizing predictions of species distributions and projecting potential future shifts under global change. *Global Change Biology* **9**: 1353–1362.

Thuiller, W., S. Lavorel, M. T. Sykes and M. B. Araujo (2006). Using niche-based modelling to assess the impact of climate change on tree functional diversity in Europe. *Diversity and Distributions* **12**: 49–60.

Trivedi, M. R., P. M. Berry, M. D. Morecroft and T. P. Dawson (2008). Spatial scale affects bioclimate model projections of climate change impacts on mountain plants. *Global Change Biology* **14**: 1089–1103.

Vinebrooke, R. D., K. L. Cottingham, J. Norberg, M. Scheffer, S. I. Dodson, S. C. Maberly and U. Sommer (2004). Impacts of multiple stressors on biodiversity and ecosystem functioning: the role of species co-tolerance. *Oikos* **104**: 451–457.

Williams, J. W., S. T. Jackson and J. E. Kutzbach (2007). Projected distributions of novel and disappearing climates by 2100 AD. *Proceedings of the National Academy of Sciences of the USA* **104**: 5738–5742.

Yao, J., D. P. C. Peters, K. M. Havstad, R. P. Gibbens and J. E. Herrick (2006). Multi-scale factors and long-term responses of Chihuahuan Desert grasses to drought. *Landscape Ecology* **21**: 1217–1231.

Zavaleta, E. S. and K. B. Hulvey (2004). Realistic species losses disproportionately reduce grassland resistance to biological invaders. *Science* **306**: 1175–1177.

Zavaleta, E. S., M. R. Shaw, N. R. Chiarello, H. A. Mooney and C. B. Field (2003). Additive effects of simulated climate changes, elevated CO_2, and nitrogen deposition on grassland diversity. *Proceedings of the National Academy of Sciences of the USA* **100**: 7650–7654.

SCOTT M. PEARSON AND JENNIFER M. FRATERRIGO

6

Habitat quality, niche breadth, temporal stochasticity, and the persistence of populations in heterogeneous landscapes

Summary

Spatial heterogeneity in habitat quality creates variation in demographic performance among subpopulations and results in source–sink dynamics. We extend this idea to explore the effects of within-patch heterogeneity on population persistence in a simulation model. Spatial heterogeneity, niche breadth, and temporal stochasticity in the environment are widely recognized as important drivers of population structure, yet few studies have examined the combined influence of these factors. Simulated populations had life-history traits resembling perennial forest herbaceous plants, and simulated landscapes were based on forests of the southern Appalachian Mountains. Habitat quality varied continuously within and between habitat patches using realistic patterns based on topographic gradients. Temporal stochasticity in survival was implemented to simulate interannual climatic variation, and levels of stochasticity were varied to reflect different frequencies of extreme events. The effects of habitat fragmentation, spatial variation in habitat quality, and niche breadth resulted in differential demographic performance among habitat patches of similar size and shape. These effects overshadowed the influences of temporal stochasticity on population persistence. The results suggest that populations of forest perennials may be more sensitive to habitat fragmentation and variation in habitat quality than to temporal stochasticity due to climate. Specialist species will be more sensitive than generalists to such changes.

Sources, Sinks and Sustainability, ed. Jianguo Liu, Vanessa Hull, Anita T. Morzillo and John A. Wiens. Published by Cambridge University Press.

Background

Understanding the response of species under scenarios of concurrent land cover and climate change remains a significant challenge for ecologists and natural resource managers. Land use drives changes in land cover and is responsible for endangering native biological diversity (Walker 1992; Tracy and Brussard 1994; Pearson *et al.* 1999). Land cover changes alter both the abundance and spatial patterns of habitat, and may affect the population dynamics of native species (Lindenmayer and Fischer 2006), particularly if these changes alter the abundance of source and sink populations (Pulliam 1988). Individual species, which differ in their patterns of life history and habitat needs, will decline from habitat loss and fragmentation to differing degrees. Species having specialized habitat needs, requiring large areas for home ranges, or having limited vagility will be impacted more than generalist species that disperse well and can live in small isolated patches (Terborgh 1974; Dale *et al.* 1994). The spatial distribution of land cover changes is often correlated with natural gradients in soil productivity, topography, and/or soil moisture (Turner *et al.* 2003), so habitat losses or gains are non-random with respect to regional biota. The consequences of landscape change will involve an interaction between species' life histories, the spatial distribution of habitat quality, and complicating factors such as demographic variability.

Mountainous terrain strongly influences the spatial distribution of habitat types in the southern Appalachians. Topographic variability creates gradients in site moisture, temperature (Bolstad *et al.* 1998b), annual solar radiation, and soil fertility. In turn, these gradients affect the location of vegetation community types (Whittaker 1956; Day and Monk 1974), rates of ecosystem processes (Elliott *et al.* 1999), and human land uses (Turner *et al.* 2003). By influencing these abiotic gradients, topography creates a mosaic of habitat types within broad land cover types such as forests. Correlations between microclimate and terrain have been successfully exploited to produce vegetation maps from spatial data on elevation and terrain shape (e.g., McNab 1996; Bolstad *et al.* 1998a; Simon *et al.* 2005). These environmental gradients likewise affect habitat quality at the patch level and at specific locations (i.e., sites) within patches. Collectively, within-patch heterogeneity determines the average quality of a patch. But variance in quality between sites, and the spatial distribution of high- and low-quality sites may also be important for persistence in patches of marginal or average quality. Habitat quality will be expressed by differences in demographic rates and probabilities of population persistence within and between patches.

If land cover patterns remain relatively stable, climate change may affect demographic rates and the quality of habitats. Climate change can be expressed as trends in the mean and variability of temperature and moisture availability. While increases in mean conditions have received much attention, climate change scenarios also predict increased variability in seasonal and interannual conditions which may be experienced as increased frequency of extreme events (Boer *et al.* 2000; Waterson 2005; IPCC 2007). Variability in moisture and temperature affects growth and mortality (Olano and Palmer 2003), population dynamics (Adler *et al.* 2006; Levine *et al.* 2008), and community composition (Adler and HilleRisLambers 2008). Variability can result in more species turnover and an increase in generalist species (Gonzalez-Megias *et al.* 2008). Life-history strategies, including niche specialization and longevity, will affect sensitivity to climatic variation (Ibáñez *et al.* 2007). Interannual variability in plant demography due to climate may be modulated by spatial variability in habitat quality.

At the patch level, the source/sink status of a patch is determined by within-patch demography (*sensu* Pulliam 1988) which is related to habitat quality. When recruitment rates are low across the landscape – during a drought for example – source patches may act as refugia. Source patches of moderate habitat quality may become sinks if recruitment rates drop below replacement levels. In contrast, high-quality patches may experience reduced recruitment but remain above replacement levels, thus retaining their status as sources. If high-quality patches persist but do not produce demographic surpluses during the drought, they stand ready to provide a source of dispersers to recolonize sink patches that have experienced extinctions when the drought subsides. This scenario plays out at multiple spatial scales (see Diez and Giladi, Chapter 14, this volume). As the source–sink status of entire patches changes, so will the survival and reproductive contributions of individual sites within patches.

Habitat heterogeneity within patches creates a mosaic of sites having higher and lower rates of survival, reproduction and recruitment (Meekins and McCarthy 2001). Within-patch heterogeneity can be especially high when "patches" are defined by a broad criterion such as land cover type. Fine-scale heterogeneity in temperature, moisture, soil nutrients, etc. may be correlated with occurrence (Palmer 1990; Fraterrigo *et al.* 2006) and demographic rates (Meekins and McCarthy 2001). Although not separated into discrete patches, source–sink dynamics can occur between the sites within patches. The demographic rates associated with an entire patch will be a function of the average habitat quality of its embedded sites. During periods of reduced survival and reproduction, clusters of high-quality sites within patches should promote population persistence by providing refugia at a fine spatial scale.

This study investigated interactions between spatial patterns of habitat, gradients in habitat quality, niche breadth, and temporal stochasticity representative of climatic variation. We used a spatially explicit model to simulate population dynamics on virtual maps representative of real landscapes in our study area. The ecological motivation for this analysis was to understand the joint implications of these factors on the spatiotemporal dynamics of herbaceous plants found in the mesic deciduous forests of the southern Appalachian Mountains. Many plant species decline when their habitat is fragmented and disturbed (Pearson *et al.* 1998; Lennartsson 2002; Vellend *et al.* 2006), and life-history traits affect a species' ability to tolerate environmental changes. For example, long-lived species are more tolerant of stochastic variation than short-lived species (Morris *et al.* 2008). Within this general context, we addressed the following questions:

- How do habitat size and fragmentation, gradients in habitat quality, and niche specialization interact with temporal stochasticity to affect the persistence of populations?
- How do these factors rank in importance with respect of population persistence and occupancy in heterogeneous landscapes?

Research methods

Model description

The spatially explicit model simulated the survival, reproduction and dispersal of populations on maps that varied in the amount and spatial pattern of habitat. The purpose of this model was to compare responses of species with different life-history strategies to the effects of habitat fragmentation, environmental gradients, and temporal stochasticity in demographic rates. The model tracks population dynamics (growth vs. decline; range expansion vs. constriction) of classes of species that differ in aspects of their life history (long-lived vs. short-lived; good dispersal vs. poor dispersal), but it cannot precisely simulate the population dynamics of any *specific* species. Thus its formulation sacrifices realism and precision for generality. Rather than providing predictions about the population dynamics of specific species, the model can be used to determine which life-history strategies (e.g., high survivorship, poor dispersal vs. low survivorship, good dispersal) are successful on a particular landscape under a given level of interannual temporal stochasticity.

The model simulated the changes in the occupation of suitable habitat through time. The basic structure of the model is similar to a spatially explicit, cellular automaton implementation of the Levins model (*sensu* Bascompte and Sole 1996; Matlack and Monde 2004; Fraterrigo *et al.* 2009). Rather than estimating abundances, this style of model simulates and tracks whether cells of suitable habitat are occupied or empty. The model interfaced with maps in a geographic information system (GIS) which contained cells of three classes:

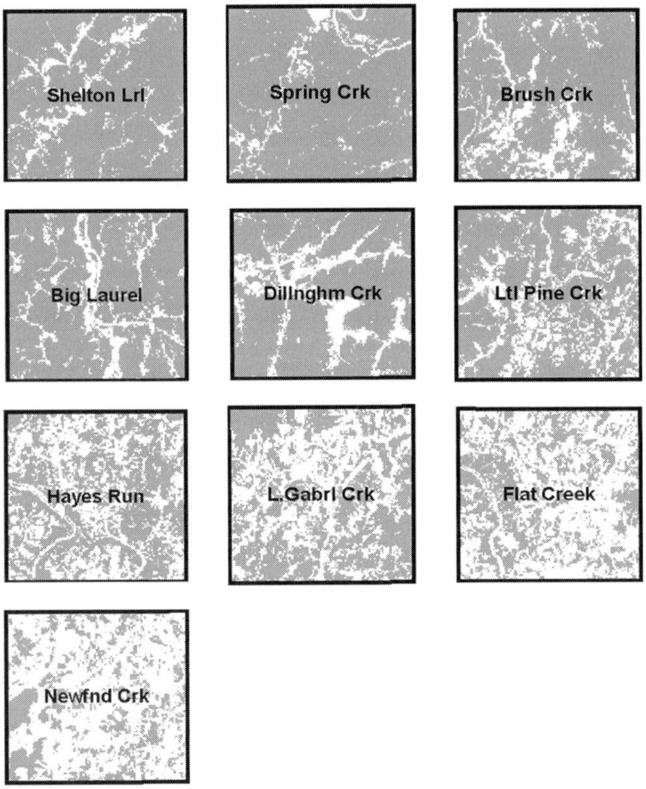

FIGURE 6.1.
Forest cover (gray) maps used in simulation experiment. Each map is 8 km × 8 km in extent with 50 m cells. See Table 6.1 for landscape metrics.

(a) suitable habitat occupied by the species, (b) unoccupied suitable habitat, and (c) unsuitable habitat. Its basic functional unit was a map cell, rather than an individual organism. While we did not simulate individual organisms and their propagules, demographic properties, such as rates of survival and fecundity, were applied to occupied cells. Depending on the organism of interest, an occupied cell could represent the presence of an individual, a breeding pair, or a subpopulation.

The habitat maps incorporated patterns of land cover and topography sampled from the French Broad River watershed in western North Carolina, USA (Fig. 6.1). The habitat maps were 8 km × 8 km in extent with a cell size of 50 m, and varied with respect to the abundance and fragmentation of forest cover (Table 6.1). Forested land cover was classified as potentially suitable habitat, while non-forest covers were classified as unsuitable. A forest patch was defined as a set of contiguous forested cells using an eight-cell adjacency rule. Fragmentation was defined as an increase in the number of patches with a decrease in patch size, given the same amount of habitat. Habitat amount and

TABLE 6.1. *Landscape metrics for forest cover in ten maps used for simulations. Each map was 8 km × 8 km in extent with cell size of 50 m. Patch area and core area are measured in hectares (ha). Metrics were calculated using FRAGSTATS (McGarigal and Marks 1995).*

Map	Percent cover	Number of patches	Mean (SD) patch area	Mean patch shape[a]	Mean (SD) core area[b]
shltnlrl	87.2	6	900.3 (2,012.0)	8.33	622.3 (1,391.6)
sprngcrk	91.2	10	562.0 (1,676.3)	6.44	423.5 (1,269.6)
brshcrk	80.8	32	155.7 (861.9)	10.95	94.6 (526.7)
biglrl	82.3	35	143.9 (576.0)	10.20	89.6 (364.2)
dillghm	78.6	41	119.1 (706.9)	8.89	82.1 (497.2)
ltpine	71.0	64	68.7 (528.0)	17.29	29.7 (234.1)
hayesrun	55.1	134	25.8 (173.9)	24.10	5.8 (43.8)
gabcrk	54.3	139	24.0 (171.4)	21.61	6.8 (52.1)
fltcrk	39.5	249	9.8 (66.6)	25.66	1.5 (14.8)
newfnd	29.9	282	6.7 (20.2)	22.92	1.2 (9.1)

[a] Patch perimeter divided by minimum perimeter of compact patch of equal size.
[b] Core areas included cells ≥ 100 m from patch edge.

fragmentation were correlated in these maps (Table 6.1). The effects of fragmentation, while controlling for habitat amount, have previously been analyzed using this model (Fraterrigo *et al.* 2009). In this study, we did not separate the effects of habitat amount and fragmentation but considered them jointly by ranking the study landscapes by number of patches and mean patch area (Table 6.1). Within forest patches, habitat quality varied with respect to site moisture as influenced by topographic gradients. Given our interest in herbaceous species, site moisture was assumed to be correlated with survival.

Population responses to variation in site moisture were adjusted to create three levels of niche breadth, hereafter referred to as niche specialization. An index of habitat quality (*QUAL*) was calculated from a topographic relative moisture index (TRMI), a spatially explicit index of site moisture incorporating the effects of landform, terrain curvature, and patterns of overland flow (Parker 1982). TRMI values from maps for our study areas (Simon *et al.* 2005) were rescaled to a minimum of 0 and maximum of 100. To incorporate niche breadth, three different functions were composed to relate *QUAL* to the rescaled TRMI (Fig. 6.2). These functions allowed us to describe the moisture response of species with three levels of niche specialization: *low* (a moisture generalist), *moderate*, and *high* (a specialist). The *QUAL* variable ranged from 0% to 100%. Cells with *QUAL* = 100% conveyed the highest habitat quality, and

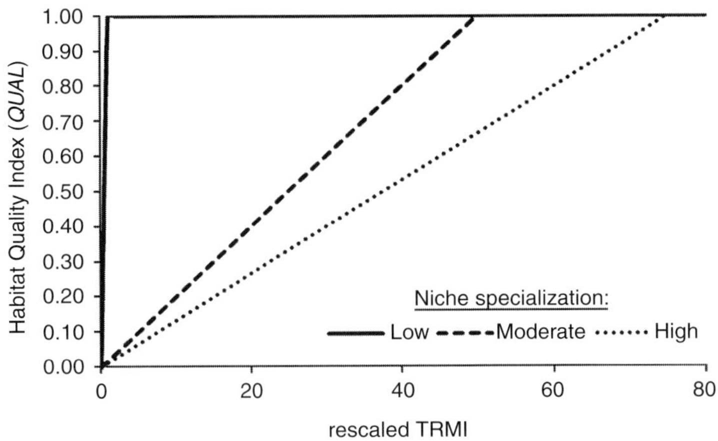

FIGURE 6.2.
Habitat quality index (*QUAL*) based on terrain relative moisture index (TRMI) scaled to 0–100. Generalist species (low niche specialization; solid line) had a low degree of niche specialization and experienced high-quality habitat over a broad range of TRMI values. In contrast, habitat specialists (high niche specialization; dotted line) experienced high *QUAL* values only at high values of TRMI. *QUAL* affected survival rates in the population model.

QUAL = 0% indicated unsuitable cells. Cells with intermediate values of *QUAL* experienced reduced levels of survival commensurate with habitat quality.

The life-history strategies of individual species were described by three parameters: survival probability (*SURV*), fecundity (*FEC*), and dispersal (*DISP*). Survival was simply the probability of a cell remaining occupied from one time step to the next. The survival rate for each specific cell was discounted by habitat quality; an occupied cell remained occupied until the next time step with a probability of *SURV***QUAL*. A given cell that was unoccupied, because it was previously empty or because the previous occupant(s) died, could be colonized by receiving a propagule from an adjacent or nearby cells within the maximum dispersal distance (*DISP*). Fecundity (*FEC*) was the probability that an occupied cell produced propagules that colonized an adjacent suitable, unoccupied cell. Dispersal was modeled using a function to discount the fecundity parameter with increasing distance from an occupied cell (i.e., propagules were more likely to colonize adjacent cells than cells farther away). The function for this distance decay coefficient (*DIST*) is:

$$DIST = 1 - [(distance - 1)/(DISP)]^S \quad (6.1)$$

where *distance* is the Euclidean distance in cell lengths, measured from cell centroids, from the focal cell to an occupied cell. *DISP* is the maximum distance (in cells) at which the distance decay coefficient is greater than zero. In the model programming, the values of *DIST* were constrained to be ≥0. The shape of this decay coefficient function was affected by the parameter S, and a negative exponential shape ($S = 0.5$) resembling a seed shadow was used for these simulations (*sensu* Fahrig 1992; Levin *et al.* 2003). Thus, an empty cell had the following probability of being colonized:

$$\text{Probability of being colonized} = \Sigma\, FEC^{*}DIST_i^{*}K \qquad (6.2)$$

where $DIST_i$ is the coefficient of decay with distance, from Eq. (6.1), and K is a constant to normalize $DIST_i$ for different values of $DISP$. The product $FEC^{*}DIST_i^{*}K$ was summed for all i cells of occupied habitat within the maximum dispersal distance of a given unoccupied cell.

Collectively, *SURV*, *FEC*, and *DISP* affect the ability of a population to spread within a map by colonizing empty cells. In order to compare the impacts of changing a life-history trait on population performance, it is necessary to compare virtual "species," as defined by a combination of life-history parameters and niche specialization, which have comparable rates of spread. Real-world species experience life-history tradeoffs in rates of survival, fecundity and dispersal (Silvertown *et al.* 1997; Franco and Silvertown 2004) which affect population growth rates. In this, and similar cellular automaton models, population growth is manifest as a diffusive rate of spread (Clark *et al.* 2001; Matlack and Monde 2004). For these simulations, *FEC* and *K* were adjusted so that all four species had similar rates of spread on a homogeneous map of cells with maximum suitability. This adjustment permitted a more direct examination of the effects of life history on habitat occupancy.

Temporal environmental stochasticity was introduced by changing the probability of survival at each time step, under the assumption that environmental stochasticity causes density-independent fluctuation in some demographic parameters. To add temporal variability to survival, we stochastically varied *SURV* at each time step by drawing a random number from a beta distribution. The beta distribution, which is bounded by 0 and 1, was parameterized by solving for α and β for the desired mean and variance (see Mood *et al.* 1974 for equations). For example, if $SURV = 0.2$ and survival variability was 50%, we set the mean to 0.2 and variance to 0.01 (standard deviation = 0.1) and solved for α and β. The new value of *SURV* was applied to all occupied cells in the landscape for the current time step. Thus, temporal variability in these parameters was synchronous across space and was analogous to regional-scale variation in the climate regime.

Simulation experiments

We conducted a factorial experiment to measure the effects of landscape characteristics (i.e., habitat fragmentation and environmental gradients), niche specialization, species' life histories, and temporal stochasticity. We specified six levels of stochasticity: 0%, 2.5, 5, 10, 30, and 50%. There were four "species" as specified by combining two levels of survival (*SURV* = 0.2 and 0.4) and maximum dispersal (*DISP* = 2 and 3 cells) each. To maintain similar rates of spread among these species, *FEC* was set to 0.09 when *SURV* = 0.2, and *FEC* = 0.08 when *SURV* = 0.4; *K* = 1.0 when *DISP* = 2, and *K* = 0.67 when *DISP* = 3. There were three levels of niche specialization, as described above, and ten habitat maps.

The parameters of these virtual species were inspired by the variety of life-history traits found in the herbaceous community of our study area, which vary considerably in their survivorship and fecundity and are sensitive to edaphic conditions (Pearson *et al.* 1998; Gilliam and Roberts 2003). For example, Jack-in-the-pulpit (*Arisaema triphyllum* (L.) Schott) is a habitat generalist found in a wide range of conditions, whereas large-flowered bellwort (*Uvularia grandiflora* Sm.) is a habitat specialist which is limited to moist rich soils. Both of these species are long-lived perennials. In contrast, touch-me-not (*Impatiens capensis* Meerb.) is an annual and a habitat specialist requiring mesic conditions. Many of the native forest herbs of the eastern temperate forest are perennials with limited dispersal ability. For example, foam flower (*Tiarella cordifolia* L.) has small seeds that fall directly below the plant. These seeds may be dispersed a short distance by gravity or by overland flow of water. Other herb species (e.g., *Trillium* (L.) spp., *Disporum* (Salisb.) spp., and *Viola canadensis* L.) are dispersed by ants (Beattie and Culver 1981; Smith *et al.* 1989).

At the beginning of each run of the model, the demographic parameters were specified and 33% of the habitat cells were randomly occupied. We chose this level of occupancy so that landscape occupancy could initially change in either direction. Simulations run with other values of initial occupancy indicated that initial conditions had no effect on model behavior (data not shown). During a given time step, changes in the occupancy of habitat cells were determined by first evaluating the probability of survival for occupied cells and then evaluating the probability of colonization for unoccupied habitat cells using Eqs. (6.1) and (6.2). During reproduction, the pattern of occupancy from the previous time step was used; this sequence was analogous to using seeds produced during the prior growing season to colonize empty sites. When a given cell becomes unoccupied due to mortality (i.e., dies), propagules from that cell (from the previous time step) contribute to its probability of being recolonized. The model was run for 100 time steps on each map (dynamics typically stabilized after 80 time steps) and each run was replicated ten times. This design involved 10 maps × 3 niches × 2 survival × 2 dispersal × 6 stochasticity levels × 10 replicates = 7,200 runs.

We recorded the proportion of suitable habitat cells occupied and the proportion of habitat patches occupied at the end of each time step, and these two parameters were our primary response variables. We wanted to detect when the experimental treatments (e.g., habitat pattern, niche specialization, stochasticity) resulted in reduced levels of habitat occupancy. By taking proportions, we were able to compare maps with different amounts of forest cover (Table 6.1) and correct for differences in the abundance of forest habitat prior to statistical analysis. To correct for any boundary effects, cells within the maximum dispersal distance (*DISP*) of the edge of the map were excluded from the analysis.

At the patch level, mean habitat quality and the distribution of *QUAL* values among patch cells may affect population persistence and the source/sink status of patches. To quantify these potential effects, we conducted a second simulation experiment using a set of 12 artificial patches in which mean habitat quality ranged from 25% to 100%. For seven patches, the statistical distribution of *QUAL* was varied around a mean *QUAL* of 75%. The highest-quality cells were clustered in the middle of the patch; cells of progressively lower quality surrounded the high-quality cluster in a pattern similar to that observed in the maps of real landscapes (Fig. 6.3B). We used $SURV = 0.2$ and $DISP = 3$ for this experiment, which included ten replicate runs for each patch. Patch size was fixed at 2,500 ha (100 × 100 cells). The proportion of suitable habitat cells occupied was recorded after 100 time steps in each simulation and compared with the initial conditions of 33% occupancy.

Data analysis

Response variables were arcsine square root transformed (appropriate for proportional data, Sokal and Rohlf 1995) to meet normality assumptions, and analysis of variance (ANOVA) was used to test for main effects and interactions. A single factor ("Map") represented the joint influence of habitat amount and fragmentation associated with the ten landscapes. Effects were ranked based on their corresponding *F* values. The global occupancy of suitable cells and the occupancy of individual patches were analyzed separately.

Logistic regression was used to model the probability of persistence at the patch level using the main effects of niche specialization, patch size, habitat quality, and temporal stochasticity, as well as their interactions. Patch size was measured as the log-transformed area of each patch measured in hectares. Habitat quality was calculated by taking the mean *QUAL* of all cells in the patch. Each patch was considered as an observation, which created a large sample size. Therefore, we only considered effects with $Z > 3.0$ or $P < 0.02$ to be statistically significant. To visualize the interaction between patch size and habitat quality, separate models were estimated for the three levels of niche specialization and plotted. All analyses were conducted in R (version 2.7; R Development Core Team 2009).

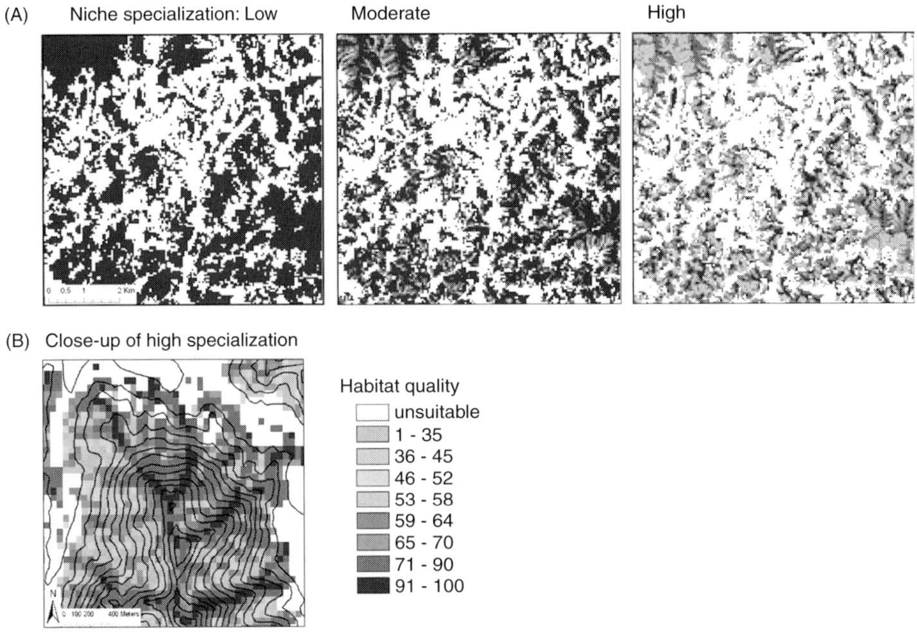

FIGURE 6.3.
Habitat quality (*QUAL*) map for the Lower Gabriels Creek study area (L. Gabrl Crk in Fig. 6.1). The abundance and spatial distribution of habitat quality varied for species with low, moderate, and high levels of niche specialization (A). A close-up view with 25-m elevation contours shows that quality was affected by aspect and terrain shape (B). Quality values ranged from 1% (low) to 100% (high). Color version available online at: www.cambridge.org/9780521199476.

Results

The analysis of variance revealed the relative influence of habitat amount and fragmentation, niche specialization, and temporal stochasticity. For the proportion of cells occupied, niche specialization had the strongest effect, followed by differences between maps, mean survival rate, and dispersal distance based on the relative magnitude of the *F* statistics (Table 6.2). The effects of temporal stochasticity were weaker than these other effects. There was an interaction between stochasticity and life-history strategy (i.e., "Surv × disp × stoch" in Table 6.2). For the proportion of patches occupied, the ranking of effects was similar, except that differences between maps had the strongest effect.

Niche specialization and the spatial pattern of habitat had strong effects on habitat occupancy. On average, the increasing levels of niche specialization reduced the occupancy of suitable cells and entire patches (Fig. 6.4). At the end of the simulations, mean cell occupancy rates were 0.278, 0.118, and 0.062 for low, moderate, and high levels of specialization, respectively, across

TABLE 6.2. *Analysis of variance of simulation results of cell and patch occupancy after 100 time steps. The relative influence of each factor was compared using F values.*

Proportion of cells occupied					
Source	df	SS	MSS	F	P
Map	9	150.83	16.76	2,788.4	<0.001
Niche specialization	1	99.51	99.51	16,557.4	<0.001
Survival	1	8.79	8.79	1,462.8	<0.001
Dispersal	1	1.35	1.35	224.9	<0.001
Stochasticity	1	0.30	0.30	49.6	<0.001
Niche sp × stoch	1	0.01	0.01	0.9	0.336
Surv × disp × stoch	1	0.10	0.10	15.9	<0.001
Proportion of patches occupied					
Map	9	77.38	8.60	8,030.5	<0.001
Niche specialization	1	2.03	2.03	1,894.0	<0.001
Survival	1	0.94	0.94	876.0	<0.001
Dispersal	1	0.02	0.02	17.2	<0.001
Stochasticity	1	0.03	0.03	23.4	<0.001
Niche × stoch	1	0.00	0.00	2.2	0.139
Surv × disp × stoch	1	0.01	0.01	9.2	0.002

combinations of the other factors. Likewise, cell occupancy declined on landscapes with reduced amount and increasingly fragmented habitat (Fig. 6.4A). Cell occupancy was particularly low for maps in which forest coverage was less than 70% (see Table 6.1 and Fig. 6.4A). These maps also had patches which tended to be smaller in total area, have more complex shapes, and smaller core areas (Table 6.1, Fig. 6.1). Patch occupancy rates were greatest for two maps, Shelton Laurel (labeled "shtnlrl") and Spring Creek ("springcrk"), which had much larger mean patch areas and core areas than other maps (Table 6.1, Fig. 6.4B). The pattern of declining occupancy with fragmentation was similar among the three levels of niche specialization (nonsignificant interaction term, $P > 0.20$).

Source–sink theory predicts demographic differences between patches related to habitat quality. The habitat occupancy was affected by the patch structure of the maps and variations in habitat quality among patches. The logistic regression analysis revealed that patch-level persistence was affected by niche specialization, patch size, and habitat quality (Table 6.3). Persistence of populations in patches was positively correlated with patch size (Figs. 6.4 and 6.5). Moreover, niche specialization and mean habitat quality, at the patch level, modulated the effects of patch size (Figs. 6.4 and 6.5) and had significant

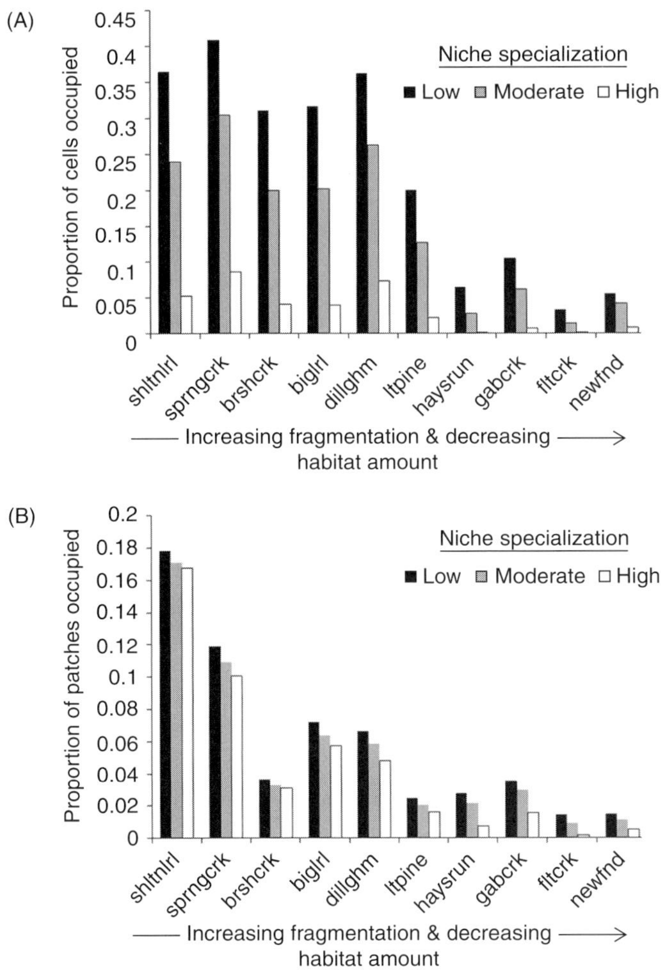

FIGURE 6.4.
Mean proportion of cells (A) and patches (B) occupied after 100 time steps for each map and each level of niche specialization. Maps are arranged in order of increasing habitat fragmentation (see Table 6.1 for landscape pattern metrics).

statistical interactions with this factor (Table 6.3). Patches with greater than average habitat quality had higher occupancy rates for moderate and high specialists.

Higher levels of specialization reduced the occupancy rate of small patches (Fig. 6.5), and fragmented landscapes experienced lower occupancy at both the patch and cell levels. Generalist species, by definition, were not sensitive to habitat quality. However, increasing specialization was correlated with increases in the size and average quality of patches necessary to sustain populations (Fig. 6.5). To achieve a patch occupancy rate of 60%, a specialist required a patch size of ≥1,100 ha for the highest-quality patches. When habitat quality

TABLE 6.3. *Logistic regression analysis of patch occupancy after 100 time steps. Summary analysis of deviance table is presented. Given the large sample size, only terms with Z > 3.0 should be considered statistically significant. Only significant interaction terms are listed.*

Source	df	Deviance	Residual deviance	Z	P
Species[a]	3	333	72,655	12.14	<0.0001
Niche specialization	2	473	72,182	9.82	<0.0001
Stochasticity	2	6	72,176	1.91	0.0560
Patch size[b]	1	52,854	19,322	3.16	0.0016
Habitat quality[c]	1	168	19,154	10.85	<0.0001
Niche × pat_size	1	57	19,096	4.54	<0.0001
Niche × habquality	1	138	18,958	7.84	<0.0001
Pat_size × habquality	1	18	18,939	5.58	<0.0001
Niche × pat_size × habquality	1	12	18,927	3.30	0.0010
Null model	357,479	72,989			

[a] Combinations of survival and dispersal define four species.
[b] Log-transformed patch area.
[c] Mean of quality index for all cells in patch.

was reduced by 50%, a patch size of >3,800 ha was needed. Small patches often had too few cells of high quality to sustain populations of specialists (see below). Populations of specialist species were largely confined to the largest patches by the end of their simulations (Fig. 6.5).

The second simulation experiment using 12 artificial patches revealed how mean habitat quality and the distribution of habitat quality among cells affected demography within patches. In patches with no habitat heterogeneity (e.g., patches A–D; Table 6.4), occupancy increased by >100% over initial levels when mean habitat quality was >80%. These populations could be considered source populations because their populations are expected to spread to all suitable cells in the patch and have the potential to contribute colonists to nearby patches. Patches with increase rates of <100% were classified as sink habitat. Those populations would eventually go extinct given enough time steps. When mean habitat quality was fixed at 75%, variance in quality among the patch's cells affected occupancy increase. Occupancy rates grew as the proportion of *QUAL* = 100 cells increased (Table 6.4). Notably, patch L (Table 6.4) was classified as a source, with an increase of >100%, while the homogeneous patch B was a sink, having an increase of 89% despite having the same mean habitat quality.

Temporal stochasticity negatively affected population persistence but only at high levels of this parameter; moreover, the effects varied among life histories.

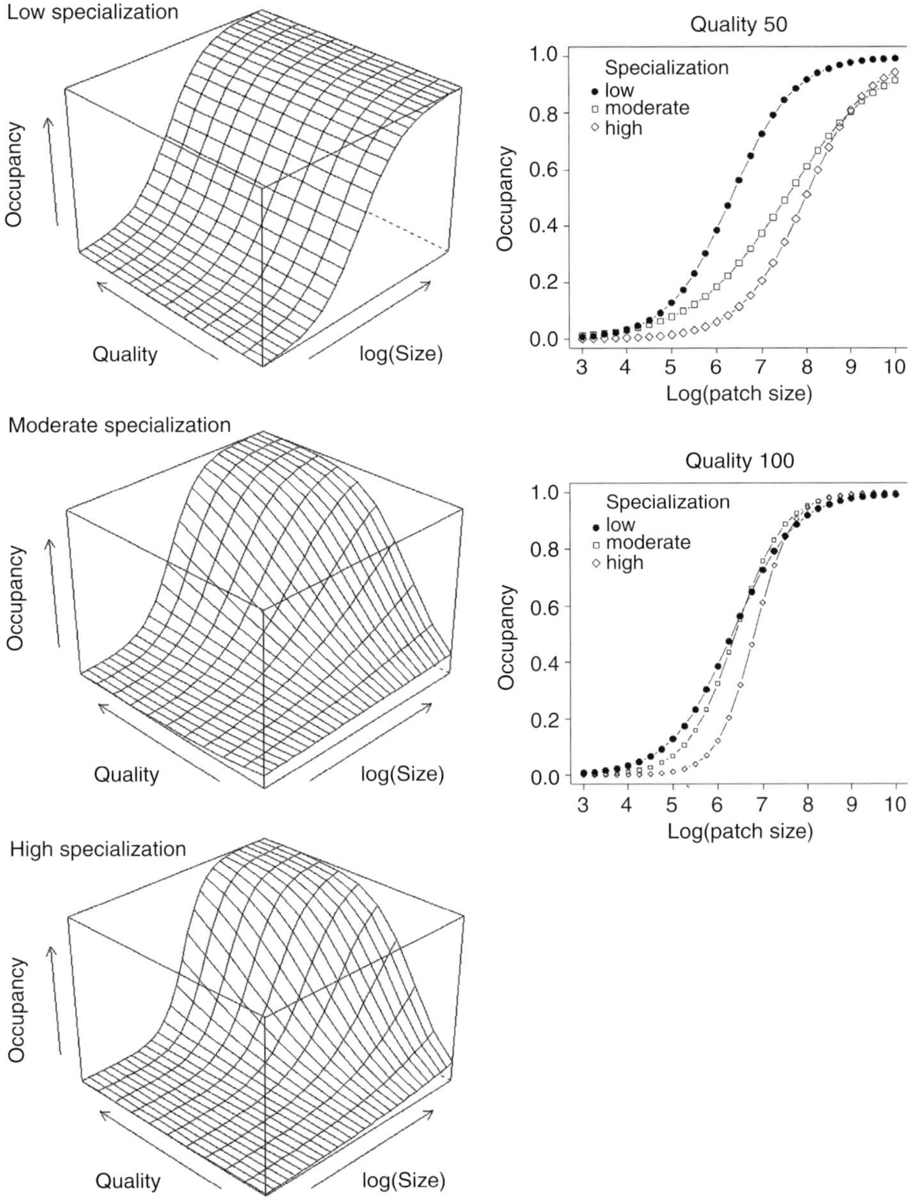

FIGURE 6.5.
Changes in frequency of patch occupancy with mean habitat quality and patch size. Surfaces derived from logistic regression models. Line graphs show change in occupancy relative to patch size at moderate (50%) and maximum (100%) habitat quality for three levels of niche specialization

Higher levels of stochasticity (>10%) reduced the cell occupancy rates, while lower rates of stochasticity had little effect on habitat occupancy (Fig. 6.6). Species with higher survival and dispersal were more sensitive to stochasticity than species with lower survival and dispersal values. Populations with high survival and

TABLE 6.4. *Effects of habitat quality on source/sink status of patches. Proportion of suitable habitat occupied (occupancy) was recorded after 100 time steps using species with survival = 0.2 and dispersal = 3 cells. Patches with increase of occupancy of >100% were demographic sources, while patches with lower rates were demographic sinks.*

	Proportion of cells in four quality categories						
Patch	*QUAL*=25.0	50.0	75.0	100.0	Mean quality	Occupancy	Increase (%)[a]
A	0.0	0.0	0.0	1.0	100.0	0.495	149
B	0.0	0.0	1.0	0.0	75.0	0.298	89
C	0.0	1.0	0.0	0.0	50.0	0.049	15
D	1.0	0.0	0.0	0.0	25.0	0.001	0
E	0	0	0.4	0.6	90.0	0.419	126
F	0	0.2	0.3	0.5	82.5	0.372	112
G	0.00	0.10	0.80	0.10	75.0	0.300	90
H	0.00	0.20	0.60	0.20	75.0	0.305	92
I	0.00	0.22	0.56	0.22	75.0	0.298	89
J	0.10	0.10	0.50	0.30	75.0	0.321	96
K	0.00	0.30	0.40	0.30	75.0	0.311	93
L	0.20	0.00	0.40	0.40	75.0	0.341	102

[a] Percent increase in occupancy. At beginning of simulation, 33.3% of suitable cells were occupied.

dispersal experienced an 8% reduction in cell occupancy at a stochasticity level of 50% compared with simulations with no (0%) stochasticity. Survival stochasticity had an insignificant main effect in the logistic regression analysis of patch occupancy, due to the strength of the other factors and their interactions (Table 6.3). Marginally significant interactions indicated that stochasticity may interact synergistically with other factors to reduce population persistence, but this phenomenon was a relatively minor source of variation in these data.

Discussion

The spatial pattern of habitat and niche specialization had stronger effects than temporal stochasticity at both the landscape- and patch-level scales. In this model, environmental gradients altered survival rates and the spatial arrangement of high-quality habitat within and between patches. Local population persistence was positively correlated with mean habitat quality of patches and negatively correlated with niche specialization. For generalist species, all forested cells were considered high-quality habitat, and any large forested patch of a minimum size represented a sustainable population.

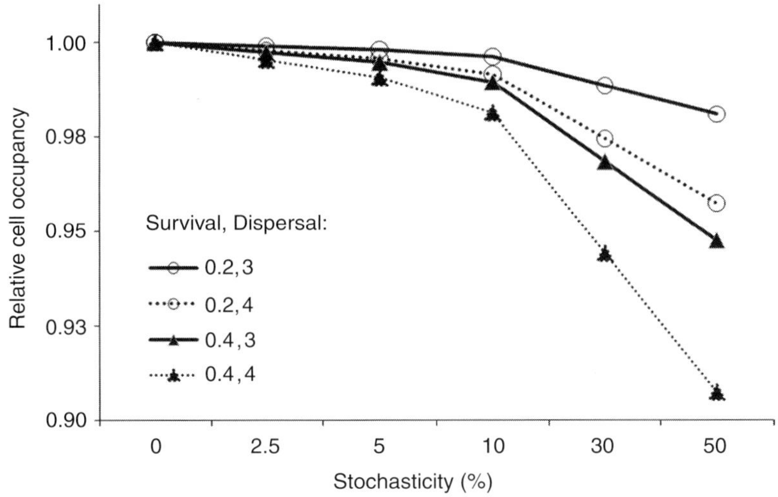

FIGURE 6.6.
Reduction in proportion of cells occupied with increasing levels of temporal stochasticity. Occupancy levels for stochastic simulations were standardized against occupancy for simulations with no stochasticity (0%).

For moderate and high specialists, persistence within a patch depended on patch size and within-patch heterogeneity in the habitat quality. Low-quality patches experienced high probabilities of extinction. Within large patches, clusters of low-quality cells were occupied, but only in a transitory fashion. Environmental gradients and habitat fragmentation interacted to influence the demographic potential of these study landscapes. Niche specialization exacerbated the negative effects of fragmentation and reductions in habitat quality. The interaction between these factors can be understood by applying the general concept of source and sink populations.

The concept of source and sink populations (Pulliam 1988) has raised awareness of how variation in demographic performance between patches affects the occupancy of individual patches as well as the landscape-wide distribution of a population. The spatial arrangement of patches is important because sink patches may persist only if they are within the dispersal window of at least one demographic source (Foppen *et al.* 2000). Local population dynamics depend on the within-patch balance between mortality and fecundity as well as the demographic subsidies from nearby sources.

In the original models of source/sink metapopulations, within-patch conditions were considered homogeneous, but this situation is seldom typical in real landscapes. If habitat quality is strongly correlated with fine-scale edaphic conditions (or some other measure of resource abundance such as food availability; e.g., Nagy and Holmes 2004), which – in turn – affect the demographic performance of sedentary species, then edaphic conditions may be used to estimate the demographic potential of a given site (Seydack *et al.* 2000; Ecke *et al.* 2002; Gonzalez-Megias *et al.* 2005; Van De Pol *et al.* 2006; Flinn 2007; Phillips 2007). Within-patch variation in edaphic parameters may be analyzed to estimate the mean productivity or quality of small patches. For patches larger than the dispersal range of an individual organism, spatial heterogeneity in habitat quality may create source–sink dynamics among sites within individual patches. Environmental gradients can create continuous variation in habitat quality among and within patches.

Increasing niche specialization shifts more of the habitat from source to sink status. A critical number of high-quality source cells (or sites) are required for population persistence (Table 6.4). Reducing patch size may reduce the number of high-quality cells below this critical threshold. For example, Bender *et al.* (1998) found that forest-interior, specialist species were more sensitive to patch size effects than generalists were. This sensitivity may be partly explained by edge effects, which is one manifestation of species sensitivity to within-patch variation in habitat quality. Sites near edges are of lower quality, due to altered microclimate (Matlack 1993), than sites in the patch interior. Generalists, due to broad tolerances, may be insensitive to such edge effects. Our results suggest that specialist species may suffer disproportionately greater habitat losses if within-patch variation in habitat quality exists.

Habitat fragmentation and quality affect the persistence of populations as well as the recovery of forest plant populations after disturbance. Once fragmented, forest patches may gradually lose species over time. With long-lived herbaceous species, several decades or a century or more may be required to pay off this extinction debt (Vellend *et al.* 2006). Our simulation results suggest that the rate of decline will be correlated with patch size and mean habitat quality. Pearson *et al.* (1998) found reduced diversity of herbaceous species in small forest fragments in the southern Appalachians, and patch size effects were further influenced by land use history. Prior land uses are known to affect edaphic characteristics of forest stands such as the abundance and heterogeneity of soil nutrients (Fraterrigo *et al.* 2005, 2006) and rates of nutrient cycling (Compton and Boone 2000), which affect habitat quality. In our simulations, some forest patches did not support populations because (a) patch size was so small as to be subject to stochastic extinctions and/or (b) the habitat quality of suitable cells was too low to support population persistence (e.g., a sink patch).

In our simulations, dispersal limitations left many suitable patches unoccupied because they were effectively too isolated to be colonized once they went extinct. Likewise, localized extinctions could occur within large patches. Even small clusters of high-quality cells could remain unoccupied if those cells were separated from occupied cells by a band of low-quality cells. A stepping-stone type of recolonization to these high-quality clusters was possible, but its likelihood was correlated with the quality and number of the intervening cells. In large patches with an abundance of large clusters of high-quality habitat, a sustainable extinction–recolonization dynamic was achieved within the patch.

In parts of Europe and North America, forests are expanding following agricultural abandonment, and the spatial arrangement of old forest patches is of great importance. At the landscape scale, the abundance and proximity of "ancient forests" (not cleared for agriculture) is correlated with the patterns of diversity among recovering forest patches (Matlack 1994a, 1994b). Moreover, life-history traits, such as colonizing ability, influence the establishment and persistence of populations (Vellend 2003; Flinn and Vellend 2005). Verheyen *et al.* (2003) found that habitat fragmentation negated any effects of varying habitat quality for dispersal-limited species in these post-agricultural forests in the initial phase of recovery. They argue that recovery after disturbance is a two-stage process in which colonizing ability is initially most important in establishing populations in new forest patches. Next, habitat quality influences the abundance and persistence within patches. This process takes many years to play out and will affect forest biodiversity for a century or more of recovery (Vellend *et al.* 2007). While those studies stress the importance of colonizing ability, our simulations suggest that niche specialization is also important for population persistence if the abundance and pattern of habitat changes.

Prior work with this model (Fraterrigo *et al.* 2009) investigated the consequences of temporal stochasticity using binary habitat maps in which the abundance of habitat was held constant but the level of fragmentation was varied. Survival, fecundity, and dispersal were varied, and stochasticity was added to the first two of these parameters. Survival was a key influence on population responses; it interacted with both temporal stochasticity and habitat fragmentation. When survival probability was high, temporal stochasticity had a large effect on population dynamics, resulting in decreased landscape-wide occupancy; whereas habitat fragmentation had little influence. However, habitat fragmentation was crucial when survival probability was low, reflecting the importance of colonization dynamics in organisms with short life spans. Survival stochasticity tended to exacerbate the negative impacts of habitat fragmentation and reproductive stochasticity. Stochasticity increased the number of local extinctions in isolated patches that could not be repopulated due to low local densities and inaccessibility to colonists. These findings suggest that greater environmental variability, such as might arise due to climate change, will compound population losses due to habitat fragmentation. The simulations discussed herein differed by incorporating variation in habitat quality (as opposed to binary maps) and in levels of niche specialization.

For populations, the ability to tolerate climatic variability and change will be complicated by the availability, fragmentation, and quality of their habitats (see also Andersen, Chapter 5, this volume). Individuals living at marginal sites may not survive occasional extreme conditions. If temporal stochasticity affects demographic rates, high-quality source patches – or source sites within patches – may act as refugia and reduce the chance of extinction during periods of low survival or reproduction (Murphy 2001; Magoulick and Kobza 2003). In addition, the source/sink status of sites may change within or between seasons (Fauth 2001; Frouz and Kindlmann 2001; Altwegg *et al.* 2003; Schmidt 2003). Therefore, the persistence and productivity of populations will depend on the abundance and spatial distribution of high-quality sites (Bonesi *et al.* 2002) as well as temporal variation in the site quality. Sink habitats may also promote persistence by providing the occasional colonist to re-establish a source population which has experienced a stochastic extinction (Frouz and Kindlmann 2001; Schmidt 2003) and provide the opportunity for stepping-stone colonization of unoccupied source sites and patches. However, synchronized temporal fluctuations across the landscape may limit the persistence of sink populations during periods of reduced survival or fecundity (Matthews and Gonzalez 2007).

Conclusions

Our results suggest that habitat loss and fragmentation are a greater threat to populations of native species than the type of temporal stochasticity employed in our model. We expected increased stochasticity to dramatically reduce population persistence, as has been found in previous work with this model and in other studies, especially for populations at the edge of their geographic ranges (e.g., Williams *et al.* 2003). However, our simulations suggest either that the effects of habitat quality, patch size, and niche breadth are much stronger than stochasticity, or that the simulated populations are tolerant of stochasticity as implemented in our model. Pulliam (1988) argued that knowledge of source–sink dynamics is critical for effective conservation, and the strong effects of patch-level habitat quality in our study support his assertion. Because land use/land cover changes occur more rapidly than climatically driven changes, habitat loss and fragmentation have been identified as a greater threat than climate change for some regions (see review by Dale 1997) and geographic species (Stevens and Baguette 2008). Moreover, landscape change may be correlated with environmental gradients (Wear and Bolstad 1998) and affect high- and low-quality habitats to different degrees (Turner *et al.* 2003). Our results suggest that dispersal-limited specialist species are actually more sensitive to fragmentation and gradients in habitat quality than to temporal stochasticity. The concept of source–sink dynamics is critical to understanding the consequences of these changes for native species. This concept may be applied to entire patches, as originally conceived, or extended to fine-scale heterogeneity within land cover patches.

Acknowledgments

Ideas for these modeling efforts were generated in discussions with Monica G. Turner and Alan B. Smith, and students at Mars Hill College and the University of Wisconsin-Madison. Two anonymous reviewers provided useful comments for improving this work. Financial support was provided by a Charles Bullard Fellowship from Harvard University, Mars Hill College, and a grant from the Long-Term Ecological Research (LTER) Program of the National Science Foundation (Grant No. DEB-0218001, Coweeta LTER).

References

Adler, P. B. and J. HilleRisLambers (2008). The influence of climate and species composition on the population dynamics of ten prairie forbs. *Ecology* **89**: 3049–3060.

Adler, P. B., J. HilleRisLambers, P. C. Kyriakidis, Q. Guan and J. M. Levine (2006). Climate variability has a stabilizing effect on the coexistence of prairie grasses. *Proceedings of the National Academy of Sciences of the USA* **103**: 12793–12798.

Altwegg, R., A. Roulin, M. Kestenholz and L. Jenni (2003). Variation and covariation in survival, dispersal, and population size in barn owls *Tyto alba*. *Journal of Animal Ecology* **72**: 391–399.

Bascompte, J. and R. V. Sole (1996). Habitat fragmentation and extinction thresholds in spatially explicit models. *Journal of Animal Ecology* **65**: 465–473.

Beattie, A. J. and D. C. Culver (1981). The guild of myrmecochores in the herbaceous flora of West Virginia forests. *Ecology* **62**: 107–115.

Bender, D. J., T. A. Contreras and L. Fahrig (1998). Habitat loss and population decline: a meta-analysis of the patch size effect. *Ecology* **79**: 517–533.

Boer, G. J., G. Flato and D. Ramsden (2000). A transient climate change simulation with greenhouse gas and aerosol forcing: projected climate to the twenty-first century. *Climate Dynamics* **16**: 427–450.

Bolstad, P. V., W. T. Swank and J. M. Vose (1998a). Predicting southern Appalachian overstory vegetation with digital terrain data. *Landscape Ecology* **13**: 271–283.

Bolstad, P. V., L. Swift, F. Collins and J. Régnière (1998b). Measured and predicted air temperatures at basin to regional scales in the southern Appalachian mountains. *Agricultural and Forest Meteorology* **9**: 161–176.

Bonesi, L., S. Rushton and D. Macdonald (2002). The combined effect of environmental factors and neighboring populations on the distribution and abundance of *Arvicola terrestris*: an approach using rule-based models. *Oikos* **99**: 220–230.

Clark, J. S., M. Lewis and L. Horvath (2001). Invasion by extremes: population spread with variation in dispersal and reproduction. *American Naturalist* **157**: 537–554.

Compton, J. E. and R. D. Boone (2000). Long-term impacts of agriculture on soil carbon and nitrogen in New England forests. *Ecology* **81**: 2314–2330.

Dale, V. H. (1997). The relationship between land-use change and climate change. *Ecological Applications* **7**: 753–769.

Dale, V. H., H. Offerman, S. Pearson and R. V. O'Neill (1994). Effects of forest fragmentation on neotropical fauna. *Conservation Biology* **8:** 1027–1036.
Day, F. P. and C. D. Monk (1974). Vegetation patterns on a southern Appalachian watershed. *Ecology* **34:** 329–346.
Ecke, F., O. Lofgren and D. Sorlin (2002). Population dynamics of small mammals in relation to forest age and structural habitat factors in northern Sweden. *Journal of Applied Ecology* **39:** 781–792.
Elliott, K. J., J. M. Vose, W. T. Swank and P. V. Bolstad (1999). Long-term patterns in vegetation–site relationships in a southern Appalachian forest. *Journal of the Torrey Botanical Society* **126:** 320–334.
Fahrig, L. (1992). Relative importance of spatial and temporal scales in a patchy environment. *Theoretical Population Biology* **41:** 300–314.
Fauth, P. T. (2001). Wood thrush populations are not all sinks in the agricultural midwestern United States. *Conservation Biology* **15:** 523–527.
Flinn, K. M. (2007). Microsite-limited recruitment controls fern colonization of post-agricultural forests. *Ecology* **88:** 3103–3114.
Flinn, K. M. and M. Vellend (2005). Recovery of forest plant communities in post-agricultural landscapes. *Frontiers in Ecology and the Environment* **3:** 243–250.
Foppen, R. P. B., J. P. Chardon and W. Liefveld (2000). Understanding the role of sink patches in source–sink metapopulations: reed warbler in an agricultural landscape. *Conservation Biology* **14:** 1881–1892.
Franco, M. and J. Silvertown (2004). A comparative demography of plants based upon elasticities of vital rates. *Ecology* **85:** 531–538.
Fraterrigo, J. M., M. G. Turner, S. M. Pearson and P. Dixon (2005). Effects of past land use on spatial heterogeneity of soil nutrients in southern Appalachian forests. *Ecological Monographs* **75:** 215–230.
Fraterrigo, J. M., M. G. Turner and S. M. Pearson (2006). Interactions between past land use, life-history traits and understory spatial heterogeneity. *Landscape Ecology* **21:** 777–790.
Fraterrigo, J. M., S. M. Pearson and M. G. Turner (2009). Joint effects of habitat configuration and temporal stochasticity on population dynamics. *Landscape Ecology* **24:** 863–877.
Frouz, J. and P. Kindlmann (2001). The role of sink to source re-colonisation in the population dynamics of insects living in unstable habitats: an example of terrestrial chironomids. *Oikos* **93:** 50–58.
Gilliam, F. S. and M. R. Roberts (2003). *The Herbaceous Layer in Forests of Eastern North America*. Oxford University Press, New York.
Gonzalez-Megias, A., J. M. Gomez and F. Sanchez-Pinero (2005). Regional dynamics of a patchily distributed herbivore along an altitudinal gradient. *Ecological Entomology* **30:** 706–713.
Gonzalez-Megias, A., R. Menendez, D. Roy, T. Brereton and C. D. Thomas (2008). Changes in the composition of British butterfly assemblages over two decades. *Global Change Biology* **14:** 1464–1474.
Ibáñez, I., J. S. Clark, S. LaDeau and J. HilleRisLambers (2007). Exploiting temporal variability to understand tree recruitment response to climate change. *Ecological Monographs* **77:** 163–177.
IPCC (2007). *Climate Change 2007: The Physical Science Basis*. Contribution of Working Group I to the Fourth Assessment Report of the Intergovernmental Panel on Climate Change. Cambridge University Press, Cambridge, UK.
Lennartsson, T. (2002). Extinction thresholds and disrupted plant–pollinator interactions in fragmented plant populations. *Ecology* **83:** 3060–3072.
Levin, S. A., H. C. Muller-Landau, R. Nathan and J. Chave (2003). The ecology and evolution of seed dispersal: a theoretical perspective. *Annual Review of Ecology, Evolution, and Systematics* **34:** 575–604.
Levine, J. M., A. K. McEachern and C. Clark (2008). Rainfall effects on rare annual plants. *Journal of Ecology* **96:** 795–806.

Lindenmayer, D. B. and J. Fischer (2006). *Habitat Fragmentation and Landscape Change: An Ecological and Conservation Synthesis*. Island Press, Washington, DC.
Magoulick, D. D. and R. M. Kobza (2003). The role of refugia for fishes during drought: a review and synthesis. *Freshwater Biology* **48**: 1186–1198.
Matlack, G. R. (1993). Microenvironment variation within and among deciduous forest edge sites in the eastern United States. *Biological Conservation* **66**: 185–194.
Matlack, G. R. (1994a). Plant species migration in a mixed-history forest landscape in eastern North America. *Ecology* **75**: 1491–1502.
Matlack, G. R. (1994b). Vegetation dynamics of the forest edge: trends in space and successional time. *Journal of Ecology* **82**: 113–123.
Matlack, G. R. and J. Monde (2004). Consequences of low mobility in spatially and temporally heterogeneous ecosystems. *Journal of Ecology* **92**: 1025–1035.
Matthews, D. P. and A. Gonzalez (2007). The inflationary effects of environmental fluctuations ensure the persistence of sink metapopulations. *Ecology* **88**: 2848–2856.
McGarigal, K. and B. J. Marks (1995). *FRAGSTATS: Spatial Pattern Analysis Program for Quantifying Landscape Structure*. USDA-Forest Service General Technical Report PNW-GTR-351, Pacific Northwest Research Station, Portland, OR.
McNab, W. H. (1996). Classification of local- and landscape-scale ecological types in the southern Appalachian Mountains. *Environmental Monitoring and Assessment* **39**: 215–229.
Meekins, J. F. and B. C. McCarthy (2001). Effect of environmental variation on the invasive success of a nonindigenous forest herb. *Ecological Applications* **11**: 1336–1348.
Mood, A. M., F. A. Graybill and D. C. Boes (1974). *Introduction to the Theory of Statistics*, 3rd edition. McGraw-Hill, New York.
Morris, W. F., C. A. Pfister, S. Tuljapurkar, C. V. Haridas, C. L. Boggs, M. S. Boyce, E. M. Bruna, D. R. Church, T. Coulson, D. F. Doak, S. Forsyth, J.-M. Gaillard, C. C. Horvitz, S. Kalisz, B. E. Kendall, T. M. Knight, C. T. Lee and E. S. Menges (2008). Longevity can buffer plant and animal populations against changing climatic variability. *Ecology* **89**: 19–25.
Murphy, M. T. (2001). Source–sink dynamics of a declining Eastern kingbird population and the value of sink habitats. *Conservation Biology* **15**: 737–748.
Nagy, L. R. and R. T. Holmes (2004). Factors influencing fecundity in migratory songbirds: is nest predation the most important? *Journal of Avian Biology* **35**: 487–491.
Olano, J. M. and M. W. Palmer (2003). Response of an Appalachian old-growth forest to a severe drought episode. *Forest Ecology and Management* **174**: 139–148.
Palmer, M. W. (1990). Spatial scale and patterns of vegetation, flora and species richness in hardwood forests of the North Carolina Piedmont. *Coenoses* **5**: 89–96.
Parker, A. J. (1982). The topographic relative moisture index: an approach to soil-moisture assessment in mountain terrain. *Physical Geography* **3**: 160–168.
Pearson, S. M., A. B. Smith and M. G. Turner (1998). Forest patch size, land use and mesic forest herbs in the French Broad River Basin, North Carolina. *Castanea* **63**: 382–395.
Pearson, S. M., M. G. Turner and J. B. Drake (1999). Landscape change and habitat availability in the Southern Appalachian Highlands and Olympic Peninsula. *Ecological Applications* **9**: 1288–1304.
Phillips, N. E. (2007). A spatial gradient in the potential reproductive output of the sea mussel *Mytilus californianus*. *Marine Biology* **151**: 1543–1550.
Pulliam, H. R. (1988). Sources, sinks, and population regulation. *American Naturalist* **132**: 652–661.
R Development Core Team (2009). *R: A Language and Environment for Statistical Computing*. R Foundation for Statistical Computing, Vienna, Austria.
Schmidt, K. A. (2003). Linking frequencies of acorn masting in temperate forests to long-term population growth rates in a songbird: the veery (*Catharus fuscescens*). *Oikos* **103**: 548–558.
Seydack, A. H. W., C. Vermeulen and J. Huisamen (2000). Habitat quality and the decline of an African elephant population: implications for conservation. *South African Journal of Wildlife Research* **30**: 34–42.

Silvertown, J., M. Franco and J. L. Harper (eds.) (1997). *Plant Life Histories: Ecology, Phylogeny and Evolution*. Cambridge University Press, Cambridge, UK.

Simon, S. A., T. K. Collins, G. L. Kauffman, W. H. McNab and C. J. Ulrey (2005). *Ecological Zones in the Southern Appalachians: First Approximation*. Forest Service Research Paper SRS-41. US Department of Agriculture, Southern Research Station, Asheville, NC.

Smith, B. H., P. D. Forman and A. E. Boyd (1989). Spatial patterns of seed dispersal and predation of two myrmecochorous forest herbs. *Ecology* **70**: 1649–1656.

Sokal, R. R. and F. J. Rohlf (1995). *Biometry: The Principles and Practice of Statistics in Biological Research*, 3rd edition. W. H. Freeman, New York.

Stevens, V. M. and M. Baguette (2008). Importance of habitat quality and landscape connectivity for the persistence of endangered natterjack toads. *Conservation Biology* **22**: 1194–1204.

Terborgh, J. (1974). Preservation of natural diversity: the problem of extinction prone species. *Bioscience* **24**: 715–722.

Tracy, C. R. and P. R. Brussard (1994). Preserving biodiversity: species in landscapes. *Ecological Applications* **4**: 205–207.

Turner, M. G., S. M. Pearson, P. Bolstad and D. N. Wear (2003). Effects of land-cover change on spatial pattern of forest communities in the southern Appalachian Mountains (USA). *Landscape Ecology* **18**: 449–464.

Van De Pol, M., L. W. Bruinzeel, D. I. K. Heg, H. P. Van Der Jeugd and S. Verhulst (2006). A silver spoon for a golden future: long-term effects of natal origin on fitness prospects of oystercatchers (*Haematopus ostralegus*). *Journal of Animal Ecology* **75**: 616–626.

Vellend, M. (2003). Habitat loss inhibits recovery of plant diversity as forests regrow. *Ecology* **84**: 1158–1164.

Vellend, M., K. Verheyen, H. Jacquemyn, A. Kolb, H. Van Calster, G. Peterken and M. Hermy (2006). Extinction debt of forest plants persists for more than a century following habitat fragmentation. *Ecology* **87**: 542–548.

Vellend, M., K. Verheyen, K. M. Flinn, H. Jacquemyn, A. Kolb, H. V. A. N. Calster, G. Peterken, B. J. Graae, J. Bellemare, O. Honnay, J. Brunet, M. Wulf, F. Gerhardt and M. Hermy (2007). Homogenization of forest plant communities and weakening of species–environment relationships via agricultural land use. *Journal of Ecology* **95**: 565–573.

Verheyen, K., G. R. Guntenspergen, B. Biesbrouck and M. Hermy (2003). An integrated analysis of the effects of past land use on forest herb colonization at the landscape scale. *Journal of Ecology* **91**: 731–742.

Walker, B. H. (1992). Biodiversity and ecological redundancy. *Conservation Biology* **6**: 18–23.

Waterson, I. G. (2005). Simulated changes due to global warming in the variability of precipitation, and their interpretation using a gamma-distributed stochastic model. *Advances in Water Resources* **28**: 1368–1381.

Wear, D. N. and P. Bolstad (1998). Land-use changes in southern Appalachian landscapes: spatial analysis and forecast evaluation. *Ecosystems* **1**: 575–594.

Whittaker, R. H. (1956). Vegetation of the Great Smoky Mountains. *Ecological Monographs* **26**: 1–80.

Williams, C. K., A. R. Ives and R. D. Applegate (2003). Population dynamics across geographical ranges: time-series analyses of three small game species. *Ecology* **84**: 2654–2667.

MATTHEW R. FALCY AND BRENT J. DANIELSON

7

When sinks rescue sources in dynamic environments

Summary

Many species of conservation concern occur in spatially heterogeneous landscapes composed of different patches that function as population sources and population sinks. Temporal variation in habitat quality, due to a cycle of habitat disturbance and subsequent recovery, can create relatively underappreciated source–sink dynamics. A cycle of disturbance and recovery can cause a given patch to alternate between functioning as a population source and a population sink. During "good" years, this patch can conceivably sustain another nearby patch that is always sink habitat. However, after a disturbance, the putative source patch may then depend upon individuals from that sink for recolonization. Thus, the metapopulation can depend upon the presence of sink populations for long-term persistence, provided that the sink is relatively unaffected by disturbance. We developed a simple, two-patch model of source–sink dynamics in order to explore the sensitivity of long-term metapopulation persistence to the temporal scale of habitat disturbance, recovery in a putative source patch, and the rate of population decline in the sink. We found that management directed at decelerating the rate of population decline in the sink can have a much greater affect on metapopulation persistence than management targeted at increasing the rate of habitat recovery in the source. This result is magnified as disturbance frequency increases. It is hoped that sinks crucial to metapopulation survival are given appropriate conservation status.

Sources, Sinks and Sustainability, ed. Jianguo Liu, Vanessa Hull, Anita T. Morzillo and John A. Wiens. Published by Cambridge University Press.

Introduction

Modeling population dynamics has played a prominent role in the history of ecology (Kingsland 1995). Many early models assume, for convenience only, that space is homogeneous. This began to change as the theory of island biogeography and metapopulation concepts grew. By the mid-1980s, many ecologists were concerned with spatial variation in habitat at the landscape scale. Thus, ecologists quickly embraced source–sink dynamics, which is predicated on the fact that landscapes are composed of heterogeneous habitats and that demographic parameters are habitat-specific. The original formulations of source–sink dynamics (Holt 1985; Pulliam 1988) make a clear and convincing case for the consequences of spatial variation of habitat quality on population dynamics without invoking temporal variability. These models illustrate that spatial differences in demographic rates, when coupled with immigration and emigration, lead to density-dependent limits to population growth. This is important. Source–sink models essentially use space rather than time to explain how populations self-regulate. But the exclusion of temporal variation in the early models is somewhat unrealistic because conditions clearly vary across seasonal, yearly, and decades-long intervals of disturbance and succession.

Source–sink models have been developed that incorporate temporal variation of the underlying spatial differences in demographic parameters among habitats. Holt (1997; Chapter 2, this volume) showed that dispersal into sink habitats is an evolutionarily stable strategy if source fitness periodically falls below sink fitness, and descendants of dispersers can move back to the source. Holt (1997) further demonstrated that utilization of sink habitat increases with increasing sink fitness and deviation from carrying capacity in the source. Gonzalez and Holt (2002) and Holt *et al.* (2003) demonstrated that autocorrelation of temporal variation in the growth rate of a sink population increases the average sink population size, a phenomenon known as the "inflationary effect." This finding was extended to an ensemble of connected sink patches, where temporal autocorrelation of the growth rates within patches was shown to permit metapopulation persistence (Roy *et al.* 2005). Foppen *et al.* (2001) used simulations to show that the resilience of a source after a catastrophe was enhanced by the presence of a sink, and empirical work on temporal variation in source–sink beetle dynamics has demonstrated that floods can transform sources into transient sinks (Johnson 2004). Gyllenberg *et al.* (1996) and Morris (Chapter 3, this volume) have also shown that a sink can rescue a source from extinction. In this chapter, we extend this idea to an analysis of management options for scenarios with different disturbance frequencies, rates of source habitat recovery, and population decline in the sink.

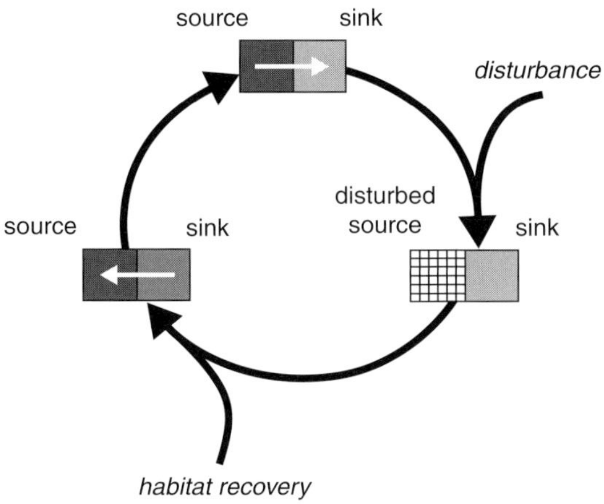

FIGURE 7.1.
Source–sink dynamics where disturbances affect one of two populations. Before disturbances occur, a typical source–sink relationship exists between two populations (topmost portion of figure). Movement of individuals from the source to the sink (white arrow) maintains the sink population. A disturbance degrades the source, putting the system in a critical period where both populations are at risk. Once the source habitat has recovered it is repopulated by individuals dispersing from the sink, eventually restoring the system to the original source–sink scenario.

In this chapter, we explore the effects of a particular form of temporal variability in source–sink dynamics. Specifically, we are concerned with the effects of recurrent disturbances that devastate a source but leave an associated sink relatively undisturbed. We model a simple system of two populations linked by dispersal. In the absence of disturbances, one population occurs in a habitat that is a permanent source while the other population occurs in a sink. Disturbances reduce population size in the source and destroy source habitat, thereby placing the source population in jeopardy until the habitat is sufficiently recovered. This process of habitat destruction and renewal is depicted in Figure 7.1. Note that the system enters a critical period after disturbances, wherein metapopulation persistence crucially depends on the recovery and recolonization of source habitat before the sink population goes extinct. Thus habitat availability, not population size, limits population growth and recovery. Here we explore the interactions among three rates that are critical to long-term metapopulation viability in this system:

1. the rate of habitat recovery in the source
2. the rate of population decline in the sink
3. the frequency of disturbances affecting the source.

The interactions between these factors are important because they have direct relevance to the identification of best management actions for the following dilemma: Should conservation and recovery efforts be devoted to increasing the rate of post-disturbance habitat recovery in the source or decreasing the rate of population decline in the sink? Which of these two rates has the greatest impact on population viability? We can add another layer of complexity by asking: How does the relative influence of these rates on population viability change as the disturbance frequency changes? Our objective in this chapter is to provide an answer to these questions.

Example systems

In order of increasing generality, we provide three examples of systems that may exhibit the source–sink dynamics described in this chapter. A common feature of these systems is that all sources are destroyed at the same time and sinks are relatively unharmed. A system with just one source is also relevant (indeed, we model a system with one source and one sink). If at least one source was unaffected by disturbance, then recovery of the metapopulation would be largely driven by dispersal out of that source. Thus, a key component of the systems we describe is temporal synchrony of source destruction. The systems we describe are likely to occur where characteristics conferring enhanced productivity upon a habitat are correlated with the occurrence of disturbances.

Alabama beach mouse and hurricanes

Our formulation of this particular source–sink dynamic is motivated by our involvement with the Alabama beach mouse (*Peromyscus polionotus ammobates*). This subspecies of the oldfield mouse became a federally listed endangered species in 1985. Its entire historic range is limited to an approximately 48 km strip of Alabama's coveted beachfront real estate. Less than 40% of the 2,830 hectares (ha) of pre-development habitat remains (USFWS 2005). Habitat loss and fragmentation certainly threaten the Alabama beach mouse's existence, and the severity of this problem is dramatically magnified when hurricanes strike.

Beach mice make burrows and forage for seeds and insects in sandy dunes lightly vegetated by sea oats (*Uniola paniculata*), blue stem (*Andropogon maritimus*), and beach grass (*Panicum amarum*). The dunes gradually increase in height (and age) until a well-defined dune apex is reached at roughly 250 meters (m) from the open ocean. On the other side of this dune ridge is a very distinct habitat type: soils are more compact and the vegetation is dominated by sand live oak (*Quercus geminata*), saw palmetto (*Serenoa repens*), and rosemary (*Ceratiola ericoides*). This area is referred to as "scrub." Beach mice were once thought to inhabit dunes exclusively (Howell 1909, 1921; Ivey 1949), but it is now known that beach mice also occur at low densities in scrub adjacent to the dunes (Sneckenberger 2001).

In addition to differences in substrate density, vegetation type, and beach mouse density between the dunes and scrub, the two habitat types also differ in susceptibility to hurricane-induced inundation. On September 16, 2004, Hurricane Ivan was at the boundary of a category 3–4 storm when it made landfall directly over the Alabama beach mouse's entire range. The storm created a 3.0–4.5 m surge ("surge" is quite literally a moving hill of water created by hurricane-strength wind) that obliterated the dunes but left the scrub relatively intact. At the time of writing this, beach mice are largely absent from dunes that do not have scrub behind them, suggesting that the scrub provides refugia from storms and acts as a source of colonists for the recovering dune habitat. Indeed, 47% of the mice captured at a site in the scrub after Hurricane Opal in 1995 had been previously captured in the dunes before the storm (Swilling *et al.* 1998), suggesting that mice take refuge in the scrub. Since all of the mice inhabiting the dunes before the storm could not be marked, it is not known how many of the mice found in the scrub after the storm originated in the dunes; 47% is a very conservative minimum estimate.

It is a likely possibility that when there has been a long period of time without a hurricane, dunes function as a population source and scrub functions as a sink. After a major hurricane, however, dune habitat is destroyed and beach mouse persistence critically depends on the dune habitat recovering before the populations taking refuge in the scrub decline to extinction. If dune habitat does indeed recover before the mice in the scrub are gone then they can recolonize the dunes, and populations will persist in relative safety until the next hurricane occurs.

Increasing the rate of recovery of dune habitat is a seemingly effective method of conserving beach mice. Planting native vegetation and establishing sand fences (sand fences are structures that intercept drifting grains of sand, thereby hastening dune formation) on beaches affected by hurricanes are effective methods of accelerating the recovery of dune habitat. Dune restoration is practiced both inside and outside of the range of beach mice. It is somewhat ironic that beach mice benefit from a practice partially motivated by attracting tourism, which increases the demand for hotels and condominiums constructed upon their habitat. Nonetheless, it is not apparent that increasing the rate of dune recovery is more effective than decreasing the rate of population decline in the scrub. The latter could be accomplished through supplemental feeding or vegetation management. The most effective conservation strategy will depend on the relative contribution of source habitat recovery rate and the rate of population decline in the sink to population viability, as well as the relative cost of implementing each strategy. Another part of this management dilemma concerns the frequency of hurricanes. Again, it is not apparent whether the relative effectiveness of managing dunes, rather than scrub, changes as hurricane frequency changes. The effects of global climate change on hurricane frequency and intensity is a very active area of research (Emanuel 2005; Landsea 2005; Trenberth 2005; Webster *et al.* 2005).

Riparian vegetation and floods

The banks of rivers and streams have unique substrate, moisture, nutrients, and light availability compared with the landscapes in which they are embedded. These physical conditions permit vegetative communities that are often quite distinct from their upland counterparts (Malanson 1993). Indeed, riparian zones are often afforded special legal protection from activities such as logging and farming that would otherwise alter plant composition. Such regulations are intended to conserve biodiversity and provide cleaner water.

We can readily imagine the source–sink dynamics of riparian plants. Species that are well adapted to the physical conditions adjacent to rivers and streams thrive; riparian zones function as sources for these plants. Propagules from such plants are dispersed into the uplands where they are ultimately outcompeted by other species. Thus, upland areas function as sinks for riparian-adapted plant species.

This source–sink relationship can be severely disrupted by a flood. Plants in the "good" habitat along the river bank will be killed by a flood and the habitat structure (substrate, moisture, and nutrients) will be temporarily altered. Plants in the "poor" upland habitat will be relatively unaffected by the flood, and can be the source of immigrants to the recovering riparian habitat. When viewed over a time scale of a couple of hundred years, the presence of some riparian-adapted plants along a particular river may be credited to upland "sinks."

Parks and fires

If unprotected habitats surrounding a park are sufficiently degraded, then such areas may function as population sinks. Habitats inside the protected area may normally function as population sources, but if fires have historically been suppressed, then potential fuel materials can accumulate, and the ensuing fires can be catastrophic. In this case, the putative source may be destroyed and long-term metapopulation persistence will be contingent upon recolonization from fire-resistant sinks outside the park into the recovering sources.

Model

Various model formulations could have been used to explore the interactions among (1) the rate of habitat recovery in the source, (2) the rate of population decline in the sink, and (3) the frequency of disturbances affecting the source. We chose to use population viability (i.e., the probability of persisting over a given time period) as a response variable because of its (sensible) appeal to conservation biologists. Modeling population viability typically requires explicit consideration of demographic and/or environmental stochasticity because these often play a prominent role in extinctions. We also wanted a demographic model that tracks population size so that we could efficiently simulate immigration and emigration of individuals between populations. We further sought a model that would allow us to "manipulate" carrying capacity in order to simulate and evaluate the effectiveness of active (i.e., human-mediated) habitat recovery. Given these constraints, we elected to use a demographic submodel described by Morris and Doak (2002: 127–133) but modified to suit our scenarios of interest. The demographic model simulates population size through time with both demographic and environmental stochasticity.

The mean growth rate over each time step is computed using a familiar logistic equation for the source population:

$$\mu_{\lambda}^{source} = e^{r\left(1-\frac{N_t^{source}}{K_t}\right)} \tag{7.1}$$

where r is the intrinsic growth rate of the source population, which must be positive because births must exceed deaths if the habitat is a source. For all simulations, we fixed this value at 0.1. N_t is the current population size in the source and K_t is the current carrying capacity. With every time step, K increases by a given amount, hereafter denoted ΔK_{source}. Thus, for the sake of simplicity, the value of K is assumed to increase monotonically through time, and the rate of change is given by ΔK_{source}. The value of ΔK_{source} is systematically varied because the rate of habitat recovery in the source is one of three parameters being explored in this study. ΔK_{source} is varied from 25 to 250 individuals/year in increments of 25. There is no need to specify an upper limit to K because it is constrained by disturbances: K is reset to 500 individuals every time a disturbance occurs.

The sink population is not modeled with density dependence; rather, the population experiences exponential decline

$$\mu_{\lambda}^{sink} = e^{r} \tag{7.2}$$

when r, the growth rate of the sink, is negative. Since the population growth rate in the sink is another parameter under examination, it is also systematically varied. The growth rate in the sink varies from 0 to –0.1 in increments of 0.01.

Note that Eqs. (7.1) and (7.2) yield the mean growth rate for source and sink populations, respectively. Since we wish to simulate stochasticity (in order to generate a probability of persistence), it is necessary to also compute the variance around these means. This is done by computing both demographic and environmental contributions to the variance of mean growth rate using:

$$\sigma_\lambda^2 = \frac{1}{N_t} + 0.05. \tag{7.3}$$

Note that the first term on the right side of this equation contains population size in the denominator, indicating increased variance as population size decreases. This simulates demographic stochasticity. The second term on the right side of Eq. (7.3) is a density-independent constant that simulates environmental stochasticity.

With target values for the mean and variance of growth rates in both source and sink populations on hand, it is necessary to convert these quantities to an exponential scale because this is how random deviations subsequently enter the model. The mean and variance of a log-normal scale are related to the mean and variance of a normal variable being exponentiated by

$$\mu_r = \log(\mu_\lambda) - \frac{1}{2}\log\left(\frac{\sigma_\lambda^2}{\mu_\lambda^2} + 1\right) \tag{7.4}$$

$$\sigma_r = \sqrt{\log\left(\frac{\sigma_\lambda^2}{\mu_\lambda^2} + 1\right)}. \tag{7.5}$$

Projecting population size one time step forward is performed by

$$N_{t+1}^{source} = N_t^{source} e^{(\mu_r + (\sigma_r \varepsilon))} + \left(\alpha N_t^{sink}\right) - \left(\beta N_t^{source}\right) \tag{7.6}$$

$$N_{t+1}^{sink} = N_t^{sink} e^{(\mu_r + (\sigma_r \varepsilon))} - \left(\alpha N_t^{sink}\right) + \left(\beta N_t^{source}\right). \tag{7.7}$$

Since ε is a zero-mean, unit-variance random normal deviate in an exponent, the growth rates are log-normal (a desired outcome because this is frequently found in nature). μ_r is used in the exponent so that the mean growth rate matches μ_λ. Migration is simulated by moving a fixed proportion of individuals from the sink into the source (α), and from the source to the sink (β).

These procedures produce floating point variables for population size, which is biologically unrealistic (populations are composed of discrete individuals, not fractions thereof) and can result in a significant underestimation of the probability of extinction. Thus, a "floor" command is used at every iteration that eliminates all fractional individuals and restores population size to an integer value.

We simulated disturbances with a Poisson process. The mean return interval of the disturbance was systematically varied from 10 to 20 years in 1-year increments. When a disturbance occurs, 80% of the source population is immediately lost and the carrying capacity of the source is set to 500 individuals (thereafter increasing according to ΔK_{source}). The sink population is unaffected by disturbances. Both populations were initiated with 200 individuals and simulated for 100 time steps (years). If no individuals remained at the end of the simulation, an extinction event was recorded. The probability of extinction was estimated by repeating each simulation 1,000 times.

We ran two sets of simulations with different source–sink migration rates in order to assess the robustness of our results. Our focal simulations were run with 6% of the sink population moving into the source (α, Eqs. 7.6 and 7.7), and 8% of the source population moving into the sink (β, Eqs. 7.6 and 7.7). These migration rates were increased to 10% and 12%, respectively, in the second set of simulations. We also performed separate simulations of systems both with and without an associated sink in order to provide a first approximation of the importance of sinks in the simulated environment. Unlike all other simulations that return the probabilities of persisting for 100 "years," these populations were simulated until extinction occurred, thereby yielding persistence time.

Results

In our model, the presence of a sink population greatly enhances the persistence of the source population (Fig. 7.2). Indeed, median source population persistence time increases from 38 "years" (SD = 14, N = 1,000) when there is no associated sink (but individuals continue to emigrate from the source) to 135 "years" (SD = 102, N = 1,000) when a sink is present. These results are qualitatively consistent with previous findings that sinks enhance metapopulation persistence (Gyllenberg *et al.* 1996; Foppen *et al.* 2001). Indeed, since our putative source population cannot persist without the presence of the sink, these results agree with other findings that two sinks can sustain one another (Jansen and Yoshimura 1998). Results from the systematic variation of three model parameters are shown in Figure 7.3. These results were obtained with $\alpha = 6\%$ and $\beta = 8\%$. Not surprisingly, the probability of metapopulation extinction decreases as

1. the rate of habitat recovery in the sink increases
2. the population growth rate of the sink increases
3. the frequency of disturbances decreases.

For all levels of disturbance frequency, the effect of increasing the rate of population growth in the sink from −0.1 to 0 had a much greater effect on the probability of extinction than increasing the carrying capacity (rate of habitat

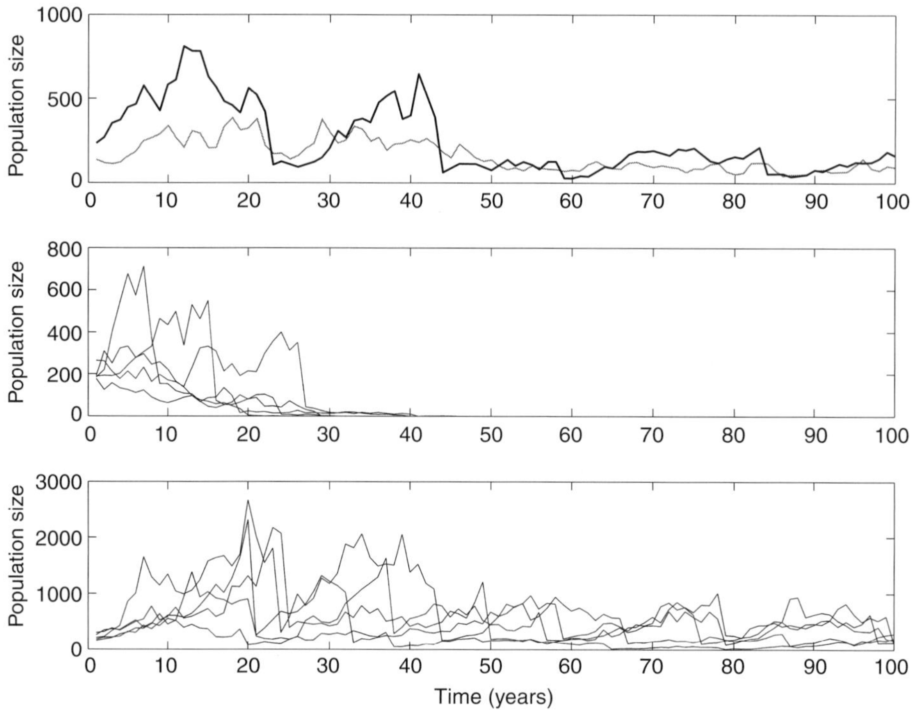

FIGURE 7.2.
Population trajectories when mean disturbance intervals is 20 years, habitat recovery rate in the source (ΔK_{source}) is 250, and population growth rate in the sink is −0.02. All populations were initiated with 200 individuals. The top panel shows a single trajectory of the source population (heavy line) and the sink population (light line). The middle panel shows five replicate source populations when 8% of animals emigrate from the source, but the sink is not present (all emigrants die and immigration from a sink does not occur). The bottom panel shows five replicate source populations that are connected to a sink. Here, 8% of animals emigrate from source to sink and 6% of animals in sink emigrate to source. Note different scales on the *y*-axis for the three panels.

recovery) of the source from 25 to 250 individuals/year. Indeed, the contour lines in Figure 7.3 are almost horizontal, indicating that the effect of the recovery rate of the source is small compared with the effect of the sink population's growth rate. The contour lines in Figure 7.3 become increasingly horizontal when moving from the upper-left panel to the bottom-right panel. This indicates that as disturbances become more frequent, the relative importance of source habitat recovery rate declines.

Figure 7.4 displays two extinction isosurfaces that differ only in the given source–sink dispersal rates. The surfaces depict the combination of three parameter values that result in a 90% chance of extinction (an isosurface can be thought of as a sheet inserted into a stack of panels from Figure 7.3, with the sheet connecting equal probabilities of extinction). Isosurfaces for smaller

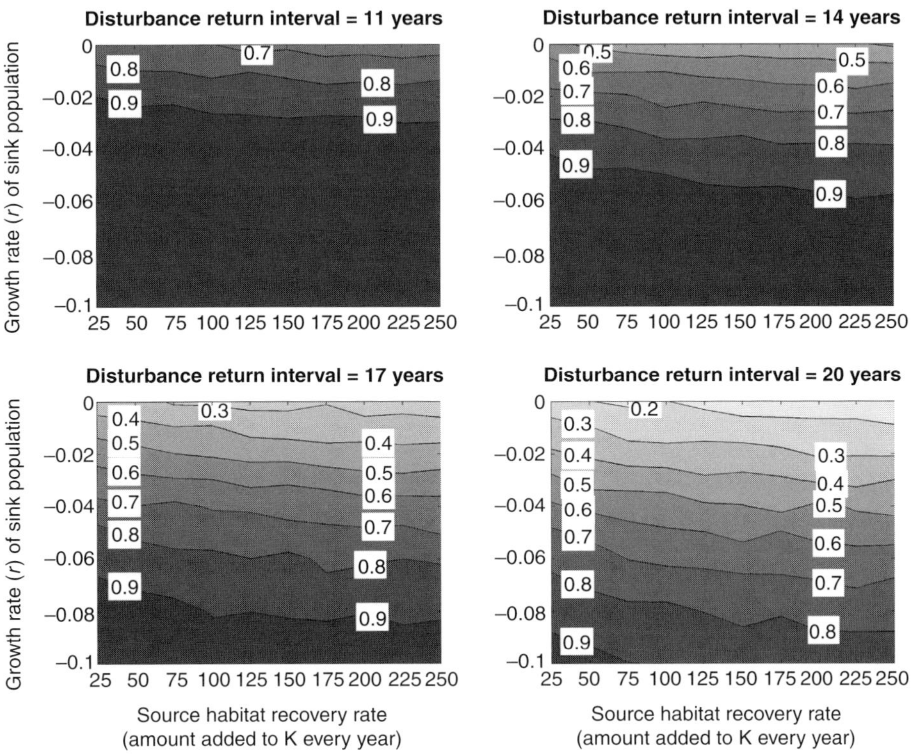

FIGURE 7.3.
Probability of extinction of a source–sink metapopulation.

extinction probabilities have similar shapes but occupy smaller slices of the parameter space, and are therefore not shown. Since the two surfaces differ only in the given dispersal rates between source and sink populations, their similar shape indicates that dispersal rates do not have a significant qualitative effect on the interactions between the parameters. Indeed, the position of the surfaces within the parameter space is very similar, indicating a small quantitative effect of dispersal rate.

Observing the bottom edge of a surface (i.e., a 10-year disturbance return interval) reveals that decreasing the rate of habitat recovery in the source can be "compensated" by a very minor increase in population growth rate in the sink in order to maintain a constant probability of extinction. The top edge of the surface suggests the same qualitative relationship. Quantitatively, however, there is a difference. The corner farthest from the viewer touches the edge of the box at approximately an 18-year disturbance return interval. As the recovery rate of the source decreases, a comparatively larger "compensation" in sink growth rate is needed at the 18-year mark. Stated simply, the surface is slightly twisted inside the box, revealing the slight three-way interaction between the factors. A very salient feature of Figure 7.4 is the approximately straight-line

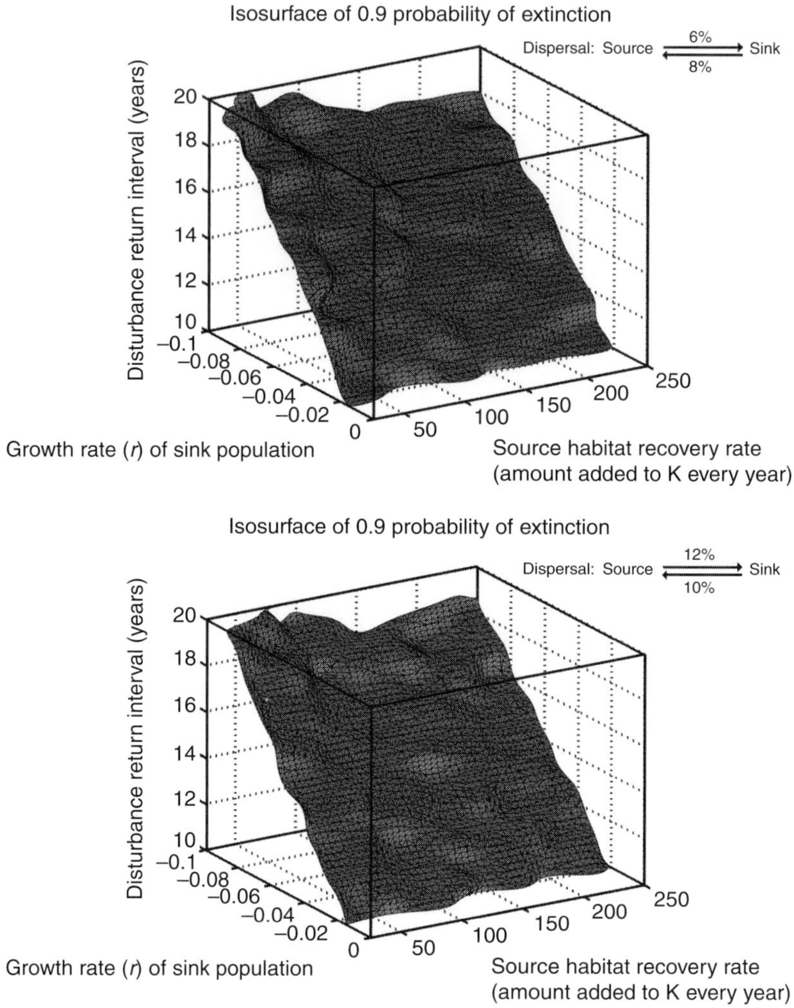

FIGURE 7.4.
The surfaces constitute combinations of three factors (model parameter values) that result in a 90% chance of extinction. Top and bottom panels differ in given dispersal rates between source and sink habitats.

relation between disturbance return interval and population growth rate in the sink. This is a significant interaction.

Discussion

The results presented here were obtained from a model of population dynamics that does not incorporate spatially correlated environmental stochasticity between source and sink populations. This is somewhat unrealistic in that both populations are within dispersal distance of one another and are therefore likely to experience similar environmental phenomena such as droughts, seed masts, etc. Therefore, we ran other simulations with varying degrees of correlated environmental stochasticity to determine whether this created qualitatively different results from those reported here. It did not. Thus, for clarity and simplicity, we focus on output derived from a model with uncorrelated environmental stochasticity.

The rate of habitat recovery in the source (ΔK_{source}) and the rate of population decline in the sink (r_{sink}) have different units, so asking which has a greater impact on the probability of extinction is, to borrow a colloquialism, a bit like comparing apples and oranges. Transforming these rates to a common currency facilitates management decisions. One could then ask whether a given sum of money will "buy more persistence" if it is spent on improving the source habitat recovery rate versus increasing the population growth rate in the sink habitat. However, since the relative amounts of ΔK_{source} and r_{sink} that can be purchased with a given amount of money will vary from system to system (and possibly the amount of money to be spent within a particular system), we will not endeavor to transform ΔK_{source} and r_{sink} to a monetary scale. Instead, we limit our discussion to the effects of varying the model parameters, and petition managers to consider how much a given sum of money can change the rate of habitat recovery in a source versus how much the same amount of money can decrease the rate of population decline in the sink.

Our analysis demonstrates that increasing the rate of population growth in the sink has a much larger effect on population persistence than increasing the rate of post-disturbance habitat recovery in the source. This occurs over a range of values that managers are likely to encounter and potentially manipulate. Although a final decision to invest in source management versus sink management will depend on how many units of ΔK_{source} and r_{sink} can be purchased, sink management seems prudent, since even a modest improvement of r_{sink} has a larger effect on population viability than a big improvement in ΔK_{source}.

The precise frequency distribution of disturbances is often very difficult to estimate, especially when disturbances occur on a multidecadal time scale. To further complicate matters, global warming is suspected to augment some disturbance regimes. Our model shows that, as disturbance frequency increases, the relative effectiveness of increasing source habitat recovery rate decreases. Thus, management aimed at increasing the population growth rate of the sink will become even more prudent if disturbance frequency increases.

Conclusions

The values of biological definitions go far beyond their mere aesthetic appeal. Scientific progress is contingent upon our ability to communicate with precision and robustness. Evolutionary biologists have been trying to define "species" for over a century, and unanimous acceptance of a definition does not seem forthcoming (Hey 2006). Ecologists face a similar predicament with "habitat." The difficulty in arriving at an agreed-upon definition may have more to do with the progress of science itself than differences in the preferences of individual scientists. The target of a scientific definition is constantly moving as new discoveries add nuance to existing ideas.

Pulliam (1996) noted that the theory of island biogeography and metapopulation concepts helped legitimize the notion of "unoccupied habitat," and that progress in source–sink dynamics is legitimizing notions that organisms can occupy "unsuitable habitat." We hope our results will help legitimize the notion of "critical sink habitat." Since proper management of imperiled species and associated legal conflicts can hinge on what is labeled "habitat," we hope that the definition of habitat will include the types of sinks we described. Morris (2003) provides an excellent definition of habitat that can facilitate the conservation of species in complex landscapes with source–sink dynamics.[1]

In this chapter, we described scenarios where sink populations rescue source populations, thereby saving a metapopulation from extinction. We demonstrated how improvement of sink habitat can be more effective in enhancing persistence than increasing the rate of recovery of source habitat. If society desires sustainable landscapes then we must be able to identify and protect sinks that rescue sources (see also Gimona *et al.*, Chapter 8, this volume). In some instances, this may not be achieved unless we adopt a definition of habitat that accommodates the critical role played by some sinks.

The source–sink dynamics described in this chapter occur over a time scale sufficiently long to include multiple disturbances. Thus, the consequences of degrading the quality of a sink that rescues a source may not become fully apparent for a very long time. The time lapse between sink degradation and metapopulation extinction constitutes a special form of the so-called "extinction debt" (Tilman *et al.* 1994), a term referring the time lag between habitat loss and fragmentation and ensuing extinction. Here, however, payment of the extinction debt occurs over a multiple-disturbance time period, which may be a substantially longer period than that caused by habitat loss and fragmentation. It is, therefore, conceivable that some recent extinctions and current metapopulation declines may be the result of events even more distant than commonly thought. As the first decade of the twenty-first century draws to a close with the release of this volume, it seems timely to note that it remains a matter of mere speculation to envisage the number of twenty-second century

[1] **Habitat:** A spatially bounded area, with a subset of physical and biotic conditions, within which the density of interacting individuals, and at least one of the parameters of population growth, is different than in adjacent areas (Morris 2003).

metapopulation extinctions that will ultimately be caused by events occurring in this decade. To end on a less macabre note, we offer hope in the form of a management directive: identify, protect and enhance sinks that rescue sources in dynamic environments.

Acknowledgments

This work was supported with the assistance of a grant to B. J. Danielson from the United States Fish and Wildlife Service. The ideas developed here were stimulated by discussions with a long line of beach mouse biologists including D. LeBlanc, W. Lynn, L. McNeese, J. Phillips, S. Sneckenberger, R. Tawes, and M. Wooten, and a number of other ecologists, most notably B. Watts. Four anonymous reviewers provided helpful comments on a previous version of this manuscript.

References

Emanuel, K. (2005). Increasing destructiveness of tropical cyclones over the past 30 years. *Nature* **436**(7051): 686–688.
Foppen, R. P. B., J. P. Chardon and W. Liefveld (2001). Understanding the role of sink patches in source–sink metapopulations: reed warbler in an agricultural landscape. *Conservation Biology* **14**(6): 1881–1892.
Gonzalez, A. and R. D. Holt (2002). The inflationary effects of environmental fluctuations in source–sink systems. *Proceedings of the National Academy of Science* **99**(23): 14872–14877.
Gyllenberg, M., A. V. Osipov and G. Södervacka (1996). Bifurcation analysis of a metapopulation model with sources and sinks. *Nonlinear Science* **6**: 329–366.
Hey, J. (2006). On the failure of modern species concepts. *Trends in Ecology and Evolution* **21**(8): 447–450.
Holt, R. D. (1985). Population dynamics in two-patch environments: some anomalous consequences of an optimal habitat distribution. *Theoretical Population Biology* **28**: 181–208.
Holt, R. D. (1997). On the evolutionary stability of sink populations. *Evolutionary Ecology* **11**: 723–731.
Holt R. D., M. Barfield and A. Gonzalez (2003). Impacts of environmental variability in open populations and communities: "inflation" in sink environments. *Theoretical Population Biology* **64**: 315–330.
Howell, A. H. (1909). Notes on the distribution of certain mammals of the southeastern United States. *Proceedings of the Biology Society of Washington* **22**: 55–68.
Howell, A. H. (1921). A biological survey of Alabama. *North American Fauna* **45**: 1–88.
Ivey, R. D. (1949). Life history notes on three mice from the Florida east coast. *Journal of Mammalogy* **30**: 157–162.
Jansen, V. A. A. and J. Yoshimura (1998). Populations can persist in an environment consisting of sink habitats only. *Proceedings of the National Academy of Science* **95**: 3696–3698.
Johnson, D. (2004). Source–sink dynamics in a temporally heterogeneous environment. *Ecology* **85**(7): 2037–2045.
Kingsland, S. E. (1995). *Modeling Nature: Episodes in the History of Population Ecology*. University of Chicago Press, Chicago, IL.
Landsea, C. (2005). Hurricanes and global warming. *Nature* **438**: E11–E13.

Malanson, G. P. (1993). *Riparian Landscapes*. Cambridge University Press, New York.
Morris, D. W. (2003). Toward an ecological synthesis: a case for habitat selection. *Oecologia* **136**: 1–13.
Morris, W. F. and D. F. Doak (2002). *Quantitative Conservation Biology: Theory and Practice of Population Viability Analysis*. Sinauer Associates, Sunderland, MA.
Pulliam, H. R. (1988). Sources, sinks, and population regulation. *American Naturalist* **132**(5): 652–661.
Pulliam, H. R. (1996). Sources and sinks: empirical evidence and population consequences. In *Population Dynamics in Ecological Space and Time* (O. E. Rhodes Jr., R. K. Chesser and M. H. Smith, eds.). University of Chicago Press, Chicago, IL: 45–70.
Roy, M., R. D. Holt and M. Barfield (2005). Temporal autocorrelation can enhance the persistence and abundance of metapopulations comprised of coupled sinks. *American Naturalist* **166**(2): 246–161.
Sneckenberger, S. I. (2001). Factors influencing habitat use by the Alabama beach mouse (*Peromyscus polionotus ammobates*). MS thesis, Auburn University, Auburn, AL.
Swilling, W. R. Jr., M. C. Wooten, N. R. Holler and W. J. Lynn (1998). Population dynamics of Alabama beach mouse (*Peromyscus polionotus ammobates*) following Hurricane Opal. *American Midland Naturalist* **140**: 287–298.
Tilman, D., R. M. May, C. L. Lehman and M. A. Nowak (1994). Habitat destruction and the extinction debt. *Nature* **371**: 65–66.
Trenberth, K. (2005). Uncertainty in hurricanes and global warming. *Science* **308**(5729): 1753–1754.
USFWS (US Fish and Wildlife Service) (2005). *Draft Revised Habitat Recovery Plan for the Alabama Beach Mouse (Peromyscus polionotus ammobates)*. US Fish and Wildlife Service, Atlanta, GA.
Webster, P. J., G. J. Holland, J. A. Curry and H. R. Chang (2005). Changes in tropical cyclone number, duration, and intensity in a warming environment. *Science* **309**(5742): 1844–1846.

ALESSANDRO GIMONA, J. GARY POLHILL AND BEN DAVIES

8

Sinks, sustainability, and conservation incentives

There is always an easy solution to every human problem – neat, plausible, and wrong. H. L. Mencken (*The Divine Afflatus*)

Summary

Sustainability of agro-ecosystems can be achieved if farming systems are both ecologically sound and economically viable. Therefore, it is critically important for conservation scientists to see wide-scale biodiversity policy as only one aspect of a complex socio-ecological system, in which independent land managers, subject to financial constraints, make choices subject to a range of objectives, most of which are only tangentially influenced by considerations of nature conservation. Conservation incentives are a policy instrument to reconcile conservation and land managers' objectives. Two broad approaches – payment for specific conservation actions (payment-for-activities), and payment for specific environmental outcomes (payment-for-results) – warrant particular attention. We investigate how undetected sinks might influence species persistence and richness in different policy and socio-economic contexts. To this end, we used a spatially explicit agent-based model of land use decision making, coupled with a spatially explicit metacommunity model. Our results show that, except when land managers are satisfied by low financial returns, the assumptions made by policy makers regarding habitat suitability of target species can have serious consequences on species' persistence when sinks are present but not detected. Sinks are more influential for species associated with habitat that does not tend to become rare, due to the profitability associated with land use conversion under free-market conditions. For other habitat types, habitat turnover due to market-driven land use change is more important for conservation.

Sources, Sinks and Sustainability, ed. Jianguo Liu, Vanessa Hull, Anita T. Morzillo and John A. Wiens. Published by Cambridge University Press.

Background

Sustainable agriculture must be ecologically sound, economically viable, and socially responsible (Ikerd 2006). In many parts of the world, low-intensity farming has been ecologically sound, and important for the conservation of valuable habitats and species. A drive to increase economic returns, however, has led to the expansion and increasing intensity of modern agriculture, which has profoundly influenced landscapes and biodiversity worldwide (e.g., Mattison and Norris 2005; Secretariat of the Convention on Biological Diversity 2006). In most developed countries, intensive agriculture has become the dominant land use, landscapes in high-intensity areas have been greatly simplified in structure, and most habitats have been replaced by rather uniform arable fields or improved grassland (e.g., Robinson and Sutherland 2002). Also, agriculture has become more specialized and spatially segregated (Warren *et al.* 2008). As a consequence, habitats have been lost, or their quality degraded, and functional landscape connectivity compromised, in large areas, due to fragmentation (e.g., Fischer and Lindenmayer 2007). At the landscape scale, the decline of farmland species such as birds (Gregory *et al.* 2004), bees (Kwaiser and Hendrix 2008), and plants (Hald 1999) is due to the combined effect of this marked loss of functional heterogeneity and of more intensive land management practices (e.g., Benton *et al.* 2003).

Given the extent of the areas involved, reserve-based conservation needs to be integrated with conservation of the "wider landscape." From a theoretical point of view, there are two principal reasons why a reserve-based conservation paradigm might not be sufficient to save a large number of species from extinction.

1. Firstly, a static conservation paradigm, based on saving some important areas, might – by itself – not be effective in the long run, due to the dynamic nature of landscapes.
2. Secondly, there is a risk that such areas might not be able to accommodate viable populations, because habitat area is an important functional property of landscapes, to which species richness is related (e.g., Rosenzweig 2003).

The conservation status of many species could be improved by spatially targeting conservation measures in agricultural landscapes, and by recognizing their highly dynamic state.

Conservation requirements, however, need to be reconciled with the fact that land managers' decisions are mainly financially – rather than biodiversity – oriented. This can be addressed through public policy, whose fundamental purpose is to resolve conflicts between interests of individuals and the goals of society (Ikerd 2006). Ecological sustainability can, in principle, be achieved, but farming systems must also be made economically viable if they are to be sustainable. The public ultimately pays for the cost of conservation-friendly policies, either through availability and prices of agricultural products, or through government fiscal policy and expenditure for conservation incentives. Therefore, it is critically important for conservation scientists to see wide-scale biodiversity policy as only one aspect of a complex socio-ecological system, in which independent land managers make choices subject to a range of objectives, most of which are only tangentially influenced by considerations of nature conservation. Conservation incentives are therefore aimed at enhancing the sustainability of agro-ecosystems by paying for activities and land use practices that are thought to enhance the provision of biodiversity and other ecosystem services. However, for reasons which are often unclear, incentive schemes are not always effective, and the response varies among taxa (Kleijn and Sutherland 2003).

Particular challenges come from the fact that many populations have spatial dynamics at scales wider than the local management area (e.g., farm), so habitat value might be context-dependent (e.g., Robinson *et al.* 2001; Concepción *et al.* 2008), and connectivity time-dependent (Clergeau and Burel 1997). Also, some land uses can constitute a demographic sink (Pulliam 1988) for organisms such as birds (e.g., Hatchwell *et al.* 1996; Chamberlain and Fuller 2000; Arlt and Pärt 2007), small mammals (Tattersall *et al.* 2004), butterflies (Boughton 1999; Ockinger and Smith 2007), and bees (Ockinger and Smith 2007).

The effect of sinks

The source–sink concept has been incorporated into conservation literature and management for two decades (e.g., Meffe and Carroll 1997) and has been used to explain the presence of species in low-quality habitat (Duguay *et al.* 2001; Tittler *et al.* 2001). Spatial linkages between local communities are thought to have strong effects on species sorting and coexistence (see also Benkman and Siepielski, Chapter 4, this volume). This is made more complex by the directionality of fluxes, as in source–sink cases.

Theory regarding sink effects in simple systems is freely available (e.g., Dunning *et al.* 1992; Doebli and Ruxton 1998; Amarasekare and Nisbet 2001; Gundersen *et al.* 2001; Namba and Hashimoto 2004), and the role of environmental variation and its temporal pattern has been recognized (Gonzalez and Holt 2002; Gonzalez and De Feo 2007). However, there has been no theoretical attempt (as far as we are aware) to explore the dynamic effect of sinks in systems where demographic processes are interacting in rather complex ways with socio-economic processes which drive the temporal availability of habitat. Therefore for human-influenced systems (an increasing proportion of the Earth's surface), the theoretical long-term consequences of the existence of sinks have not yet been explored adequately because most models ignore the socio-economic factors that provide the context in which population dynamic processes that drive habitat dynamics. This makes it more difficult to generate hypotheses explaining why policy measures, such as incentive schemes, are effective or otherwise.

A theoretical question that has received virtually no attention is what happens if conservation incentives – due to imperfect knowledge – promote the creation of habitats or the adoption of land uses that are actually sinks for the (set of) species of interest and whether this can be avoided by rewarding outcomes (species) rather than activities.

Studying this problem is interesting for the application of source–sink theory to conservation in managed landscapes, because it can help focus both theoretical and empirical efforts on situations in which sink effects are likely to be relatively important compared with other factors. We have developed tools to begin such an exploration.

Research methods

A coupled human–natural system model

In agro-ecosystems, the landscape structure, which influences species diversity, emerges from the interaction of biophysical constraints and individual decisions influenced by factors such as crop prices, management input costs, and economic aspirations. Agent-based modeling seems a natural tool with which to model the human portion of such systems, as analytical models would be much more difficult to formulate and solve. Arthur *et al.* (1997) have listed several properties of complex adaptive systems, of which we see this as an example, that pose a challenge for traditional mathematical modeling techniques.

It has further been argued that agent-based modeling is particularly well suited to studying coupled human–natural systems (Hare and Deadman 2004). Boulanger and Bréchet (2005), highlighting the promise of agent-based modeling in the study of sustainable development, note that it allows an intuitive representation of the environment and the embedding of agents within it. Bousquet and Le Page (2004) conclude that researchers in ecology and the social sciences can use agent-based modeling to study the interactions between spatial, network and hierarchical levels of organization, a view supported by Huigen (2004).

FEARLUS (Polhill *et al.* 2001) is an agent-based modeling system designed to build models for studying land use change. This is a modeling tool flexible enough to capture differences between individual land managers but still able to produce relatively simple general models. It has been used to study various aspects of boundedly rational land use decision-making algorithms and their interaction with differing degrees of spatiotemporal heterogeneity in factors influencing economic returns, including imitation (Polhill *et al.* 2001) and aspiration (Gotts *et al.* 2003). We have coupled this model with a metacommunity model which is an extension of the stochastic patch occupancy model (SPOM) framework (Moilanen 1999, 2004). See Box 8.1 for details.

Box 8.1

The sequence of events in FEARLUS is depicted on the left-hand side in Figure 8.B1.1, each cycle of which is intended to represent a year. (In what follows, we adopt the convention of giving entities in the model upper-case initial letters.) Starting from the top, Land Managers use their Land Use Selection algorithm to decide the Land Use of each Land Parcel they own. The Economy and Climate for the Year are then obtained (these are effectively exogenous time series) and, together with the Biophysical Characteristics of the Land Parcels (also exogenous, but varying spatially rather than temporally), are used to compute the Yield and Economic Return to the Land Manager (the latter in the "Harvest" step). A Government Agent (an optional component of the model) may then make some observations and issue grants or fine Land Managers according to Government Policy. After the Harvest, an optional Approval phase takes place, in which Managers may use rules to Approve or Disapprove of their neighbors for various reasons. Managers then learn from their experience of different Land Uses. At the end of the Year, those Managers with negative accumulated wealth in their Account are regarded as being bankrupt, and must sell all their Land Parcels to solvent neighbors or to in-migrant Managers. Parcels are sold in an auction, and Managers have rules determining how much to bid and which Parcels to bid for (Polhill *et al.* 2008).

The overall effect is to create an evolutionary environment in which Managers using more successful decision-making algorithms tend to accumulate more Land Parcels, and those using less successful algorithms tend to go bankrupt. These dynamics need not necessarily apply, however. If the amount of money required to prevent loss (a parameter of the model) is too low, Managers with even relatively poor decision-making algorithms will stay in business. Likewise, if this parameter is too high, it is not possible to

Box 8.1 (*Cont.*)

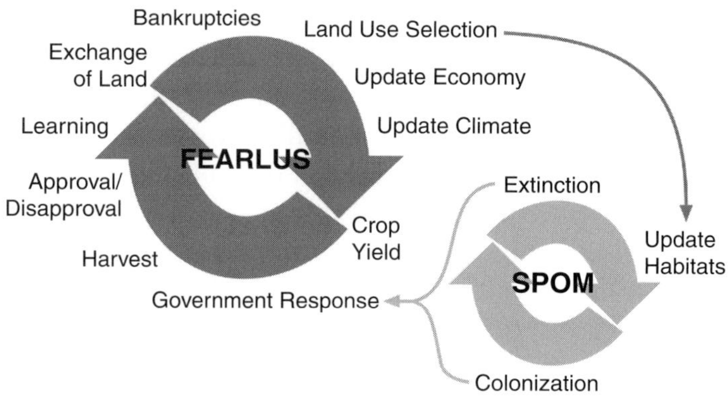

FIGURE 8.B1.1.
The sequences of events in the agent-based model of land use change (FEARLUS) and the species metacommunity model (SPOM), and how each influences the other.

make enough money to stay in business, no matter how good the algorithm. Rapidly changing Climate or Economy can also make it difficult to stay in business, particularly if conditions in one Year offer no predictive basis for conditions in the following Year.

The right-hand side of Figure 8.B1.1 depicts the sequence of events in the SPOM. The SPOM is a metacommunity ecological model, which simulates populations on a lattice (in this case toroidal) in which each cell can be thought of as a landscape Patch. The SPOM models the presence or absence of Species on Patches of land, rather than recognizing individuals (as in classic individual-based ecological modeling). This is similar to Moilanen's (2004) SPOMSIM, in that it uses the same equations to model extinction and colonization, but has a number of extensions, the most significant of which are to allow for multiple Species and interactions between them, to allow for Habitat preferences, and modeling the effect of sink Habitat. There is a many-to-many relationship between Species and Habitats in the model; a Species may be configured to survive on several Habitats if it is adaptable, and a Habitat may be suitable for several Species. Each Year, the Species occupying a Patch compute a Local Extinction Probability, which is the chance of the Species ceasing to occupy the Patch. Patches providing Habitat for – but not occupied by – a Species, also represent a Colonization Probability, which determines the chance that the Patch will become occupied by the Species in that Year. The SPOM also provides the (optional) possibility that the availability of Habitats on each Patch may change once per

Year, before Extinctions and Colonizations are computed again in the next cycle.

Whilst the Local Extinction Probability depends on parameters of the Species (and, optionally, also on the presence of competitor Species) and the amount of Habitat on the Patch, the Colonization Probability depends also on the occupancy of the Species in other Patches in the Environment. As a consequence, the long-term survivability of a Species depends on the pattern of Habitats at the landscape scale.

The fact that FEARLUS and the SPOM operate at the same spatial and temporal scale, and at similar levels of abstraction, allows them to be integrated more easily. Land Uses chosen by Land Managers in FEARLUS translate into Habitats in the SPOM.

A feedback to Land Managers is provided through the Government Agent, which pays for conservation incentives.

Scheduling is indicated by the two sets of arrows in Figure 8.B1.1. The Update Habitats step in the SPOM takes place after the Land Use Selection step in FEARLUS, and the Extinction and Colonization steps in the SPOM following the Habitat change take place before the Government Response step in FEARLUS.

Together, FEARLUS and the SPOM present a socio-ecological system in which Species distributions adjust to changes in Land Use arising from Land Managers' decisions and any demographic changes in Land Manager populations. Land Managers' decisions, insofar as they depend on Government grants or fines, are affected in turn by Species distributions. This creates a modeling system in which some explorations of biodiversity Policy can be made.

In the description below we use capital initial letters to indicate entities that are part of our model (e.g., Land Managers).

Land Managers' decisions, incentive strategies and sinks

This coupled modeling tool permitted us to build stylized models to investigate the space-time dynamics of a socio-economic system. In this particular application, the objectives were to investigate the interaction between undetected sink Habitats and incentive-based Policies.

We set up simulations of a relatively simple system, where some Species of conservation interest share Habitat with other "less interesting" Species and in which conservation incentives reward either the choice of Land Uses providing the appropriate Habitat, or the occurrence of target Species on a Parcel of land.

Species, Habitats and landscape

Land Use here incorporates notions both of crop and management *practices* (as these partly determine the level of intensity) which have an effect on Habitat suitability for Species of interest. In this model system, we made available two main land cover types "G" and "A," at three levels of intensity (from "1" – low, to "3" – high) giving Land Uses labeled GL1, GL2, GL3, AL1, AL2, and AL3. Six corresponding Habitats were specified in the SPOM model: GH1, GH2, GH3, AH1, AH2, and AH3. The more intense the Land Use, the fewer Species were able to use the corresponding Habitat.

One possible interpretation of this model configuration (although by no means the only one) is a landscape where arable and grassland patches coexist due to the land use system, with the grassland patches hosting a wider pool of species. Practices of increasing intensity are different in the two types of patches but both have the effect of excluding some of the species.

Ten Species were specified, G1–6, A1–3, and a competitor Species C1, which was able to outcompete and exclude some of the Species (G1–3) if present on the same Patch. This was intended to simulate a situation in which a complete lack of management would result in lower alpha diversity (patch species richness) with respect to a moderately intense regime. This is the case for many grassland systems, where grazing can promote diversity (e.g., Wallis de Vries *et al.* 1998). The Species parameters were such that they represent functional groups tolerating an increasing amount of land use intensity.

All Species were characterized by a Dispersal Distance, a Probability of Extinction, given Habitat occupied, and a Probability of Colonization which depended on the configuration of occupied Patches in the landscape (see Box 8.2). Table 8.1 shows which Species could live on which Habitats. To create a potential refuge from competition, Land Uses GL1, GL3, AL1, AL2 and AL3 provided Habitats GH1, GH3, AH1, AH2 and AH3, respectively, while Land Use GL2 provided two Habitat types: GH1 (20% of the Patch area) and GH2 (80% of the Patch area). Only GH1 was available to the superior competitor C1. The relatively more vulnerable Species were G5, G6 and A2, A3, having more specialized Habitat requirements and shorter average dispersal distances.

Box 8.2

Dispersal, Colonization, and Local Extinction are modeled as in Moilanen (2004). Local Extinction can also be caused by competition (see below). All Patches have unit area.

For each Species *s* and Patch *i* the following equations were used:

Connectivity

$$S_{is}(t) = A_{is}^{c}\sum_{j\neq i} O_{js}(t)D(d_{ij},\alpha_s,\ldots)A_{js}^{b}$$

where j denotes Patches that are not the focal Patch I, c and b are parameters, A_{is} is the available area (i.e., the amount of Habitat area made available by the present Land Use for Species s), $O_{js}(t)$ is an indicator variable assuming the value 1 if the Patch is occupied by the Species, 0 otherwise. $D_s(d_{ij}, \alpha_s) = \exp(-\alpha_s d_{ij})$, and d_{ij} is the distance between two Patches.

Colonization

At each time step t

$$C_{is}(t)=\frac{[S_{is}(t)]^2}{[S_{is}(t)]^2+y^2}$$

where $S_{is}(t)$ is the connectivity of Patch i at time t, for Species s and y is a parameter.

Local Extinction

$$E_{is}=\frac{\mu_s}{A_{is}^{x}}$$

where μ is the Extinction Probability of a Patch of unit size, A is the available area of the Patch, and x is a scaling parameter (always set to 1 in our simulations).

In addition (in this exercise) when a competitor is present this can cause Local Extinction of inferior competitors within n time steps (here $n = 3$).

The parameter values used in the simulation are shown in Table 8.B2.1.

TABLE 8.B2.1. Parameter values used in the simulation.

Species	c	B	α	μ	β
G1	1.0	1.0	0.8	0.1	1.0
G2	1.0	1.0	0.9	0.1	1.0
G3	1.0	1.0	1.1	0.1	1.0
G4	1.0	1.0	1.3	0.1	1.0
G5	1.0	1.0	1.3	0.1	1.0
G6	1.0	1.0	1.3	0.1	1.0
A1	1.0	1.0	1.3	0.1	1.0
A2	1.0	1.0	0.9	0.1	1.0
A3	1.0	1.0	0.8	0.1	1.0
C1	1.0	1.0	1.3	0.05	1.0

TABLE 8.1. *Species–Habitat matrix used in the demonstration experiments. Habitats are the rows and Species are the columns. A letter "Y" in a cell indicates that the Species can occupy a Patch having the Habitat in the row, and disperse from it. An "S" in a cell indicates that the Habitat in the row can be configured to be a sink Habitat; open to occupation by the column Species, but the Species cannot disperse from it. In runs where sink Habitats were not specified, "S" behaves as "Y."*

	Species									
Habitat	G1	G2	G3	G4	G5	G6	A1	A2	A3	C1
GH1	Y	Y	Y	Y	Y	Y	–	–	–	Y
GH2	Y	Y	Y	Y	Y	–	–	S	S	–
GH3	Y	Y	Y	–	–	–	–	–	–	–
AH1	–	–	–	–	S	S	Y	Y	Y	–
AH2	–	–	–	–	–	–	Y	Y	–	–
AH3	–	–	–	–	–	–	Y	–	–	–

All Species were assumed to have relatively short dispersal distances due to the nature of the study, which was aimed at investigating the vulnerability of Species to fragmentation. An average Dispersal Distance (i.e., $1/\alpha$, see Table 8.B2.1) between 0.8 and 1.3 cells was therefore assumed.

For Species with higher dispersal distance, 99% of dispersal events were set within five cells, while they were within three cells for Species with the shorter dispersal distance. This choice of dispersal parameters is therefore oriented to represent species such as herbaceous plants (excluding weeds), small mammals and less mobile invertebrates, especially non-flying taxa.

Local (intrinsic) Extinction rate was set to 10% per Year. This is within the range of values reported by reviews from, e.g., Fahrig and Merriam (1994) and Schoener (1983). The first study, regarding plants and animals, reported rates between 5% and 30%; the second reported rates of 1–10% for vertebrates and plants, and 10–100% for invertebrates. Given that all Patches had the same (unit) area, we set parameters scaling immigration and emigration with area equal to 1 (c and b in Table 8.B2.1). Landscape structure was determined at each time step by the collective decisions of Land Managers, given their objectives.

All runs were initialized with a random distribution of 50% AL1 and 50% GL1, and maximum Species occupancy (this was to maximize the probability of Species surviving while the initial Land Managers were still learning), and then run for 300 time steps.

Sinks

Additional Habitat was provided for the vulnerable Species, simulating a situation in which the populations of some A*x* Species are able to survive on

Habitat types provided by GL*y* Land Uses, and vice versa. In half of the simulation runs, however, these Habitat types were sinks for those Species. This is indicated as S in Table 8.1.

Because our population model has no internal Patch dynamics, but only tracks occupancy, sinks are simulated as Patches that do not contribute to colonization of other Patches in the next time step. They are therefore "black-hole" sinks, simulating a situation in which individuals do not contribute to the next generation. Neither the Government Agent issuing a financial reward, nor the Land Managers were aware of the existence of sinks.

Land Managers

For these experiments, Land Managers were implemented with a satisficing approach (Simon 1955) to decision making (rather than aiming at making the maximum possible profit). Satisficing is a commonly used heuristic approach to representing human decision making. Departure from profit maximizing is known to occur. Parker *et al.* (2007) cite evidence of various factors that lead to farmers not making fiscally optimal decisions, such as meeting subsistence requirements and cultural norms. To this may be added questions of identity as a farmer, from qualitative social research (Burton and Wilson 2006), in which "keeping the name on the farm" and being recognized by one's peers as a "good farmer" (Burton 2004) are also motivating influences on decision making orthogonal to purely pecuniary concerns.

Land Managers reviewed their choice of Land Uses on all their Parcels if the mean Profit per unit area did not meet their financial Aspirations for a specified number of consecutive Years (this number was taken from a uniform distribution in the range 0–9). When deciding whether to change Land Use, Managers consulted their experience, i.e., employed case-based reasoning (Aamodt and Plaza 1994) to choose a Land Use based on their expectations of the Climate and Economy in the coming Year, and their experience of the Land Use in the past, which includes its Economic Return. Managers with no experience of a Land Use were given the opportunity to ask neighbors for their experience of it, and use that as a basis for decision making. If neighbors had no experience of a Land Use either, then Managers assumed that that Land Use would meet their Aspirations; when other Land Uses had poorer expected outcomes, this allowed the Land Managers to experiment. Expected outcomes (Profit) were obtained for each Land Use, and a selection made at random from those Land Uses with equal maximum expected Profit. Land Managers are therefore satisficing regarding the decision to change Land Use, but maximizing once they have decided to change. Since Profit includes any subsidies from the Government, Policy has an influence on Land Use Selection by Managers.

TABLE 8.2. *Gross Economic Return per unit area to Land Managers for each Land Use under an unchanging Economy (denoted by U).*

Land Use	Yield	Price U	Return U
GL1	4	5.5	22
GL2	5	5.5	27.5
GL3	6	5.5	33
AL1	4.5	5.0	22.5
AL2	5.5	5.0	27.5
AL3	6.5	5.0	32.5

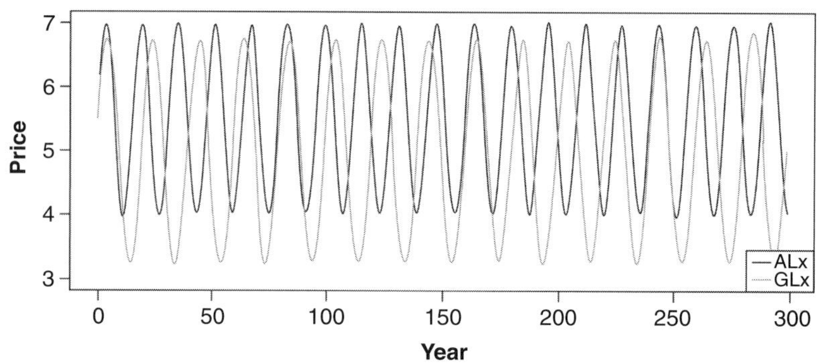

FIGURE 8.1.
Time series of price fluctuations per unit Yield in the Economy for the "G" (light curve) and "A" (dark curve) Land Uses with a variable Market.

The Economic Returns for Land Uses depend on their Yield per unit area (which increases with intensity), and Price per unit Yield. We simulated two contrasting Price time series for the Economy: one unchanging, the other variable. The settings for the unchanging Economy are shown in Table 8.2. For the variable Market, we used an approximately sinusoidal time series with a period of 16 Years, an amplitude of 1.5 and a mean of 5.5 for the "G" Land Uses, and a period of 20 Years, an amplitude of 1.75 and a mean of 5.0 for the "A" Land Uses. Figure 8.1 shows the time series of price fluctuations per unit Yield in the Economy, with a variable Market, and Table 8.3 shows the minimum and maximum Gross Economic Returns. The Climate and Biophysical Characteristics were kept constant.

Box 8.3 briefly characterizes FEARLUS and shows a summary of the parameterization used to characterize Land Managers' behavior, and their consequences for the landscape-scale abundance of Habitats, as well as its justification.

TABLE 8.3. *Gross Economic Returns per unit area to Land Managers for each Land Use under a variable Economy (denoted by V).*

Land Use	Yield	Price V (min)	Return V (min)	Price V (max)	Return V (max)
GL1	4	3.0	12.0	7.0	28.0
GL2	5	3.0	15.0	7.0	35.0
GL3	6	3.0	18.0	7.0	42.0
AL1	4.5	3.25	14.625	6.75	30.375
AL2	5.5	3.25	17.875	6.75	37.125
AL3	6.5	3.25	21.125	6.75	43.875

Box 8.3

FEARLUS has been classed as a "typification" by Boero and Squazzoni (2005), i.e., a model focused on a particular class of phenomena, in contrast to "case-based models" fitted to a specific scenario. In such a modeling paradigm, one is concerned more with stylized scenarios of "life as it could be" than with particularities of "life as it is." Thus the absolute values of the parameters are less important than the dynamics caused by the relationships between them. Before running the experiments, exploratory runs were made to find sets of parameters covering the continuum of dynamics in the simulated social and ecological systems: from situations in which there is no economic pressure on Land Managers – and so no Land Use change, to those in which the pressure is so great that it is impossible for Managers to stay in business; from cases where all Species survive, to cases where all rapidly become extinct. From these explorations, we derive parameters for the reported experiments trading off covering the full range of dynamics with demands on computational power by eliminating those generating particularly unrealistic outcomes (e.g., where the bankruptcy rate is too high).

The Profit returned to Land Managers in each time step is given by the Gross Economic Return per unit Yield, less a Break-Even Threshold (representing input costs), on each Land Parcel they own:

$$R_{m,t} = \Sigma_{p \in} P_m \left[g(E_t, U_{p,t}) y(U_{p,t}) - b \right]$$

where $R_{m,t}$ is the profit of Manager m at time step t, P_m is the set of Parcels owned by Manager m (iterated over by p), E_t is the state of the Economy at time step t (determined from the exogenous time series as input to the model: "flat" or "var2"), $U_{p,t}$ is the Land Use applied by Manager m to Parcel p in time step t, $g()$ is a lookup table returning the gross income per unit Yield

Box 8.3 (*Cont.*)

for the Economy state and Land Use (see Tables 8.2 and 8.3), $y()$ is a lookup table returning the Yield per Parcel of the Land Use (see Tables 8.2 and 8.3), and b is the Break-Even Threshold.

The Aspiration Threshold (ASP) was set to 0.5, 1, and 5 units. The Break-Even Threshold (BET) per unit area was set at 25 and 30. This number is subtracted from a Land Manager's Gross Economic Return. Thus, from Table 8.2, when there is an unchanging Economy ("flat"), GL1 and AL1 are never profitable for either BET used (i.e., they are less than 25), and for the higher BET, only the most intensive Land Uses, GL3 and AL3 are profitable (i.e., more than 30). By contrast, when the variable Economy ("var2") is used, from Table 8.3 we see that all of the Land Uses are sometimes unprofitable. Thus, the less intensive Land Uses, in particular, require Government incentives if Land Managers are to deploy them. The parameters used in the simulations are listed and described in Table 8.B3.1.

TABLE 8.B3.1. *Parameters explored in the simulation experiments.*

Parameter	Description	Values
Government	Specifies the set of rules used to reward Land Managers for biodiversity.	*RewardSpecies*: Give a reward to each Land Manager for the presence of any awardable Species on each Land Parcel; *RewardActivity*: Give a reward to each Land Manager for using any awardable Land Use on each Land Parcel.
Sink	Specifies whether or not certain Habitats are sinks for some Species (i.e., the Species can be present on Patches with this Habitat, but cannot disperse from them).	Yes; No
Market	Exogenous time series providing the level of Economic Return to Land Managers per unit Yield of each Land Use.	*Flat*: The Economic Returns for each Land Use do not change, and are 5.5 for the "G" Land Uses and 5.0 for the "A" Land Uses; *Variable*: The Economic Returns for each Land Use have a sinusoidal time series with a period of 16 Years,

Parameter	Description	Values
		amplitude of 1.5 and mean of 5.5 for the "G" Land Uses, and a period of 20 Years, amplitude of 1.75 and mean of 5.0 for the "A" Land Uses.
Break-Even Threshold	The amount of Economic Return per unit area that a Land Manager needs to make to avoid making a loss.	25; 30
Aspiration Threshold	The amount of Profit per unit area that the Land Manager hopes to make. If this is not achieved, the Land Manager will review the Land Uses allocated to all Parcels they own.	0.5; 1.0; 5.0
Reward	The amount the Government gives to Land Managers per awardable Species or Land Use.	0.0; 5.0; 10.0
Ratio	An amount by which to divide the Reward when the Government awards by Species.	1; 2; 3
Stop C1	Designs the incentives to stop C1 from causing extinction of non-target species. In activity-based policies, this amounts to not rewarding for Land Use GL1; in outcome-based policies, this amounts to rewarding for Species G3. Since this had no effect on the reported results, it is not discussed in the main text.	Yes; No

Conservation incentives

When an activity-based Policy was simulated, Managers received a payment for each Parcel in which they deployed GL1, GL2, or AL1. When a results-based Policy was in place, Managers received a payment for *each* occurrence of

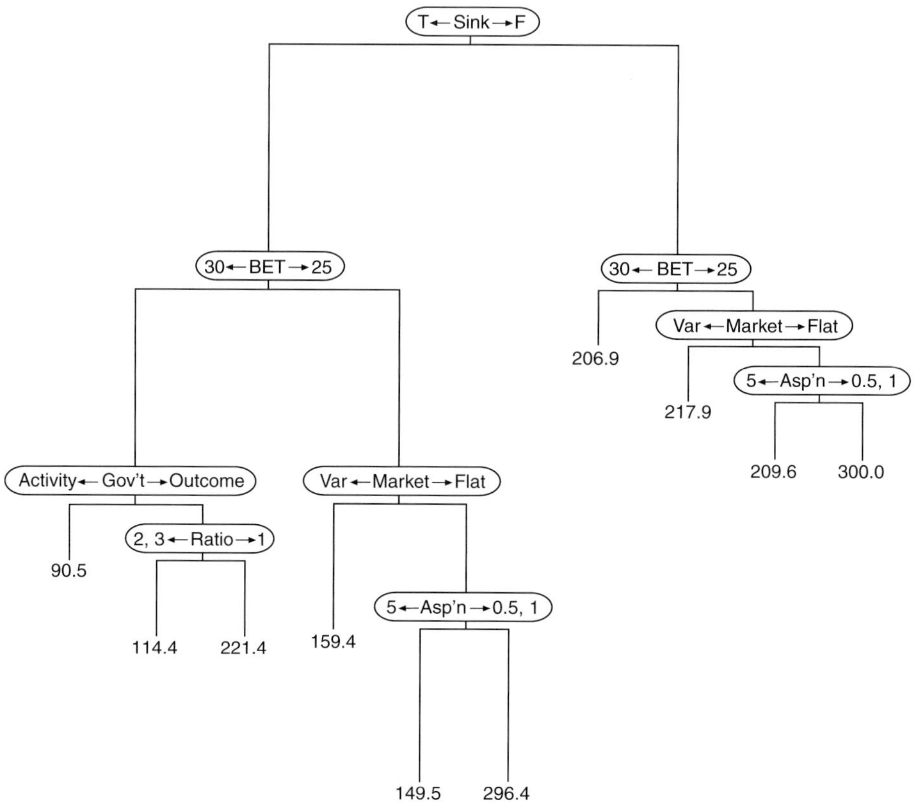

FIGURE 8.2.
A regression tree relating model parameters to extinction time of Species A3 in all the 16,800 runs. The "leaves" of the tree show the (fitted) average extinction time.

Species G5, G6, A2, or A3 on a Parcel they owned. Three Rewards levels were implemented: 0 (as a control), and 5 and 10 income units.

Analysis of results

The results are summarized in the regression trees (Breiman *et al.* 1984) in Figures 8.2, 8.3 and 8.4, which are based on 20 replicate runs (using different seeds for the pseudo-random number generator) for each combination of parameters, resulting in 16,800 runs. We analyzed persistence time and total species richness at the landscape level as a function of parameter values. Only variables useful for explaining the data appear in a tree. The "leaves" of each tree are the fitted values, while the "knots" show which variable best explained a particular split in the dataset. In the simulations, 70% of the runs were used to build statistical models, and 30% of the runs were used as a validation set. We ensured that the fitted values (e.g., persistence time) predicted by a tree, using

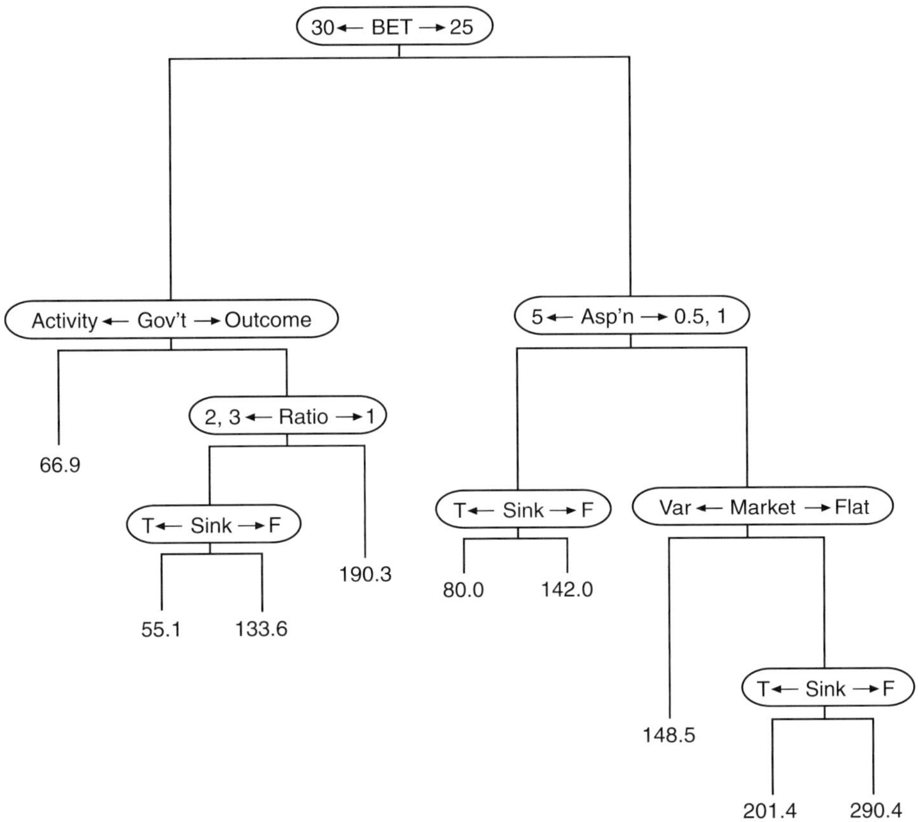

FIGURE 8.3.
A regression tree relating model parameters to extinction time of Species G6 in all the 16,800 runs. The "leaves" of the tree show the (fitted) average extinction time.

70% of the cases, correctly predicted the values in the validation set ($R^2 > 0.8$). All the regression trees presented are also tenfold cross-validated. This means that the observations were split into ten groups and, recursively, one group was left out while the other nine groups were used to grow trees of various sizes. The final tree is the one which gives the minimum cross-validation error.

Results and discussion

We concentrate mainly on the effect of sinks, and will report more fully on the effects of policy strategies in future work. We concentrate on species with narrower habitat preference and affected by sinks, which better serve to illustrate how various factors vary in importance according to species traits.

The regression trees show, in order of strength of influence (measured by decrease in variance) starting at the top, the parameters that affect average persistence time of the most vulnerable Species types in AL*x* (Fig. 8.2: Species

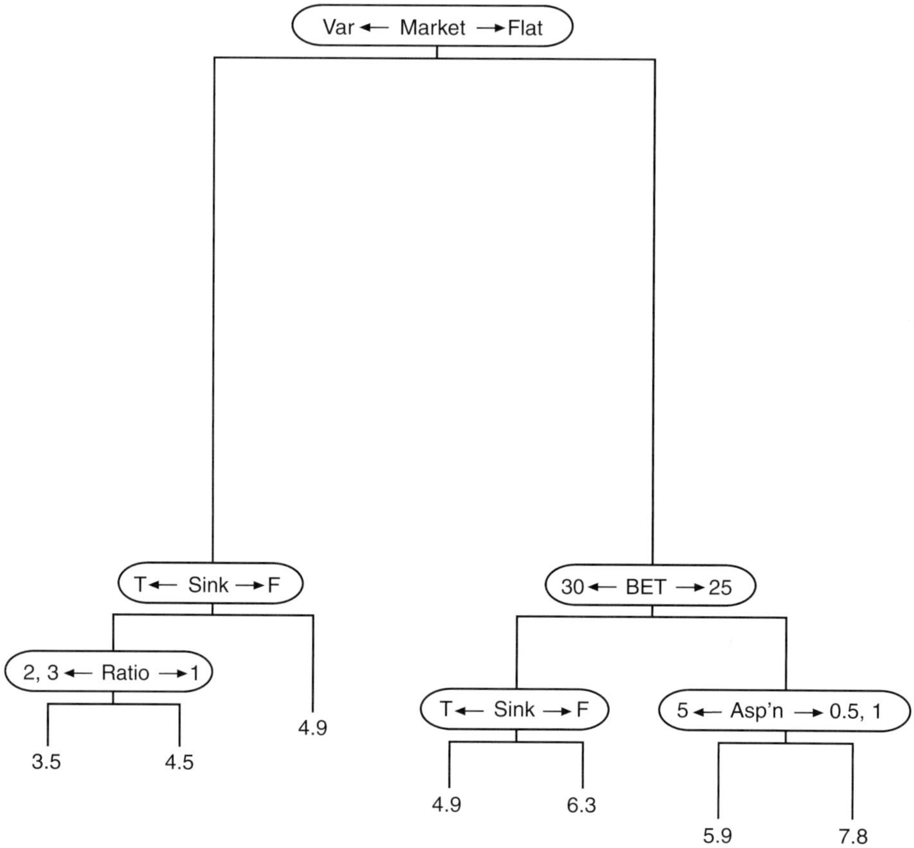

FIGURE 8.4.
A regression tree relating model parameters to species richness in all the 16,800 runs. The "leaves" of the tree show the (fitted) average landscape-scale species richness.

A3) and GL*x* (Fig. 8.3: Species G6) Land Uses, and landscape species richness (Fig. 8.4). Although the importance of source–sink dynamics for biodiversity conservation has been widely recognized, the results confirmed our hypothesis that its practical significance must be understood in the context of Land Use decisions.

In general, rewarded sinks compete with sources for Land Use allocation. Sinks are the most important factor affecting persistence time for a Species, such as A3 (Fig. 8.2), living in Habitats which do not tend to become very rare because they are associated with Land Uses of moderate profitability (including any Government incentive). In this case, GL2, which is more profitable than AL1 (the Land Use providing A3's primary Habitat), is also rewarded when activity rather than outcome qualifies, hence Land Managers are more likely to adopt GL2 than AL1. Sinks (left-hand sub-tree of Fig. 8.2) have an effect because they "outcompete" sources in the marketplace, lowering the effective number of Patches and therefore persistence time. In this situation, BET = 30 (left-hand sub-tree of the left-hand sub-tree of Fig. 8.2). Here, rewarding for outcome allows A3 to persist, on average, for longer than when rewarding per activity. However, since multiple Species can survive on some Parcels, this effect could be due entirely to larger total subsidies. To test this, we introduced the "ratio" parameter (see Table 8.B3.1), which is aimed at making absolute rewards for outcome and activity comparable when set to 1, and observed that for higher values of ratio explored, with consequent lower per-Species Rewards, the effect is reduced (see Fig. 8.2; average persistence for 114.4 with ratios 2 and 3 versus 221.4 steps with ratio 1).

For Species such as G6 (Fig. 8.3), living in Habitats associated with Land Uses having higher opportunity costs (AL1 and GL1), the effect of sinks is dwarfed by other factors. In these cases, persistence time is not influenced *principally* by sinks, although these play a role, but the combination of input costs (BET, root node of the tree in Fig. 8.3) and Land Managers' Aspirations (right-hand sub-tree of Fig. 8.3) led to different levels of Land Use intensity, and therefore to different relative abundances of source Habitat (GH1, principally from GL1, but to a small extent from GL2). In other words, Species living on Habitats associated with less profitable Land Uses, when these can easily be converted into more profitable ones, are vulnerable, independently of the presence of sinks. Their persistence is influenced more by Land Manager's Aspirations and Market fluctuations. If Land Managers have high Aspirations relative to the Returns from such Land Uses (left sub-tree of the right sub-tree of Fig. 8.3), the Habitat becomes rare, sometimes disappearing altogether, and the Species is likely to become extinct. With high input costs (left-hand sub-tree of Fig. 8.3), the outcome-based strategy works better. Though, again, this is partly because the outcome Policy leads to higher overall expenditure and reward, the fact that sinks have a significant effect only when there is a lower per-Species Reward ("ratio" 2 or 3) suggests that large incentives, or regulation, can preserve these Habitats for longer. Even though the activity-based Policy rewards for GL1, the main provider of the source Habitat for G6, it also rewards for the more profitable GL2, and Managers have no incentive to adopt GL1. Here too, there are good reasons to believe that there are contexts in which outcome-based incentive schemes would be more successful.

Habitat availability over time is therefore the main driver of persistence in the modeled system. Sinks are more influential for Species living in Habitats associated with sufficiently profitable or incentivized Land Uses (which do not tend to become very rare). In this case Land Managers allocate to sinks when the combination of incentives and profitability makes them attractive. For these Species, Policy incentives improve persistence only when there are high input costs to be repaid, meaning that non-awardable Land Uses become less attractive. In such cases, landscape-level persistence time is enhanced only when enough subsidies are provided, while "black-hole" sinks, as expected, shorten this time.

As far as species richness is concerned (Fig. 8.4):

1. market variability, by inducing land use (and hence habitat) turnover on the land parcels, has the highest effect on richness;
2. when the market is variable, the effect of incentives is subordinate to that of sinks (left sub-tree of left sub-tree, Fig. 8.4);
3. when the market is invariant the effect of sinks is subordinate to that of input costs (left sub-tree of right sub-tree, Fig. 8.4), which, together with profit orientation, drive intensification of land use.

Finally, it is worth noting that extinction of some non-target species, and thus decline in diversity, could not be avoided because the target species are not acting as "umbrella" species. This means that if landscape-scale species richness is the policy goal, incentive schemes will have to be designed carefully.

To summarize, sinks are an important issue, and worth considering for the conservation of species associated with habitats that remain moderately abundant. In other cases the disappearance of sources due to opportunity cost and land managers' profit orientation is the main cause of concern.

The consequences of modeling species abundance rather than occupancy have not been explored in this framework. Models show that sinks can stabilize population dynamics, thus avoiding density-dependent crashes in patches where there is a high growth rate (e.g., Kawecki 2004) by acting as a sort of buffer (see also Morris, Chapter 3, this volume). In the context of agro-ecosystems, this situation is likely to be relevant for populations that thrive and rapidly grow in local patches, e.g., of pests, but less so for most species of conservation concern. Also, this study was concerned with regional rather than local effects. Our results show conditions in which sinks matter, and it is here that the results might provide a lower limit to persistence times.

It is worth noting that in models tracking abundance, when dispersal involves individuals that do not contribute to the next generation in source communities, local coexistence is not influenced by dispersal (Amarasekare and Nisbet 2001). This is implicitly assumed by our model. Also, our model did not examine the effects of density dependence on dispersal rates (Amarasekare 2004a, 2004b).

Finally, as in population viability analysis, persistence time and richness should be considered as currencies used to compare scenarios rather than as stand-alone quantities.

Conclusions

Source–sink theory (Shmida and Ellner 1984; Pulliam 1988) is supported by empirical evidence (e.g., Thomas *et al.* 1996; Boughton 1999; Cousins and Lindborg 2008), and is considered to be one of the theories at the foundation of landscape ecology (Wiens *et al.* 1993). This work contributes to understanding the details of its relevance to managed landscapes.

We have explored the effect of sinks in situations that have received virtually no attention so far, namely in a system in which economic factors, policy factors, and land managers' aspirations are allowed to interact and to influence landscape structure, and therefore population persistence.

Our results show that the assumptions made by policy makers regarding habitat suitability can have serious consequences on species' persistence. If many of the lower land use intensity patches (e.g., fields with field margins) are undetected sinks, these might lead to the "wrong" policy. The problem facing policy makers, however, is not simple. Because persistence depends on landscape-scale attributes of a whole population, and is realized (or not) over a relatively long time period, it cannot be measured or predicted easily. Often, the presumed habitat quality (*sensu* Van Horne 1983) of individual patches composing the landscape mosaic is what is used to decide whether a habitat type or land use practice should be incentivized. This might result in perverse incentives, causing land managers to adopt land uses that are supposedly conservation-friendly but which might compete with the source habitat of target species for financial support.

Our results are consistent with studies suggesting that environmental variation plays an important role in determining community composition. The temporal structure of variation, for example, can influence extinction risks (Heino 1998), population dynamics (e.g., Gonzalez and Holt 2002), and coexistence (Holt *et al.* 2003).

In the system simulated, as in many semi-natural landscapes, local communities are assembled through dispersal, and species are "filtered out" by local environmental conditions and competition in some habitats. However, the disturbance regime can be severe, especially when market-driven signals translate into frequent land use change, and often local disappearance of habitat. Market-driven landscapes therefore appear to be prone to biodiversity crises when market conditions change. If we consider such disturbance as a type of "environmental variation" which is known to impact appreciably on population processes (e.g., Gonzalez and De Feo 2007), we can notice that, in this system too, the internal structure of environmental variation is a major driver of diversity. Such disturbance depends on market prices and land managers' attitudes, and is altered by incentives, which interact with "black-hole" sinks.

To understand how present and future (e.g., due to climate change) natural environmental variation is likely to impact on communities in managed landscapes, and what role sinks might have in future species extinction or persistence, it is necessary to devote attention to how socio-economic factors are likely to drive the dynamics of habitat availability and quality at the landscape level.

Acknowledgments

We acknowledge the Scottish Government Rural and Environmental Research and Analysis Directorate for financial support.

References

Aamodt, A. and E. Plaza (1994). Case-based reasoning: foundational issues, methodological variations, and system approaches. *AI Communications* **7**: 39–59.

Amarasekare, P. (2004a). The role of density-dependent dispersal in source–sink dynamics. *Journal of Theoretical Biology* **226**: 159–168.

Amarasekare, P. (2004b). Spatial variation and density-dependent dispersal in competitive coexistence. *Proceedings of the Royal Society of London, Series B* **271**: 1497–1506.

Amarasekare, P. and R. M. Nisbet (2001). Spatial heterogeneity, source–sink dynamics, and the local coexistence of competing species. *American Naturalist* **158**: 572–584.

Arlt, D. and T. Pärt (2007). Nonideal breeding habitat selection: a mismatch between preference and fitness. *Ecology* **88**: 792–801.

Arthur, W. B., S. Durlauf and D. Lane (1997). Introduction. In *The Economy as a Complex Evolving System II* (W. B. Arthur, D. Durlauf and S. Lane, eds.). Addison-Wesley, Reading, MA: 1–14.

Benton, T. G., J. A. Vickery and J. D. Wilson (2003). Farmland biodiversity: is habitat heterogeneity the key? *Trends in Ecology and Evolution* **18**: 182–188.

Boero, R. and F. Squazzoni (2005). Does empirical embeddedness matter? Methodological issues on agent-based models for analytical social science. *Journal of Artificial Societies and Social Simulation* **8**(4): http://jasss.soc.surrey.ac.uk/8/4/6.html.

Boughton, D. A. (1999). Empirical evidence for complex source–sink dynamics with alternative states in a butterfly metapopulation. *Ecology* **80**: 2727–2739.

Boulanger, P.-M. and T. Bréchet (2005). Models for policy-making in sustainable development: the state of the art and perspectives for research. *Ecological Economics* **55**: 337–350.

Bousquet, F. and C. Le Page (2004). Multi-agent simulations and ecosystem management: a review. *Ecological Modelling* **176**: 313–332.

Breiman, L., J. H. Friedman, R. A. Olshen and C. J. Stone (1984). *Classification and Regression Trees*. Chapman and Hall, New York.

Burton, R. J. F. (2004). Seeing through the "good farmer's" eyes: towards developing an understanding of the symbolic value of "productivist" behaviour. *Sociologia Ruralis* **44**: 195–215.

Burton, R. J. F. and G. A. Wilson (2006). Injecting social psychology theory into conceptualisations of agricultural agency: towards a post-productivist farmer self-identity. *Journal of Rural Studies* **22**: 95–115.

Chamberlain, D. E. and R. J. Fuller (2000). Local extinctions and changes in species richness of lowland farmland birds in England and Wales in relation to recent changes in agricultural land-use. *Agriculture, Ecosystems and Environment* **78**: 1–17.

Clergeau, P. and F. Burel (1997). The role of spatio-temporal patch connectivity at the landscape level: an example in a bird distribution. *Landscape and Urban Planning* **38**: 37–43.

Concepción, E. D., M. Díaz and R. D. Baquero (2008). Effects of landscape complexity on the ecological effectiveness of agri-environment schemes. *Landscape Ecology* **23**: 135–148.

Cousins, S. A. O. and R. Lindborg (2008). Remnant grassland habitats as source communities for plant diversification in agricultural landscapes. *Biological Conservation* **141**: 233–240.

Doebeli, M. and G. D. Ruxton (1998). Stabilization through spatial pattern formation in metapopulations with long-range dispersal. *Proceedings of the Royal Society of London, Series B* **265**: 1325–1332.

Duguay, J. P., P. B. Wood and J. V. Nichols (2001). Songbird abundance and avian nest survival rates in forests fragmented by different silvicultural treatments. *Conservation Biology* **15**: 1405–1415.

Dunning J. B., B. J. Danielson and H. R. Pulliam (1992). Ecological processes that affect populations in complex landscapes. *Oikos* **65**: 169–175.

Fahrig, L. and G. Merriam (1994). Conservation of fragmented populations. *Conservation Biology* **8**: 50–59.

Fischer, J. and D. B. Lindenmayer (2007). Landscape modification and habitat fragmentation: a synthesis. *Global Ecology and Biogeography* **16**: 265–280.

Gonzalez, A. and A. De Feo (2007). Environmental variability modulates the insurance effects of diversity in non-equilibrium communities. In *The Impact of Environmental Variability on Ecological Systems* (D. Vasseur and K. McCann, eds.). Springer, Dordrecht, The Netherlands: 159–178.

Gonzalez, A. and R. D. Holt (2002). The inflationary effects of environmental fluctuations in source–sink systems. *Proceedings of the National Academy of Sciences of the USA* **99**: 14872–14877.

Gotts, N. M., J. G. Polhill and A. N. R. Law (2003). Agent-based simulation in the study of social dilemmas. *Artificial Intelligence Review* **19**: 3–92.

Gregory, R. D., D. G. Noble and J. Custance (2004). The state of play of farmland birds: population trends and conservation status of lowland farmland birds in the United Kingdom. *Ibis* **146**(Suppl. 2): 1–13.

Gundersen, G., E. Johannesen, H. P. Andreassen and R. A. Ims (2001). Source–sink dynamics: how sinks affect demography of sources. *Ecology Letters* **4**: 14–21.

Hald, A. B. (1999). The impact of changing the season in which cereals are sown on the diversity of the weed flora in rotational fields in Denmark. *Journal of Applied Ecology* **36**: 24–32.

Hare, M. and P. Deadman (2004). Further towards a taxonomy of agent-based simulation models in environmental management. *Mathematics and Computers in Simulation* **64**: 25–40.

Hatchwell, B. J., D. E. Chamberlain and C. M. Perrins (1996). The demography of blackbirds *Turdus merula* in rural habitats: is farmland a sub-optimal habitat? *Journal of Applied Ecology* **33**: 1114–1124.

Heino, M. (1998). Noise colour, synchrony and extinctions in spatially structured populations. *Oikos* **83**: 368–375.

Holt, R. D., M. Barfield and A. Gonzalez (2003). Impacts of environmental variability in open populations and communities: inflation in sink environments. *Theoretical Population Biology* **64**: 315–333.

Huigen, M. (2004). First principles of the MameLuke multi-actor modelling framework for land use change, illustrated with a Philippine case study. *Journal of Environmental Management* **72**: 5–21.

Ikerd, J. (2006). On defining sustainable agriculture [available at www.sustainable-ag.ncsu.edu/onsustaibableag.htm].

Kawecki, T. J. (2004). Ecological and evolutionary consequences of source–sink population dynamics. In *Ecology, Genetics, and Evolution of Metapopulations* (I. Hanski and O. E. Gaggiotti, eds.). Elsevier, Amsterdam: 387–414.

Kleijn, D. and W. J. Sutherland (2003). How effective are European agri-environment schemes in conserving and promoting biodiversity? *Journal of Applied Ecology* **40**: 947–969.

Kwaiser, K. S. and S. D. Hendrix (2008). Diversity and abundance of bees (Hymenoptera: Apiformes) in native and ruderal grasslands of agriculturally dominated landscapes. *Agriculture Ecosystems and Environment* **124**: 200–204.

Mattison, E. H. A. and K. Norris (2005). Bridging the gaps between agricultural policy, land-use and biodiversity. *Trends in Ecology and Evolution* **20:** 610–616.
Meffe, G. K. and C. R. Carroll (eds.) (1997). *Principles of Conservation Biology*, 2nd edition. Sinauer Associates, Sunderland, MA.
Moilanen, A. (1999). Patch occupancy models of metapopulation dynamics: efficient parameter estimation using implicit statistical inference. *Ecology* **80:** 1031–1043.
Moilanen, A. (2004). Spomsim: software for stochastic patch occupancy models of metapopulation dynamics. *Ecological Modelling* **179:** 533–550.
Namba, T. and C. Hashimoto (2004). Dispersal-mediated coexistence of competing predators. *Theoretical Population Biology* **66:** 53–70.
Öckinger, E. and H. G. Smith (2007). Semi-natural grasslands as population sources for pollinating insects in agricultural landscapes. *Journal of Applied Ecology* **44:** 50–59.
Parker, D. C., A. Hessl and S. C. Davis (2007). Complexity, land-use modeling, and the human dimension: fundamental challenges for mapping unknown outcome spaces. *Geoforum* **39:** 789–804.
Polhill, J. G., N. M. Gotts and A. N. R. Law (2001). Imitative versus nonimitative strategies in a land use simulation. *Cybernetics and Systems* **32:** 285–307.
Polhill, J. G., D. C. Parker and N. M. Gotts (2008). Effects of land markets on competition between innovators and imitators in land use: results from FEARLUS-ELMM. In *Social Simulation Technologies: Advances and New Discoveries* (C. Hernandez, K. Troitzsch and B. Edmonds, eds.). Information Science Reference, Hershey, PA: 81–97.
Pulliam, H. R. (1988). Sources, sinks, and population regulation. *American Naturalist* **132:** 652–661.
Robinson, R. A. and W. J. Sutherland (2002). Post-war changes in arable farming and biodiversity in Great Britain. *Journal of Applied Ecology* **39:** 157–176.
Robinson, R. A., J. D. Wilson and H. Q. P. Crick (2001). The importance of arable habitat for farmland birds in grassland landscapes. *Journal of Applied Ecology* **38:** 1059–1069.
Rosenzweig, M. L. (2003). Reconciliation ecology and the future of species diversity. *Oryx* **37:** 194–205.
Secretariat of the Convention on Biological Diversity (2006). *Global Biodiversity Outlook 2*. Secretariat of the CBD, Montreal.
Schoener, T. W. (1983). Field experiments on interspecific competition. *American. Naturalist* **122:** 240–284.
Shmida, A. and S. Ellner (1984). Coexistence of plant species with similar niches. *Vegetatio* **58:** 29–55.
Simon, H. (1955). A behavioral model of rational choice. *Quarterly Journal of Economics* **69:** 99–118.
Tattersall, F. H., D. W. Macdonald, B. J. Hart and W. Manley (2004). Balanced dispersal or source–sink: do both models describe wood mice in farmed landscapes? *Oikos* **106:** 536–550.
Thomas, C. D., M. C. Singer and D. A. Boughton (1996). Catastrophic extinction of population sources in a butterfly metapopulation. *American Naturalist* **148:** 957–975.
Tittler, R., S. J. Hannon and M. R. Norton (2001). Residual tree retention ameliorates short-term effects of clear-cutting on some boreal songbirds. *Ecological Applications* **11:** 1656–1666.
Van Horne, B. (1983). Density as a misleading indicator of habitat quality. *Journal of Wildlife Management* **47:** 893–901.
Wallis de Vries, M. F., J. P. Bakker and S. E. Van der Wieren (1998). *Grazing and Conservation Management*. Kluwer Academic Publishers, Dordrecht, The Netherlands.
Warren, J., C. Lawson and K. Belcher (2008). *The Agri-Environment*. Cambridge University Press, Cambridge, UK.
Wiens, J. A., N. C. Stenseth, B. Van Horne and R. A. Ims (1993). Ecological mechanisms and landscape ecology. *Oikos* **66:** 369–380.

Part III

Progress in source–sink methodology

Just as source–sink and related theories have developed rapidly over the past decade, the toolbox for estimating and modeling sources and sinks has also expanded considerably. Effective analyses of source–sink systems require accurate estimates of key demographic parameters of local populations such as reproductive rates, survival, and dispersal. These estimates provide the foundation for making assumptions, developing models, and drawing conclusions from source–sink studies. To understand population dynamics at a broader scale, it is also important to quantify demographic relationships among different local source or sink populations as well as their contributions to a larger population network. This section presents several important aspects of progress in methods for estimating population demographics with respect to source–sink status (Chapters 9–11), in techniques for modeling source–sink dynamics across complex landscapes (Chapters 12–14), and in experimentation on source–sink dynamics (Chapter 15).

Of the demographic parameters modeled within the framework of source–sink dynamics, dispersal is especially problematic, since it is often difficult to observe and measure directly. In Chapter 9, Pulliam and colleagues illustrate that ignoring or discounting dispersal can result in inaccurate estimation of population parameters, as well as misclassification of sources and sinks. They propose and test two new maximum likelihood approaches for estimating dispersal, which allow for isolation of the relative contributions of fecundity and dispersal to overall population change. Such an approach to parameter estimation may greatly improve future parameterization of source-sink models and also have broad implications for the field testing of ecological theory.

Although it is common that populations are spatially structured as a population network, much more research is needed to differentiate demographic contributions of local populations to a population network and to the growth rate of such a network. Using data on individually marked reed buntings (*Emberiza schoeniclus*) from 21 local populations in northeastern Switzerland, Pasinelli and colleagues (Chapter 10) demonstrate that accurate estimation of demographic parameters is important when considering the relative contributions of the local populations to a population network. They test a recently developed metric (the C^r metric) that simultaneously estimates the relative demographic contributions of different local populations to the population network by including emigration estimates. Their results show that metrics including emigration more comprehensively describe the source–sink status of local populations than metrics on processes within local populations. Furthermore, immigration is crucial to the viability of the network of local populations, and small local populations can make contributions to a population network that are as valuable as those of large local populations.

While sources and sinks exist in a wide variety of systems, it is difficult to detect them in some systems because of problems in accurately estimating essential demographic parameters. These problems pose challenges for source–sink management that make the development of new methods for identifying sources and sinks essential. This topic is discussed in Chapter 11 with respect to migratory bird species that have complex annual cycles involving dispersal to vastly distant locations. In this chapter, Dunning and colleagues discuss the lack of relevant demographic measurements for quantifying sources and sinks when studying the dynamics of migratory grassland birds. They propose alternative measures such as territory occupancy and age distribution of successful dispersers as useful indicators of the status of grassland breeding bird populations.

In addition to the estimation of demographic parameters, another important aspect of source–sink research is the development of models to capture the spatial dynamics of sources and sinks. The next three chapters present approaches that are applied explicitly to source–sink systems: network analysis, range-limit models, and hierarchical models. These approaches add depth to the understanding of the complexity of dynamic populations distributed across landscapes of varying habitat quality.

The first approach is network analysis (Chapter 12), a tool that uses the properties of graphs to describe the topology of a landscape and to quantify connectivity between landscape elements (patches and corridors). Jordán illustrates how network analysis can be used to assign weighted rankings to sources and sinks based on their positions in the landscape relative to other local populations. Using two case studies on landscape graphs of flightless bush cricket and seven forest-living carabid species in Hungary, he demonstrates methods of determining quantitative priority ranks of landscape elements and procedures for pairing suitable techniques with particular problems. Furthermore, he argues that the efficiency of conservation efforts can be enhanced by setting quantitative priority ranks for landscape elements.

Many species' range-limit models have been developed to predict shifts in species' distribution ranges due to human impacts such as land use and climate change. To make these models more useful, however, it is necessary to incorporate information about the impacts of human activities and environmental variation on the demographic parameters that dictate the geographic distribution of species. In Chapter 13, Etterson and colleagues develop a range-limit model (CPSS) that predicts the distributional range of the coastal plain swamp sparrow (*Melospiza georgiana nigrescens*) and embed the CPSS fecundity gradient in Pulliam's source–sink model. The results show reasonable predictions of the current CPSS range limits, although there are underlying uncertainties.

Population dynamics and underlying processes (e.g., birth, death, migration) differ between spatial and temporal scales. Scale is a central tenet in the ecological literature, but integrating scale with quantitative studies of population dynamics is not easy, so it is often not explicitly integrated into population modeling. In Chapter 14, Diez and Giladi show how scale has affected studies of source–sink dynamics and illustrate the utility of hierarchical models for understanding source–sink dynamics by capturing the dynamics of demographic parameters of local populations at different temporal and spatial scales. Using a case study of the forest herb *Hexastylis arifolia*, they demonstrate how demographic rates and population growth rates may be quantified at different scales.

While most source–sink studies have been based on observations or modeling, experimentation – especially long-term experimentation – is a useful approach to understanding the outcome and processes underlying source–sink dynamics. Matter and Roland (Chapter 15) use a long-term experimental removal study on alpine butterflies to explore circumstances under which recovery from disturbance can occur. Their results indicate that immigration was sufficient to allow most local populations to recover from a sudden population removal, highlighting the resilience of this particular population network.

The chapters in this section make it clear that testing the theory and ideas of source–sink dynamics ultimately requires detailed information about the demography of subpopulations and how they are linked by dispersal. Such information is often not easy to obtain for any one population, much less for multiple populations in a spatial context. Yet it must be done. There are good suggestions here about how to proceed. They show that, with careful thought and perseverance, even difficult things are possible.

H. RONALD PULLIAM, JOHN M. DRAKE AND JULIET R. C. PULLIAM

9

On estimating demographic and dispersal parameters for niche and source–sink models

Summary

Demography plays a central role in ecology, and accurate estimation of demographic parameters is essential to testing basic theories of life-history evolution, population regulation, and species coexistence, as well as applying ecological theory to a broad range of practical issues including species conservation, predicting shifts in species distribution due to climate change, and understanding the emergence and spread of new diseases. Our goals in this chapter are twofold: first, to demonstrate a problem with estimating demographic parameters that stems from ignoring dispersal, and second, to propose a solution to this problem. We illustrate the problem with a simulation model that shows how ignoring dispersal may lead to the misclassification of sources and sinks; we attempt to solve the problem by using generalized linear models to differentiate population change due to fecundity from population change due to dispersal, thereby helping to improve source and sink classification. We believe that the particular example discussed is only one of a number of parameter-estimation problems limiting our ability to test and apply ecological theories in field systems and that an approach to parameter estimation similar to that used here may be broadly useful in narrowing the gap between ecological theory and field testing of that theory.

Introduction

The successful application of source–sink theory to field studies requires that investigators accurately assess individual fitness and properly assign

Sources, Sinks and Sustainability, ed. Jianguo Liu, Vanessa Hull, Anita T. Morzillo and John A. Wiens. Published by Cambridge University Press.

populations to source and sink categories. In general, this requires identifying and tracking individuals or groups of individuals to determine their survival and reproductive success, and performing statistical analyses to estimate vital rates and population growth rates. Because dispersal is rarely observed directly, it is often even more difficult to estimate than survival and reproduction. Nonetheless, dispersal is an important part of population dynamics, especially in the heterogeneous environments where metapopulation and source–sink dynamics may prevail, and erroneous assumptions about dispersal can lead to systematic errors in the estimation of vital rates. For example, in some studies, fecundity is estimated by assuming that all juveniles found in a prescribed area are the offspring of the adults occupying the same area. This assumption ignores the possibility that some of the juveniles found in the area may have been produced elsewhere or that some of the juveniles born in the area may have dispersed elsewhere before the census was conducted. As shown later in this chapter, ignoring or otherwise not properly accounting for dispersal may result in underestimating the productivity of sources and overestimating the productivity of sinks and, in many cases, this leads to the misclassification of sources and sinks.

With increasing attention to climate change, ecologists have developed a wide array of "climate envelope" or "niche models" aiming to relate habitat suitability and population dynamics to biophysical and climatic variables, or proxies of such variables. Because both demographic and detailed environmental data are hard to come by, investigators often attempt to relate population density, presence–absence data, or even presence-only data, to readily available geographic data such as topography or estimated rainfall (Elith and Burgman 2002; Guisan and Thuiller 2005). Even when both detailed demographic and bioclimatic information is available, dispersal is often ignored, and this may lead to serious problems in parameter estimation and ultimately in the prediction of species ranges and the potential impacts of climate change.

Detailed niche models may require estimates of survival and fecundity in order to specify the range of suitable environmental conditions that allow a species to persist and spread (Kadmon 1993; Urban *et al.* 2007). Some models go further and attempt to specify a species-specific optimum and niche width for multiple environmental variables (Pulliam 2000; Diez 2006; Warren 2007). In most field studies that provide data for niche models, the spatial scales of dispersal and environmental heterogeneity overlap. For example, niche models may attempt to estimate fecundity and survival as functions of environmental variables such as temperature and moisture (Diez and Pulliam 2007; Warren 2007; Pulliam and Waser 2010), but if juveniles found on a study site are not produced on the site, parameter estimates may be severely biased when dispersal is ignored. For example, seeds produced in a moist microhabitat might fall into a nearby dry patch and the resulting seedlings might thus be attributed incorrectly to the less productive adults in the drier patch. In this case, a regression model fitting parameters to an equation describing fecundity as a function of moisture, but ignoring dispersal, will give biased estimates of the influence of moisture on seedling production. Although problems with parameter estimation related to ignoring dispersal may be especially acute at small spatial scales, landscape-scale models are not immune to problems associated with dispersal because, regardless of scale, dispersal leads to a mismatch between where individuals are found and where they do well (Pulliam 2000; Kearney 2006; Kearney *et al.* 2008). We believe that the solution to these problems requires a disentanglement of the roles of dispersal and demography in producing observed patterns of distribution (Kadmon and Tielborger 1999).

In this chapter, we ask how much the failure to account for dispersal biases the estimation of vital rates and niche parameters, and how likely such biases are to lead to errors in estimating population growth rates. We do this with a simple, though not trivially simple, simulation model in which survival, reproduction, and dispersal of hypothetical plant populations are determined by drawing samples from appropriate probability distributions with known parameters and then using statistical models to reconstruct the input parameters. We believe that this medium-complexity model is useful as a demonstration of concept because it manifests a bias that we believe is common in population models, while remaining simple enough to understand.

To demonstrate the problem, we first ignore dispersal and use generalized linear models to estimate parameters. As expected, some parameters are consistently overestimated or underestimated. We then address the problem of dispersal by using a maximum likelihood approach to estimate demographic variables and niche parameters, taking dispersal into account. By comparing the estimated demographic and dispersal parameters to the "true" parameter values used to generate the data, we show that the maximum likelihood methods lead to more accurate estimates of demographic parameters and are more likely to lead to the correct interpretation of the dependence of population growth on environmental factors.

The models: *nichemaker* and *nicheobserver*

We developed two linked models to simulate and analyze population dynamics. The first, *nichemaker*, simulates population dynamics of hypothetical plant populations and generates data similar to that a field researcher might collect. The second, *nicheobserver*, contains a collection of regression and maximum likelihood statistical routines to analyze the data output of *nichemaker* and to estimate the "true" parameter values originally used to generate the data. In short, *nichemaker* simulates population dynamics of hypothetical plant populations and generates data that is, in turn, analyzed by *nicheobserver*. Both *nichemaker* and *nicheobserver* are written in R, a free statistical and programming software package available at http://cran.r-project.org/. Shortened versions of the code for both *nichemaker* and *nicheobserver* are presented as an Appendix to this chapter and the code used to simulate all results presented in the chapter is available by request from the senior author (H. R. Pulliam).

The *nichemaker* model simulates the dynamics of interacting plant populations on a grid. In all examples presented in this chapter, there are eight plant species interacting on a grid of 16 × 16 cells, although the number of species and size of the grid may be easily modified. All of the hypothetical plant species have the same simple life cycle. Seeds are produced in the spring and disperse prior to an annual census, which occurs shortly after germination. During the census, an observer counts all seedlings and adults on all cells. To keep the model simple, the following additional assumptions are made: all seeds germinate, any plant that has survived for one or more years is considered an adult, and all adults of a given species have the same probability of survival and the same fecundity function. In this hypothetical setting, as in many real ones, seedlings are not censused until after dispersal; therefore, locally produced seedlings cannot be distinguished from immigrants.

Nichemaker incorporates two levels of stochastic variation (Table 9.1). For a given replicate of the model, each species is first assigned a set of randomly determined demographic and dispersal parameters drawn from a uniform distribution. (The use of the uniform distribution ensures that probabilities are bounded and that parameters sample the full range of possible values within the chosen bounds.) Once the demographic and dispersal parameters are assigned, the actual values of survival, reproduction, and dispersal are drawn from a second set of distributions using the previously chosen parameters. For example, each species has a probability of adult survival (p_{Ai}) drawn from a uniform distribution with a minimum of 0.8 and a maximum of 0.9. If the draw for a species happens to be 0.87, this does not mean 87% of the adults always survive, but rather the number surviving is then a binomially distributed random variable with parameter $p_{Ai} = 0.87$ and order $n_{i,x,y,t}$ (the number of adults of species i on cell x,y at time t). Table 9.1 defines the parameters and environmental characteristics and specifies the probability distributions that *nichemaker* uses to generate them. An example of the actual niche and demography parameter values generated by a particular replicate of the *nichemaker* model is shown in Table 9.2.

In *nichemaker*, only adults reproduce and each individual's fecundity, measured as the number of viable seeds produced, depends on environmental conditions in the grid cell where the adult resides. For concreteness, we think of

TABLE 9.1. *Demographic parameters and environmental characteristics are randomly assigned by* nichemaker *using specified probability distributions.* Nicheobserver *estimates parameters from data using either generalized linear regression models* (R) *or maximum likelihood models* (L). *Survival probabilities and dispersal fraction have only one likelihood estimate each but other parameters have two different versions of maximum likelihood estimates denoted by subscripts* L1 *and* L2; *the former incorporating the regression estimate of the dispersal fraction and the latter based on the simultaneous estimation of dispersal and fecundity parameters.*

Parameter definition	Distribution	Parameter estimates: regression (*R*) and maximum likelihood (*L*)
Probability that a seedling of species *i* survives from time *t* to $t+1$	$p_{Ji} \sim$ Uniform(0.1, 0.2)	p_{JiR}, p_{JiL}
Number of seedlings surviving on grid cell *x,y* from time *t* to $t+1$	$s_{J,i,x,y,t} \sim$ Binomial(p_{Ji}, $J_{i,x,y,t}$)	–
Probability of an adult of species *i* surviving from time *t* to $t+1$	$p_{Ai} \sim$ Uniform(0.8, 0.9)	p_{AiR}, p_{AiL}
Number of adults surviving on grid cell *x,y* from time *t* to $t+1$	$s_{A,i,x,y,t} \sim$ Binomial(p_{Ai}, $n_{i,x,y,t}$)	–
Probability that a seed disperses to an adjacent grid cell	$p_{Di} \sim$ Uniform(0.05, 0.15)	p_{DiR}, p_{DiL}
Log of the maximum expected fecundity of an individual of species *i*	$r_{Oi} \sim$ Uniform(1.0, 2.0)	$r_{OiR}, r_{OiL1}, r_{OiL2}$
Optimum pH for species *i*	$opt_i \sim$ Uniform(2.0, 9.0)	$opt_{iR}, opt_{iL1}, opt_{iL2}$
Niche width for species *i*	$w_i \sim$ Uniform(1.0, 10.0)	w_{iR}, w_{iL1}, w_{iL2}
Density dependence for species *i*	$a_i \sim$ Uniform(0.04, 0.06)	a_{iR}, a_{iL1}, a_{iL2}
pH of grid cell *x,y*	$pH_{x,y} \sim$ Uniform(5.0, 9.0)	–
Mean fecundity of an adult of species *i* on grid cell *x,y* at time *t*	$f_{i,x,y,t} = \exp(r_{oi} - (opt_i - pH_{x,y})^2/w_i + a_i N_{x,y,t})$	–
Number of seeds of species *i* produced on grid cell *x,y* at time *t*	$S_{i,x,y,t} \sim$ Poisson($n_{i,x,y,t}$ * $f_{i,x,y,t}$)	–

Notes. $J_{i,x,y,t}$: number of seedlings of species *i* on grid cell *x,y* at time *t*; $n_{i,x,y,t}$: number of adults of species *i* on grid cell *x,y* at time *t*; $N_{x,y,t}$: population density (all species combined) on grid cell *x,y* at time *t*.

TABLE 9.2. *An example of the parameters generated by the* nichemaker *model for eight hypothetical plant species. Each replicate model generates a unique set of model parameters. The parameter values in this table were generated by a single run of the* nichemaker *model and these same parameter values are used for Figures 9.1, 9.2, and 9.3.*

Species	1	2	3	4	5	6	7	8
Adult survival (p_{Ai})	0.87	0.81	0.88	0.88	0.84	0.85	0.82	0.82
Seedling survival (p_{Ji})	0.12	0.16	0.18	0.16	0.12	0.19	0.18	0.17
Optimum pH (opt_i)	8.53	7.32	8.22	6.33	7.71	6.89	7.54	8.91
Niche width (w_i)	4.82	5.68	1.54	4.06	5.68	3.69	2.51	3.14
Log max fecundity (r_{Oi})	1.09	1.19	1.91	1.52	1.83	1.30	1.37	1.17
Density dependence (a_i)	0.050	0.051	0.055	0.049	0.057	0.055	0.050	0.046
Dispersal fraction (p_{Di})	0.099	0.062	0.112	0.060	0.137	0.053	0.100	0.123

the environmental variable as pH, although it could just as easily be moisture or another environmental factor. Fecundity depends on pH and population density experienced by individuals at the cell level. Each grid cell is assigned a randomly drawn pH at the start of each replicate of the model, and the pH of that cell is then assumed to be constant throughout the duration of that replicate. For the examples presented, the mean pH of all cells on the grid is 7.0 and the pH of each individual cell is drawn from a uniform distribution with a minimum of 5 and a maximum of 9.

In real populations, the response of population growth rate to environmental variables such as pH and moisture is often unimodal (Lawesson and Oksanen 2002), with the highest population growth occurring at intermediate values of the environmental variables. For biological realism, we require a relationship between pH and fecundity that is unimodal, smooth, positive, and defined for all real positive numbers. Following convention (e.g., MacArthur 1969; May and MacArthur 1972), we adopt a bell-shaped function for the response of fecundity to pH, so that the mean number of seeds produced by each adult of species i at time t on the grid cell with coordinates x,y is given by

$$f_{i,x,y,t} = \exp(r_{Oi} - (opt_i - pH_{x,y})^2/w_i - a_i N_{x,y,t}) \qquad (9.1)$$

where r_{0i} is log maximum per capita fecundity, opt_i is the optimum pH for species i, $pH_{x,y}$ is the assigned pH of grid cell x,y. Niche width (w_i) is defined as the mean square deviation (variance) of the fecundity curve for species i and, as such, provides a measure of the range of pH values tolerated by each species. Population growth is density dependent and the parameter a_i measures the per capita decline in fecundity as density increases. The density dependence parameter a_i is drawn from a Uniform(0.04, 0.06) distribution; a value of $a_i = 0.05$ results in a maximum density per grid cell on the order of 30–40 individuals. Note that the similarity between Eq. (9.1) and the normal distribution does not imply that f is a probability distribution. Also, note that a biotic interaction between species is incorporated into the assumption that fecundity declines proportionally to the sum of the abundance of all adults of all individuals of all species on the grid cell, i.e., $N_{x,y,t} = \sum_{i=1,8} n_{i,x,y,t}$ where $n_{i,x,y,t}$ is the number of adults of species i on the cell x,y at time t.

The number of adults in year $t+1$ is the sum of the number of adults present at time t that survive to be counted again at time $t+1$ and the number of seedlings found on the cell at time t that survive to be counted at time $t+1$. In particular, if there are $n_{i,x,y,t}$ adults of species i on a grid cell x,y at time t then the number $s_{A,i,x,y,t}$ surviving to be counted in the next census is a random number drawn from a binomial distribution with parameter p_{Ai} and order $n_{i,x,y,t}$. Similarly, if there are $J_{i,x,y,t}$ seedlings on a cell at time t, the number $s_{J,i,x,y,t}$ of these that survive until the next census is a binomially distributed random variable with parameter p_{Ji} and order $J_{i,x,y,t}$. However, it is important to note that $s_{J.i,x,y,t}$ is not the number of surviving seedlings produced by the adults on the cell but rather a number that depends on fecundity, dispersal, and survival in a three-step process. First, the number of viable seeds produced on a particular grid cell is a random variable ($S_{i,x,y,t}$) drawn from a Poisson distribution with mean $n_{i,x,y,t} {}^* f_{i,x,y,t}$, where $n_{i,x,y,t}$ is the number of adults of species i present on the cell and $f_{i,x,y,t}$ is the mean per capita fecundity as determined by local environmental conditions (see Eq. 9.1). Second, the fraction of seeds dispersing from the cell is a binomially distributed random variable with parameter p_{Di} and order $S_{i,x,y,t}$. For simplicity, dispersal is assumed to occur only between adjacent cells and p_{Di} is the fraction of all seeds produced on a focal cell that disperse to one of its neighbor cells. Apparent fecundity $J_{i,x,y,t}$ is the number of seedlings actually counted on the cell after dispersal has taken place (recall that 100% of seeds are assumed to germinate) and, finally, $s_{J,x,y,t}$ is the number of these seedlings that survive until the following year.

We initialized *nichemaker* with one adult of each species on each grid cell. (We also explored the alternative of initializing the simulations by randomly placing adults on a grid with similar results.) Since demographic parameters are randomly chosen, some species are better adapted to conditions on the grid and increase in abundance, especially on grid cells with pH close to the species optimum. Other species, not as well suited for local conditions, decline. To represent that the grid is embedded in a larger landscape, new immigrants randomly arrive according to a species-specific immigration rate I_i, chosen from an exponential distribution. Thus, immigration rates vary among species and, on average, one new immigrant of each species arrives once every $1/I_i$ years. When an immigrant does arrive, it is randomly assigned to one of the 256 grid cells.

Parameter estimates for adult and seedling survival

The data analyzed by *nicheobserver* are similar to those a field investigator might collect by visiting a real population once each year and counting the number of adults and seedlings on each grid cell. The question we ask is what inferences the investigator can draw from the data concerning demographic and dispersal parameters and to what extent such census and environmental data can be used to determine how demographic parameters depend on pH and density. Since our focus is on parameter estimation, and specifically how dispersal affects parameter estimation, we assume that the investigator knows the appropriate model to use but does not know the appropriate model parameters. (Of course, model selection is itself an important issue, but one largely beyond the scope of this chapter.) We are particularly interested in how dispersal complicates estimates of population growth rate, because the proper classification of sources and sinks requires accurate assessment of population growth rates.

For each time step, *nichemaker* simulates reproduction, dispersal, and survival and stores the results of the annual census of seedlings and adults for statistical analysis by *nicheobserver*. Typically, a replicate is run for 60+ years and the data may be analyzed for all years or for some chosen subset of years. For our analysis, we choose blocks of 5 or 10 years' duration. Among other questions, we ask whether parameters estimated for early sample blocks, i.e., years 1–5, differ in quality from estimates made for a later time block (61–65) from the same replicate. Because the model is initialized far from stationarity, the early blocks have more occurrences of rare species and greater ranges of density for all species, perhaps making parameter estimation somewhat easier, but the later blocks are presumably more like the data that a field researcher would encounter for most real populations. We also ask how much increasing sample size, by increasing duration of the study from 5 to 10 years (i.e., years 61–65 to years 61–70), increases the accuracy of parameter estimation.

Nicheobserver uses the data generated by *nichemaker* and generalized linear models (GLMs) to estimate demographic and dispersal probabilities. The GLMs used to estimate survival probability are logistic regression models, hereafter referred to simply as regression models. Within a replicate, survival probability of a species is constant and does not depend on pH or population density, so the appropriate regression model for adult or seedling survival is as follows:

$$\text{glm}(Y \sim \textit{spfactor}\text{-}1, \text{family} = \text{binomial}(\text{link} = \text{logit})) \quad (9.2)$$

The regression models are written in a form recognized by the R statistical software. *Y* represents the data to be explained, consisting of two columns. For seedling survival, the first column is the number of "successes" (number of seedlings that survive until the next census) and the second column is "failures" (number of seedlings that do not survive). Likewise, for adult survival, the first column is the number of surviving adults and the second is the number of adults that do not survive. The term *spfactor* refers to species number (1–8) and inclusion of -1 after *spfactor* instructs R to estimate a separate survival probability for all eight species. (Without this term, R would estimate a mean survival probability plus survival probabilities for only seven of the eight species.) As shown in Figure 9.1, the regression estimates of adult and seedling survival match the true values used to generate the data fairly well. Statistical analyses of the match between true and estimated parameter values from a large number of replicates are presented in Table 9.3.

Parameter estimates for fecundity

Given the assumptions we have made, an investigator does not have direct access to fecundity data, *per se*, but rather must make an inference about fecundity based on post-dispersal data. The data available are what we call "apparent fecundity," defined as the number of seedlings per adult found on each grid cell during the annual census. Based on the assumptions of the *nichemaker* model, we expect to find that pH, population density, and dispersal all influence apparent fecundity. To illustrate the effect of pH on fecundity and the type of data an investigator might be faced with, we plot the data for each species in Figure 9.2. The figure shows the number of seedlings for each species, cell and time plotted against pH. With close inspection, the influence of pH on seedling recruitment can be discerned, and comparison of Figure 9.2 with Table 9.2 shows that, despite considerable scatter, each species tends to have greater fecundity when pH is near the optimum pH for the species.

Since the annual census takes place after dispersal, estimating fecundity based on the number of seedlings per adult on a cell ignores dispersal. Our approach is to first estimate fecundity with a regression model ignoring dispersal and then to compare this approach to a method that estimates fecundity

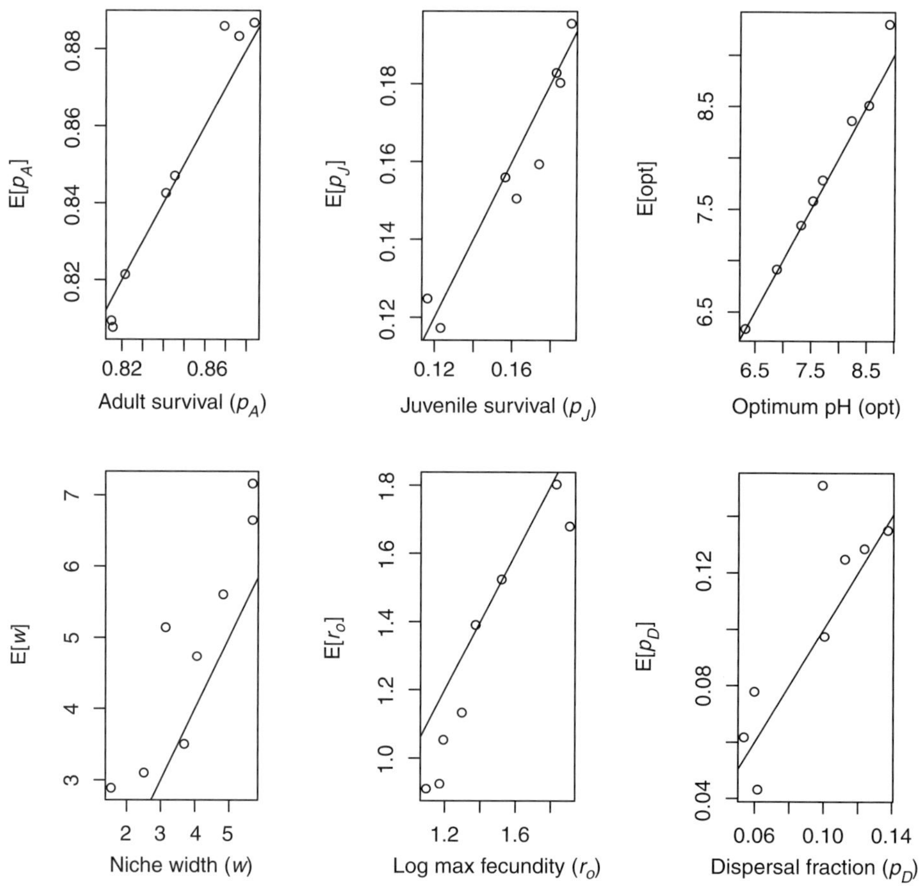

FIGURE 9.1.
An example of parameter estimates using generalized linear models compared with the true parameters used to generate the data. The estimates for some parameters (e.g., seedling and adult survival) closely approximate the true parameter values, shown by the 1:1 diagonal line. However, the regression models may overestimate or underestimate the true values of other parameters (e.g., niche width (w) is overestimated and log maximum fecundity (r_O) is underestimated in this example).

taking dispersal into account. To estimate fecundity parameters, we first reparameterize Eq. (9.1) as a quadratic:

$$f_{i,x,t} = \exp(b_{1i} + b_{2i}\,pH_{x,y} + b_{3i}\,pH^2_{x,y} + a_i N_{t,x,y}). \tag{9.3}$$

Taking the derivative of f with respect to pH and setting it equal to zero, we see that the optimum pH for species i can be estimated by the equation

$$opt_{iR} = -b_{2i}/2b_{3i} \tag{9.4a}$$

where the subscript R indicates that the estimate was made using a regression model. Furthermore, comparing Eqs. (9.1) and (9.3), we see that niche width and maximum log fecundity are estimated by

TABLE 9.3. *Mean values of Spearman's ρ based on 100 replicates of* nichemaker *and the correlations between estimated and true parameter values. Data were analyzed separately for three different time periods (years 61–65, 61–70, and 1–5) for each replicate run. Population growth rates (λ_0 and λ_N) were estimated using Eq. (9.7) and either regression or likelihood parameter estimates and then compared using Spearman's ρ to the growth rates calculated using the true parameter values. Wilcoxon rank sum tests (two-tailed) were used to compare the means of the regression correlations with the means of the likelihood correlations. The numbers in parentheses after the means are the number of replicates (out of a total of 100) for which Spearman's ρ is significantly different from zero.*

		Spearman's ρ and the percentage significant τ values ($P < 0.05$)					
		Years 61–65		Years 61–70		Years 1–5	
Adult survival (p_A)	Regression	0.769	(69)	0.824	(83)	0.907	(100)
	Likelihood	0.769[b]	(69)	0.824[b]	(83)	0.907[b]	(100)
Juvenile survival (p_J)	Regression	0.755	(63)	0.825	(81)	0.937	(100)
	Likelihood	0.755[b]	(64)	0.825[b]	(81)	0.937[b]	(100)
Dispersal fraction (p_D)	Regression	0.707	(60)	0.749	(61)	0.825	(93)
	Likelihood	0.717[b]	(60)	0.734[b]	(62)	0.776[a]	(75)
Optimum pH (*opt*)	Regression	0.693	(62)	0.726	(69)	0.980	(97)
	Likelihood 1	0.879[a]	(88)	0.899[a]	(94)	0.972[b]	(99)
	Likelihood 2	0.884[a]	(91)	0.896[a]	(93)	0.957[a]	(97)
Niche width (*w*)	Regression	0.288	(12)	0.297	(11)	0.851	(90)
	Likelihood 1	0.525[a]	(31)	0.577[a]	(39)	0.851[b]	(88)
	Likelihood 2	0.337[b]	(19)	0.330[b]	(15)	0.570[a]	(24)
Log max fecundity (r_O)	Regression	0.479	(27)	0.520	(30)	0.910	(91)
	Likelihood 1	0.503[b]	(38)	0.548[b]	(33)	0.900b[2]	(91)
	Likelihood 2	0.490[b]	(26)	0.519[b]	(30)	0.835[a]	(72)

TABLE 9.3. (*cont.*)

		Spearman's ρ and the percentage significant τ values ($P < 0.05$)					
		Years 61–65		Years 61–70		Years 1–5	
Density dependence (*a*)	Regression	0.444	(21)	0.466	(21)	0.591	(34)
	Likelihood 1	0.438[b]	(23)	0.488[b]	(23)	0.623[b]	(49)
	Likelihood 2	0.407[b]	(24)	0.483[b]	(24)	0.571[b]	(24)
Population growth rate (λ_0)	Regression	0.861	(100)	0.874	(100)	0.992	(100)
	Likelihood 1	0.897[a]	(100)	0.920[a]	(100)	0.993[b]	(100)
	Likelihood 2	0.889[a]	(100)	0.905[a]	(100)	0.987[a]	(100)
Population growth rate (λ_N)	Regression	0.904	(100)	0.809	(100)	0.994	(100)
	Likelihood 1	0.946[a]	(100)	0.956[a]	(100)	0.997[a]	(100)
	Likelihood 2	0.941[a]	(100)	0.948[a]	(100)	0.994[b]	(100)

[a] Difference between regression estimate and likelihood estimate significant by two-tailed Wilcoxon test ($P < 0.01$).

[b] Difference not significant ($P > 0.01$).

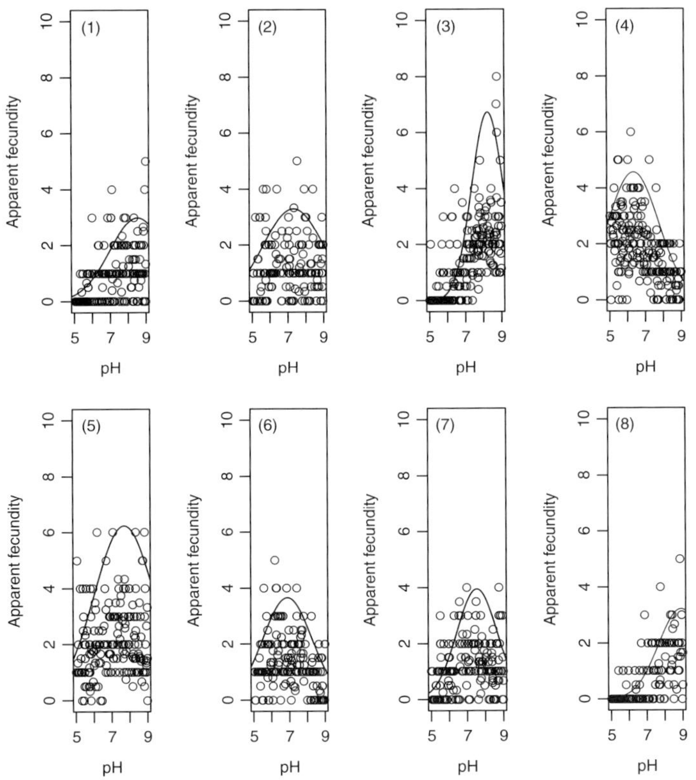

FIGURE 9.2.
Apparent fecundity is the number of seedlings per adult found during the annual census of a cell. Each panel shows the data for a single species and year (year 5). The solid line shows the expected fecundity for a cell with no competition ($N_t = 0$). The scatter of points reflects both variation between cell pH and the species optimum pH, and variation in actual population density. Despite this variation, each species tends to have its highest apparent fecundity when pH is near the optimum for the species. For example, as can be seen in both this figure and Table 9.2, species 4 has a relatively low pH optimum (6.3) and species 8 has a relatively high pH optimum (8.9) and the scatter of points reflects this difference. Color version available online at: www.cambridge.org/9780521199476.

$$w_{iR} = -1/b_{3i} \tag{9.4b}$$

and

$$r_{OiR} = b_{1i} + opt^2/w = b_{1i} - b_{2i}^2/4b_{3i}. \tag{9.4c}$$

For now, ignoring dispersal, we assume that the number of seedlings observed on a cell is the same as the number of seedlings produced by the adults present on that cell. The number of observed seedlings on a cell with known pH and adult density is assumed to be a Poisson-distributed random variable with mean

$$n_{i,x,y,t}{}^{*} f_{i,x,y,t} = n_{i,x,y,t} \exp(b_{0i} + b_{1i} pH_{x,y} + b_{2i} pH_{x,y}^2 - aN_{t,x,y}) \tag{9.5}$$

where $f_{i,x,y,t}$ is the per capita fecundity and $n_{i,x,y,t}$ is the number of adults on cell x,y at time t. Taking the log of the right side of this equation, we see that the corresponding statistical model in R is

$$\text{glm}(J \sim logn + PH + I(PH^2) + N, \text{family} = \text{Poisson}(\text{link} = \log)). \tag{9.6}$$

Because slopes and intercepts all depend on species identity, we chose to do a separate regression for each species rather than treating species as a factor in a single regression model including all species. Accordingly, prior to statistical analysis, we sorted the data by species and selected only those grid cells and times for which adults are present. In Eq. (9.6), J is a vector containing the number of seedlings and *logn* is the natural logarithm of the number of adults of the focal species for each cell and year. The terms $PH + I(PH^2)$ instruct R to include both first- and second-order effects of pH.

Although our emphasis is parameter estimation, not model selection, for several replicates we compared models with different combinations of parameters and used AIC for model selection. As expected, the best model (lowest AIC) always retained both first- and second-order pH effects and a density effect for all species. A positive first-order pH effect (b_{2i}) and a negative second-order effect (b_{3i}) indicated a bell-shaped (or Gaussian) response of fecundity to pH and, in all cases examined, first-order coefficients were found to be significantly greater than zero and second-order coefficients were significantly less than zero for all species. The regression coefficients were transformed to estimates of fecundity parameters (opt_{iR}, w_{iR}, and r_{OiR}) using Eqs. (9.4a–c), and the results are shown in Figure 9.1, where the estimated parameters are compared to the true parameters used to generate the data in the first place. In this particular example, the parameter estimates for optimal pH are close to the true values but niche width is overestimated. Later, we present the analysis of 100 separate replicates, each with a unique set of randomly selected demographic and dispersal parameters, and test the generality of this bias.

Estimating dispersal

The simplest way to estimate dispersal is to focus on those grid cells that do not have any adults on them and assume that all seedlings found on these cells must be immigrants from neighboring cells. Accordingly, *nicheobserver* sorts the data and creates two new vectors, one containing the number of seedlings, $J_{i,x,y,t}^{A=0}$, of species i on every cell x,y with no adults of species i present, and the second specifying the number of seedlings, $J_{i,x,y,t}^{neigh}$, on the four neighboring cells ($x-1,y$; $x+1,y$; $x,y-1$; and $x,y+1$). *Nicheobserver* uses the logistic regression model

$$glm(D \sim spfactor - 1, family = binomial(link = logit)) \tag{9.7}$$

to estimate the dispersal fraction (p_{Di}) for each species. In the regression model, the first column of D is $J^{A=0}$ and the second is J^{neigh}. As before, the inclusion of −1 after *spfactor* yields a regression with one coefficient (d_i) for each species. Since we used the logit link and immigrants come from any of four neighboring cells, the dispersal fraction is estimated by $p_{DiR} = 4\exp(d_i)/(1+\exp(d_i))$, where again the subscript R indicates that the parameter was estimated with a regression model. As illustrated in panel 6 of Figure 9.1 and later in Table 9.3, this estimate is strongly correlated with the true dispersal fraction.

A maximum likelihood approach to parameter estimation

So far we have estimated demographic parameters ignoring dispersal. We now ask "Can we obtain a better estimate of population growth rate by incorporating an estimate of dispersal into the estimation of fecundity?" To get a better estimate of fecundity, we need to know how many seedlings present on a cell were produced by the adults on that cell and how many immigrated from elsewhere. Our general approach was to develop a model incorporating both fecundity and dispersal and to obtain parameter estimates using maximum likelihood techniques. We developed two versions of such a model. In version 1, we use the dispersal fraction estimated by regression p_{DiR} to inform the maximum likelihood estimation of fecundity parameters. In version 2, we use maximum likelihood methods to *simultaneously* estimate the fecundity and the dispersal fraction (p_{DiL}). We expected version 2 to yield better parameter estimates because the estimate of dispersal is informed by demographic information and vice versa; however, as we show later, this was not always the case.

For maximum likelihood parameter estimation we find the set of parameters that maximizes the likelihood L – or, equivalently, minimizes $-log(L)$ – of obtaining the observed data. In our case, the data to be explained are the numbers of post-dispersal seedlings found on each cell. To implement the maximum likelihood approach, we need to specify an appropriate probability distribution for the data. In our model, the mean number of seedlings per cell is $\mu = B + I - E$, where B is the expected number of seeds produced or "born" on each cell, E is the expected number emigrating from the cell, and I is the expected number immigrating into the cell. One approach would be to model the probability of finding J seedlings on a given cell as a Poisson random variable using μ as the Poisson parameter. However, for the Poisson distribution, the variance equals the mean, and in our data, the variance is typically larger than the mean. This elevated variance is not surprising because the number of seedlings observed on a cell results from at least three different processes: seed production, emigration, and immigration, each of which has a random component.

The *negative binomial probability distribution* is more appropriate for our data than the Poisson because, like the Poisson, the negative binomial takes on discrete, positive values but, unlike the Poisson, the negative binomial permits a variance larger than the mean (i.e., "overdispersed" data). The negative binomial has two parameters, referred to as r – the size parameter, and p_{NB} – the probability parameter. Using this notation, the mean of a negative binomial is $r(1-p_{NB})/p_{NB}$ and the variance is $r(1-p_{NB})/p_{NB}^2$. Since the mean divided by the variance equals p_{NB}, the parameter p_{NB} can be thought of as the mean to variance ratio, and the variance is greater than the mean as long as $p_{NB} < 1$. In the limit as p_{NB} approaches 1, the mean equals the variance and the negative binomial converges to a Poisson distribution. Applying the negative binomial to our data, the mean number of seedlings per cell is $\mu = r(1-p_{NB})/p_{NB}$ and the variance in the number of seedlings per cell is μ/p_{NB}. Thus the probability of observing J seeds on a cell is the probability density of J, given a negative binomial with parameters r and p_{NB}, or in R notation *dnbinom*(J, r, p_{NB}).

Our maximum likelihood approach to parameter estimation is to find demographic and dispersal parameters that maximize the probability of the data given a negative binomial distribution with mean $\mu = r(1-p_{NB})/p_{NB}$ and variance μ/p_{NB}. Thus, to apply the negative binomial to our data, we need to express the mean μ of the *negative binomial distribution* as a function of the factors that determine the number of seedlings per cell. The mean number of seedlings per cell is $\mu = B - E + I$, so we need to express B, E, and I as functions of cell pH and population density. The per capita estimated fecundity (mean seed production) for species i on cell (x,y) is

$$E[f_{i,x,y,t}] = \exp(r_{0i} - (opt_i - pH_{x,y})^2/w_i - a_i N_{x,y,t}). \tag{9.8}$$

Nicheobserver uses a function called *like*( ) to calculate the probability of the data assuming a negative binomial probability distribution with parameters r and p_{NB}. The function *like*( ) uses Eq. (9.8) in its matrix form to calculate expected per capita fecundity. The expected number of seeds of species i produced on a cell x,y with $n_{i,x,y,t}$ adults is $B_{i,x,y,t} = n_{i,x,y,t}E[f_{i,x,y,t}]$. Having thus calculated B, the function *like*( ) proceeds to calculate the expected number of emigrants E and the expected number of immigrants I. The function *like*( ) uses B and a maximum likelihood estimate p_{DL} of the dispersal fraction to calculate the expected number of emigrants, $E = p_{DL}B$. Finally, in order to calculate I, *like*( ) requires an estimate of the numbers of seeds produced on neighboring cells, because cells adjacent to a focal cell are the source of immigrants into the focal cell. To calculate the number of seeds produced on neighboring cells, *like*( ) calls another function, *neighbor.calc*( ), which adds up the number of seeds produced on the appropriate neighboring cells, taking into account edge effects. Since, on average, ¼ of the seeds produced on neighboring cells immigrate to a given focal cell, the expected number of immigrants is ¼ of the dispersal fraction times the number of seeds produced on neighboring cells.

Nicheobserver uses an R-supplied function *optim(init,like)* to call the function *like( )* and calculate the probability that J_i seedlings of species i are observed, assuming a *negative binomial* distribution. *Optim(init,like)* requires a set of initial values (*init*) of the parameters to be estimated and returns to the main program the maximum likelihood parameter values (i.e., those that minimize the negative log likelihood of the data). As mentioned, we developed two versions of the likelihood model. In version 1, we used the regression estimate of dispersal fraction, p_{DiR}, and maximum likelihood estimates of the demographic parameters (opt_{iL1}, w_{iL1}, r_{OiL1}, a_{iL1}) and the mean to variance ratio p_{NBiL1}. In version 2, all six parameters (opt_{iL2}, w_{iL2}, r_{OiL2}, a_{iL2}, p_{NBiL2}, p_{DiL2}) are simultaneously estimated by the maximum likelihood function *like( )*.

Estimates of population growth (λ) and identification of sources and sinks

Our primary concern is not estimating individual parameters, such as niche width and dispersal fraction, but instead estimating population growth rates and identifying sources and sinks. Population growth rates depend on the conditions experienced by individuals in the population and these vary across space and time. Environmental conditions are experienced locally, and conditions may vary substantially from one grid cell to the next. Obviously, the individuals on a single grid cell do not constitute an entire population; nonetheless, we can estimate each cell's contribution to population growth. For the simple life cycle presented in this chapter, we estimate a cell's contribution to population growth of species i by

$$\lambda_{i,x,y,t} = p_{Ai} + p_{Ji}\exp(r_{Oi} - (opt_i - pH_{x,y})^2/w_i - a_i N_{x,y,t}). \qquad (9.9)$$

The value of $\lambda_{i,x,y,t}$ thus calculated is equivalent to the growth rate of a population all individuals of which experience the same conditions (pH and population density) as experienced by those individuals on cell x,y at time t.

To evaluate how well different statistical models estimate population growth rates, we calculated "true" growth rates using Eq. (9.9), incorporating the true parameter values used to generate the data, and compared these to growth rate estimates based on either regression or maximum likelihood models. We calculated λ for each cell and species, and both for $N = 0$ and using $N_{x,y,t}$ at each time step. In the sense of Hutchinson's classic distinction between fundamental and realized niche, the calculation at $N = 0$ can be thought of as a *fundamental lambda*, λ_0, inasmuch as it depends on abiotic conditions (pH) on a

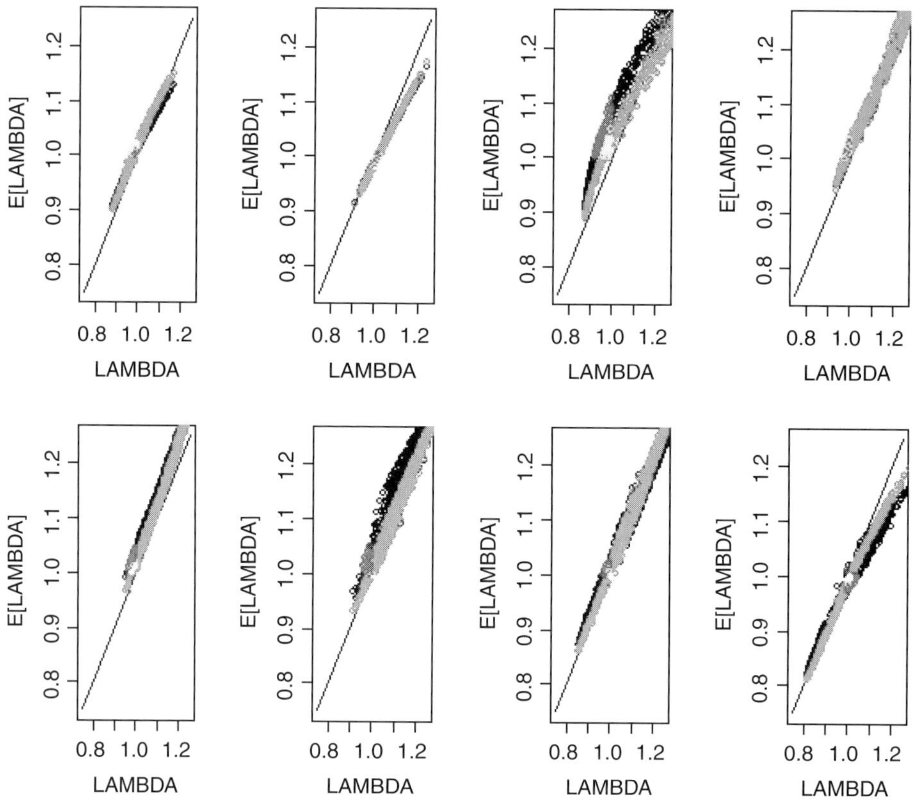

FIGURE 9.3.
Population growth rates (λ_N) estimated for eight hypothetical plant species from a single run of *nichemaker* and *nicheobserver* compared with rates based on the true parameter values used to generate the dataset. Black symbols indicate growth rate estimates based on regression parameters from generalized linear models. Color version available online at: www.cambridge.org/9780521199476.

grid cell but not on biotic conditions ($N_{x,y,t}$). Calculated in this manner, λ_0 measures the ability of a species to invade unoccupied habitat. By contrast, λ_N, the *realized lambda*, measures population growth when population density (N) is in the vicinity of a stochastic equilibrium and may, therefore, more closely resemble population growth rates measured from field data. We compare regression and maximum likelihood estimation methods both by calculating correlation coefficients between true values of λ and estimates of both λ_0 and λ_N. We also asked how often the estimated values of λ equal or exceed 1.0 when the true value is less than 1.0 and, conversely, how often the estimated value is less than 1.0 when the true value is equal to or greater than 1.0. Taking only a slight liberty with the definitions of source and sink, we will refer to these two cases, respectively, as "misclassifying sinks as sources" and "misclassifying sources as sinks" (Fig. 9.3).

Comparison of the regression and maximum likelihood parameter estimates

To evaluate the various methods of parameter estimation and return to the original question "Does ignoring dispersal result in poor parameter estimation?" we ran 100 replicates of *nichemaker*, each generating both a unique set of demographic and dispersal parameters and a unique dataset providing the number of adults and seedlings on each grid cell at each time step. We then used *nicheobserver* to analyze each dataset produced by *nichemaker* to estimate demographic and dispersal parameters using both regression and likelihood methods.

Nichemaker generates seven parameters (p_J, p_A, opt, w, r_O, a, p_D) for each species and each replicate of the model. Furthermore, all parameters are estimated for each replicate by *nicheobserver* using both regression (R) and maximum likelihood (L) methods. Thus, there are two estimates of adult survival (p_{AR} and p_{AL}), two estimates of juvenile survival (p_{JR} and p_{JL}), and two estimates of dispersal fraction (p_{DR} and p_{DL}). Estimates of survival do not require any assumption about dispersal, but the four parameters that determine fecundity (opt, w, r_O, and a) cannot be estimated without making some assumption about dispersal. The regression estimates of fecundity parameters (opt_R, w_R, r_{OR}, a_R) ignore dispersal and thereby, implicitly, assume that dispersal is not important. We developed two versions of likelihood models to estimate fecundity parameters and the two versions incorporate different assumptions about dispersal. For version 1, the fecundity parameters (opt_{L1}, w_{L1}, r_{OL1}, a_{L1}) were estimated by maximizing a likelihood function that was conditioned on the regression estimate of the dispersal fraction (p_{DR}). For version 2, the dispersal fraction (p_{DL}) and the fecundity parameters (opt_{L2}, w_{L2}, r_{OL2}, a_{L2}) were simultaneously estimated with the same likelihood function.

Table 9.3 summarizes the parameter estimates using both regression and the two versions of likelihood methods. We used Spearman's correlation (ρ) to measure how well the estimated parameters matched the true parameter values. For each of the seven parameters (p_J, p_A, opt, w, r_O, a, p_D) and for each of the 100 replicates, we calculated both Spearman's ρ and the associated probability that ρ was not different from zero (p-value). The table presents the mean values of ρ for each parameter and the number of times, out of 100, that the associated p-value was less than 0.05. We also used Spearman's ρ to evaluate how well the estimates of population growth rates (λ_0 and λ_N) based on either regression or likelihood methods matched population growth rates calculated using the true parameter values (see Fig. 9.3).

For each replicate, we estimated all parameters and performed the statistical test for three separate time periods, years 1–5, 61–65, and 61–70. Since the simulations were initialized far from equilibrium and quasi-stationary distributions were only reached after about 30 years, we take the later years (>60) to represent something akin to the data that a field observer would typically encounter, while the data from earlier years (1–5) are more like the data that might result from a field manipulation that reduced population sizes well below their steady-state values.

Both regression and likelihood methods provided good estimates of adult and juvenile survival (all ρ values were above 0.75). The correlations for both estimation methods were excellent ($\rho > 0.9$) when using data for the early years (years 1–5). Survival correlations were lower, but still relatively high ($\rho \sim 0.8$), for years 61–65, and using 10 years of data (years 61–70) resulted in only a modest increase in correlation coefficients. In addition to comparing correlation coefficients, parameter estimates can be compared in terms of parameter "bias"; that is, by the extent to which parameters tend to be systematically overestimated or underestimated. A simple measure of parameter bias is the ratio of the median parameter estimate to the median true parameter value. Parameter bias, using this measure, was very low (<1%) for all survival estimates, regardless of the estimation method or years used.

In general, the parameters that contribute to fecundity proved more difficult to estimate than survival parameters. Three of the fecundity parameters, niche width (w), log maximum fecundity (r_O), and the density dependence parameter (a), were especially difficult to estimate, and in years 61–65 less than half of the correlation coefficients (ρ) were significantly different from 0. (Note, however, that only about 5 in 100 ρ values would be greater than zero for uncorrelated data, so even these low values indicate some ability of the statistical methods to recover meaningful parameter estimates.) To determine whether or not the likelihood methods performed better than the regression methods when estimating fecundity parameters, we compared the means estimated by regression to the corresponding means estimated by likelihood methods using a two-sided paired Wilcoxon rank sum test. As expected, likelihood methods generally outperformed regression methods; however, contrary to our expectation, version 1 of the likelihood method performed slightly better than version 2. Furthermore, when using data from years 1–5, regression estimates performed about as well as, or better than, likelihood version 1, and both of these methods tended to outperform likelihood version 2.

As illustrated in Table 9.3 and Figure 9.3, both regression and likelihood methods performed well when estimating population growth rates (λ_0 and λ_N); nonetheless, likelihood methods accounting for dispersal generally outperformed regression methods (Table 9.3). As illustrated in Figure 9.3, estimates of population growth rate were strongly correlated with growth rates calculated from true population parameters. Likelihood estimates accounting for dispersal generally resulted in substantially fewer misclassifications (sources classified as sinks or sinks classified as sources). Averaging over all replicates, years, and grid cells, regression estimates ignoring dispersal resulted in misclassification in 16% of all cases compared with only 5% misclassification using likelihood methods (version 2) that accounted for dispersal. Overall, sinks misclassified as sources outnumbered sources misclassified as sinks by more than 3:1 and, as illustrated in Figure 9.3, most misclassifications occurred when the true λ was very close to 1.0.

For the most part, parameter bias was not strong, regardless of the estimation method used, inasmuch as the estimated values were typically within 2–3% of the true values. The largest bias by far was for niche width with regression methods, where the median parameter estimate was 29.1% larger than the true parameter value (years 61–65). Other noteworthy biases were the regression estimates of the density dependence parameter (underestimated by 11.3%) and dispersal fraction (underestimated by 6.1%). The biases using likelihood methods were consistently smaller than those using regression methods. The largest likelihood biases were underestimates of niche width (−2.0% with likelihood version 1 and −4.3% with version 2) and dispersal fraction (−6.1% with version 1 and −4.2% with version 2). In general, average population growth rates were very close to average values estimated by regression but were slightly overestimated by regression methods. Regression estimates of λ_0 exceeded true rates by 1.6% and regression estimates of λ_N exceeded true values by 2.6%. Likelihood estimates of λ_0 and λ_N were both within 0.5% of true values.

Discussion

In this chapter, we have shown that ignoring dispersal can result in biased estimation of demographic parameters. For example, ignoring dispersal may result in overestimating fecundity in sink habitat and underestimating it in source habitat. This is a direct result of ignoring dispersal because post-dispersal censuses discover fewer juveniles in the patches where they were born and more in the adjacent, often poorer-quality patches, to which they have dispersed. We believe that similar errors in parameter estimation will occur whenever dispersal rates and dispersal distances are large enough that substantial numbers of individuals emigrate from good patches into poor patches of habitat. Given the heterogeneity of natural and human-altered landscapes, this is likely to be true for most species most of the time.

Errors in estimation of demographic parameters may lead to even larger errors in predictions of population dynamics. Therefore, given the widespread interest in applying population and niche models to predict the consequences of climate change (Guisan and Thuiller 2005; Kearney *et al.* 2008), it is imperative that we develop better statistical methods for estimating parameters and accounting for parameter uncertainty in model prediction. Errors introduced by ignoring dispersal may present an especially difficult problem with regard to using niche models to predict changes in species distributions under various scenarios of climate change because such models are often parameterized with presence–absence data, or even presence-only data, with no information about habitat suitability.

To deal with the problem of ignoring dispersal when estimating demographic parameters, we have proposed two maximum likelihood approaches to estimate demographic parameters that take dispersal into account. In version 1, the dispersal fraction was first estimated by regression and this estimate was then used to adjust the data for emigration and immigration before fecundity parameters were estimated by likelihood methods. In the second version, we estimated dispersal and fecundity parameters simultaneously, using the same data and a single likelihood function. In cases involving quasi-stationary data (e.g., >year 60 in our examples), both likelihood approaches significantly improved on parameter estimates made with regression methods that ignore dispersal. We were surprised, however, to find that both the regression method ignoring dispersal and version 1 of the likelihood model performed better than version 2 of the likelihood model when analyzing data from the very early years (1–5) of the simulations. In hindsight, we think the primary reason for this is that in the very early years, when population densities are low, population growth is very high and increases in number due to demography may completely swamp increases or declines due to dispersal. Looking at it another way, dispersal interferes with the estimation of demographic parameters only when dispersal is relatively high compared with fecundity.

The observation that regression methods ignoring dispersal lead to acceptable parameter estimates in the early years when populations are sparse suggests that the experimental reduction of population size and subsequent monitoring of population dynamics during a growth phase could lead to better parameter estimates than might be obtained in non-experimental studies. Investigators could also take advantage of natural disturbances to estimate demographic parameters while monitoring the recovery of populations perturbed from stationary conditions. Of course, some of the problems of parameter estimation can be avoided just by recording additional kinds of data or taking censuses at different times. In the hypothetical example presented in this chapter, the confounding of dispersal and demography resulted from not being able to distinguish seeds that were produced locally from those that dispersed to the site from elsewhere. This problem is common and can be avoided by a number of techniques that allow local and imported propagules to be distinguished from one another. For example, a census of seed production before dispersal followed by a second census of seedling abundance just after dispersal would greatly enhance the ability to distinguish fecundity and dispersal. Similarly, a host of techniques – ranging from banding nestlings before they leave the nest to allow an avian researcher to distinguish local juveniles from those dispersing onto a site, to fin or toe clips and fluorescent dyes to genetic markers and stable isotopes – may help field researchers distinguish locally produced individuals from those coming from afar.

In some cases, dispersal can be observed and measured directly and dispersal distance and frequency can be directly estimated from field observations. An example of this approach is presented by Diez and Giladi (Chapter 14, this volume). These authors estimated demographic parameters by following the survival and reproduction of marked individuals and estimating dispersal distances by direct observations of ants carrying seeds on the same study plots. Similar information will often be difficult to obtain, for example, when seeds are dispersed by wind or mammals and, in the absence of direct measurements, simultaneous statistical estimation of dispersal and demography, as done in this chapter, may be the best available alternative.

Finally, we point out that while the population model developed and statistical methods suggested here are relatively easy to implement and lead to tractable results, we do not claim our methods are the only, or even the best, approach to the problem of accounting for dispersal in parameter estimation. Rather our intent is to illustrate the problem, to point out one approach to solving it, and to invite the community to offer other approaches and solutions. We made a number of relatively arbitrary choices regarding the probability distributions, the demographic and dispersal characteristics of species, and the spatial scale of the analysis. While we think these choices are both interesting and defensible, further analysis using alternative species and grid characteristics is fully warranted. For example, treating space as a continuous, rather than a discrete variable, and exploring alternative dispersal characteristics would add to the applicability of the results. Furthermore, we suggest that Bayesian parameter estimates of survival, reproduction, and population growth rates that are fully conditioned on knowledge of dispersal would lead to better parameter estimates for population models.

In summary, we hope that our initial exploration of this problem will lead to population models that more fully account for dispersal, and new statistical methods that better incorporate dispersal into parameter estimation.

Acknowledgments

The authors are grateful to the organizers of the symposium that led to this book and to the editors of the book. The senior author (H. R. Pulliam) especially wishes to thank the primary conference organizer and editor of the book,

Jianguo Liu, who was a model student during his graduate school years and has now become a distinguished scientist of whom the senior author is very proud.

References

Diez, J. M. (2006). Hierarchical patterns of symbiotic orchid germination linked to adult proximity and environmental gradients. *Journal of Ecology* **95**: 159–170.
Diez, J. M. and H. R. Pulliam (2007). Hierarchical analysis of species distributions and abundance across environmental gradients. *Ecology* **88**: 3144–3152.
Elith J. and M. A. Burgman (2002). Predictions and their validation: rare plants in the Central Highlands, Victoria, Australia. In *Predicting Species Occurrences: Issues of Accuracy and Scale* (J. M. Scott, P. Heglund and M. L. Morrison, eds.). Island Press, Washington, DC: 303–314.
Guisan, A. and W. Thuiller (2005). Predicting species distributions: offering more than simple habitat models. *Journal of Vegetation Science* **9**: 65–74.
Kadmon, R. (1993). Population dynamic consequences of habitat heterogeneity: an experimental study. *Ecology* **74**: 816–825.
Kadmon, R. and K. Tielborger (1999). Testing for source–sink dynamics: an experimental approach exemplified with desert annuals. *Oikos* **86**: 417–429.
Kearney, M. (2006). Habitat, environment and niche: what are we modeling? *Oikos* **115**: 186–191.
Kearney, M., B. L. Phillips, C. R. Tracy, G. Betts and W. P. Porter (2008). Modelling species distributions without using species distributions: the cane toad in Australia under current and future climates. *Ecography* **31**: 423–434.
Lawesson, J. E. and J. Oksanen (2002). Niche characteristics of Danish woody species as derived from coenoclines. *Journal of Vegetation Science* **13**(2): 279–290.
MacArthur, R. H. (1969). Species packing and what interspecific competition minimizes. *Proceedings of the National Academy of Sciences USA* **64**: 1369–1371.
May, R. M. and R. H. MacArthur (1972). Niche overlap as a function of environmental variability. *Proceedings of the National Academy of Sciences USA* **69**: 1109–1113.
Pulliam, H. R. (2000). On the relationship between niche and distribution. *Ecology Letters* **3**: 168–175.
Pulliam, H. R. and N. M. Waser (2010). Ecological invariance and the search for generality in ecology. In *The Ecology of Place: Contributions of Place-Based Research to Ecological Understanding* (I. Billick and M. V. Price, eds.). University of Chicago Press, Chicago, IL: 69–92.
Urban, M. C., B. L. Phillips, D. K. Skelly and R. Shine (2007). The cane toad's (*Chaunus* [*Bufo*] *marinus*) increasing ability to invade Australia is revealed by a dynamically updated range model. *Proceedings of the Royal Society – Biological Sciences – Series B* **274**: 1413–1419.
Warren, R. J. (2007). Linking understory evergreen herbaceous distributions and niche differentiation using habitat-specific demography and experimental common gardens. PhD dissertation, University of Georgia, Athens, GA.

Appendix

This abbreviated version of the *nichemaker* and *nicheobserver* models can be used to reproduce Table 9.2 and Figures 9.1, 9.2, and 9.3. The parameters of *nichemaker* can be easily changed but the user must check the values of the variables ATOTAL, JTOTAL, and EMPTY to ensure that sample sizes are large enough for analysis by *nicheobserver*. A longer version, which embeds this code in a replication loop and automatically checks sample size, is available from the senior author (H. R. Pulliam).

```
# nichemaker
SEED<-2107; set.seed(SEED) # seed random number generator
years=c(1:5); yrs=length(years) # choose the years on which analysis is performed
tmax=max(years)+1 # number of time steps in the simulation
sp=8; # size of the species pool
aveE=7; # mean pH of environment
gx=16; gy=16; m=c(gx,gy) # grid dimension gx by gy; x represents rows and y represents columns
vS=2; minS=aveE-vS; maxS=aveE+vS # min and max species pH optima
vE=2; minE=aveE-vE; maxE=aveE+vE # min and max environmental pH
# Generate new species characteristics
pA=runif(sp,0.8,0.9); pJ=runif(sp,0.1,0.2); # prob of adult and seedling (juvenile) survival
opt=runif(sp,minS,maxS); # generates optimal pH vales from Uniform distribution
w=runif(sp,1,10); # pH niche width
r=runif(sp,1,2); # max log fecundity
a=runif(sp,0.04,0.06); # density dependece of fecundity
pD=runif(sp,0.05,0.15); # fraction of seeds dispersing to one of 4 adjacent cells
IM=rgamma(sp,1,1) # immigration (from outside the grid) rate vector
# Generate new landscape characteristics
pH<-matrix(runif(gx*gy,minE,maxE),m); # generates pH vales on landscape
n<-array(0,c(sp,m,tmax)); n[1:sp,1:m[1],1:m[2],1]=1 # number adults species i, cell x,y, time t
sds<-array(0,c(sp,m,tmax)) # number of viable seeds produced
juv<-array(0,c(sp,m,tmax)) # number of seedlings
juvsurv<-array(0,c(sp,m,tmax)) # surviving seedlings
adultsurv<-array(0,c(sp,m,tmax)) # surviving adults
for (t in 1:(tmax-1))    # BEGIN TIME LOOP for simulation of population dynamics
{N<-matrix(data=0,nrow=m[1],ncol=m[2]) # matrix for cell-level density (all species combined)
invisible(sapply(1:sp,function(i){N<<-N+n[i,,,t]}))
for (i in 1:sp) # BEGIN SPECIES LOOP
{F<-exp(r[i]-(opt[i]-pH)^2/w[i]-a[i]*N) # calculate pH adjusted mean seed production for all cells
```

```
# choose total number of seeds produced by species i for all grid cells at
time t
sds[i,,,t]<-t(sapply(1:m[1],function(x){sapply(1:m[2],function(y){return(rpoi
s(1,n[i,x,y,t]*F[x,y]))})}))
# choose the number of dispersing seeds of those produced by sp i on each
grid cell
nd<-t(sapply(1:m[1],function(x){sapply(1:m[2],function(y){return(rbinom(1,
sds[i,x,y,t],pD[i]))})}))
# determine where dispersing individuals move to
dispersed<-disp.left<-disp.right<-disp.up<-disp.down<-matrix(0,gx,gy) #
initialize dispersal matrices
if (any(nd>0))
{# determine number dispersing to the left for each grid cell
disp.left<-t(sapply(1:m[1],function(x){sapply(1:m[2],function(y){return(rbin
om(1,nd[x,y],0.25))})}))
dispersed<-disp.left
}
if (any(nd-dispersed>0))
{# determine number dispersing to the right for each grid cell
disp.right<-t(sapply(1:m[1],function(x){sapply(1:m[2],function(y)
{return(rbinom(1,nd[x,y]-dispersed[x,y],1/3))})}))
dispersed<-dispersed+disp.right
}
if (any(nd-dispersed>0))
{# determine number dispersing up for each grid cell
disp.up<-t(sapply(1:m[1],function(x){sapply(1:m[2],function(y)
{return(rbinom(1,nd[x,y]-dispersed[x,y],1/2))})}))
dispersed<-dispersed+disp.up
}
if (any(nd-dispersed>0))
{# determine number dispersing down for each grid cell
disp.down<-nd-dispersed
}
# Add up totals to obtain post-dispersal juvenile distribution
juv[i,,,t]<-sds[i,,,t]-nd # seeds that didn't disperse
juv[i,,1:(m[2]-1),t]<-juv[i,,1:(m[2]-1),t]+disp.left[,2:m[2]] # add seeds disper-
sing left
juv[i,,2:m[2],t]<-juv[i,,2:m[2],t]+disp.right[,1:(m[2]-1)] # add seeds disper-
sing right
juv[i,1:(m[1]-1),,t]<-juv[i,1:(m[1]-1),,t]+disp.up[2:m[1],] # add seeds disper-
sing up
```

```
juv[i,2:m[1],,t]<-juv[i,2:m[1],,t]+disp.down[1:(m[1]-1),] # add seeds dispersing down
# Add Immigration from outside grid
I<-rpois(1,IM[i]) # number of immigrants from outside grid of each species
if (I>0) # assign each immigrant to a randomly chosen cell (x,y)
{cell.x<-ceiling(runif(I,0,gx)) # x coordinate
cell.y<-ceiling(runif(I,0,gy)) # y coordinate
invisible(sapply(1:I,function(cell){juv[i,cell.x[cell],cell.y[cell],t+1]<<-
n[i,cell.x[cell],cell.y[cell],t]+1}))
}
# Determine number of individuals surviving until NEXT census
adultsurv[i,,,t]<-t(sapply(1:m[1],function(x){sapply(1:m[2],function(y){retur
n(rbinom(1,n[i,x,y,t],pA[i]))})}))
juvsurv[i,,,t]<-t(sapply(1:m[1],function(x){sapply(1:m[2],function(y){return(r
binom(1,juv[i,x,y,t],pJ[i]))})}))
  # Update total adult population size
n[i,,,t+1]<-adultsurv[i,,,t]+juvsurv[i,,,t] # new pop total is surviving adults
plus surviving juveniles
} # END SPECIES LOOP
} # END TIME LOOP
# Store the output: each line has year(t), species(i), xaxis(x),
yaxis(y),PH(pH),
# number of adults(A), surviving adults(AS), juveniles(J), surviving
juveniles(JS), population density (N)
yrs=length(years)
species<-species<-rep(rep(1:sp,each=gx*gy),yrs)
time<-rep((years),each=sp*gx*gy)
xaxis<-rep(rep(rep(1:gx,each=gy),sp),yrs); yaxis<-rep(1:gy,gx*sp*yrs)
PH<-rep(as.vector(t(pH)),sp*yrs)
A<-as.vector(sapply(years,function(t){sapply(1:sp,function(i){t(n[i,,,t])})}))
AS<-as.vector(sapply(years,function(t){sapply(1:sp,function(i)
{t(adultsurv[i,,,t])})}))
J<-as.vector(sapply(years,function(t){sapply(1:sp,function(i)
{t(juv[i,,,t])})}))
JS<-as.vector(sapply(years,function(t){sapply(1:sp,function(i)
{t(juvsurv[i,,,t])})}))
N<-as.vector(sapply(years,function(t){rep(as.
vector(t(sapply(1:gy,function(y){sapply(1:gx,function(x){sum(n[1:sp,x,y,t])})}
))),sp)}))
# Check to see if sample sizes are adequate for statistical analysis to
proceed
```

```
ATOTAL<-matrix(0,sp); JTOTAL<-matrix(0,sp); # total numbers of adults and juveniles of each species over all years
EMPTY<-matrix(0,sp) # number of cells with no adults of species i
for (i in 1:sp)
{ATOTAL[i]=sum(A[species==i])
JTOTAL[i]=sum(J[species==i])
EMPTY[i]=length(A[A==0 & species==i])
}
min(ATOTAL,JTOTAL) # If <=10 too few individuals of some species for analysis to proceed
min(EMPTY) # If <=5 there are too few empty cell to calculate dispersal
# END of nichemaker
######################################################################
###########################################
#START of nicheobserver
# arrays to hold parameter estimates
PJ<-array(0,c(4,sp)); PA<-array(0,c(4,sp)); OPT<-array(0,c(4,sp));
W<-array(0,c(4,sp)); R<-array(0,c(4,sp)); ALPHA<-array(0,c(4,sp)); PNB<-array(0,c(4,sp))
PD<-array(0,c(4,sp)); METHOD<-array(0,c(4,sp)); REPLICATE<-array(0,c(4,sp));
LAMBDA0<-array(0,c(4,sp,m)); LAMBDAN<-array(0,c(4,sp,m,yrs))
# function to add up seeds or seedlings on neighboring cell
neighbor.calc <-function(t,seeds,up=matrix(0,gy,gx),down=matrix(0,gy,gx),right=matrix(0,gy,gx),left=matrix(0,gy,gx))
{sk<-c(((t*gx*gy)-(gx*gy-1)):(t*gx*gy))
s<-matrix(seeds[sk],gy,gx)
up[2:gy,1:gx]<-s[1:(gy-1),1:gx] # seeds on cell immediately above focal cell
down[1:(gy-1),1:gx]<-s[2:gy,1:gx] # seeds on cell immediately below focal cell
right[1:gy,1:(gx-1)]<-s[1:gy,2:gx] # seeds on cell to right of focal cell
left[1:gy,2:gx]<-s[1:gy,1:(gx-1)] # seeds on cell to left of focal cell
return(up+down+left+right)
}
# likelihood function to calculate survival parameters
likeS<-function(initS)
{pA=initS[1]; pJ=initS[2] # initial values for search
if (pA<0||pA>1||pJ<=0||pJ>=1)
{return(1.7*10^308)} # Return (effectively) infinite neg log like value for impossible parameter
```

```
P1=dbinom(AS[species==i],A[species==i],pA,log=T) # likelihood for adult survival
P2=dbinom(JS[species==i],J[species==i],pJ,log=T) # likelihood for juvenile survival
return(-sum(P1+P2))
}
#likelihood function to calculate fecundity parameters
likeF<-function(param)
{opt=param[1]; w=param[2]; r=param[3]; a=param[4]; pNB=param[5]
if (versn==1) {pd=PD[2,i]} # version 1 uses regression estimate of pD
if (versn==2) {pd=param[6]} # version 2 estimates pD by maximum likelihood methods
Ai=A[species==i] # number of adults of species i
Ji=J[species==i] # number of juveniles of species i
size=length(Ji) # number of obervations
if (pd<0||pd>1||pNB<=0||pNB>=1)
{return(1.7*10^308)} # Return (effectively) infinite neg log like value for impossible parameter
# Design matrix Xi with covariates pH, pH square, and pop density N
Ef=exp(r-(opt-PH[species==i])^2/w-a*N[species==i])
B=Ai*Ef # expected number of seeds produced
SN<-as.vector(sapply(c(1:length(years)),neighbor.calc,seeds=B))
I=0.25*pd*SN # expected number of immigrants
E=pd*B # expected number of emigrants
mu=B-E+I # expected number of juveniles
r=(mu*pNB)/(1-pNB) # negative binomial size parameter
P1=dnbinom(Ji[Ai>0],r[Ai>0],pNB,log=T) # neg bino likelihood for cells w adults
P2=dnbinom(Ji[Ai==0 & SN>0],r[Ai==0 & SN>0],pNB,log=T) # for cells w/o adults
return(-sum(P1,P2)) # negative log likelihood (minimized by optim)
}
# Vectors to store parameters estimates
PA[1,]=pA; PJ[1,]=pJ; PD[1,]=pD # 'METHOD 1' refers to true parameter values
OPT[1,]=opt; W[1,]=w; R[1,]=r
ALPHA[1,]=a;
# Estimate and display probability of JUVENILE survival # 'METHOD 2' refers to regression (glm) estimates
YJ<-cbind(JS,J-JS) # data
spfactor=factor(species)
```

```
sJ<-glm(YJ~spfactor-1,family=binomial (link=logit)) # general linearized regression model
PJ[2,1:sp]=exp(sJ$coef)/(1+exp(sJ$coef)) # inverse logit transformation
# Estimate and display probability of ADULT survival
YA=cbind(AS,A-AS)
sA<-glm(YA~spfactor-1,family=binomial (link=logit))
PA[2,1:sp]=exp(sA$coef)/(1+exp(sA$coef))
# ESTIMATE OVERALL AND CELL-SPECIFIC FECUNDITY
for (i in 1:sp) # Separate regression for each species
{As=A[species==i & A>0] # data for sp i when at least 1 adult on cell
Js=J[species==i & A>0]
Xs=PH[species==i & A>0]
logn=log(As)
Ns=N[species==i & A>0]
fs<-glm(Js~logn+Xs+I(Xs^2)+Ns,family=poisson(link=log))
OPT[2,i]=-fs$coef[3]/(2*fs$coef[4])
W[2,i]=-1/fs$coef[4]
R[2,i]=fs$coef[1]-fs$coef[4]*OPT[2,i]^2
ALPHA[2,i]=-fs$coef[5]
}
# Estimate DISPERSAL parameters based on assumption all dispersal is to one of 4 neighbor cells
nJ<-as.vector(sapply(years,function(t){as.vector(sapply(c(1:sp),neighbor.calc,seeds=J[time==t]))}))
nJD=nJ[A==0]
JD=J[A==0]
YD=cbind(JD,nJD)
spfactorD=factor(species[A==0])
sD<-glm(YD~spfactorD-1,family=binomial (link=logit))
PD[2,1:sp]=4*exp(sD$coef)/(1+exp(sD$coef))
# Max likelihood parameter estimates # METHODS 3 & 4 refer to max likelihood estimates L1 & L2
initS=c(0.8,0.2); # initial values to start max likelihood search for survival parameters
initF=c(7,5,2,0.05,0.1,0.1); # initial values to start max likelihood search for fecundity parameters
for (i in 1:sp)
{LS=optim(initS,likeS)
PA[3,i]=LS$par[1]; PA[4,i]=LS$par[1]
PJ[3,i]=LS$par[2]; PJ[4,i]=LS$par[2]
```

```
versn=1
LF=optim(initF,likeF)
OPT[3,i]=LF$par[1]
W[3,i]=LF$par[2]
R[3,i]=LF$par[3]
ALPHA[3,i]=LF$par[4]
PNB[3,i]=LF$par[5]
PD[3,i]=PD[2,i] # version 1 uses regression estimate of pD
versn=2
LF=optim(initF,likeF)
OPT[4,i]=LF$par[1]
W[4,i]=LF$par[2]
R[4,i]=LF$par[3]
ALPHA[4,i]=LF$par[4]
PNB[4,i]=LF$par[5]
PD[4,i]=LF$par[6] # version 2 estimates pD by maximum likelihood
methods
}
# Compare true fitness and regression-estimated fitness
for (i in 1:sp) # fitness calculation for empty cell (N=0)
{for (x in 2:(gx-1))
{for (y in 2:(gy-1))
{for (M in 1:4)
{LAMBDA0[M,i,x,y]=PA[M,i]+PJ[M,i]*exp(R[M,i]-(OPT[M,i]-
pH[x,y])^2/W[M,i])
for (tau in 1:yrs) {LAMBDAN[M,i,x,y,tau]=PA[M,i]+PJ[M,i]*exp(R[M,i]-
(OPT[M,i]-pH[x,y])^2/W[M,i]-ALPHA[M,i]*sum(n[1:sp,x,y,years[tau]]))}
}
}
}
}
# SUMMARY STATISTICS
PARAM<-array(0,c(4,sp,8)) # 8 parameters, 4 estimation methods
PARAM[,,1]=PA; PARAM[,,2]=PJ;
PARAM[,,3]=OPT; PARAM[,,4]=W; PARAM[,,5]=R;
PARAM[,,6]=ALPHA; PARAM[,,7]=PD
rho<-array(1,c(4,9)) # correlation coefficients
pr<-array(0,c(4,9)) # 1-tailed significance test
for (M in 2:4)
{for (p in 1:7)
```

```
{rho[M,p]=cor(PARAM[1,,p],PARAM[M,,p],use="complete.obs")
g=cor.test(PARAM[1,,p],PARAM[M,,p],use="complete.obs")
pr[M,p]=g$p.value
}
rho[M,8]=cor(LAMBDA0[1,,,],LAMBDA0[M,,,],use="complete.obs")
g=cor.test(LAMBDA0[1,,,],LAMBDA0[M,,,],use="complete.obs")
pr[M,8]=g$p.value
rho[M,9]=cor(LAMBDAN[1,,,,],LAMBDAN[M,,,,],use="complete.obs")
g=cor.test(LAMBDAN[1,,,,],LAMBDAN[M,,,,],use="complete.obs")
pr[M,9]=g$p.value
}
names<-c("pA","pJ","opt pH","niche width (w)","maxfecundity (r)","alpha (a)","pr dispersal (d)","lambda0","lambdaN")
data.frame(names,rho.reg=rho[2,1:9],rho.L1=rho[3,1:9],rho.L2=rho[4,1:9])
# Figure 1
M=2
par(mfrow=c(2,3),cex.axis=1.2,cex.lab=1.6,mar=c(5,6,2,1.5)+.1,omd=c(.01,1,.01,1),mgp=c(3.7,1,0))
plot(pA,PA[M,],xlab=expression(paste("Adult Survival (",p[A],")")),ylab=expression(paste("E[",p[A],"]")))
abline(a=0,b=1)
plot(pJ,PJ[M,],xlab=expression(paste("Juvenile Survival (",p[J],")")),ylab=expression(paste("E[",p[J],"]")))
abline(a=0,b=1)
plot(opt,OPT[M,],xlab="Optimum pH (opt)",ylab="E[opt]")
abline(a=0,b=1)
plot(w,W[M,],xlab="Niche Width (w)",ylab="E[w]")
abline(a=0,b=1)
plot(r,R[M,],xlab="Log Max Fecundity (r)",ylab="E[r]")
abline(a=0,b=1)
plot(pD,PD[M,],xlab=expression(paste("Dispersal Fraction (",p[D],")")),ylab=expression(paste("E[",p[D],"]")))
abline(a=0,b=1)
# Figure 2
par(mfrow=c(2,4),cex.lab=1.6,cex.axis=1.3,omd=c(.01,1,.01,1),mgp=c(3.6,1,0),mar=c(5,5.5,2,1.5)+.1)
for (i in 1:sp)
{plot(c(5,9),c(0,10),pch=" ",xlab="pH",ylab="Apparent Fecundity")
  use.vals<-which(A>0&time==5&species==i)
  points(PH[use.vals],J[use.vals]/A[use.vals],col=i+1)
```

```
    curve(exp(r[i]-(opt[i]-x)^2/w[i]),add=T)
  }
  # Figure 3
  # Display estimated lambdas and show misclassification of sources and sinks
  par(mfrow=c(2,4),cex.lab=1.6,cex.axis=1.3,omd=c(.01,1,.07,1),mgp=c(3.6,
1,0),mar=c(5,5.5,2,1.5)+.1)
  for (i in 1:sp)
  {   plot(c(0.75,1.25),c(0.75,1.25),type="l",xlab=expression(lambda[N]),ylab
=expression(paste("E[",lambda[N],"]")))
    abline(v=1,lty=2,col="dark grey");abline(h=1,lty=2,col="dark grey")
  for (x in 1:gx)
  {for (y in 1:gy)
  {points(LAMBDAN[1,i,x,y,],LAMBDAN[2,i,x,y,],col=1)
  points(LAMBDAN[1,i,,,],LAMBDAN[3,i,,,],col=3)
  }
  }
  for (x in 1:gx)
  {for (y in 1:gy)
  {for (tau in 1:5)
  {if (!is.na(LAMBDAN[2,i,x,y,tau]))
  {if (LAMBDAN[1,i,x,y,tau]>=1 & LAMBDAN[2,i,x,y,tau]<1)
  { points(LAMBDAN[1,i,x,y,tau],LAMBDAN[2,i,x,y,tau],col=2)} # misclassi-
fied source by regression
  if (LAMBDAN[1,i,x,y,tau]<1 & LAMBDAN[2,i,x,y,tau]>=1)
  { points(LAMBDAN[1,i,x,y,tau],LAMBDAN[2,i,x,y,tau],col=2)} # misclassi-
fied sink by regression
  if (LAMBDAN[1,i,x,y,tau]>=1 & LAMBDAN[3,i,x,y,tau]<1)
  { points(LAMBDAN[1,i,x,y,tau],LAMBDAN[3,i,x,y,tau],col=7)} # misclassi-
fied source by likelihood
  if (LAMBDAN[1,i,x,y,tau]<1 & LAMBDAN[3,i,x,y,tau]>=1)
  {points(LAMBDAN[1,i,x,y,tau],LAMBDAN[3,i,x,y,tau],col=7)} # misclassi-
fied sink by likelihood
  }}}}}
```

GILBERTO PASINELLI, JONATHAN P. RUNGE AND KARIN SCHIEGG

10

Source–sink status of small and large wetland fragments and growth rate of a population network

Summary

Many organisms persist in populations that are spatially structured by human-induced loss and fragmentation of their native habitats. Despite this, the demographic contributions of local populations to a population network and to the growth rate of such a network are still largely unexplored. Using data on individually marked young and adult female reed buntings (*Emberiza schoeniclus*) from 21 local populations studied over six years in northeastern Switzerland, we examined the source–sink status of small and large local populations with recently developed metrics. We hypothesized that including emigration to the population network (the C^r metric) would classify more local populations as sources than when only focusing on the ability of local populations to maintain themselves (the R^r metric). We further tested the hypothesis that the relative contribution of small and large local populations to the population network does not differ. The inclusion of emigration to the population network resulted in significantly higher values than when only considering the contribution of local populations to themselves, the difference between the metrics averaging 30%. Despite this, most local populations in our study turned out to be sinks (C^r value <1), suggesting that substantial immigration is required for maintaining local populations as well as the entire population network (growth rate of network always <1). Both large and small populations contributed equally to the population network. We conclude that (a) the source–sink status of local populations is more comprehensively described by metrics including emigration (such as C^r) than by metrics focusing on processes within local populations

Sources, Sinks and Sustainability, ed. Jianguo Liu, Vanessa Hull, Anita T. Morzillo and John A. Wiens. Published by Cambridge University Press.

(such as R^r); (b) the network of local populations studied here is not viable without immigration; and (c) small local populations can be as valuable as large local populations in their contribution to a population network.

Background

Many plant and animal species persist as local populations in fragmented habitats throughout human-influenced landscapes. The habitat fragments hosting local populations typically differ in size, quality and connectivity, all of which may affect the fate of individuals and ultimately the persistence of a potential network of local populations (Hanski 2005). Such networks, often referred to as metapopulations (Hanski 1999) or patchy populations (Harrison 1991; Matthysen 1999), have been shown to be less extinction-prone than isolated populations, for example because local extinctions are prevented through immigration (the rescue effect; Brown and Kodric-Brown 1977; Stacey and Taper 1992) or because empty fragments can be recolonized from nearby local populations (Hanski 2005).

Studies on networks of local populations have often addressed presence/absence patterns of single or multiple species as well as composition and persistence of species communities. Far less common are studies examining the demography of species occurring in population networks and the consequences of temporal and spatial variation in demographic rates across local populations for the persistence of population networks (see Runge *et al.* 2006 for a recent review). From ecological and evolutionary perspectives, local populations contributing disproportionately to the entire set of local populations are particularly important, because such local populations may be driving the dynamics of the network and may influence the potential for adaptation (Kawecki 2004). From a conservation point of view, local populations contributing most to a population network may be the key targets of management strategies, particularly in times of limited funding for the application of large-scale conservation measures.

In his seminal paper, Pulliam (1988) used habitat-specific demography to distinguish source from sink habitats. Based on reproductive rate and adult and juvenile survival probabilities, the metric λ (lambda) was calculated, with source habitats having $\lambda > 1$ and sinks $\lambda < 1$. The relative ease of calculation and the increasing need to distinguish source and sink habitats for conservation purposes resulted in a plethora of studies addressing some aspects of source–sink dynamics (Runge *et al.* 2006). However, Pulliam's λ exclusively focuses on processes within habitats or local populations, which is insufficient for assessing source–sink status because emigration is not taken into account. The C^r

metric recently proposed by Runge *et al.* (2006) overcomes this shortcoming by simultaneously assessing the relative demographic contribution of a focal local population to itself and to all other local populations in the network of interest.

We here present results on the demographic status and contributions of local populations of the reed bunting (*Emberiza schoeniclus*) inhabiting wetland remnants varying in size from 2 to 250 ha scattered throughout a highly human-modified landscape in Switzerland. Based on data from six years and 21 local populations, we first examine whether the C^r metric and a metric closely related to Pulliam's λ termed R^r ("self-recruitment rate"; Runge *et al.* 2006) differ in their assessment of the source–sink status of local populations. The R^r metric "represents the ability of a local population to maintain itself through retention and self-recruitment (i.e., in the absence of immigration)" (Runge *et al.* 2006). Thus, while R^r describes net gains of the focal population to itself, C^r describes net gains to all local populations collectively (Runge *et al.* 2006). We hypothesized that emigration was an important component of the population network and thus the C^r metric would classify more local populations as sources than the R^r metric. Secondly, using the C^r metric, we assessed the relative demographic contributions of large and small local populations to the population network. Our initial hypothesis was that local populations inhabiting large fragments would act as demographic sources by producing more offspring and contributing more emigrants to the population network than local populations in small fragments. An alternative hypothesis was that the relative contribution of small and large local populations does not differ, because we have found no differences in reproductive performance and recruitment of reed buntings between small and large fragments in previous research (Pasinelli *et al.* 2008). Finally, we calculated the growth rate of the entire population network per year to see whether the network is self-sustainable.

Methods

Study species

The reed bunting is a migratory, ground-nesting passerine with a transpalearctic distribution. In Switzerland, reed buntings exclusively breed in wetlands, and the presence of old reed *Phragmites* sp. is the most important cue for territory establishment (Surmacki 2004) by males returning from the wintering grounds in southeastern France (Glutz von Blotzheim and Bauer 1997). If old reed habitat (see below for an explanation) is missing, reed buntings will not settle, and the wetland will remain unoccupied throughout the breeding season. Reed buntings defend small nesting territories, while foraging takes place in undefended areas of wetland vegetation adjacent to the territory.

Radio-tracking revealed that reed buntings do not leave wetlands while foraging, a result independent of wetland fragment size (Silvestri 2006).

Reed buntings are reproductively mature in the first breeding season after the birth year. Nests are usually placed in tussocks, heaps of old grass, or under broken, horizontal old reed stems within old reed habitat, but as the breeding season progresses, nests are also placed along ditches or in sedge meadows adjacent to old reed patches (35% of a total of 416 nests; Pasinelli *et al.* 2008). In Central Europe, birds may make up to five breeding attempts per season, but more than two successful nests are rare. Clutch size is 2–6 eggs and generally declines with season. Nesting success is highly variable and is strongly affected by both predation and the occurrence of floods.

Study area and local populations

From 2002 to 2005 we recorded reproductive performance of reed buntings in 21 wetland nature reserves scattered over an area of 200 km^2 in southeastern Canton Zurich (47°16′/08°47′), Switzerland (Fig. 10.1). Based on land-use maps from the Cantonal Office for Nature Conservation, these 21 nature reserves represent all the fragments potentially suitable for reed buntings in the 200 km^2 area. The fragments range in size from 1.9 to 247.2 ha (median 10.5 ha, interquartile range 4.2–16.7 ha). All the wetlands in the Canton Zurich are nature reserves and are managed in late summer/fall to prevent natural succession and to fight exotic plant species. As a consequence, reed re-grown during spring and summer is annually cut, except for reed along water bodies. Hence, reed along water bodies has not been cut for several years and has been allowed to build bands or patches of old reed habitat, the preferred breeding habitat of the reed bunting.

We selected our study area to contain several wetland fragments with sufficient variation in size and degree of isolation on a logistically manageable spatial scale. In addition, the set of wetland fragments is relatively isolated, as there are no fragments hosting more than 20 reed buntings within 20 km of our study area, which is 20 times the natal dispersal distance of the reed bunting as known at the beginning of this study (0.95 km; Paradis *et al.* 1998). The closest extensive breeding areas of the species are approximately 50 km away along Lake Constance, collectively hosting an estimated number of more than 1,000 breeding pairs (Bauer *et al.* 2005).

The scattered distribution of the reed bunting reflects the historical, natural patchiness of the preferred habitat (see above), but it is also a consequence of the dramatic landscape alterations that have occurred during the past two centuries throughout Central Europe. Many of the formerly extensive wetlands have been drained in the twentieth century, resulting in substantial loss of reed habitats. In Switzerland, the reed bunting is currently confined to nature reserves typically surrounded by agricultural landscapes, which do not act as barriers

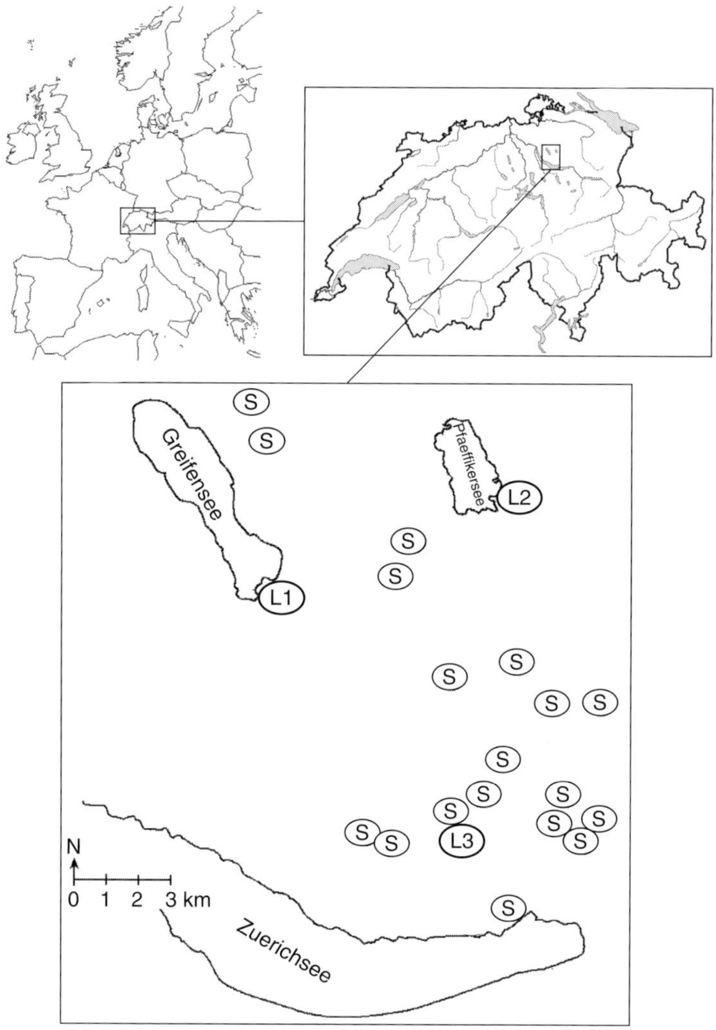

FIGURE 10.1.
Location of the local populations of the reed bunting studied in northeastern Switzerland from 2002 to 2007. L1, L2, L3 are the three large populations, S indicates the 18 small local populations (see text for further details). Words indicate names of lakes.

to dispersing individuals (Mayer *et al.* 2009), but result in a highly patchy distribution, with habitat fragments hosting 1–200 breeding pairs (Schmid *et al.* 1998). For further details on the study area, see Pasinelli and Schiegg (2006), Pasinelli *et al.* (2008), and Mayer *et al.* (2009).

We defined as a local population the breeding pairs within each fragment. In the three largest local populations, 20–60 pairs of reed buntings bred annually (Orniplan, unpublished report; G. Pasinelli, unpublished data). Within the three largest local populations, reproductive performance of at least ten breeding pairs per local population was annually monitored in randomly selected study plots along the lakefront. The study plots had been selected at the beginning of the study in 2002, and the same plots were monitored in all years. In the other 18 local populations, all breeding pairs present were annually monitored, with the annual number of breeding pairs ranging from 0 to 5. The three study plots in the large local populations as well as the 18 small local populations are referred to as the "intensively monitored study area."

Field procedures

From mid-March to early August 2002–2005, each local population was visited at least twice per week by two observers. Nests were located by standing on ladders and observing females building the nest, leaving the nest, and returning to it during incubation or when the parents were feeding the young. The young were individually color-banded between nestling days 6 and 9. Given that partial brood loss is rare in the nestling stage (18 of 296 nests; G. Pasinelli, unpublished data), we considered the number of nestlings banded as equal to the number of fledglings, because fledging usually occurs from day 10 onward (Glutz von Blotzheim and Bauer 1997). We did not check nests after young had been banded, in order to avoid premature fledging (Glutz von Blotzheim and Bauer 1997). Adults were caught with mist nets using playback tapes (males) or at the nest when feeding the young (males and females). After capture, birds were individually color-banded. Over 90% of the study population was color-banded in all study years.

Two blood samples (each max. 50 μl) per individual were taken by puncturing the brachial vein (permission number from the Cantonal Veterinary Office, Zurich: 169/2001) to sex nestlings in order to obtain the number of female fledglings for the demographic analysis (see below). Blood was absorbed with heparinized microcapillaries. Samples were either stored in microcapillaries directly or blown into APS-buffer (Arctander 1988) and stored at −20°C. Sexing of nestlings followed the PCR-based method of Griffiths *et al.* (1998).

From May to July 2003–2007, identity of territory owners in the intensively monitored study area was assessed to infer dispersal patterns. Further, we systematically searched for banded birds outside the monitored study plots in the three large local populations and opportunistically in wetlands outside of our 200 km^2 study area. This area outside the intensively monitored study area is below referred to as "outside area." We focused our search for banded reed buntings on wetlands because the species does not use habitats other than wetlands during breeding time in the Canton Zurich. The period between May and July corresponds to the breeding season of the reed bunting in our study area; individuals observed during that time are considered breeding birds. Non-breeding territorial individuals were extremely rare (G. Pasinelli, unpublished data).

Calculations of source–sink status of local populations and population growth rate

Extra-pair matings are a general feature of the reed bunting's social system, with 54–86% of the nests containing extra-pair young and 30–55% of all nestlings produced being the result of extra-pair fertilizations (Dixon *et al.* 1994; Bouwman *et al.* 2005; Kleven and Lifjeld 2005; Keiser 2007). Frequencies of extra-pair fertilizations in the reed bunting are among the highest reported to date (Griffith *et al.* 2002). A few males could therefore, in theory, father most of the offspring, making females rather than males the limiting sex for population growth. We therefore calculated demographic rates, contributions of local populations, and population growth rate of the overall network based on females only.

The source–sink status of local populations was assessed in three ways as follows. First, because the reed bunting population was stage-structured (i.e., consisting of two stages: first-year and adult individuals), an expression similar to the one detailed in Appendix A of Runge *et al.* (2006) was used to calculate C^r, which is the per capita demographic contribution of local population r to the entire population network in the next time step (here the next year). First-year individuals were birds banded as fledglings in our study area, with "first-year" referring to the time period between fledging and the end of the subsequent breeding season. Adult individuals were birds first captured and banded as adults in a breeding season. Although Appendix A in Runge *et al.* (2006) presented a pre-breeding matrix to calculate C^r, we used a post-breeding matrix (see below), with numbers of fledglings (i.e., the first-year birds) and adults counted at the end of the breeding season (see Caswell 2001 and Morris and Doak 2002 for the difference between pre-breeding and post-breeding censuses). Incorporating demographic rates per stage class in the calculation of C^r increases realism and should be considered whenever such rates are available. The stage structure required demographic parameters of the stage classes (k) to be multiplied by their relative abundance (w_k^r, see below) (Runge *et al.* 2006: Appendix A). Population projection matrices provide a useful way of presenting stage structure (Caswell 2001). Because a population projection matrix for the many local populations in this study would be too complex to portray, a matrix for two local populations is presented:

$$\left[\begin{array}{cc|cc} \phi_J^{11}\beta_A^1 & \phi_A^{11}\beta_A^1 & \phi_J^{21}\beta_A^1 & \phi_A^{21}\beta_A^1 \\ \phi_J^{11} & \phi_A^{11} & \phi_J^{21} & \phi_A^{21} \\ \hline \phi_J^{12}\beta_A^2 & \phi_A^{12}\beta_A^2 & \phi_J^{22}\beta_A^2 & \phi_A^{22}\beta_A^2 \\ \phi_J^{12} & \phi_A^{12} & \phi_J^{22} & \phi_A^{22} \end{array}\right]$$

with β_A^1 and β_A^2, respectively, being the per capita adult (A) female reproduction in local populations 1 and 2, ϕ_A^{11} and ϕ_J^{11}, respectively, being the probability that an adult (A) and juvenile (J) female in local population 1 in one breeding season was alive and in local population 1 in the next breeding season, and ϕ_A^{12} and ϕ_J^{12}, respectively, being the probability that an adult (A) or juvenile (J) female in local population 1 in one breeding season was alive and in local population 2 in the next breeding season. Note that in the above matrix the ϕ_j parameters are multiplied by β_A parameters, because per capita reproductive rate of first-year (β_J) and older females (β_A) did not differ (G. Pasinelli, unpublished data). The ϕ parameters are transition probabilities, i.e., the product of the apparent survival probability (S) and the movement probability (ψ), respectively, per stage class (see below for estimation of S and ψ). To obtain C^1, i.e., the contribution of local population 1 to the entire population network in the next time step (here the next year), the columns pertaining to local population 1 are added and multiplied by their relative stage-class abundances:

$$C^1 = w_J^1(\phi_J^{11}\beta_A^1 + \phi_J^{11} + \phi_J^{12}\beta_A^2 + \phi_J^{12}) + w_A^1(\phi_A^{11}\beta_A^1 + \phi_A^{11} + \phi_A^{12}\beta_A^2 + \phi_A^{12}). \tag{10.1}$$

Local population 1 is classified as a source if $C^1 > 1$ and as a sink if $C^1 < 1$. Note that Eq. (10.1) is for two local populations, but it can be generalized to C^r with many local populations.

Second, calculating C^r as just explained can reflect the contribution of different local populations in two years. For example, in the lower-left submatrix of the matrix shown above, the fecundity terms (i.e., $\phi_J^{12}\beta_A^2$ and $\phi_A^{12}\beta_A^2$) consist of the transition probabilities (ϕ_J^{12} and ϕ_A^{12}) from local populations 1 to 2 between year 1 and 2, and of the per capita female reproduction (β_A^2) from local population 2 in year 2, a consequence of the post-breeding approach applied. Hence, the C^r value of local population 1 in one year is affected by the reproductive rate of local population 2 in the next year. This problem is universal to all analyses based upon post-breeding matrices. To avoid the mixing of contributions from different local populations and years, we also calculated C^r based on equation 6 in Runge *et al.* (2006), reminiscent of a pre-breeding approach, as

$$C^1 = \phi_A^{11} + \phi_A^{12} + \beta_A^1(\phi_J^{11} + \phi_J^{12}). \tag{10.2}$$

Third, by removing all terms dealing with transition (i.e., emigration) probabilities from one local population to another, Eqs. (10.1) and (10.2) simplify to:

$$R^1 = w_J^1(\phi_J^{11}\beta_A^1 + \phi_J^{11}) + w_A^1(\phi_A^{11}\beta_A^1 + \phi_A^{11}) \tag{10.3}$$

$$R^1 = \phi_A^{11} + \beta_A^1\phi_J^{11} \tag{10.4}$$

with R^1 being the "self-recruitment rate" of local population 1 (Runge *et al.* 2006). R^1 is conceptually similar to Pulliam's (1988) λ, in that it focuses on the demographic processes within a local population (or habitat patch, fragment, etc.). Local population 1 is classified as a source if $R^1 > 1$ and as a sink if $R^1 < 1$. The structured post-breeding approach to estimating R^1 (Eq. 10.3) is novel, and to make comparisons with C^1 from Eq. (10.1) valid, we used the following year's reproductive rate, as we did with post-breeding C^1.

Growth rate of the entire population network per year (λ^T) was calculated by summing the annual C^r values of the local populations, with C^r values weighted by the relative abundance of the respective local populations (counted either pre- or post-breeding, see above). Note that λ^T calculated in this way is identical to deriving population growth from a structured population matrix model (see Runge *et al.* 2006 for details). Like population growth derived from matrix models, λ^T does not account for the effects of immigration.

Estimating demographic rates

Multistate capture–mark–recapture models were used to estimate probabilities of encounter (p), apparent survival (S) and movement (ψ) between local populations. Since individual identification occurred by resighting rather than recapture, the models in fact are capture–mark–resighting models, but were parameterized statistically as the multistate models detailed in Hestbeck *et al.* (1991). An important point is that S reflects apparent survival (combined probability of true survival and fidelity to the study area), because some emigrants will always be missed (unless the study species can be perfectly surveyed). Thus, S is less than the rate of true survival, which is true both for every local population and for entire patchy populations or metapopulations. In consequence, transition probabilities (ϕ) represent minimum estimates, because ϕ parameters are products of S and ψ.

Multistate capture–mark–resighting histories were constructed for each banded female (fledglings and adults), with the state referring to the three large local populations (L1, L2, L3), the 18 small local populations pooled into one category (S) and the outside area (X). Small local populations had to be pooled because it was not possible to reliably estimate p, S and ψ for local populations hosting only very few breeding pairs. The category X was included to improve estimation of transition probability (and hence survival probability), because some fledglings recruited to territories outside the intensively monitored study area. The multistate analyses were based on the fates of 126 adult females and 374 fledgling females followed from 2002 to 2007 (i.e., five resighting occasions).

We used the program MARK 5.0 to estimate p, S and ψ (White and Burnham 1999). Seven models with different structures for S and ψ (constant or variable across states and time) were evaluated (Table 10.1). In contrast, p was allowed to vary with age only because we saw no reason why p would vary across states, and insufficient data precluded estimating p across time. Owing to the small sample size, only additive models in terms of S and ψ were examined. Because we were interested in separately evaluating the source–sink status of each local population category as accurately as possible, ϕ parameters were then calculated by model-averaging (see below) across the top three models, for which the sum of the Akaike weights was 99.9%. Akaike weights indicate the level of support for a given model by the data (Burnham and Anderson 2002). Goodness-of-fit (GOF) of the fully time-dependent model was determined with the program U-CARE (Pradel *et al.* 2003), which indicated no lack of fit ($\chi^2 = 22.1$, df $= 38$, $P > 0.98$). This implies no lack of fit of reduced-parameter models either, such as the models detailed above in which we were interested.

The annual per capita reproductive rate was obtained by summing the female fledglings produced per local population category per year and dividing by the number of breeding females in the respective local population category in that year. 38 fledglings from 13 nests (out of a total of 214 nests with fledglings) could not be sexed, so we assumed half of those nestlings per nest to be females, based on a 50:50 primary offspring sex ratio in reed buntings in our study area (own unpublished data) and elsewhere (Bouwman *et al.* 2007; Keiser 2007).

The ϕ parameters were calculated based on model-averaging S and ψ values across the top three models in program MARK. Model-averaged variance–covariance matrices were also obtained from MARK; these were then used according to the Delta method (Powell 2007) to estimate variance. Resampling ($n = 1{,}000{,}000$) from the beta distribution, based upon the point estimate and variance, was then used to estimate the 95% confidence intervals (CI) for the ϕs to avoid CIs containing values less than 0. The Delta method was also applied to estimate variances for C and R values, with 95% CIs estimated by 1.96*SE (see Fig. 10.4).

Results

Demographic rates

Of the seven models examined, the best-supported one had time-constant survival S and movement ψ probabilities (Table 10.1). Based on this model, encounter probabilities (p) over all the years and local populations were 0.421 (95% CI 0.235–0.634) for first-year females and 0.736 (0.583–0.847) for adult females. Model-averaged apparent annual first-year female survival (S) was 0.103 (95% CI 0.045–0.162). Survival rates differed neither across years nor fragments (Fig. 10.2, Table 10.1). Apparent adult female survival was

TABLE 10.1. *Multistate model results for female reed buntings in northeastern Switzerland from 2002 to 2007. Models sorted according to decreasing support.* S = *apparent survival probability,* p = *encounter probability,* ψ = *movement probability, a = age (first-year vs. adults),* t = *time (five periods),* g = *groups (i.e., the five states L1, L2, L3, S, X, representing the four local populations plus the outside area).* K = *number of parameters,* ΔAICc = *difference in AICc to the best model,* A_weights = *Akaike weights, indicating support for a model.*

Model	K	−2log(L)	AICc	ΔAICc	A_weights
$\{S_a p_a \psi_a\}$	7	741.6	755.7	0.000	0.773
$\{S_{a+t} p_a \psi_{a+t}\}$	15	727.4	758.3	2.534	0.218
$\{S_a p_a \psi_{a+g}\}$	26	710.5	765.0	9.211	0.008
$\{S_{a+t+g} p_a \psi_{a+g}\}$	34	696.1	768.3	12.586	0.001
$\{S_{a+g} p_a \psi_{a+g}\}$	30	706.5	769.8	14.044	0.001
$\{S_{a+t+g} p_a \psi_{a+t+g}\}$	50	682.1	791.5	35.744	0.000
$\{S_{a+g} p_a \psi_{a+t+g}\}$	46	692.3	792.2	36.419	0.000

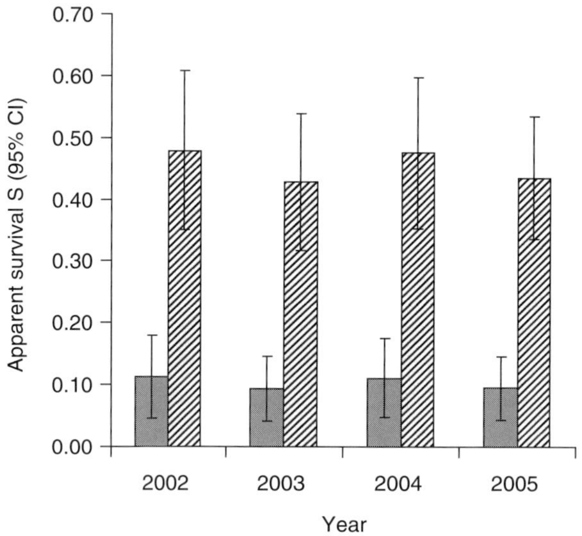

FIGURE 10.2.
Apparent survival *S* of first-year (filled bars) and adult female reed buntings (hatched bars). Estimates and error bars (95% confidence intervals) calculated by model-averaging over the three top models in Table 10.1.

consistently higher than first-year survival, when compared within the same local population categories (Fig. 10.2). Model-averaged apparent annual adult survival was 0.455 (0.340–0.570). No differences were apparent between years and local populations (Fig. 10.2, Table 10.1).

Model-averaged survival/emigration (ϕ^{rs}) and survival/philopatry (ϕ^{rr}) rates of first-year females across years and local populations were in the

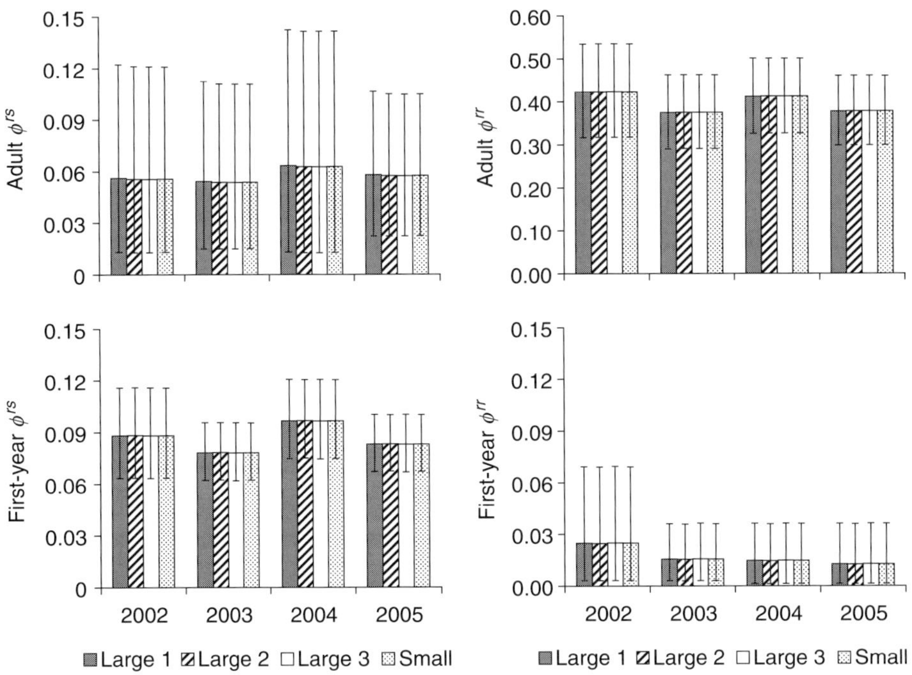

FIGURE 10.3.
Annual survival/emigration (ϕ^{rs}) and survival/philopatry rates (ϕ^{rr}) of first-year and adult female reed buntings per local population. ϕ^{rs} rates are the products of apparent annual survival and probabilities of moving from local population r to the other local populations s (diagrams on the left); ϕ^{rr} rates are products of apparent annual survival and probabilities of staying within a local population (diagrams on the right). ϕ^{rs} and ϕ^{rr} rates are based on model-averaging S and ψ values across the top three models in Table 10.1; error bars (95% confidence intervals) calculated with the Delta method (see Methods). Note different scaling of y-axes.

ranges 0.078–0.097 and 0.013–0.025, respectively (Fig. 10.3). In 2003, 2004, and 2005, ϕ^{rs} were larger than ϕ^{rr} of the respective local population category, as indicated by the non-overlapping 95% confidence intervals in Figure 10.3. In 2002, the 95% confidence intervals slightly overlapped, although ϕ^{rs} and ϕ^{rr} rates were quite different. Values of both rates varied little across fragments and years (Fig. 10.3). For adult females, model-averaged ϕ^{rs} rates ranged from 0.054 to 0.063 and ϕ^{rr} rates from 0.375 to 0.424. ϕ^{rr} rates of adults within a local population category and year were always substantially higher than ϕ^{rs} rates (Fig. 10.3). Note that the ϕ rates of both first-year and adult females were very similar across all local population categories, since the two best models did not support differences between local populations (Table 10.1).

Over all years and local population categories, per capita production of female fledglings averaged 1.79 (range 0.73–3.0, n = 16). We found no difference in per capita production across years (Friedman test, $\chi^2 = 2.1$, df = 3, $P = 0.557$; Table 10.2) or between the local populations ($\chi^2 = 1.8$, df = 3, $P = 0.622$).

Comparison between R^r and C^r

Based on both R^r metrics, all local populations were sinks in every year (Fig. 10.4). Two possible exceptions were local populations L1 and L2 in 2004, when 95% confidence intervals included 1 for structured R^r values (calculated with Eq. 10.3). The inclusion of emigration rates did not change the general source–sink assessment, because most C^r metric values and associated 95% confidence intervals were also <1 (Fig. 10.4). The only exception was local population L2 with a C^r of 1.01 in 2004, when calculated with Eq. (10.1). Further, 95% confidence intervals included 1 for structured C^r values (Eq. 10.1) of L1 and L3 in 2004, and for non-structured C^r values (Eq. 10.2) of L1, L2 and L3 in 2002. Pooled over the years, the C^r metric based on Eq. (10.1) classified one local population as source and 11 as sinks, while the R^r metric based on Eq. (10.3) classified all 12 local populations as sinks (Fisher exact test; $P > 0.49$, $n = 24$). Based on Eqs. (10.2) and (10.4), the respective numbers were 16 sinks for C^r and 16 sinks for R^r ($P = 1.0$, $n = 32$). Thus, the C^r metrics did not classify more local populations as sources than the R^r metrics did. Nevertheless, population-specific C^r metrics were significantly larger than the respective R^r metrics in every year and with both approaches to calculate C^r and R^r (Wilcoxon tests, $P < 0.045$ in each case). C^r values calculated with Eq. (10.1) were on average 28.9% (±5.5, SD) larger than the respective R^r values. C^r values calculated with Eq. (10.2) were on average 32.7% (±3.8, SD) larger than the respective R^r values.

Temporal and spatial variation in C^r

The C^r values calculated with Eq. (10.2) did not significantly differ across years (Friedman test; $\chi^2 = 5.1$, df = 3, $P = 0.165$) (Fig. 10.4). Likewise, C^r values did not differ among the four categories of local populations ($\chi^2 = 1.8$, df = 3, $P = 0.622$). Thus, on a per capita basis, the relative contribution of the small local populations to the population network was not statistically different from that of the larger local populations (Fig. 10.4). Note that the respective tests were not possible with C^r values calculated with Eq. (10.1), owing to insufficient sample size.

Growth rate of the entire population network

In every year, growth rate λ^T of the population network, calculated by summing the weighted annual C^r values of the local populations (see Methods), was <1 (Fig. 10.5). That the overall size of the population network (as measured by the number of breeding females) did not decline over the study period implies substantial immigration. The closest large breeding area (>1,000 breeding pairs), which may act as both source of immigrants to and receiver of emigrants from our population network, is approximately 50 km away along Lake Constance.

TABLE 10.2. *Per capita female fledgling production across years and local populations. Per capita production calculated as the sum of female fledglings produced per local population category per year divided by the number of breeding females in the same local population category in that year. N = number of nests.*

	Local population category						
Year	Large 1	Large 2	Large 3	Small	$\bar{X}$	SD	N
2002	3.0	1.9	2.2	1.6	2.2	0.6	110
2003	2.0	1.6	2.0	1.8	1.8	0.2	114
2004	1.1	0.7	1.4	1.9	1.3	0.5	99
2005	2.0	2.4	1.6	1.6	1.9	0.4	117
X	2.0	1.7	1.8	1.7			
SD	0.8	0.7	0.4	0.2			
N	103	93	80	164			

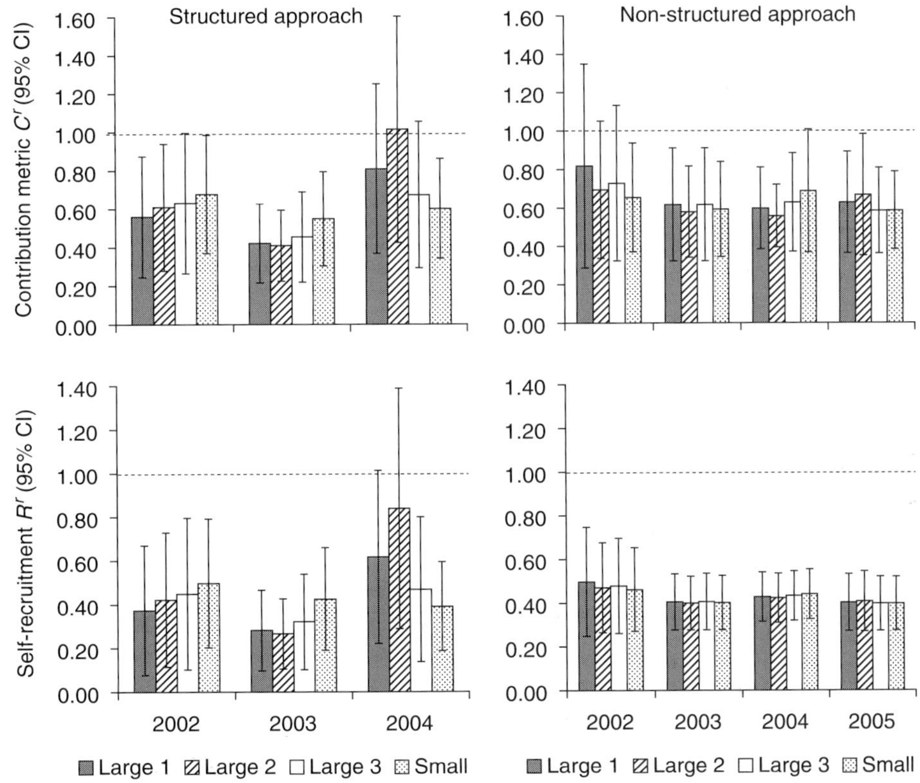

FIGURE 10.4.
C^r and R^r values per local population category and year for the reed bunting in northeastern Switzerland, calculated with structured (Eqs. 10.1 and 10.3) and non-structured approaches (Eqs. 10.2 and 10.4). Dashed horizontal lines: $C^r = 1$ and $R^r = 1$; values above or below dashed lines indicate sources or sinks, respectively. Note that in the structured approach, no estimates were possible for the year 2005, which would have required reproductive data from 2006. Note different scaling of y-axis.

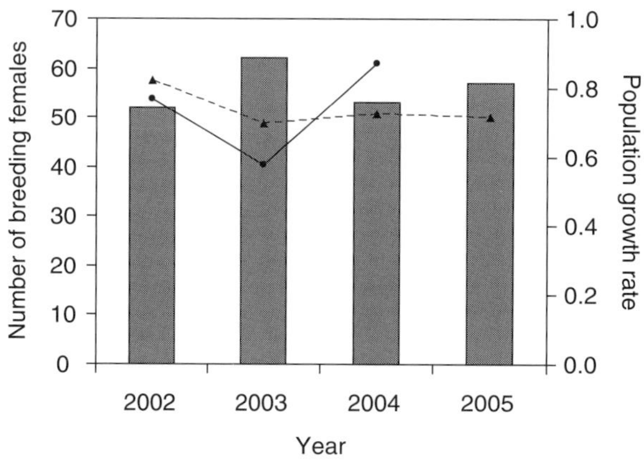

FIGURE 10.5.
Relation between annual number of female breeders (left axis, bars) and growth rate λ^T of the population network (right axis, symbols and lines) from one year to the next. Dots and solid line = structured approach (Eq. 10.1), triangles and dashed line = non-structured approach (Eq. 10.2). λ^T could not be calculated from 2005 to 2006 with the structured approach (no reproductive data from 2006).

Discussion and conclusions

Methodological issues

To assess the source–sink status of local populations, we have used both structured (Eqs. 10.1 and 10.3) and non-structured approaches (Eqs. 10.2 and 10.4) to calculate the metrics C^r and R^r, the latter being conceptually similar to Pulliam's λ (1988) (Runge *et al.* 2006). A structured approach allows examination of the contribution and elasticity of the different stage classes to the overall population growth of the network, which may yield important insights for species conservation (e.g., Caswell 2001). A structured approach should thus be used for any population that has observable age or stage classes, but most source–sink studies have so far neglected this point. An operational definition of age or stage structure is the observation of more than one age or stage class in a population. In age-classified models, individuals always advance to the next age class. In stage-classified models, individuals may advance, stay in the same stage class, or even regress to previous stage classes (Caswell 2001). In this study, both adults and newborns were observed every year. Because adults stayed in the same class from year to year, this is a stage-class model. Observations occurred post-breeding and thus two stage classes define the population.

One difference between the R^r presented in Eq. (10.3) and Pulliam's λ (1988) is the inclusion of population structure. When populations are structured, the different classes need to be weighted by relative stage-specific abundance if metrics such as C^r or R^r are to be accurately calculated. As an extreme example, suppose that due to prior overharvesting of adults, a population consists of 99% juveniles and 1% adults. Even if adult survival were 1.0 (after harvest ceased), adults would contribute proportionally little to population growth. Hence the need to account for the different proportions of stage classes in a population.

Despite the advantages outlined above, applying a post-breeding structured approach to C^r can be conceptually challenging, because C^r reflects the contribution of the focal population r to the population network through the reproductive contribution of philopatric individuals the next year (β_A), the dispersal contribution (ϕ^{rs} values) and, in addition, the reproductive contribution of emigrants from r in their target populations s (β_A^s) the next year. Thus, C^r calculated with the post-breeding structured approach consists of both contributions from r and all the local populations containing emigrants from r. This makes it difficult to examine the potential influence of certain factors characterizing local population r, for example habitat quality or genetic composition, on C^r particularly if dispersal is high. Using a non-structured approach in the form of Eq. (10.2) partially circumvents this complication, at the cost of being unable to assess the contribution and elasticities of stage classes and of overestimating C^r in organisms that have different vital rates in different stage classes. Note that Eq. (10.2) is non-structured because observations (e.g., population counts) would occur just prior to breeding, when adults and 1-year-olds are inseparable (cf., Morris and Doak 2002; Runge *et al.* 2006). Equation (10.2) in this study represents a "pseudo-pre-breeding" approach, because the data were collected according to a post-breeding sampling protocol.

Whether to use a structured or non-structured approach depends on the goals of the analysis. If the contributions and the elasticities of stage classes are of interest, then a structured approach is recommended. If the focus is on the relative contribution of local populations to a network, as in our example, then a non-structured approach in the form of Eq. (10.2) may be appropriate, so long as its use in a structured population does not bias estimates of C^r. Note that such bias is circumvented only when these three conditions are met: there are no more than two stage classes (or if more than two stage classes occur, the adult stage classes have identical survival and reproductive rates), reproductive rates between the two stage classes are equal, and stable age distribution is the weight determined to be most appropriate (the use of Eq. (10.2) essentially assumes a stable age distribution).

Ecological conditions of the study system, as well the ecology and life history of the study species, may affect the approach of choice. For example, differences between post- and pre-breeding approaches will increase with temporal variation in vital rates: if a good year for reproduction follows a bad one, the differences between the two approaches can be stark. High emigration rates combined with high spatial variation in reproductive rates will also accentuate the differences between the two approaches. Thus, (detailed) knowledge of the species' ecology and the study system should assist the choice of approach.

By and large, the structured and non-structured approaches gave similar results in our study, in that local populations were almost always classified as sinks. The few possible discrepancies illustrate the complications pointed out above. For example, local population L2 was classified as source in 2004 by the structured C^r, but as sink by the non-structured C^r. This is because reproduction in 2005 was higher than in 2004 (Table 10.2; Pasinelli *et al.* 2008), resulting in a proportionately larger reproductive contribution of emigrants from L2 in 2005 than in 2004.

Both C^r contribution metrics (Runge *et al.* 2006) resulted in significantly higher values than the respective R^r metrics, with the differences averaging 30%. Depending on the metric used (i.e., C^r or R^r), different assessments of the conservation value of local populations or fragments may result, which in turn may have profound consequences for the allocation of resources for conservation and management, and actions taken. This is particularly important when local populations have C^r values close to 1, in which case the use of R^r may lead to an underestimation of the value of a fragment, if only those with $R^r > 1$ are deemed valuable. Local populations with $C^r > 1$ and $R^r < 1$ depend on immigrants for their continued existence, but at the same time provide emigrants for other local populations. Such situations have been termed "dependent sources" (Hixon *et al.* 2002) and may be very common in highly mobile organisms such as birds (e.g., Sillett and Holmes 2002). We therefore strongly advocate use of the C^r metric in addition to R^r (perhaps along with other metrics; see Ovaskainen and Hanski 2003; Runge *et al.* 2006) in studies attempting to assign source–sink status to local populations.

Source–sink status of local reed bunting populations

We found no support for our first hypothesis that C^r would classify more local populations as sources than would R^r, as all local population categories turned out to be primarily sinks, irrespective of the metric used to assess source–sink status. One possible explanation for this finding is that estimated transition rates were biased low due to emigration from the sampled area, resulting in incorrect estimates of C^r. In mobile organisms, such as birds, it is notoriously difficult to adequately estimate dispersal and emigration (Walters 1998; Dunning *et al.*, Chapter 11, this volume). Based on ring recoveries in the UK, mean natal dispersal distance was previously estimated to be 0.95 km (Paradis *et al.* 1998). Median natal dispersal distance in our study was 4 km (interquartile range: 0.98–5.22 km, $n = 30$), and the longest natal dispersal event was 12 km (G. Pasinelli, unpublished data). However, the suggestion that young reed buntings may settle much farther from their birthplaces than suggested by our natal dispersal data is supported by a male nestling banded in a local population 150 km away, who settled and bred in the local population L2. Such long-distance dispersal events are probably not uncommon in the reed bunting, because population genetic data indicate high levels of gene flow both within our study area (Mayer *et al.* 2009) and between our study area and other local populations throughout Europe (Mayer 2009). Thus, it is possible that the scale of our study area was insufficient to accurately assess C^r, because dispersal occurs over much larger distances than those possible within our study area and expected at the beginning of our study. The problem associated with accurately estimating dispersal in spatially limited study sites has long been recognized (e.g., van Noordwijk 1984) and may be accentuated in highly mobile organisms such as migratory bird species, in which propensity of natal dispersal is greater than in resident species (Weatherhead and Forbes 1994). Assigning source–sink status to local populations by means of C^r and similar metrics requires taking into account the scale of the study area relative to the dispersal capacity of the species; that is, in species with high dispersal capacities, such as the reed bunting, estimation of C^r is less accurate than in species with low dispersal capacities. In widely dispersing species, a population classified as a sink within the study area may therefore be a source range-wide.

On the other hand, ϕ^{rs} rates of first-year and adult females from the network would need to be, when averaged over years and local populations, 0.32 (assuming adult ϕ^{rs} as estimated above) and 0.42 (assuming juvenile ϕ^{rs} as estimated above), respectively, to arrive at a C^r of 1 for the entire network. In other words, ϕ^{rs} rates of first-year and adult females would need to be, respectively, over 365% and 725% larger than the estimated ϕ^{rs}. Most dispersal events are due to juveniles in their first year of life, while adult dispersal is rare in most species (but see Dale *et al.* 2005 for an exception), so that underestimation of natal dispersal may be more severe than adult dispersal in our study. Further, assuming ϕ^{rs} of adult females to be zero (i.e., no emigration of adult females within and from the population network), first-year ϕ^{rs} would need to be large to turn the different local populations into sources (Fig. 10.6). The same holds true for adult female ϕ^{rs} (i.e., setting first-year ϕ^{rs} to zero) (Fig. 10.6). Thus, emigration rates from the population network surveyed, as well as dispersal among local populations within the network, would need to be much larger than the observed emigration within the network, suggesting that underestimation of total emigration is unlikely to be the only reason for the sink status of our local populations.

Another explanation for the sink status of most local populations in our study may be low first-year and/or adult female survival. Apparent annual survival of females in their first year of life varied from 0.05 to 0.16 in our study.

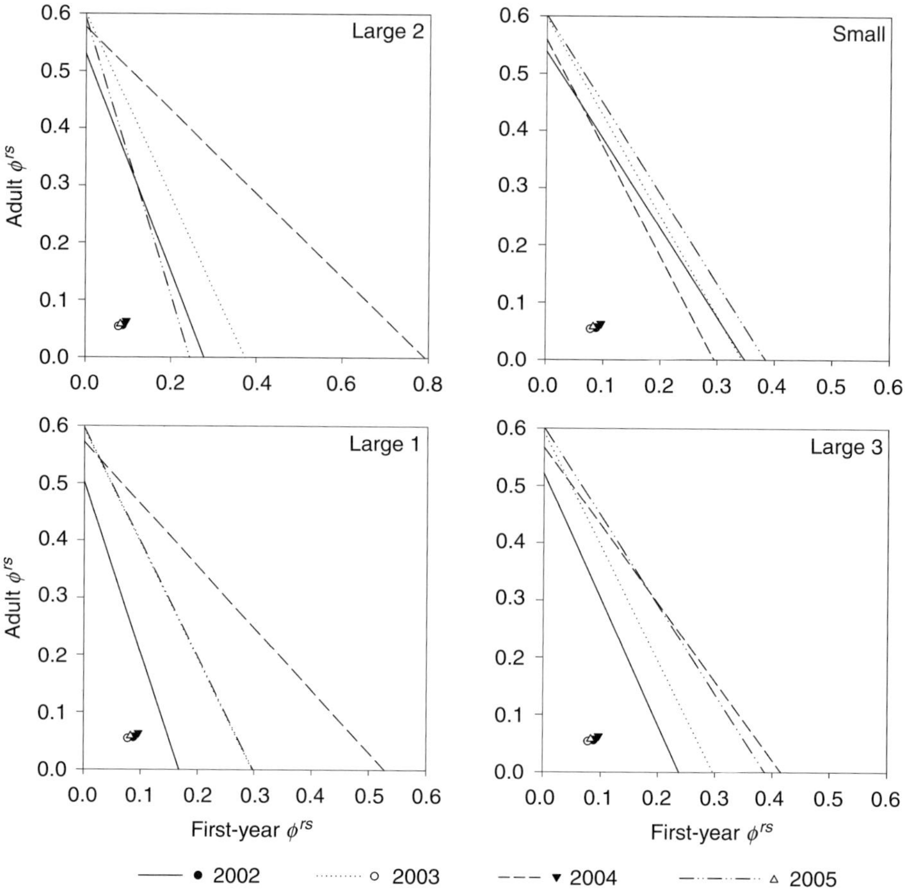

FIGURE 10.6.
Source–sink status of local populations in relation to first-year and adult survival/emigration (ϕ^{rs}) rates of female reed buntings. Lines indicate values needed for $C^r = 1$ in each local population in the different years, with estimates of ϕ_J^{rr}, ϕ_A^{rr} and β_A^r as calculated above and assuming no emigration from the network. Symbols give observed ϕ^{rs} rates. Any point above or below a given line would indicate that the respective local population was a source or sink, respectively. Calculations are based on the non-structured approach (Eq. 10.2). Note that in population Large 1, the lines for 2003 and 2005 strongly overlap. Note different scaling of x-axes.

Compared with first-year survival rates (both sexes combined) in British reed buntings, our values are on the lower end of those reported by Peach *et al.* (1999) and well below those reported by Siriwardena *et al.* (1998). However, the comparison is problematic for at least two reasons. First, both Siriwardena *et al.* (1998) and Peach *et al.* (1999) based their assessments on juveniles caught during summer in mist nets; that is, they considered only young that had already survived the immediate post-fledgling period, when juvenile mortality has been shown to be highest in several songbird species (Krementz *et al.* 1989; Thomson *et al.* 1999; Naef-Daenzer *et al.* 2001; Yackel Adams *et al.* 2006).

Second, reed buntings in Britain are resident (Glutz von Blotzheim and Bauer 1997) and hence do not experience mortality during migration as the reed buntings studied here may do. Compared to other passerines, apparent annual first-year survival rates of female reed buntings in our study were rather low (citril finch *Serinus citrinella* averages around 0.28–0.37, Senar *et al.* 2003; lark bunting *Calamospiza melanocorys* 0.19–0.25, Yackel Adams *et al.* 2006). Reduced first-year survival had been implicated in the decline of British reed bunting populations (Peach *et al.* 1999), and thus low first-year survival may be responsible for the sink status of the local populations studied here.

In contrast, adult survival rates estimated in our study (0.38–0.59) were in the range of values reported from British reed buntings (adults of both sexes) by Siriwardena *et al.* (1998) and Peach *et al.* (1999) and from other passerines based on mark–recapture analyses (e.g., Lebreton *et al.* 1992; Brawn *et al.* 1995; Powell *et al.* 2000; Peach *et al.* 2001; Senar *et al.* 2003). We therefore conclude that adult survival was not the primary reason for the sink situation observed.

Finally, reproductive output may have been insufficient. Per capita production of female fledglings ranged from 0.73 to 3.0 per year. Elsewhere, reproductive performance values of the reed bunting were very similar (Glutz von Blotzheim and Bauer 1997; Keiser 2007), suggesting that per capita reproductive performance was not causing the sink status of the local populations.

Even though most C^r values were below 1, the C^r metric still provides a measure of relative worth for each local population. For example, all local populations provided emigrants to other local populations, even when classified as sinks. Although we do not know whether some of the emigrants leaving our study area ended up in source local populations, the situation (dispersal away from sinks) is reminiscent of a study on citril finches where higher emigration was observed from sinks to sources than vice versa (Senar *et al.* 2003).

Our results further imply that small local populations need not be less valuable than large ones, given the very similar C^r values of S to L1, L2 and L3 (Fig. 10.4). This contrasts with the general notion that large habitat fragments, patches, local populations, etc. are more valuable than small ones from a conservation point of view (see also Robinson and Hoover, Chapter 20, this volume). In fact, small and large wetland fragments have previously been shown to be equally suited in terms of reproduction and recruitment for the reed bunting (Pasinelli *et al.* 2008). The joint consideration of reproductive, survival, and emigration rates in the C^r metric again supports the idea that small local populations can be important components of population networks as well. Consequently, conservation efforts in favor of relatively small local populations may be as valuable as measures taken for larger populations (see also Falcy and Danielson, Chapter 7, this volume). Finally, simulation studies have consistently shown population networks to be more viable than single populations or networks consisting of relatively isolated local populations (e.g., Stacey and Taper 1992; Hanski 2005). Many small local populations may thus be at least as important as a few large populations in terms of long-term persistence of species, for example by providing emigrants to other similar populations, even though each single small local population may be classified as sink.

Acknowledgments

Access to nature reserves was granted by the Cantonal Office for Nature Conservation, which also provided us with maps and data on our study sites. Capturing and handling of birds was made possible thanks to permits from the Swiss Ornithological Institute, Sempach, and the Federal Office for the Environment, Berne. We thank K. Blum, C. Elmiger, P. Frei, E. Glaus, A. Gouskov, P. Halder, M. Keiser, B. Kurt, C. Mayer, and S. Michler for their great help during fieldwork. Genetically sexing the fledglings was done by A. Gouskov and C. Mayer. M. Kneubühler, B. Miranda, and P. Thee provided us with GPS equipment and invaluable advice. M. Schaub provided advice with the program MARK, F. Korner-Nievergelt and H. Schmid helped with the maps in Figure 10.1, and three anonymous reviewers contributed helpful comments to previous versions of the manuscript. For financial support, we are grateful to the Swiss National Science Foundation (grant no. 3100A0–100150/1 to K. Schiegg and G. Pasinelli), Stiftung Zürcher Tierschutz, Fachstelle Naturschutz des Kantons Zürich, Georges und Antoine Claraz-Schenkung, Ala-Fonds zur Förderung der Feldornithologie, Ornithologische Gesellschaft Zürich, and Graf Fabrice von Gundlach und Payne Smith-Stiftung.

References

Arctander, P. (1988). Comparative studies of avian DNA by restriction fragment length polymorphism analysis: convenient procedures based on blood samples from live birds. *Journal of Ornithology* **129**: 205–216.

Bouwman, K. M., C. M. Lessells and J. Komdeur (2005). Male reed buntings do not adjust parental effort in relation to extrapair paternity. *Behavioral Ecology* **16**: 499–506.

Bouwman, K. M., R. E. van Dijk, J. J. Wijmenga and J. Komdeur (2007). Older male reed buntings are more successful at gaining extrapair fertilizations. *Animal Behaviour* **73**: 15–27.

Brawn, J. D., J. R. Karr and J. D. Nichols (1995). Demography of birds in a neotropical forest: effects of allometry, taxonomy, and ecology. *Ecology* **76**: 41–51.

Brown, J. H. and A. Kodric-Brown (1977). Turnover rates in insular biogeography: effect of immigration on extinction. *Ecology* **58**: 445–449.

Bauer, H.-G., M. Peintinger, G. Heine and U. Zeidler (2005). Veränderungen der Brutvogelbestände am Bodensee: Ergebnisse der halbquantitativen Gitterfeldkartierungen 1980, 1990 und 2000. *Die Vogelwelt* **126**: 141–160.

Burnham, K. P. and D. R. Anderson (2002). *Model Selection and Inference: A Practical Information-Theoretic Approach*. Springer, New York.

Caswell, H. (2001). *Matrix Population Models: Construction, Analysis, and Interpretation*. Sinauer Associates, Sunderland, MA.
Dale, S., A. Lunde and O. Steifetten (2005). Longer breeding dispersal than natal dispersal in the ortolan bunting. *Behavioral Ecology* **16**: 20–24.
Dixon, A., D. Ross, S. L. C. O'Malley and T. Burke (1994). Paternal investment inversely related to degree of extra-pair paternity in the reed bunting. *Nature* **371**: 698–700.
Glutz von Blotzheim, U. N. and K. M. Bauer (1997). *Handbuch der Vögel Mitteleuropas*. Aula, Wiesbaden, Germany.
Griffith, S. C., I. P. F. Owens and K. A. Thuman (2002). Extra-pair paternity in birds: a review of interspecific variation and adaptive function. *Molecular Ecology* **11**: 2195–2212.
Griffiths, R., M. C. Double, K. Orr and R. J. G. Dawson (1998). A DNA test to sex most birds. *Molecular Ecology* **7**: 1071–1075.
Hanski, I. (1999). *Metapopulation Ecology*. Oxford University Press, Oxford, UK.
Hanski, I. (2005). *The Shrinking World: Ecological Consequences of Habitat Loss*. International Ecology Institute, Oldendorf/Luhe, Germany.
Harrison, S. (1991). Local extinction in a metapopulation context: an empirical evaluation. In *Metapopulation Dynamics: Empirical and Theoretical Investigations* (M. E. Gilpin and I. Hanski, eds.). Academic Press, London: 73–88.
Hestbeck, J. B., J. D. Nichols and R. A. Malecki (1991). Estimates of movement and site fidelity using mark–resight data of wintering Canada geese. *Ecology* **72**: 523–533.
Hixon, M. A., S. W. Pacala and S. A. Sandin (2002). Population regulation: historical context and contemporary challenges of open vs. closed systems. *Ecology* **83**: 1490–1508.
Kawecki, T. J. (2004). Ecological and evolutionary consequences of source–sink population dynamics. In *Ecology, Genetics and Evolution of Metapopulations* (I. Hanski and O. E. Gaggiotti, eds.). Academic Press, San Diego, CA: 387–446.
Keiser, M. (2007). Habitat occupation strategies and breeding behaviour in reed buntings (*Emberiza schoeniclus*). PhD thesis, Departement für Biologie, Abteilung Ökologie und Evolution, Universität Freiburg, Freiburg, Switzerland.
Kleven, O. and J. T. Lifjeld (2005). No evidence for increased offspring heterozygosity from extrapair mating in the reed bunting (*Emberiza schoeniclus*). *Behavioral Ecology* **16**: 561–565.
Krementz, D. G., J. D. Nichols and J. E. Hines (1989). Postfledging survival of European starlings. *Ecology* **70**: 646–655.
Lebreton, J.-D., K. P. Burnham, J. Clobert and D. R. Anderson (1992). Modelling survival and testing biological hypotheses using marked animals: a unified approach with case studies. *Ecological Monographs* **62**: 67–118.
Matthysen, E. (1999). Nuthatches (*Sitta europaea*: Aves) in forest fragments: demography of a patchy population. *Oecologia* **119**: 501–509.
Mayer, C. (2009). Living in a naturally fragmented world: from extra-pair paternity in local populations to spatial population structure of the Reed bunting (*Emberiza schoeniclus*) across Europe. PhD thesis, Institute of Zoology, University of Zurich, Switzerland.
Mayer, C., K. Schiegg and G. Pasinelli (2009). Patchy population structure in a short-distance migrant: evidence from genetic and demographic data. *Molecular Ecology* **18**: 2353–2364.
Morris, W. F. and D. F. Doak (2002). *Quantitative Conservation Biology: Theory and Practice of Population Viability Analysis*. Sinauer Associates, Sunderland, MA.
Naef-Daenzer, B., F. Widmer and M. Nuber (2001). Differential post-fledging survival of great and coal tits in relation to their condition and fledging date. *Journal of Animal Ecology* **70**: 730–738.
Ovaskainen, O. and I. Hanski (2003). How much does an individual habitat fragment contribute to metapopulation dynamics and persistence? *Theoretical Population Biology* **64**: 481–495.
Paradis, E., S. R. Baillie, W. J. Sutherland and R. D. Gregory (1998). Patterns of natal and breeding dispersal in birds. *Journal of Animal Ecology* **67**: 518–536.
Pasinelli, G. and K. Schiegg (2006). Fragmentation within and between wetland reserves: the importance of spatial scales for nest predation in reed buntings. *Ecography* **29**: 721–732.

Pasinelli, G., C. Mayer, A. Gouskov and K. Schiegg (2008). Small and large wetland fragments are equally suited breeding sites for a ground-nesting passerine. *Oecologia* **156**: 703–714.

Peach, W. J., D. B. Hanmer and T. B. Oatley (2001). Do southern African songbirds live longer than their European counterparts? *Oikos* **93**: 235–249.

Peach, W. J., G. M. Siriwardena and R. D. Gregory (1999). Long-term changes in over-winter survival rates explain the decline of reed buntings *Emberiza schoeniclus* in Britain. *Journal of Applied Ecology* **36**: 798–811.

Powell, L. A. (2007). Approximating variance of demographic parameters using the Delta method: a reference for avian biologists. *Condor* **109**: 949–954.

Powell, L. A., J. D. Lang, M. J. Conroy and D. G. Krementz (2000). Effects of forest management on density, survival, and population growth of wood thrushes. *Journal of Wildlife Management* **64**: 11–23.

Pradel, R., C. M. A. Wintrebert and O. Gimenez (2003). A proposal for a goodness-of-fit test to the Arnason–Schwarz multisite capture–recapture model. *Biometrics* **59**: 43–53.

Pulliam, H. R. (1988). Sources, sinks, and population regulation. *American Naturalist* **132**: 652–661.

Runge, J. P., M. C. Runge and J. D. Nichols (2006). The role of local populations within a landscape context: defining and classifying sources and sinks. *American Naturalist* **167**: 925–938.

Schmid, H., R. Luder, B. Naef-Daenzer, R. Graf and N. Zbinden (1998). *Schweizer Brutvogelatlas: Verbreitung der Brutvögel in der Schweiz und im Fürstentum Liechtenstein 1993–1996.* Schweizerische Vogelwarte, Sempach, Switzerland.

Senar, J. C., M. J. Conroy and A. Borras (2003). Asymmetric exchange between populations differing in habitat quality: a metapopulation study on the citril finch. *Journal of Applied Statistics* **29**: 425–441.

Sillett, T. S. and R. T. Holmes (2002). Variation in survivorship of a migratory songbird throughout its annual cycle. *Journal of Animal Ecology* **71**: 296–308.

Silvestri, G. (2006). *Do food supply and reproductive success vary with site size?* Institute of Zoology, University of Zurich, Zurich.

Siriwardena, G. M., S. R. Baillie and J. D. Wilson (1998). Variation in the survival rates of some British passerines with respect to their population trends on farmland. *Bird Study* **45**: 276–292.

Stacey, P. B. and M. Taper (1992). Environmental variation and the persistence of small populations. *Ecological Applications* **2**: 18–29.

Surmacki, A. (2004). Habitat use by reed bunting *Emberiza schoeniclus* in an intensively used farmland in western Poland. *Ornis Fennica* **81**: 137–143.

Thomson, D. L., S. R. Baillie and W. J. Peach (1999). A method for studying post-fledging survival rates using data from ringing recoveries. *Bird Study* **46**: S104–S111.

Van Noordwijk, A. J. (1984). Problems in the analysis of dispersal and a critique on its "heritability" in the great tit. *Journal of Animal Ecology* **53**: 533–544.

Walters, J. R. (1998). The ecological basis of avian sensitivity to habitat fragmentation. In *Avian Conservation: Research and Management* (J. M. Marzluff and R. Sallabanks, eds.). Island Press, Washington, DC: 181–192.

Weatherhead, P. J. and M. R. L. Forbes (1994). Natal philopatry in passerine birds: genetic or ecological influences? *Behavioral Ecology* **5**: 426–433.

White, G. C. and K. P. Burnham (1999). Program MARK: survival estimation from populations of marked animals. *Bird Study* **46**: S120–S139.

Yackel Adams, A. A., S. K. Skagen and J. A. Savidge (2006). Modeling post-fledging survival of lark buntings in response to ecological and biological factors. *Ecology* **87**: 178–188.

JOHN B. DUNNING JR., DANIEL M. SCHEIMAN AND
ALEXANDRA HOUSTON

11

Demographic and dispersal data from anthropogenic grasslands: what should we measure?

Summary

Studies of population dynamics of grassland birds have often followed the source–sink paradigm of Pulliam (1988). We present examples of demographic, dispersal, and modeling studies done with bird species found in anthropogenic grasslands of the midwestern USA. Although we believe that we have gained valuable insights into the factors that affect bird populations found in restored grasslands, hayfields and pastures, some of the demographic and dispersal processes assumed in Pulliam (1988) are difficult to measure with grassland birds. More importantly, the population dynamics of many migratory birds do not follow the structure of individual-based models used in the study of source–sink dynamics as pioneered by Pulliam (1988). We suggest measures of population stability such as territory occupancy, age distribution of successful dispersers, production of offspring, and dispersal may be useful as an alternative for assessing the health of grassland breeding bird populations.

Background

Grassland birds have shown some of the strongest declines of any group of North American birds and are increasingly considered a major conservation priority. This is true both regionally and within individual regions (Askins *et al.* 2007). These declines have resulted in considerable recent attention on grassland bird populations (Walk and Warner 1999; Herkert *et al.* 2003; Scheiman *et al.* 2003; Cunningham and Johnson 2006). Studies have included basic

Sources, Sinks and Sustainability, ed. Jianguo Liu, Vanessa Hull, Anita T. Morzillo and John A. Wiens.
Published by Cambridge University Press.

demography (e.g., Harmeson 1974; Galligan *et al.* 2006), and impacts of grazing, haying, and other agricultural practices (Temple *et al.* 1999; Sutter and Ritchison 2005).

Throughout the Midwest, most natural grasslands were converted to agriculture many decades ago (Herkert 2007). Natural grasslands currently cover <1% of their original extent, and many patches of native prairie are extremely small. Anthropogenic substitutes, such as hayfields, Conservation Reserve Program (CRP) lands, and restored grasslands on public and private lands make up the majority of available habitat for grassland birds, but even these can be small and interspersed with non-habitat (DeVault *et al.* 2002; Perlut *et al.* 2006). Many grassland-based studies of avian ecology have focused on edge effects and other aspects of fragmentation (Winter and Faaborg 1999; Fletcher and Koford 2003; Renfrew *et al.* 2005). The ultimate goal of many of these research projects is to determine how grassland birds are maintaining their populations in the artificial, highly fragmented grasslands left in the region. The source–sink model (Pulliam 1988) is often used to assess population viability in a fragmented system (McCoy *et al.* 1999; Askins *et al.* 2007; Perkins *et al.* 2008).

Using the demographic models that were the basis of Pulliam (1988), the population dynamics of a species can be determined by measuring its habitat-specific (or patch-specific) demographic rates such as adult survivorship, juvenile survivorship, and reproductive success. These variables allow one to calculate the population's finite rate of increase (λ). Source populations, the most valuable to conserve, are defined as patches or populations with $\lambda > 1$, allowing the population to produce dispersers that can rescue declining sink populations ($\lambda < 1.0$; Pulliam 1988). To determine conservation strategies following this paradigm, we need to generate habitat-specific demographic data. In addition, we require information on dispersal of individuals between patches because the production of excess individuals is useless unless those individuals can successfully disperse to new breeding locations.

In this chapter we provide examples of demographic, dispersal, and modeling studies from our research on bird populations in anthropogenic grasslands in Indiana. We feel that our research is representative of grassland bird population studies, but based on this experience, we appreciate the difficulty of applying a Pulliam (1988) source–sink paradigm to this system. We believe a different approach may be necessary to understand the population dynamics of migratory grassland birds occupying fragmented landscapes and to develop conservation strategies for these species.

One note on terminology: Pulliam (1988) discussed the situation where organisms in different patches experience different demographic rates. In some of our studies, our patches were of different habitats, thus making the term "habitat-specific demography" appropriate. In other studies, we looked at patches of the same habitat but in different landscape settings. "Patch-specific demography" is more accurate for these studies. Pulliam's general theory can be applied to both situations, and he finessed the possible confusion between "patch" and "habitat" by using the term "compartment" in the development of his theory. We will follow this convention by referring to "compartments" in the general case, and either "habitat" or "patch" in the specific cases where those terms are more accurate.

Demographic study

Many common grassland birds will colonize restored sites, and abundances can be similar between restored and natural grasslands (Askins *et al.* 2007). Fewer studies of restored grassland bird populations have quantified demography and determined whether restored habitats support source or sink populations (Millenbah *et al.* 1996; Patterson and Best 1996). Some bobolink (*Dolichonyx oryzivorus*) and dickcissel (*Spiza americana*) populations in anthropogenic grasslands were considered to be sinks, or sustainable only if annual survival was unrealistically high (McCoy *et al.* 1999, 2001; Fletcher *et al.* 2006).

A basic assumption of the Pulliam model is that organisms in different compartments may experience different demographic rates. Indeed, this assumption is critical to the source–sink concept – certain compartments support populations that are demographically successful enough to create sources, while other compartments do not allow for successful breeding and/or survival. In grassland systems, however, this assumption of compartment-specific demography has not been well studied. For example, we examined 20 published studies of reproductive success of dickcissels, bobolinks, and grasshopper sparrows (*Ammodramus savannarum*) in natural or anthropogenic grasslands (list of studies available from J. B. Dunning on request). Of these, only Zimmerman (1982, 1983, 1984) presented breeding data from natural and anthropogenic habitats simultaneously in the same studies. A few other studies examined reproduction in two categories of anthropogenic habitats: e.g., hayfields versus grazed meadows (Bollinger and Gavin 1989, 2004). But most studies looked at reproduction ecology in only one kind of grassland habitat. Therefore, we conducted a study in multiple grassland habitats to determine whether reproductive success was habitat-specific for a group of common grassland birds.

We studied habitat-specific reproductive success of grasshopper sparrows, dickcissels, common yellowthroats (*Geothlypis trichas*), field sparrows (*Spizella pusilla*), and Henslow's sparrows (*Ammodramus henslowii*) in restored grassland patches in Indiana. In particular, we examined whether populations in actively restored grasslands (sites planted with native prairie species) were more successful than those in fallowed agricultural fields in which passive plant community succession occurred. If fallowing fields results in the same or better bird populations, then the lower cost and manpower requirements of fallowing would suggest that it is a preferable restoration technique (at least from the perspective of avian conservation).

We evaluated the relationship between three habitat treatments (restored grassland, fallowed fields, and a large established grassland preserve) and the density and reproductive success of grassland birds at The Nature Conservancy's (TNC) Kankakee Sands Preserve in northwest Indiana. Kankakee Sands consists of 2,917 ha in various stages of grassland and wetland restoration – the largest private prairie restoration east of the Mississippi River. It is adjacent to several state-owned natural areas, including the 259-ha Beaver Lake Nature Preserve, an established older grassland that we used for comparison. The Beaver Lake preserve has been managed as a grassland for decades, and is dominated by native grasses and forbs. Historically, however, this site was not a true eastern tallgrass prairie remnant, therefore we use the term "established grassland" for Beaver Lake, rather than describing it as native prairie.

The Nature Conservancy purchased Kankakee Sands in 1996, and by 2002 had restored 1,040 ha of prairie and wetlands. Most of the property had been used in row-crop farming, and restoration consisted of herbicide application to eliminate exotics, and replanting of native grasses and forbs. Prescribed burning and limited use of herbicide encouraged development of native prairie. We conducted weekly line-transect surveys (100 m apart) on ten plots between mid-May and early August 2002–2003, noting breeding activity and plotting locations of all birds on site maps to delineate territories. Each plot was in a separate patch of restored grassland, fallowed field, or the established grassland. We also banded and resighted adult grasshopper sparrows and dickcissels.

Breeding status was quantified using a reproductive success index (Vickery *et al.* 1992). In this technique one assesses the status of reproduction in each territory once or twice a week by observing adult behavior. Observers looked for evidence of territory establishment (singing males), nesting (adults carrying nest materials, food), and production of independent fledglings (adults carrying food beyond the nestling stage, or visual observation of fledglings). We used the Vickery index to rank territories as to breeding success (rank 1 – territorial male; 2 – pairing; 3 – nesting; 4 – nestlings; 5 – fledglings; 6 – second brood attempted; 7 – second brood successful). A nest was deemed successful if it achieved a rank of 5–7, indicating the production of at least one set of fledglings. Use of the Vickery index is controversial for some grassland birds (Rivers *et al.* 2003) but it allowed us to quantify relative breeding success without finding nests – which is labor-intensive, and intrusive for grassland birds. In this way, we were able to track reproduction indirectly for a larger number of territories than would have been possible otherwise: a critical benefit when comparing large-scale landscapes.

Because each plot within a treatment type was measured over two years, we analyzed the data using a generalized linear mixed model (GLMM) and treated plot as a random effect, and treatment and year as fixed effects. We used a binomial distribution with a logit link function for modeling reproductive success rate, a Poisson distribution with a log link function and an offset for plot area when modeling count responses (bird abundance, reproductive success), or a Gaussian distribution for modeling vegetation variables. Analyses were done in program R using function glmmPQL in the library MASS.

We measured vegetation composition and structure at 57–200 spots per plot, depending on plot size, and then tested for variation among plots and treatments in vegetation height; vegetation volume; and volumes of individual plant community components (e.g., grass, sedge, etc.). Response variables for birds included abundance, number of successful nests, and successful nests per nest attempt. For each variable, we tested for variation due to treatment type and year (both as fixed effects) and vegetation variables, all treated as fixed effects in GLMMs. Because of limited sample size ($N = 10$ plots $\times$ 2 years), we restricted model comparison to two-variable models, did not include interaction terms, and accepted $P < 0.10$ as indicative of significant relationships to reduce Type II error. Because there are no straightforward model selection criteria currently available for random-effects models (J. Moore, personal communication), after consultation with a statistician we chose to use the P-values of individual variables to select the "best" models. If a model had two significant variables, we selected that model over single-variable models. If multiple two-variable models were significant, then we considered both models potentially important.

For all reproductive-success analyses, we defined success as any territory that received a final Vickery ranking of 5–7. Our initial analyses measured the number of successful territories per plot with an offset for plot area. We also analyzed the proportion of successful nests (ranks 5–7) per nesting attempt (final ranks 3–7), which allows our results to be compared with traditional nest-monitoring studies that do not account for territories where no nest was located (ranks 1–2).

Apart from Henslow's sparrows and dickcissels, density was more associated with plot variation than treatment types. With the common yellowthroat, the best model included only a vegetation variable: there were more yellowthroats on plots with dense vegetation. There were no significant treatment effects with this species. The best model for grasshopper sparrow included only vegetation height ($t = -3.51$, df $= 9$, $P = 0.007$), as the sparrows were not common on three restored plots with tall and dense vegetation. On the other two restored sites, grasshopper sparrow densities were similar to those on fallowed sites and the established site. The best model for field sparrow density was the null model. There was a single random plot effect: one fallowed site had the most field sparrows in both years.

Dickcissel density was affected by treatment type. Both fallowed and restored sites had more dickcissels than the established grassland ($t = 2.22$, df = 7, $P = 0.062$; and $t = 2.93$, df = 7, $P = 0.022$, respectively), and the restored sites had more dickcissels than the fallowed sites did ($t = 2.14$, df = 7, $P = 0.069$). There was also a significant year effect, with more dickcissels in the second year of the study ($t = 2.76$, df = 9, $P = 0.022$). For Henslow's sparrow, the best model included only Treatment, as there were more Henslow's sparrows on the established site than on either fallowed ($t = 3.70$, df = 7, $P = 0.008$) or restored plots ($t = 3.74$, df = 7, $P = 0.007$).

Reproductive success was affected by treatment type for several species (Fig. 11.1). There were too few Henslow's sparrow territories to analyze reproductive success. For dickcissel, the best model included both Year and Treatment. This species had a greater number of successful territories on restored sites compared with the established grassland ($t = 2.055$, df = 7, $P = 0.0789$) and fallowed sites ($t = 2.127$, df = 7, $P = 0.071$). Reproductive success in the fallowed sites did not differ from that in the established grassland ($t = 1.180$, df = 7, $P = 0.267$). The significant Year effect reflected that there were fewer successful dickcissel nests in 2003 than in 2002 ($t = -4.15$, df = 9, $P = 0.0025$).

For grasshopper sparrow, the best model included both Treatment and Vegetation Height. Sites with higher vegetation had fewer successful territories ($t = -4.278$, df = 9, $P = 0.0021$) and restored sites had increased success compared with fallowed sites ($t = 2.437$, df = 7, $P = 0.045$). The established site had reproductive success that was intermediate between the restored and fallowed sites, but was not significantly different from either. Interestingly, although Figure 11.1 seems to suggest that grasshopper sparrows had fewer successful territories per hectare on restored sites when compared with fallowed sites, when we took into account the plot-level variation in vegetation height we found that the addition of Treatment as a variable improved the model. This suggests that although the number of successful territories per hectare may be somewhat higher on fallowed sites, much of this variation is attributable to changes in vegetation height, and once we controlled for vegetation height we found that restored sites actually had increased reproductive success per hectare compared with fallowed sites.

Common yellowthroat patterns were not associated with treatment type. The best models were two models that contained two variables each: "Year ($t = -2.434$, df = 7, $P = 0.0452$) + Vegetation Height ($t = 3.937$, df = 7, $P = 0.0056$)" and "Year ($t = -8.696$, df = 7, $P = 0.0001$) + Vegetation Volume ($t = 8.834$, df = 7,

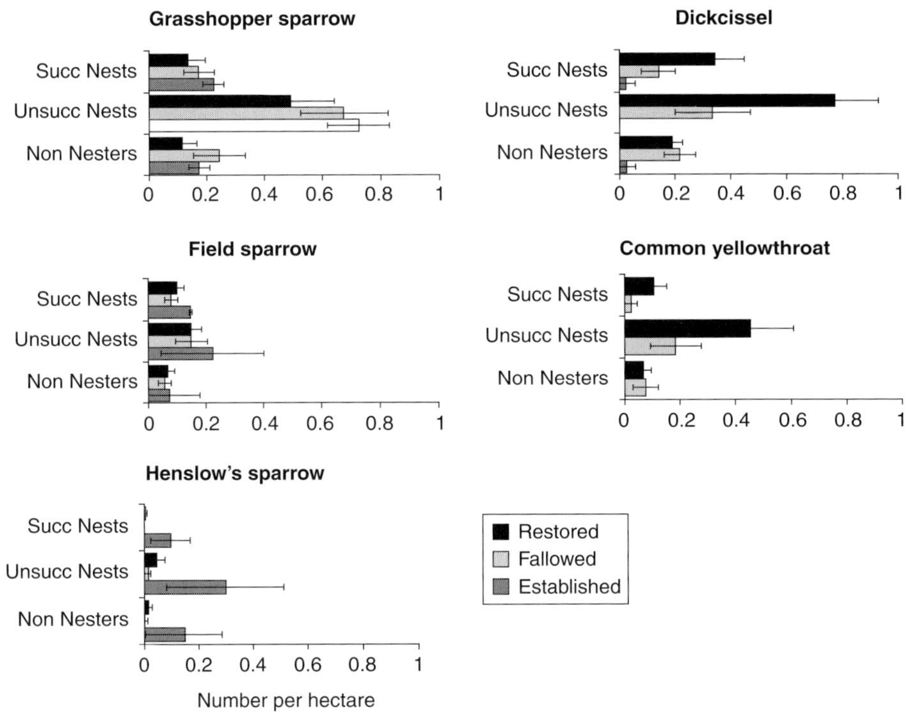

FIGURE 11.1.
Mean number of territories per hectare (and standard error) of common yellowthroat, dickcissel, field sparrow, grasshopper sparrow and Henslow's sparrow within a treatment type (Restored, Fallowed, Established) that experienced varying levels of reproductive success in 2002 and 2003 at Kankakee Sands, Indiana.

$P < 0.00001$)." This suggests that common yellowthroats had more successful nests in areas where vegetation was denser and taller.

There was no effect of Treatment on field sparrow reproductive success, although there was a significant Year effect (more successful nests in 2003; $t = 2.155$, df = 9, $P = 0.0596$). Many study plots had few or no field sparrow nests in either year, presumably because this species favors nesting in woody vegetation, which was poorly developed in most of these grassland sites.

The analysis of the proportion of nests that were successful yielded slightly different results. The best model for grasshopper sparrows included Treatment and Grass Volume, with a greater likelihood of success with less dense grass ($P = 0.0487$) and an increased reproductive success rate on the restored ($P = 0.0763$) and established sites ($P = 0.0712$) as compared with the fallowed sites. For dickcissels we did not find a Treatment effect on proportional success rate, but the Year effect remained ($P = 0.0132$), with a greater proportion of successful nests in 2002. For the common yellowthroat there was consistent evidence of a Year effect; the best model included both Year ($P = 0.0025$) and Vegetation Volume

($P = 0.0056$), with increased proportional nesting success where there was a greater overall vegetation density.

In summary, our study showed that density varied between years in some species, but did not vary between restored and fallowed sites for any species except dickcissels. Reproductive success also varied between years and was higher in restored sites than in fallowed sites for grassland specialists such as dickcissel and grasshopper sparrow, but not for more generalist species such as common yellowthroat and field sparrow. This information highlights a major component of Pulliam's (1988) model: that population demography such as reproductive success may vary between compartments even when population density does not differ: an important assumption of source–sink dynamics.

Dispersal study

Another critical assumption of source–sink dynamics concerns the movement of individuals between compartments. Although he did not explicitly model dispersal, Pulliam (1988) argued that source populations are responsible for maintaining regional populations by producing dispersers that can rescue declining sinks. In agricultural landscapes, many anthropogenic grasslands are privately owned and managed as independent entities. We were therefore interested in quantifying how populations in such landscapes were linked in a metapopulation framework via dispersal. We captured adult bobolinks breeding in a network of six hayfields and pastures in Warren County, Indiana, and monitored interpatch movements within and between years. The study plots constituted all suitable bobolink habitat within a 945 km^2 region, and were separated by distances ranging from 2.3 to 17.8 km (mean = 9.3 km). A major goal of our study was to determine whether the populations were consistent with the characteristics of different metapopulation models as defined by Stith *et al.* (1996). Details of this study have been published (Scheiman *et al.* 2007) and so are summarized here.

Over 2001–2004, we banded 201 bobolinks (143 males, 58 females) and resighted 124 birds at least once during subsequent capture occasions. Thirty birds (all males) were recaptured or resighted in subsequent breeding seasons. Twenty-four of these 30 males were site-faithful. Ten birds (also all males) moved up to 14.2 km (mean 7.4, median 7.3 km) to a different population from the one in which they were banded, either within or between years.

The dispersal and survival rates detected in this study suggest that the birds in individual hayfields and pastures were functionally linked into a single population. This fits Stith *et al.*'s (1996) definition of a "patchy metapopulation" (Scheiman *et al.* 2007). No individual population appeared to be isolated from the others, based on the maximum observed dispersal distance. Although the grassland patches were managed independently under separate landowners, the bobolink populations were not independent; instead they were interconnected through frequent dispersal.

Modeling study

Results of the dispersal study described above can be used to parameterize population models to assess the viability of specific populations, and such models are a critical step in determining the source/sink nature of specific populations (Reed *et al.* 2002). Population viability analysis (PVA) is a tool used to predict the risk of extinction under various scenarios of demographic and landscape change (Akçakaya 1991; Beissinger and Westphal 1998). We performed a PVA for the bobolink populations described above (Scheiman 2005; Scheiman *et al.* 2007) to estimate local and metapopulation extinction risk and to assess model sensitivities to a range of parameter values and the effect of a catastrophe (haying) on extinction risk.

We used the program RAMAS Metapop (Akçakaya 2002) to build a spatially explicit, stochastic metapopulation model, incorporating a polygynous mating system, ceiling-style density dependence, and age and sex structure (see Scheiman 2005 for model details). Population growth was modeled using population-specific stage matrices containing survival and fecundity. We derived adult male survival and its standard deviation (SD) from our field study using the program MARK (version 4.1; White and Burnham 1999). Adult female survival was taken from Martin (1971); juvenile survival was taken from Wittenberger (1976). Because we had only a single value for adult female survival, juvenile survival, and fecundity, we assumed that these values were constant among populations. Thus, population-specific matrices differed only in their values for adult male survival and its SD.

We also created additional extinction scenarios by varying the values of survival, fecundity, *K*, and dispersal from the baseline scenario described above, and examined the sensitivity of model results to changes in these parameters (Scheiman 2005). In addition, we added a catastrophe to the baseline scenario in the form of mid-season haying that decreased fledging rate to 0.06, based on haying effects reported elsewhere (Bollinger *et al.* 1990; Ells 1995).

We examined correlations between the population trajectories of all pairs of populations under each scenario, predicting that high correlations between populations suggest synchrony in population dynamics. Synchrony increases the risk of metapopulation extinction because if all populations go extinct simultaneously, recolonization through dispersal cannot occur (Akçakaya 2000).

Under the baseline scenario, the metapopulation grew and persisted for 50 years (Fig. 11.2). Overall metapopulation extinction probability was 0.00

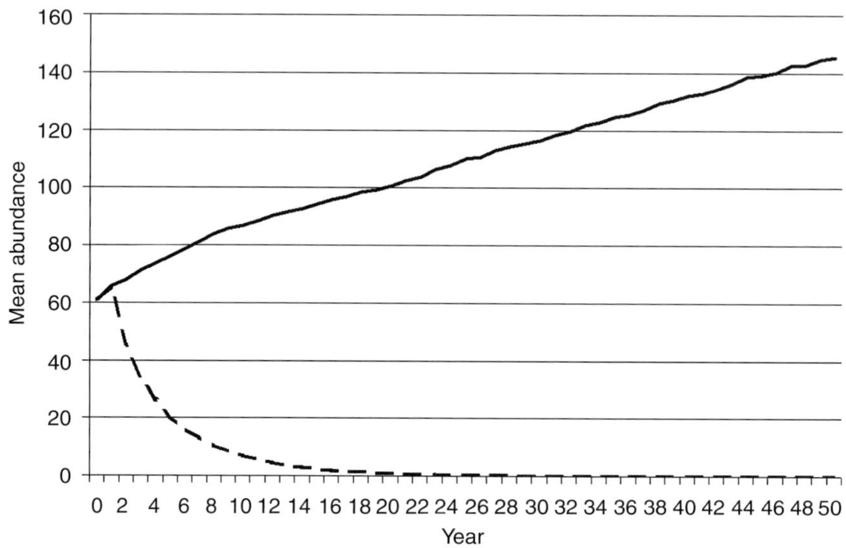

FIGURE 11.2.
Population trajectory for a bobolink metapopulation under baseline (solid line) and catastrophe (dashed line) scenarios. Population viability analysis simulations were run for 50 years and 1,000 simulations.

and local extinction risk was low. Populations went locally extinct but usually were quickly recolonized. Not surprisingly, the probability of metapopulation extinction under the scenario of mid-season haying was 100% (Fig. 11.2). Even reducing the chance of haying by 50–75% was not enough to prevent metapopulation extinction for most iterations of the model.

All pairs of populations had strongly positively correlated (r = 0.80–1.00) population trajectories under the baseline scenario. Sign and strength of correlation coefficients were inconsistent under the zero-dispersal scenario (r = −0.72–0.95). Populations were strongly positively correlated under the haying scenario (r = 0.81–1.00), with five pairs perfectly correlated.

Under baseline conditions this bobolink metapopulation appeared stable and able to persist for at least 50 years. Setting dispersal rate to zero did not appreciably affect metapopulation extinction risk, but local extinction risk increased because recolonization could not follow local extinction. Dispersal probably was largely responsible for demographic synchrony, as correlations were weaker in the absence of dispersal. This suggests that bobolinks display patchy metapopulation dynamics because dispersal can alter population persistence (Stith *et al.* 1996).

Metapopulation extinction was practically unavoidable under the catastrophe scenario. Increasing fecundity, increasing dispersal rate, or decreasing the frequency of haying was not enough to compensate for widespread reproductive failure (Scheiman 2005). Recruitment via dispersal was sufficient to offset the loss of *in situ* recruitment only if the majority of populations were not hayed and thus could contribute dispersers.

Metapopulation-wide haying imposed strong synchrony in population trajectories, contributing to local and metapopulation extinction (Hanski 1999). Even where the amount of available land is limited, collaborating with private landowners to ensure that most patches are not disturbed during peak breeding season should ensure the long-term persistence of bobolinks in the landscape.

What are we measuring?

The studies described above yielded valuable insights into the kinds of specific demographics that are the building blocks of information to understanding population viability. We feel these studies are representative of similar work on grassland birds. However, our experience with these studies also suggests that the parameters most often measured for grassland birds may be inadequate for assessing source–sink dynamics in this system, which we now explain in the following section.

The original source–sink models were envisioned for resident organisms where local population dynamics are determined by the survivorship, breeding, and movement of local populations (Pulliam 1988). But migratory birds such as those that dominate our grassland study systems are more mobile, less site-faithful, and their populations are potentially more interconnected than resident species.

Pulliam (1988) assumed that population changes reflect three demographic traits: adult survivorship, juvenile survivorship, and reproductive success. These demographic variables are projected to be compartment-specific, and differences in these traits are what define sources and sinks. We argue, however, that these three traits are rarely measured concurrently in field studies of migratory grassland birds, especially in a compartment-specific manner. Many studies measure one, or rarely two, of the traits and then depend on other published studies for values of other variables, which are often assumed or estimated in the original work. The values used are rarely compartment-specific (e.g., With *et al.* 2008). For example, our own data were insufficient for calculating independent measures of survivorship for grasshopper sparrows, dickcissels, and female bobolinks in different habitat patches. We did find, however, that reproductive success varied across three habitat types for grasshopper sparrows and dickcissels.

We relied on resighting banded birds to estimate survivorship. Adult survivorship of grassland birds is difficult to measure directly because these species exhibit low site philopatry (e.g., Vickery 1996; Temple 2002). Thus, measures of adult return rates are complicated by the movement of adults out of the study area. The apparent stability of populations in this system may not be determined by the return of territorial adults, but rather by the re-shuffling of adults between compartments within a study region (e.g., breeding dispersal). Banding returns in one or a few patches are typically inadequate for measuring this movement.

Juvenile survivorship is used in the Pulliam (1988) model to estimate how many locally produced offspring survive to disperse to breeding locations within a region and therefore contribute to population size in subsequent seasons. There are major complications in measuring juvenile survivorship directly in many habitats, not just grasslands (e.g., Rush and Stutchbury 2008). Juveniles are difficult to track after they become independent, and virtually no studies have followed young passerines long enough to determine whether they survive and where they settle to breed in later seasons. Juvenile survivorship rates are sometimes based on limited samples collected until the young birds disperse or migrate from the study region (after which they can no longer be followed). More commonly, researchers simply estimate juvenile survivorship as half of the adult survival rate, which itself may be estimated.

Perhaps more importantly, most grassland species in our region are migratory; and multiple published studies of migratory birds have shown that juveniles often do not return to the local patches in which they were raised (e.g., Nolan 1978). Researchers often band hundreds of nestlings, with only a few birds returning in subsequent seasons as breeders. Yet new breeders establish territories each year. It is not known where these new immigrants come from, but they are clearly not locally produced offspring. Therefore, it is unclear why juvenile survivorship is an important demographic for modeling these types of migratory bird populations – survival of local young does not appear to result in dispersers that colonize locally available breeding sites. As one reviewer suggested, it could be that locally produced juveniles settle within the general region, so that juvenile survivorship is relevant to the viability of the regional metapopulation, but this is also unstudied.

Reproductive success is the third component of the Pulliam (1988) source–sink model and we acknowledge that breeding success is a critical component of healthy bird populations. The breeding success variable that is of interest in the source–sink model is the production of offspring. Many breeding studies of grassland birds, however, quantify reproductive success indirectly using either daily survival rate (Mayfield-related methods; e.g., Winter and Faaborg 1999; Renfrew *et al.* 2005; Galligan *et al.* 2006) or the "proportion of nests that successfully produce at least one fledgling" (e.g., Hughes *et al.* 1999; Sutter and Ritchison 2005). Some studies report the proportions of nests that were either depredated or parasitized (Herkert *et al.* 2003; Patten *et al.* 2006). None of these measures translate easily to the production of offspring. In our studies, we also used an indirect measure of reproductive success (the Vickery method) to compensate for the difficulty of locating a sufficient sample size of nests. We gained insights from our breeding studies but, like other researchers, we did not quantify the production of offspring. Thus, it is clear to us that many grassland bird studies do not quantify reproduction in a manner that contributes to the kind of source–sink modeling proposed by Pulliam (1988).

What should we be measuring?

If the demographic traits used in Pulliam (1988) are difficult to measure or not particularly applicable to grassland bird systems, which traits could be valuable in assessing the stability of these populations? We suggest that three variables might provide equivalent information to the variables used in individual-based models but are more relevant to grassland birds and possibly other migratory species (such as the migratory songbirds discussed by Robinson and Hoover, Chapter 20, this volume). These variables are territory occupancy, age-class ratios of new breeders, and disperser production. We also need to develop improved measures of dispersal success rates to understand the health of populations in this system.

Pulliam (1988) used adult survivorship to calculate how many adults return between years to maintain population size. In grassland systems, low site philopatry means that surviving individuals of many species do not necessarily return in subsequent years. But the most critical patches in this system could be identified as patches whose territories are reliably occupied from year to year by breeders, even if the individuals that hold the territories change frequently. We (and others) have used spot-mapping and other territory-mapping techniques to monitor territory occupancy, breeder abundance, and distributions (DeVault *et al.* 2002; Cunningham and Johnson 2006). In our studies, we banded adults and observed high turnover among years, although some patches were consistently occupied by unmarked birds that likely came from outside our study areas. We argue, therefore, that a metric that focuses on consistency of territory occupation is a more relevant approach for vagile, migratory grassland birds, rather than individual-based adult survivorship. Habitat quality should be associated with the choices made by dispersing adults, so that high-quality habitat patches may be those whose territories are consistently occupied even if the breeders do not return each year, while poor-quality patches may be more likely to have unoccupied territories. One metric that focuses on territory occupation is an incidence function (Hanski 1999), although Hanski (1994) argued that the calculation of incidence of occupancy requires data from a large number of patches, suggesting that this specific metric may not be usable in our study region, where a limited number of habitat fragments remain.

Territory occupancy cannot by itself define high-quality patches as sources, because in some situations organisms may be forced to choose territories of poor quality. For instance, when sources are fully occupied, both sources and sinks may have high occupancy, with individuals moving from sink to source when possible (Pulliam and Danielson 1991). Additional information is therefore necessary in order to define high-quality habitat.

We argued above that juvenile survivorship is less relevant in this system because juveniles rarely return to breed in natal patches. The age of newly arrived breeders is still a relevant factor, though, that can be determined for many passerine species by examination of molt patterns. Birds hatched in the previous summer (e.g., inexperienced breeders) can be distinguished from older birds in many species by differences in the wear and molt of certain feather groups (Pyle 1997; Pyle *et al.* 2008). In at least some species, we can catch newly settled breeders and identify inexperienced individuals from older birds that are resettling from other areas and therefore are at least potentially experienced. As above, assuming that habitat selection is adaptive, high-quality habitat may be more likely to attract experienced breeders routinely, while poor-quality habitat may contain a higher percentage of less experienced birds. Therefore the percentage of new breeders that are older birds could be used as an indicator of potential population health.

Patch-specific reproductive success is critical for defining habitat quality in grassland systems, as envisioned by Pulliam (1988). But success must be defined as the production of dispersers, since sources are defined as populations whose production is regularly high enough to generate dispersers that colonize other areas. Thus we need to quantify breeding success using metrics that measure the production of offspring, rather than daily nest survival, or proportion of nests that fail. The Vickery index is useful because territories are scored as to whether they produce one or more sets of fledglings, but it is still an indirect estimate. A better approach might be modeled on constant-effort mist-netting programs, because these programs quantify the production of independent offspring on a study-plot scale (e.g., Krementz and Christie 1999).

Finally, dispersal rate and direction must be quantified to demonstrate the connection between subpopulations across a landscape (as underscored in Pulliam *et al.*, Chapter 9, this volume). Future studies should work to improve estimates of dispersal. Our bobolink study suggests that we need to cover a larger geographic area, encompassing a wider array of patches, to locate widely dispersing birds. Tracking a sufficient number of marked birds over a larger area will be difficult, although citizen–science networks can be employed to increase sightings. Data obtained using radio or satellite transmitters could correct dispersal distribution estimates obtained using other marking schemes. Besides expanding the search radius for banded birds, the use of stable-isotope (Hobson *et al.* 2004) and genetic techniques (Arsenault *et al.* 2005) might help to quantify dispersal ability.

We do not know the relationship between the three demographic variables that we have proposed (territory occupancy, percentage of adult breeders, number of offspring produced) and source–sink dynamic variables as modeled by Pulliam (1988). To determine this, we would need to measure each variable across populations while simultaneously measuring the components of the Pulliam model, which – as we argue above – is difficult in our system. But the alternate demographic variables could be combined to provide an assessment of the overall health of a population and the population's potential to contribute to regional conservation of a species. Since setting conservation priorities has been a primary use of the source–sink concept (indeed, this use was a prominent component of the discussion in Pulliam's original paper), measuring the population characteristics that we have proposed would allow one to assess the role of individual bird populations when the source–sink metrics are not available. Each of the metrics we proposed can be scaled from 0 to 1.0 as:

P_{terr} (territory occupancy) = proportion of territories occupied each year;
P_{age} (breeder age class) = proportion of breeders with adult plumage characteristics;
P_{breed} (offspring production) = population productivity / maximum productivity;

where the maximum productivity is defined as either the theoretical maximum (assuming each territory produces the maximum possible offspring based on clutch size and double-brooding potential) or the maximum observed among populations under study. These measurements can then be combined into an index of population health (IPH) similar to indices of biotic integrity (IBI) used in conservation of aquatic systems (Karr *et al.* 1986). In these indices, attributes of a system are measured that reflect overall system health and integrity (Karr and Chu 1999). The individual metrics are scaled relative to an expected value for an undisturbed system, and then summed to give an overall score indicative of the quality of the system. In the present case, assuming adaptive habitat selection by grassland birds suggests that the maximum expected value of each metric equals 1.0; e.g., when territories are fully occupied each year; when incoming and returning breeders are all adults; and when production of offspring equals the maximum expected or observed. Thus a potential index of population health would be:

$$IPH = \Sigma (P_{terr}, P_{age}, P_{breed})$$

with an expected range of 0.0–3.0. This index could be used to rank populations (or habitat patches) by their demographic characteristics so that conservation priorities can be set. We propose that this IPH ranking could then be combined with dispersal information to understand both the relative health of a given population and its potential for playing a supportive role in maintaining regional populations.

Conclusions

Having conducted studies that contribute to our understanding of population stability in grassland birds, we propose that our approach to examining population dynamics of these species needs to improve. It is absolutely true that just because something is easy to measure does not make it the right thing to measure. But applying the successful model of Pulliam (1988) to a system whose basic structure does not fit the model is also not productive. We argue that there are aspects of grassland bird populations that can be measured and used to assess population quality other than the components of individual-based models as developed by Pulliam (1988). We believe that the source–sink framework has been useful conceptually for conserving grassland bird populations, which are often found in fragmented landscapes in our region. It is still clear, given habitat- or patch-specific variation in demography and dispersal, that populations will vary in their importance to a conservation network. But the demographic components of the Pulliam (1988) model are difficult to measure in the field, and some (e.g., adult survivorship) may not be relevant to the dynamics of individual grassland bird populations. We may need a different set of parameters to identify healthy populations that can play a role similar to that proposed for source populations (Pulliam 1988) in maintaining regional grassland birds.

Acknowledgments

We thank Ron Pulliam for the inspiration for these and many similar studies, and for his mentoring of J. B. Dunning during his time at the Institute of Ecology, University of Georgia. Many of the ideas presented here originated during discussions of the Pulliam laboratory group during the early 1990s; it is instructive that we are still grappling with these same issues almost 15 years later. We thank the US Forest Service, US Department of Energy, National Science Foundation, The Nature Conservancy, Amos W. Butler Audubon Society, Indiana Academy of Science, Sigma Xi, the Wilson Ornithological

Society, and Purdue University for support. This chapter benefits from the comments made by three anonymous reviewers.

References

Akçakaya, H. R. (1991). Population viability analysis and risk assessment. In *Wildlife 2001: Populations* (D. R. McCullough and R. H. Barrett, eds.). Elsevier Applied Science, New York, USA: 148–157.

Akçakaya, H. R. (2000). Viability analyses with habitat-based metapopulation models. *Population Ecology* **42**: 45–53.

Akçakaya, H. R. (2002). *RAMAS Metapop: Viability Analysis for Stage-structured Metapopulations* (version 4.0). Applied Biomathematics, Setauket, NY.

Arsenault, D. P., P. B. Stacey and G. A. Hoelzer (2005). Mark–recapture and DNA fingerprinting data reveal high breeding-site fidelity, low natal philopatry, and low levels of genetic population differentiation in flammulated owls (*Otus flammeolus*). *Auk* **122**: 329–337.

Askins, R. A., F. Chavez-Ramirez, B. C. Dale, C. A. Haas, J. R. Herkert, F. L. Knopf and P. D. Vickery (2007). Conservation of grassland birds in North America: understanding ecological processes in different regions. *Ornithological Monographs* **64**.

Beissinger, S. R. and M. I. Westphal (1998). On the use of demographic models of population viability in endangered species management. *Journal of Wildlife Management* **62**: 821–841.

Bollinger, E. K. and T. A. Gavin (1989). The effects of site quality on breeding-site fidelity in bobolinks. *Auk* **106**: 584–594.

Bollinger, E. K. and T. A. Gavin (2004). Responses of nesting bobolinks (*Dolichonyx oryzivorus*) to habitat edges. *Auk* **121**: 767–776.

Bollinger, E. K., P. B. Bollinger and T. A. Gavin (1990). Effects of hay-cropping on eastern populations of the bobolink. *Wildlife Society Bulletin* **18**: 142–150.

Cunningham, M. A. and D. H. Johnson (2006). Proximate and landscape factors influence grassland bird distribution. *Ecological Applications* **16**: 1062–1075.

DeVault, T. L., P. E. Scott, R. A. Bajema and S. L. Lima (2002). Breeding bird communities of reclaimed coal-mine grasslands in the American Midwest. *Journal of Field Ornithology* **73**: 268–275.

Ells, S. F. (1995). Bobolink protection and mortality on suburban conservation lands. *Bird Observer* **23**: 98–112.

Fletcher, R. J. and R. R. Koford (2003). Spatial responses of bobolinks (*Dolichonyx oryzivorus*) near different types of edges in northern Iowa. *Auk* **120**: 799–810.

Fletcher, R. J., R. R. Koford and D. A. Seaman (2006). Critical demographic parameters for declining songbird breeding in restored grasslands. *Journal of Wildlife Management* **70**: 145–157.

Galligan, E. W., T. L. DeVault and S. L. Lima (2006). Nesting success of grassland and savanna birds on reclaimed surface coal mines of the Midwestern United States. *Wilson Journal of Ornithology* **118**: 537–546.

Hanski, I. (1994). A practical model of metapopulation dynamics. *Journal of Animal Ecology* **63**: 151–162.

Hanski, I. (1999). *Metapopulation Ecology*. Oxford University Press, Oxford, UK.

Harmeson, J. P. (1974). Breeding ecology of the dickcissel. *Auk* **91**: 348–359.

Herkert, J. R. (2007). Conservation reserve program benefits on Henslow's sparrow within the United States. *Journal of Wildlife Management* **71**: 2749–2751.

Herkert, J. R., D. L. Reinking, D. A. Wiedenfeld, M. Winter, J. L. Zimmerman, W. E. Jensen, E. J. Finck, R. R. Koford, D. H. Wolfe, S. K. Sherrod, M. A. Jenkins, J. Faaborg and S. K. Robinson (2003). Effects of prairie fragmentation on the nest success of breeding birds in the midcontinental United States. *Conservation Biology* **17**: 587–594.

Hobson, K. A., L. I. Wassenaar and E. Bayne (2004). Using isotopic variance to detect long-distance dispersal and philopatry in birds: an example with ovenbirds and American redstarts. *Condor* **106**: 732–743.

Hughes, J. P., R. J. Robel, K. E. Kemp and J. L. Zimmerman (1999). Effects of habitat on dickcissel abundance and nest success in Conservation Reserve Program fields in Kansas. *Journal of Wildlife Management* **63**: 523–529.

Karr, J. R. and E. W. Chu (1999). *Restoring Life in Running Waters: Better Biological Monitoring*. Island Press, Washington, DC.

Karr, J. R., K. D. Fausch, P. L. Angermeier, P R. Yant and I. J. Schlosser (1986). *Assessing Biological Integrity in Running Waters: A Method and its Rationale*, Illinois Natural History Survey Special Publication 5. Illinois Natural History Survey, Champaign, IL.

Krementz, D. G. and J. S. Christie (1999). Scrub-successional bird community dynamics in young and mature longleaf pine-wiregrass savannahs. *Journal of Wildlife Management* **63**: 803–814.

Martin, S. G. (1971). Polygyny in the bobolink: habitat quality and the adaptive complex. PhD dissertation, Oregon State University, Corvallis, OR.

McCoy, T. D., M. R. Ryan, E. W. Kurzejeski and L. W. Burger (1999). Conservation Reserve Program: source or sink for grassland birds in Missouri? *Journal of Wildlife Management* **63**: 530–537.

McCoy, T. D., M. R. Ryan, L. W. Burger and E. W. Kurzejeski (2001). Grassland bird conservation: CP1 vs. CP2 plantings in Conservation Reserve Program fields in Missouri. *American Midland Naturalist* **145**: 1–17.

Millenbah, K. F., S. R. Winterstein, H. Campa, L. T. Furrow and R. B. Minnis (1996). Effects of conservation reserve program field age on avian relative abundance, diversity and productivity. *Wilson Bulletin* **108**: 760–770.

Nolan, V. (1978). The ecology and behavior of the prairie warbler *Dendroica discolor*. *Ornithological Monographs* **26**.

Patten, M. A., E. Shochat, D. L. Reinking, D. H. Wolfe and S. K. Sherrod (2006). Habitat edge, land management, and rates of brood parasitism in tallgrass prairie. *Ecological Applications* **16**: 687–695.

Patterson, M. P. and L. B. Best (1996). Bird abundance and nesting success in Iowa CRP fields: the importance of vegetation structure and composition. *American Midland Naturalist* **135**: 153–167.

Perkins, D. W., P. D. Vickery and W. G. Shriver (2008). Population viability analysis of the Florida grasshopper sparrow (*Ammodramus savannarum floridanus*): testing recovery goals and management options. *Auk* **125**: 167–177.

Perlut, N. G., A. M. Strong, T. M. Donovan and N. J. Buckley (2006). Grassland songbirds in a dynamic management landscape: behavioral responses and management strategies. *Ecological Applications* **16**: 2235–2245.

Pulliam, H. R. (1988). Sources, sinks, and population regulation. *American Naturalist* **132**: 652–661.

Pulliam, H. R. and B. J. Danielson (1991). Sources, sinks, and habitat selection: a landscape perspective on population dynamics. *American Naturalist* **137**(Suppl.): S50–S66.

Pyle, P. (1997). *Identification Guide to North American Birds, Part 1*. Slate Creek Press, Bolinas, CA.

Pyle, P., S. L. Jones and J. M. Ruth (2008). *Molt and Aging Criteria for Four North American Grassland Passerines*. US Department of Interior; Fish and Wildlife Service, Biological Technical Publication FWS/BTP-R6011–2008, Washington, DC.

Reed, J. M., L. S. Mills, J. B. Dunning Jr., E. S. Menges, K. S. McKelvey, R. Frye, S. R. Beissinger, M. Anstett and P. Miller (2002). Emerging issues in population viability analysis. *Conservation Biology* **16**: 7–19.

Renfrew, R. B., C. A. Ribic and J. L. Nack (2005). Edge avoidance by nesting grassland birds: a futile strategy in a fragmented landscape. *Auk* **122**: 618–636.

Rivers, J. W., D. P. Althoff, P. S. Gipson and J. S. Pontius 2003. Evaluation of a reproductive index to estimate dickcissel reproductive success. *Journal of Wildlife Management* **67**: 136–143.

Rush, S. A. and B. J. M. Stutchbury (2008). Survival of fledgling hooded warblers (*Wilsonia citrina*) in small and large forest fragments. *Auk* **125**: 183–191.
Scheiman, D. M. (2005). Dispersal rates, extinction risk, and metapopulation dynamics of bobolinks. PhD dissertation, Purdue University, West Lafayette, IN.
Scheiman, D. M., E. K. Bollinger and D. H. Johnson (2003). Effects of leafy spurge infestation on grassland birds. *Journal of Wildlife Management* **67**: 115–121.
Scheiman, D. M., J. B. Dunning and K. A. With (2007). Metapopulation dynamics of bobolinks occupying agricultural grasslands in the Midwestern United States. *American Midland Naturalist* **158**: 415–423.
Stith, B. M., J. W. Fitzpatrick, G. E. Woolfenden and B. Pranty (1996). Classification and conservation of metapopulations: a case study of the Florida scrub-jay. In *Metapopulations and Wildlife Conservation* (D. R. McCullough, ed.). Island Press, Washington, DC: 187–215.
Sutter, B. and G. Ritchison (2005). Effects of grazing on vegetation structure, prey availability, and reproductive success of grasshopper sparrows. *Journal of Field Ornithology* **76**: 345–351.
Temple, S. A. (2002). Dickcissel (*Spiza americana*). In *The Birds of North America: No. 703* (A. Poole and F. Gill, eds.). The Birds of North America, Inc., Philadelphia, PA: 1–24.
Temple, S. A., B. M. Fevold, L. K. Paine, D. J. Undersander and D. W. Sample (1999). Nesting birds and grazing cattle: accommodating both on Midwestern pastures. *Studies in Avian Biology* **19**: 196–202.
Vickery, P. D. (1996). Grasshopper sparrow (*Ammodramus savannarum*). In *The Birds of North America: No. 239* (A. Poole and F. Gill, eds.). The Academy of Natural Sciences, Philadelphia, PA and The American Ornithologists' Union, Washington, DC: 1–24.
Vickery, P. D., M. L. Hunter and J. V. Wells (1992). Use of a new reproductive index to evaluate relationships between habitat quality and breeding success. *Auk* **109**: 697–705.
Walk, J. W. and R. E. Warner (1999). Effects of habitat area on the occurrence of grassland birds in Illinois. *American Midland Naturalist* **141**: 339–344.
White, G. C. and K. P. Burnham (1999). Program MARK: survival estimation from populations of marked animals. *Bird Study* **46**(Suppl.): 120–138.
Winter, M. and J. Faaborg (1999). Patterns of area sensitivity in grassland-nesting birds. *Conservation Biology* **13**: 1424–1426.
With, K. A., A. W. King and W. E. Jensen (2008). Remaining large grasslands may not be sufficient to prevent grassland bird declines. *Biological Conservation* **141**: 3152–3167.
Wittenberger, J. F. (1976). Habitat selection and the evolution of polygyny in bobolinks (*Dolichonyx oryzivorus*). PhD dissertation, University of California at Davis, Davis, CA.
Zimmerman, J. L. (1982). Nesting success of dickcissels (*Spiza americana*) in preferred and less preferred habitats. *Auk* **99**: 292–298.
Zimmerman, J. L. (1983). Cowbird parasitism of dickcissels in different habitats and at different nest densities. *Wilson Bulletin* **95**: 7–22.
Zimmerman, J. L. (1984). Nest predation and its relationship to habitat and nest density in dickcissels. *Condor* **86**: 68–72.

FERENC JORDÁN

12

Network analysis: a tool for studying the connectivity of source–sink systems

Summary

In order to optimize landscape conservation efforts, the most important landscape elements (patches and corridors) need to be identified quantitatively. A major aspect of functional importance is the role of a landscape element in maintaining connectivity. From a network perspective, different habitat patches and corridors can be ranked according to suitable centrality indices. Choosing and applying different indices may reflect differences in the problems studied (e.g., whether dispersal is limited) or data type (e.g., whether corridor permeability can be quantified). Within this framework, it is possible to consider the quality of patches and corridors or to make a distinction between source and sink patches. Network analytical tools can be used to quantify the local and global properties of directed landscape graphs (depicting source–sink systems). I present two case studies for illustrating how to set quantitative priority ranks of landscape elements and how to match the suitable techniques to particular problems (an undirected network of flightless bush crickets and a directed network of forest-living carabids). In the second case, I also calculate where to build a new corridor and suggest which corridor qualities to improve in the directed landscape graph. I propose a plan for redesigning a future freeway across the studied area. Finally, I argue that setting quantitative priority ranks for landscape elements could increase the efficiency of conservation efforts.

Sources, Sinks and Sustainability, ed. Jianguo Liu, Vanessa Hull, Anita T. Morzillo and John A. Wiens. Published by Cambridge University Press.

Introduction

Landscape ecology is a composite discipline integrating elements from both biology and geography. For some, mostly methodologically oriented people, it is ecological research based on maps and aerial photos. For others, landscape ecology is ecology on a large scale: they speak of the "landscape scale" of dozens or hundreds of square kilometers. For the third kind of scientists, landscape ecology is not a topic but a perspective, proposing that the ambiguous borders of ecosystems are not a problem but an exciting challenge. According to this latter view, the great problem of landscape ecology is how to better understand interactions between "neighbor" ecosystems. This is a contradiction in terms: as soon as you define two "neighbor" ecosystems, you need the border between them. So, the question is not whether a forest and a creek are neighbors – the question is to what extent they influence each other. From this very functional point of view, the core concepts of landscape ecology include migration and colonization (Dunning *et al.* 1992), spatial population dynamics (Pulliam 1988), and subsidization (Polis *et al.* 1997). Of course, an arsenal of methods is needed (for example, to serve spatially explicit models) and scales do matter (different organisms use different spatial and temporal scales; see Diez and Giladi, Chapter 14, this volume), but the essence of landscape ecological problems is much more general: both coyotes (Crooks and Soulé 1999) and continents (Brown *et al.* 1974) can become isolated, for example.

The need for the landscape view on ecological systems goes back a long way. Community and ecosystem ecology have traditionally discussed communities and ecosystems as if they were well-defined units. But where are their limits on the map and where are their limits in the interaction space? As the loss of natural habitats proceeds, as the remaining patches become fragmented and some become totally isolated, and as climate change causes various organisms to change their area, these have become issues of current conservation biology. If we do not understand how organisms use the landscape (and the seascape), how they move, how they interact in space and behave in new community contexts, we probably have no chance of achieving the right conservation practice.

In this chapter, I briefly discuss topological principles and the tools of network analysis in supporting this functional view of landscapes. I review some methods, discuss how to match suitable methods to actual problems, and present two case studies.

Network models for landscapes

A habitat can be evaluated by several kinds of information. Habitat suitability analyses can consider the soil, the vegetation cover, the distance from the nearest town, and many other characteristics of the habitat patch in question (e.g., Glenz *et al.* 2001). However, there is an important property of every habitat patch which cannot be evaluated only by looking at or studying the habitat in question. This is how it is related to other, more or less similar, habitats; for example, whether a forest patch is completely isolated in a sea of grassland or whether it is just one of several connected forest patches. While evaluating the isolated or connected character of a habitat patch, we need to think of one or more model organisms that we are mostly interested in, since these properties differ for different organisms. For example, a forest habitat patch can be considered as isolated for carabids but well connected for birds. This seemingly simple but very difficult problem lies at the heart of how to connect biological phenomena to the abiotic components of a landscape, i.e., how to add function to structure. If we have a model organism or a general rationale to help us decide what should be the basis for determining the fact and extent of connectedness, we will be able to characterize the position of the focal habitat patch within the context of other similar ones. In other words, we can build a network model for analyzing the topological and structural characteristics of the landscape.

Network analysis explores the properties of graphs: mathematical objects composed of a set of nodes, a set of links, and a relationship defined between the two sets (see Harary 1969). For example, a landscape graph is defined as (1) a set of habitat patches, (2) a set of corridors, and (3) a relation between the sets: link α exists between nodes A and B if the focal organism can disperse between the two corresponding habitat patches (A and B) through a corridor (represented by α). The tools of graph theory have been imported into many disciplines, from landscape ecology (Urban and Keitt 2001) to community ecology (Cohen 1978), and from ethology (Croft *et al.* 2004) to political science (Harary 1961). Network science is an integrative field of multidisciplinary interest (see Newman 2003).

The network perspective on landscapes works well for certain terrestrial organisms (such as carabids, see Tischendorf and Wissel 1997; Lövei *et al.* 2006) and is increasingly being used in the "seascape" analysis of marine systems (see Shimazaki *et al.* 2004; Treml *et al.* 2008; Saenz-Agudelo *et al.* 2009). The old view of spatially homogeneous marine ecosystems is no longer held, even though in some cases it might be more difficult to identify the patch–corridor pattern in wet systems (depending on scale). Network thinking can be generalized for some types of questions (such as how to consider linear and loop-like landscape elements) but it essentially remains highly species-specific (an exception is given in Rothley and Rae 2005).

In most cases, network analysis can be a helpful tool to better understand the part-to-whole relationships between habitat patches and the landscape. The question is not simply whether a particular patch is isolated. If the patchwork is connected, it is also important to understand how corridors are patterned among patches (what is the topology of the patchwork?). Certain configurations either enhance or weaken the overall connectedness of the whole landscape (for example, compare the "necklace" arrangement with the "cross" topology of Cantwell and Forman 1993). If the whole patchwork is well connected (e.g., a "cross"), we can expect that migration will be easier, gene flow will be more intense, the loss of genetic diversity will be less serious and, in general, the metapopulation inhabiting the habitat patchwork will have a higher chance of survival and will be less vulnerable. Roughly, these are the risks of living in small and isolated habitat fragments (e.g., Saccheri *et al.* 1998), driving the population below the minimal viable population size. It is important to emphasize that there can be a difference between putting emphasis on abiotic or biotic elements in the landscape. One approach concentrates on a physical entity, such as a habitat patch (like a pond or an island), and provides a top-down description of the abiotic system, independently of the inhabiting organisms. Another approach is a more biologically-based, bottom-up description of ecological systems, focusing on spatial phenomena (dispersal, migration). Network analysis works for both, but we should bear in mind that the important factors and parameters are different. For example, patch size may be more relevant in the first, while immigration rate may be more relevant in the second case. It needs be mentioned that, as in the case of communities, the borders of a "landscape" are somewhat ambiguous: hence, network analysis needs the insights and judgment of the researcher to make a sensible decision about how to construct the graph.

Fitting relevant methods to particular problems

Depending on what kind of information is represented in the network model (for example, whether the network is weighted by the permeability or the length of corridors), we need to choose suitable method(s) for network analysis. Apart from weighting links, a network can also be directed: source–sink systems can be represented by a directed network (Pulliam 1988), even if the identities of sources and sinks are highly dynamic (Dias 1996). Weighting the network is important if dispersal is strongly limited by distance, or if the permeability of corridors varies and this is important. In the case of a landscape ecological problem, one has to decide whether weights and directionality are important biological features and can be considered as relevant data for the analysis. Then, based on this, weighted (weighted distance) or binary (degree) as well as directed (reachability) or undirected (clustering coefficient) indices can be used.

A landscape graph can be studied at two levels: "global" network properties characterize the whole system (e.g., diameter), while "local" measures (e.g., node degree) quantify the neighborhood of individual landscape elements (patches or corridors) within the whole system. Beyond characterizing a landscape graph by macroscopic measures (a single number or distribution), there is a strong recent interest in quantifying the positions of its nodes and links (Ricotta *et al.* 2000; Jordán *et al.* 2003; Pascual-Hortal and Saura 2006; Bodin and Norberg 2007; Minor and Urban 2007; Estrada and Bodin 2008). These local indices may help in better understanding the internal structure of the patchwork, making landscape management more efficient, and setting quantitative conservation priorities. Some of the most basic indices are presented here, from the most local to more global ones, considering larger and larger neighborhoods of graph nodes.

The position of node *i* can be characterized by its degree (D_i), giving the number of neighboring patches directly connected to patch *i* (Wassermann and Faust 1994; Minor and Urban 2007, 2008). A D_i value can also be derived for corridor *i* by taking the average degree of its end-nodes.

Whether neighbors are connected to each other, and to what extent, can be measured by the clustering coefficient of a node (and its average for the whole graph). This is calculated as the number of corridors between neighboring patches divided by their potential maximal number. The latter is difficult to determine for landscape graphs (what is the maximal number of corridors leaving a particular habitat patch?).

The topological distance (d) of two nodes is the number of links forming the shortest path connecting them (Harary 1969; Wassermann and Faust 1994). The "topographical" distance (d_{tgr}) is a more realistic corridor index that reflects both the number of links between two nodes and their weights characterizing corridor quality (permeability, length, width). It is partly a biological question how to convert this kind of information into a single weight value (although it is possible if the nature of this weighted distance is taken into account: for instance, high permeability means low topographical distance). The connectedness of a graph node can be calculated as its average distance from other nodes. Small distance values represent the more central nodes in a network: on average, a habitat patch with low d is more important in maintaining connectivity than one with a higher d (note also that the redundancy of possible pathways needs to be considered here in future applications). Since the distance of a node from the others is taken to be infinite if it is isolated, these nodes can be ignored or, as an alternative, reciprocal distance values can be used (by convention, the reciprocal of infinite is zero). Note that several distance-related measures, borrowed from social network analysis, have recently been used for characterizing landscape graphs. These include various centrality indices (e.g., Minor and Urban 2007, 2008 note that d and closeness centrality are closely related).

As well as considering corridor quality, the quality of patches should also be taken into account. This can be characterized, for instance, by local population size. The summed local population size for the largest component (a set of connected nodes) of the landscape graph may indicate an important characteristic of the metapopulation; as population size is close to the minimal viable population size, the problem of how to keep as many individuals as possible connected may be a key issue. A patch or corridor can be characterized by how much this value will be changed after deleting the given landscape element from the network.

A combined index of several network indices can be helpful in order to unite various kinds of information (e.g., pure topology, patch quality, corridor quality). For example, Jordán *et al.* (2003) used the following combined importance index for node *i*:

$$I_i = \frac{D_i - CC_i}{d_i^{tgr} + LPS_{conn(i)}^{\max}}$$

Here, I_i is the positional importance of node *i* in maintaining connectivity, D_i is its degree, CC_i is its clustering coefficient, d_i^{tgr} is its average topographical distance from other nodes, and $LPS^{\max}_{\mathrm{conn}(i)}$ is the summed population size of the largest connected component. These values are either also computable for graph links, or the average values of end-nodes may be used in order to characterize the positional importance of links. Depending on how dense the network is (how many links it has, compared to the theoretical maximum), the clustering coefficient can be useless: in a sparse network, it can be zero for each node and there is no need to consider this index (see Jordán *et al.* 2007, where the modified I^* importance index simply lacks CC).

Quantifying source–sink landscape digraphs

The habitat of a source–sink metapopulation can be modeled by a digraph (directed graph; see e.g., Schick and Lindley 2007). In this case, only a subset of network analytical tools may be used. For example, if there is a single large source and many small sinks in the system, it is a key question how large is the total population size of patches connected to the source. This is, obviously, a subcase of the more general measure above. This measure is reasonable if dispersal is not limited by distance and it is then particularly relevant (e.g., the metapopulation is highly sensitive to the loss of genetic variability). If a species is on the brink of extinction, the total number of individuals participating in gene flow may be more important than the speed of dispersal. However, if dispersal is limited by distance, a distance-based reachability index can measure how easy dispersal is from the source to the sinks (see Jordán *et al.* 2007, based on Borgatti 2006).

As well as measuring the structural importance of landscape elements in intact landscape graphs, several other questions also arise. In a sparse, low-density network, it is a relevant question what the position of an inserted corridor in the network is if connectivity needs to be maximally increased. In this case, the hypothetical landscape element needs to be positioned in the network in a feasible way. If a corridor is inserted into a model network, its possible positions must be a plausible subset of all possibilities (e.g., inserted corridors must not cross rivers). The properties of the derived graph, containing the inserted corridor, can then be compared to those of the intact network.

In case of a metapopulation where dispersal is spatially limited (i.e., corridors are weighted by length or permeability), improving the quality of existing corridors is an option. In this case, link weights referring to topographical distance can be modified and the question is which corridor to improve (e.g., its permeability can be increased). Implementation may mean the optimization of its width or reducing predation pressure in the corridor. If the metapopulation is close to the minimal viable population size, a question is how to connect every patch in the studied landscape in order to keep local populations connected (by inserting a minimal number of corridors, maximizing reachability, and achieving the maximal summed population size).

Network analysis always offers a set of tools, but selecting the most suitable method depends on whether the quality of either patches or corridors can be measured, whether the metapopulation is a source–sink system, and whether we can consider establishing new landscape elements. Note that additional measures (Pascual-Hortal and Saura 2006) as well as software programs (Saura 2007; Saura and Pascual-Hortal 2007) have also been developed for even more problem-oriented studies. The difference between using metrics for undirected and directed landscape graphs can be best illustrated by comparing case studies.

Case studies: undirected and directed landscape graphs

Figure 12.1 shows the landscape graph of a Hungarian metapopulation of the flightless bush cricket *Pholidoptera transsylvanica*, a highly mobile predator. Nodes represent semiarid grasslands, with node size being proportional to local population size (estimated in the field, by acoustic census) and classified from 1 to 4. Links represent forest roads and similar linear unforested structures (their permeability has been classified from 1 to 4, based on corridor length and vegetation cover), while the matrix is forest (Orci 1997; Varga 1997). It is worth mentioning that determination of the weights, or quality, of both nodes and links was quite simplistic (I here focus on the methodology, and better data obviously call for a repeated analysis). Dispersal is undirected,

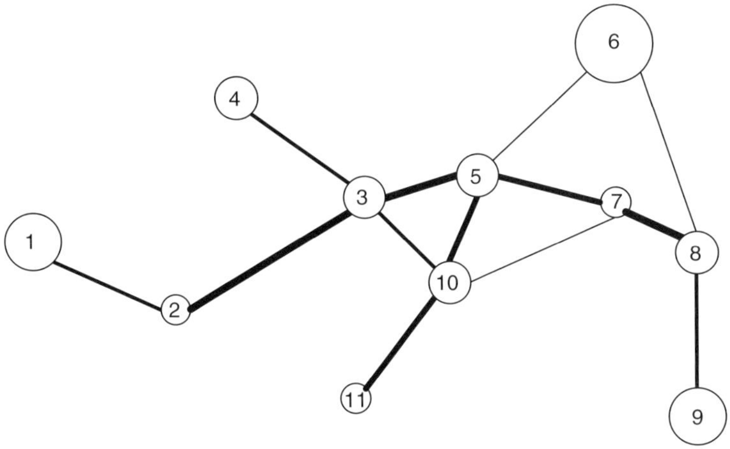

FIGURE 12.1.
Landscape graph of the Aggtelek Karst in northeastern Hungary for a flightless cricket species inhabiting grasslands (node is grassland, edge is road, matrix is forest). Node size is proportional to local population size discretized from 1 to 4, while line thickness corresponds to corridor permeability, and is similarly classified from 1 to 4 (large nodes and thick lines are more important). The graph contains 11 habitat patches (graph nodes: coded N in the text) and 13 corridors (graph links: coded L in the text) as follows: (1) Huszas töbör, (2) Kis tisztások, (3) Szilicei kaszálók, (4) U-alakú töbör, (5) Nagy-Nyilas, (6) Mogyorós-rét és tisztás, (7) Árvalányhajas, (8) Dénes töbör, (9) Nagyoldal mögötti tisztások, (10) Gyertyánsarjas, (11) Lófej-forrás alatti tisztás (Jordán *et al.* 2003).

there is no source–sink dichotomy between local populations, and the animal is a highly mobile predator. This network is a mesh-like arrangement (following Cantwell and Forman 1993). The importance rank of landscape elements can be determined, based on a combined importance index (I): patches N3 (I_{N3} = 0.2059), N5 (I_{N5} = 0.151), and N10 (I_{N10} = 0.1486) are the most important ones, while patch N11 (I_{N11} = 0.0284) is structurally the least important element in maintaining connectivity (Jordán *et al.* 2003).

Figure 12.2 shows the landscape graph of seven forest-living carabid species in the Bereg plain, Hungary. Nodes represent forest patches: node size is proportional to local population size (measured in the field by trapping, and discretized from 1 to 4). Corridors represent permeable vegetation between two patches within a distance of 1 km and the matrix is either grassland or agricultural landscape (Lövei and Sunderland 1996; Magura *et al.* 2001). In this source–sink metapopulation system, the Carpathian Mountains can be considered a large area of habitat (a source) for carabids, whereas the local populations of forest-living carabids are continuously decreasing in the other small, isolated forest patches (sinks). So the network is directed with one source and many sinks, and this feature calls for a slightly different set of network indices. Additionally, the graph is of very low density (the number of corridors is

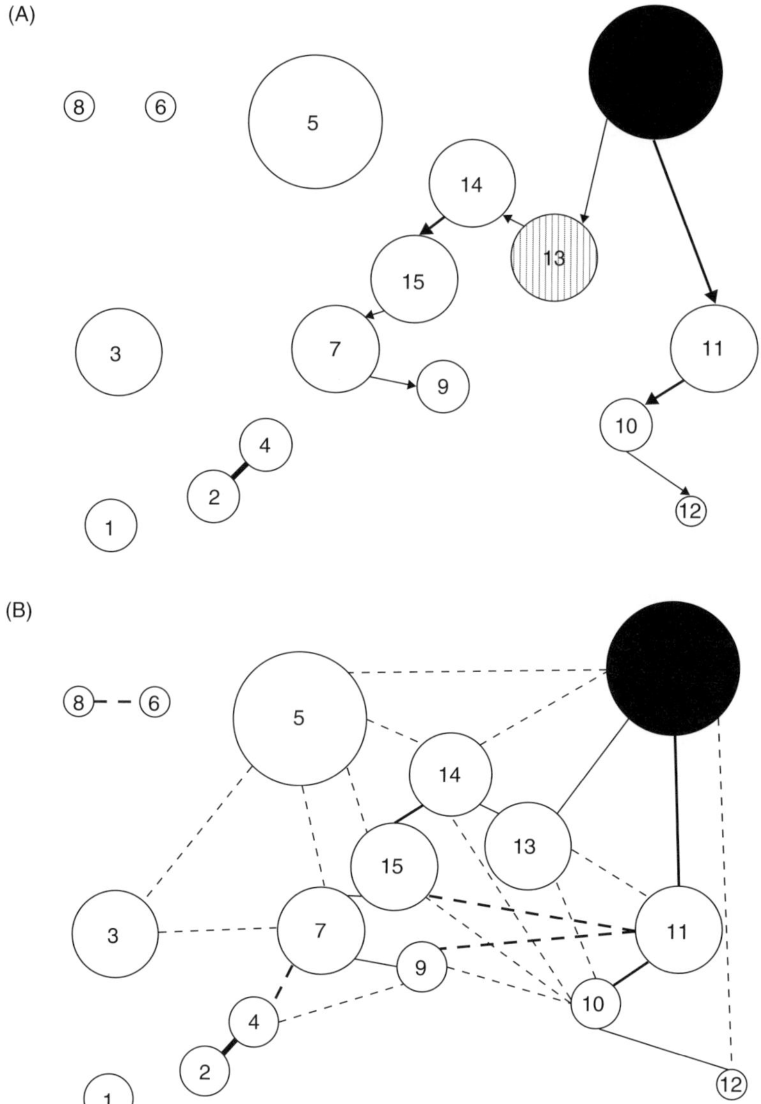

FIGURE 12.2.
Landscape graph analysis of the Bereg Plain in eastern Hungary for forest-living carabids (node is forest, edge is permeable distance, matrix is grassland; node size and line thickness are shown as in Figure 12.1). We can determine (a) which landscape element is the most important in maintaining the connectivity of this directed graph (striped graph node), (b) what are the possible positions of ecological corridors (dashed lines), (c) which corridor is the most useful to build, (d) which corridor is the best to improve, and (e) what is the most efficient way to connect all forest patches (original names of forest patches are given in Jordán *et al.* 2007). The black graph node is the only source patch (Carpathian Mountains). Direction of links is shown only in (a) for simplicity; dashed lines represent possible positions of ecological corridors.

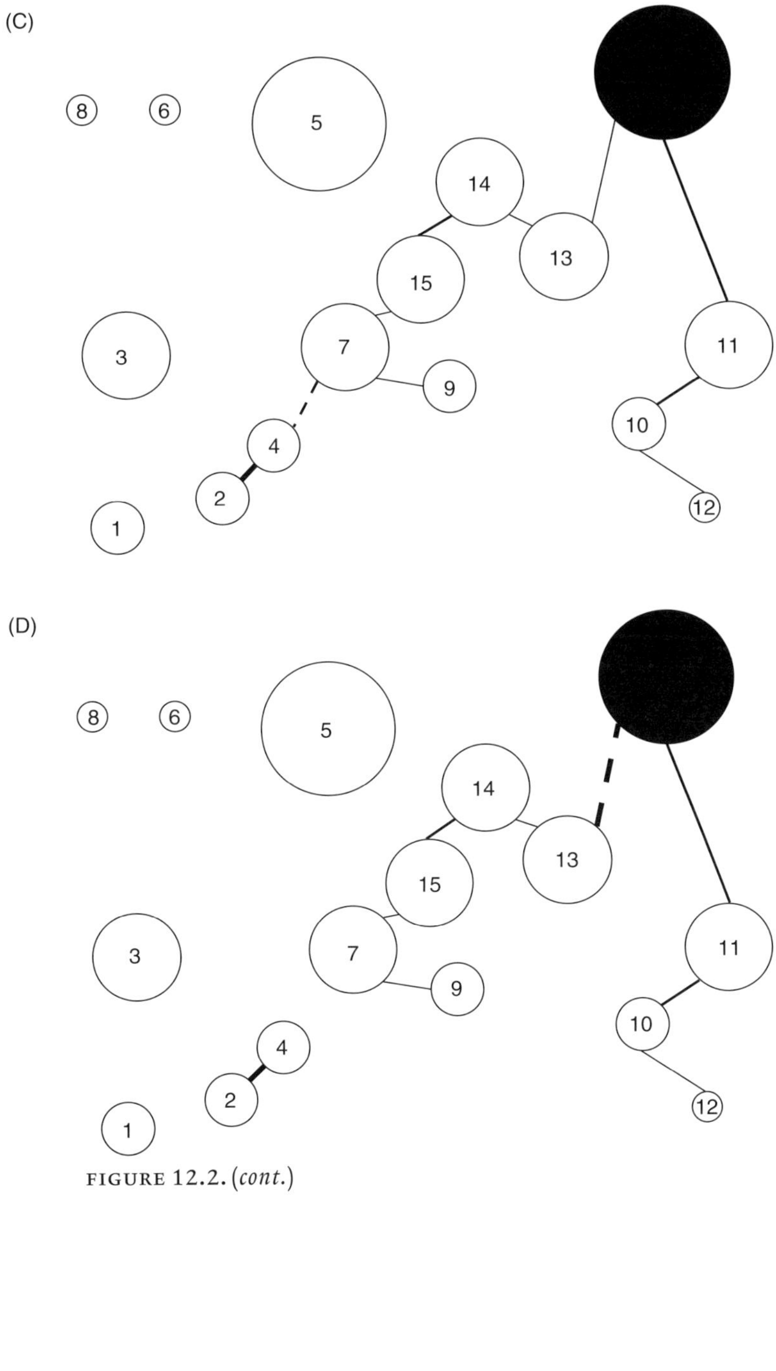

FIGURE 12.2. (*cont.*)

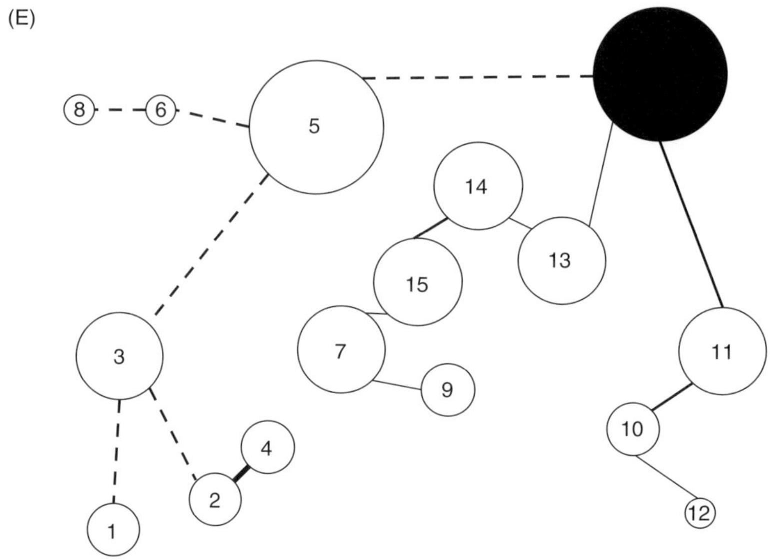

FIGURE 12.2. *(cont.)*

very small compared to the theoretical maximum), so using the clustering coefficient seems to give no information. Thus, the importance rank of landscape elements was determined by a slightly different combined importance index (I^*; Jordán *et al.* 2007; the same as the previous one discussed earlier but not containing CC). Figure 12.2a shows the most important landscape element in maintaining connectivity ($I^*_{N13} = 0.1081$) and the rank continues with corridor L13–14 ($I^*_{L13-14} = 0.0967$) and patch N14 ($I^*_{N14} = 0.0958$); the least important landscape element in this network is patch N9 ($I^*_{N9} = 0.0273$). Figure 12.2b shows the corridors that are possible to build based on various constraints (e.g., theoretically, there could be a corridor between patches N1 and N11, but it would spatially overlap with several other patches and corridors). Figure 12.2c shows which corridor to build if the sum of local population sizes linked to the source habitat is to be maximized. This particular approach is relevant only in the case of "one source–several sink" networks. Figure 12.2d shows which corridor to improve if distance-based reachability from the source node is considered, i.e., how to minimize the costs of dispersal from the source to the sinks (in terms of weights on links). Figure 12.2e shows how to connect all sinks to the source by keeping the number of inserted corridors minimal (see Jordán *et al.* 2007, for methodological details). Note that the "best" corridor (Figure 12.2c) is not part of this latter solution: this calls for holistic thinking when evaluating the role and importance of landscape elements. Since a freeway is to be built across this area, following one of three possible routes, we used the above network analytical tools to quantitatively evaluate and compare their effects on landscape connectivity (Vasas *et al.* 2009).

Discussion

The systems view on habitat patchworks means that we study both landscape elements (individual patches/corridors) and the whole system (the landscape graph), and need to understand both levels. In other words, we evaluate a patch or a corridor only in the context of the whole patchwork. This approach helps in conceptualizing and quantifying the role that single habitat patches and corridors play in maintaining landscape connectivity (as well as providing a global characterization of the landscape graph). Network analytical tools offer various possibilities for finding the most suitable and optimal methods for the study of particular problems (considering data type and quality). Local indices quantifying the network position of individual landscape elements (patches and corridors) add another aspect to the traditional parameters of habitat suitability analyses (e.g., Debeljak *et al.* 2001; Glenz *et al.* 2001) and they make it possible to set priority ranks for conservation (landscape management, landscape engineering). If these efforts are used to maintain landscape connectivity efficiently, natural processes can be kept intact and there is a higher chance of sustainability. On the other hand, if key landscape elements are lost, reduced connectivity may trigger community-wide extinctions and the chance of sustainability is seriously reduced (see Crooks and Soulé 1999).

However, before this approach can be applied, we must note that the structure of the landscape is highly species-specific and scale-dependent (as detailed in Haddad *et al.*, Chapter 22, this volume). Multiscale and multispecies landscape models are, to date, too weak to be reliably applied. Metacommunity models already exist, but metacommunity databases are very sparse (but see Melian *et al.* 2005). Still, the metacommunity context is highly important, since landscape changes (e.g., fragmentation) influence selectively different species of the local communities, e.g., mobile predators at higher trophic levels go extinct first, and altered community control may drive secondary extinctions (e.g., Crooks and Soulé 1999). If multispecies approaches are not feasible (or if they are very expensive), a compromise can be made by studying the habitat network (and the landscape graph) of keystone or umbrella species (Simberloff 1998). There is a hope that understanding the landscape ecology of keystone species could disproportionately increase our knowledge of the metacommunity – and this may make conservation efforts more efficient.

Another possibly useful extension is the multidisciplinary, comparative view. If not all kinds of networks, at least spatial networks can be analyzed in parallel; we can expect to learn something from cross-comparisons. For example, traffic networks are extensively analyzed and this may provide useful information for landscape ecologists as well (see Jordán 2008).

Acknowledgments

I am very grateful to the Pulliam Symposium Organizers for their kind invitation. Santiago Saura and Lucia Pascual-Hortal are acknowledged for interesting discussions and ideas. I also thank Professor Zoltán Varga, András Báldi, Tibor Magura, Professor Béla Tóthmérész, Vera Vasas, Kirill Márk Orci, Viktor Ködöböcz, and István Rácz for their help and for sharing field data. I gratefully thank Katie Wintermute for linguistic assistance. My research was fully supported by the Branco Weiss Fellowship of Society in Science, ETH Zürich, Switzerland.

References

Bodin, O. and J. Norberg (2007). A network approach for analyzing spatially structured populations in fragmented landscape. *Landscape Ecology* **22**: 31–44.

Borgatti, S. P. (2006). Identifying sets of key players in a social network. *Computational and Mathematical Organization Theory* **12**: 21–34.

Brown, K. S., P. M. Sheppard and J. R. G. Turner (1974). Quaternary refugia in tropical America: evidence from race formation in *Heliconius* butterflies. *Proceedings of the Royal Society of London B* **187**: 369–378.

Cantwell, M. D. and R. T. T. Forman (1993). Landscape graphs: ecological modelling with graph theory to detect configurations common to diverse landscapes. *Landscape Ecology* **8**: 239–255.

Cohen, J. E. (1978). *Food Webs and Niche Space*. Princeton University Press, Princeton, NJ.

Croft, D. P., J. Krause and R. James (2004). Social networks in the guppy (*Poecilia reticulata*). *Proceedings of the Royal Society of London Series B – Biological Sciences* **271**: S516–S519.

Crooks, K. R. and M. E. Soulé (1999). Mesopredator release and avifaunal extinctions in a fragmented system. *Nature* **400**: 563–566.

Debeljak, M., S. Dzeroski, K. Jerina, A. Kobler and M. Adamic (2001). Habitat suitability modelling for red deer (*Cervus elaphus* L.) in south-central Slovenia with classification trees. *Ecological Modelling* **138**: 321–330.

Dias, P. C. (1996). Sources and sinks in population biology. *Trends in Ecology and Evolution* **11**: 326–330.

Dunning, J. B., B. J. Danielson and H. R. Pulliam (1992). Ecological processes that affect populations in complex landscapes. *Oikos* **65**: 169–175.

Estrada, E. and O. Bodin (2008). Using network centrality measures to manage landscape connectivity. *Ecological Applications* **18**: 1810–1825.

Glenz, C., A. Massolo, D. Kuonen and R. Schlaepfer (2001). A wolf habitat suitability prediction study in Valais (Switzerland). *Landscape and Urban Planning* **55**: 55–65.

Harary, F. (1961). A structural analysis of the situation in the Middle East in 1956. *Journal of Conflict Resolution* **5**: 167–178.

Harary, F. (1969). *Graph Theory*. Addison-Wesley, Reading, MA.

Jordán, F. (2008). Predicting target selection by terrorists: a network analysis of the 2005 London underground attacks. *International Journal for Critical Infrastructures* **4**: 206–214.

Jordán, F., A. Báldi, K. M. Orci, I. Rácz and Z. Varga (2003). Characterizing the importance of habitat patches and corridors in maintaining the landscape connectivity of a *Pholidoptera transsylvanica* (Orthoptera) metapopulation. *Landscape Ecology* **18**: 83–92.

Jordán, F., T. Magura, B. Tóthmérész, V. Vasas and V. Ködöböcz (2007). Carabids (Coleoptera: Carabidae) in a forest patchwork: a connectivity analysis of the Bereg Plain landscape graph. *Landscape Ecology* **22**: 1527–1539.
Lövei, G. L. and K. D. Sunderland (1996). Ecology and behavior of ground beetles (Coleoptera: Carabidae). *Annual Review of Entomology* **41**: 231–256.
Lövei, G. L., T. Magura, B. Tóthmérész and V. Ködöböcz (2006). The influence of matrix and edges on species richness patterns of ground beetles (Coleoptera, Carabidae) in habitat islands. *Global Ecology and Biogeography* **15**: 283–289.
Magura, T., V. Ködöböcz and B. Tóthmérész (2001). Effects of habitat fragmentation on carabids in forest patches. *Journal of Biogeography* **28**: 129–138.
Melian, C. J., J. Bascompte and P. Jordano (2005). Spatial structure and dynamics in marine food webs. In *Aquatic Food Webs* (A. Belgrano, U. M. Scharler, J. Dunne and R. E. Ulanowicz, eds.). Oxford University Press, Oxford, UK.
Minor, E. S. and D. L. Urban (2007). Graph theory as a proxy for spatially explicit population models in conservation planning. *Ecological Applications* **17**: 1771–1782.
Minor, E. S. and D. L. Urban (2008). A graph-theory framework for evaluating landscape connectivity and conservation planning. *Conservation Biology* **22**: 297–307.
Newman, M. E. J. (2003). The structure and function of complex networks. *SIAM Review* **45**: 167–256.
Orci, K. M. (1997). A comparative study on grasshopper (Orthoptera) communities in the Aggtelek Biosphere Reserve. In *Research in Aggtelek National Park and Biosphere Reserve* (E. Tóth and R. Horváth, eds.). ANP Directorate, Aggtelek, Hungary: 109–116.
Pascual-Hortal, L. and S. Saura (2006). Comparison and development of new graph-based landscape connectivity indices: towards the prioritization of habitat patches for conservation. *Landscape Ecology* **21**: 959–967.
Polis, G. A., W. B. Anderson and R. D. Holt (1997). Toward an integration of landscape and food web ecology: the dynamics of spatially subsidized food webs. *Annual Review of Ecology and Systematics* **28**: 289–231.
Pulliam, H. R. (1988). Sources, sinks, and population regulation. *American Naturalist* **132**: 652–661.
Ricotta, C., A. Stanisci, G. C. Avena and C. Blasi (2000). Quantifying the network connectivity of landscape mosaics: a graph-theoretical approach. *Community Ecology* **1**: 89–94.
Rothley, K. and C. Rae (2005). Working backwards to move forwards: graph-based connectivity metrics for reserve network selection. *Environmental Modeling and Assessment* **10**: 107–113.
Saccheri, I., M. Kuussaari, M. Kankare, P. Vikman, W. Fortelius and I. Hanski (1998). Inbreeding and extinction in a butterfly metapopulation. *Nature* **392**: 491–494.
Saenz-Agudelo, P., G. P. Jones, S. R. Thorrold and S. Planes (2009). Estimating connectivity in marine populations: an empirical evaluation of assignment tests and parentage analysis under different gene flow scenarios. *Molecular Ecology* **18**: 1765–1776.
Saura, S. (2007). Evaluating forest landscape connectivity through Conefor Sensinode 2.2. In *Patterns and Processes in Forest Landscapes: Multiple Use and Sustainable Management* (R. Lafortezza, J. Chen, G. Sanesi and T. R. Crow, eds.). Springer, Berlin: 403–422.
Saura, S. and L. Pascual-Hortal (2007). *Conefor Sensinode 2.2 User's Manual: Software for Quantifying the Importance of Habitat Patches for Maintaining Landscape Connectivity through Graphs and Habitat Availability Indices*. University of Lleida, Spain.
Schick, R. S. and S. T. Lindley (2007). Directed connectivity among fish populations in a riverine network. *Journal of Applied Ecology* **44**: 1116–1126.
Shimazaki, H., M. Tamura, Y. Darman, V. Andronov, M. P. Parilov, M. Nagendran and H. Higuchi (2004). Network analysis of potential migration routes for oriental white storks (*Ciconia boyciana*). *Ecological Research* **19**: 683–698.
Simberloff, D. (1998). Flagships, umbrellas, and keystones: is single-species management passé in the landscape area? *Biological Conservation* **83**: 247–257.

Tischendorf, L. and C. Wissel (1997). Corridors as conduits for small animals: attainable distances depending on movement pattern, boundary reaction and corridor width. *Oikos* **79**: 603–611.
Treml, E. A., P. N. Halpin, D. L. Urban and L. F. Pratson (2008). Modeling population connectivity by ocean currents, a graph-theoretic approach for marine conservation. *Landscape Ecology* **23**: 19–36.
Urban, D. and T. Keitt (2001). Landscape connectivity: a graph-theoretic perspective. *Ecology* **82**: 1205–1218.
Varga, Z. (1997). Biogeographical outline of the invertebrate fauna of the Aggtelek Karst and surrounding areas. In *Research in Aggtelek National Park and Biosphere Reserve* (E. Tóth and R. Horváth, eds.). ANP Directorate, Aggtelek, Hungary: 87–94.
Vasas, V., T. Magura, F. Jordán and B. Tóthmérész (2009). Graph theory in action: evaluating planned highway tracks based on connectivity measures. *Landscape Ecology* **24**: 581–586.
Wassermann, S. and K. Faust (1994). *Social Network Analysis*. Cambridge University Press, Cambridge, UK.

MATTHEW A. ETTERSON, BRIAN J. OLSEN, RUSSELL GREENBERG AND W. GREGORY SHRIVER

13

Sources, sinks, and model accuracy

Summary

Source–sink models are a promising empirical tool for the sustainable management of animal populations across landscapes. Recent work has demonstrated a theoretical link between the demographic processes addressed in both source–sink and metapopulation models and the formation of species' range limits. In the face of large-scale anthropogenic disturbances (e.g., increasing temperature and sea level with global climate change), conceptual range-limit models that are functionally linked to these demographic mechanisms may help predict range shifts and provide insights for the management of vulnerable populations. However, the value of such models is limited by their ability to offer precise and testable predications about how demographic parameters might respond to environmental change and thus influence population dynamics. Here, we illustrate the gulf between the promise of conceptual demographic models and the difficulty of their empirical application by developing a model of range limits for a narrowly distributed tidal-marsh songbird, the coastal plain swamp sparrow (*Melospiza georgiana nigrescens*, CPSS). We first modeled a gradient in CPSS fecundity that depends on environmental factors varying with latitude. To predict the species' range limits we embed this fecundity gradient in Pulliam's (1988) source–sink model. Our resulting model predicts current CPSS range limits reasonably well. However, its predictions are also subject to substantial uncertainty. Our model framework generally conforms to the conceptual unification of source–sink theory, Hutchinson's (1957) fundamental niche, and species' range limits recently expounded by Pulliam (2000).

Sources, Sinks and Sustainability, ed. Jianguo Liu, Vanessa Hull, Anita T. Morzillo and John A. Wiens. Published by Cambridge University Press.

Background

It is now clear that sources and sinks exist (Pulliam 1995) and that anthropogenic disturbance can alter the landscape distribution of sources and sinks (Pulliam 1995; Gaona *et al.* 1998; Remes 2000; Rowe and Hopkins 2003). Pulliam (1988) presented a hypothetical case in which anthropogenic destruction of a seemingly unimportant habitat (where a small fraction of the population occurred) could have disastrous consequences if the habitat is an important source. If we are to sustainably manage landscapes with mosaics of source and sink populations, we need predictive models to quantify the effect of our actions on these populations. Thus, for management, it is not sufficient for models to simply predict the existence of sources and sinks; they must also functionally link demographic rates to specific landscape features and environmental conditions. Furthermore, predictions of demographic responses to habitat alterations must be sufficiently precise to support management actions that are designed to positively influence population stability and growth. However, spatial demographic models may be subject to considerable uncertainty, even when informed by intensive monitoring data (Etterson and Nagy 2008).

Uncertainties notwithstanding, empirical and theoretical research on source–sink dynamics continues to generate insights and novel hypotheses about the observed dynamics of spatially structured populations. As the work of Holt and colleagues has shown (Holt and Keitt 2000; Holt *et al.* 2005; Holt, Chapter 2, this volume), species' distributional limits provide a dynamic laboratory for understanding how gradients in metapopulation parameters such as occupancy, extinction, and colonization rates may limit the spatial distribution of populations. Similarly, Pulliam (2000) elaborated the relationship between resource gradients, source–sink dynamics, and a species' niche, by defining the latter as the set of environmental conditions for which $\lambda \geq 1$. By this definition, the niche is the set of all population sources ($\lambda > 1$; Pulliam 1988), although the former would also include sites for which $\lambda = 1$. This explicit and simple relationship provides a powerful conceptual approach for extending source–sink theory to investigate the relationship between environmental factors, demography, and species' range limits.

In this chapter we explore the data requirements for translating the conceptual relationship between source–sink theory and environmental factors into specific predictions about the present range limits of coastal plain swamp sparrows (*Melospiza georgiana nigrescens*, CPSS). Our primary goal is to illuminate the difficulties of using a promising conceptual model to construct an empirical model with predictions sufficiently precise to make management decisions. In keeping with classical source–sink theory, our model examines how gradients in factors affecting fecundity may create regions where productivity exceeds mortality versus regions where extinction would occur in the absence of immigration (Pulliam 1988). At the scale of a species range, we assume that a geographic transition between these two regions (from mostly sources to mostly sinks) should determine the approximate range boundary. We recognize that while we use the term boundary, metapopulation dynamics may result in a zone of populations that wink in and out of existence, as suggested by repeat surveys of the historical range of this subspecies (Beadell *et al.* 2003).

Environmental gradients and avian reproductive success in tidal-marsh habitats

We used four latitudinal gradients in avian reproductive parameters to construct our fecundity model: clutch size, nest predation rates, nest inundation rates (due to flooding), and breeding season length. Greenberg *et al.* (2006) reviewed studies of North American tidal-marsh sparrows and found that, while clutch size and flooding rates increase with increasing latitude, nest predation rates and breeding season length generally decrease with increasing latitude. Therefore, nest predation and clutch size will result in increased fecundity with increasing latitude, while flooding rates and breeding season length will result in reduced fecundity with increasing latitude.

In four broad steps we use what we know about the breeding biology of the CPSS and latitudinal gradients in tidal-marsh sparrow reproduction to predict the location of the CPSS southern range boundary.

1. We first develop a mathematical model that predicts seasonal fecundity as a function of specific CPSS traits.
2. We incorporate the four latitudinal gradients into this model to generate a fecundity curve, dependent on latitude.
3. To relate this fecundity curve to population growth rate, we embed the model within Pulliam's (1988) source–sink model to create a niche model for the CPSS.
4. Finally, we use this niche model to predict where population growth falls below replacement in the absence of immigration: the putative southern range boundary.

Our objectives are to explore the following questions.

- Given what we know about CPSS fecundity (*Seasonal fecundity model*) and latitudinal gradients in reproductive parameters for tidal-salt-marsh sparrows, what is the optimal breeding latitude for the CPSS (*Fecundity curve*)?

- Does this optimum occur within the current range of the CPSS?
- When combined with plausible values for survival probabilities, can a simple demographic model be used to predict CPSS range limits (*Niche model*)?
- If a plausible model is possible, is it feasible to estimate its parameters with sufficient precision to be able to use the model to predict range shifts resulting from natural or anthropogenic environmental change and, more importantly, for making decisions about conservation and management of this endemic subspecies (*Model precision*)?

Research methods

Study species

The coastal plain swamp sparrow (CPSS) is a recognized subspecies which is largely restricted to brackish marshes above the mean high tide line during both its breeding and non-breeding periods. Two far more widespread inland subspecies are restricted to the freshwater marshes of eastern and boreal North America. The coastal subspecies is not differentiable from those inland based on MtDNA analysis (Greenberg *et al.* 1998) or on microsatellite markers (R. Fleischer *et al.*, unpublished), but is 100% diagnosable based on its large bill and grayer/blacker coloration (Greenberg and Droege 1990): traits that the coastal endemic subspecies shares with other tidal-marsh birds (Greenberg and Droege 1990; Grenier and Greenberg 2005). Standard rearing experiments show that the morphological differences are probably genetically based (Greenberg and Droege 1990; Ballentine and Greenberg 2010). In addition to morphological differences, a number of behavioral and ecological features distinguish coastal plain birds from interior swamp sparrows (Olsen 2007), including distinctive parental behavior and vocal repertoire (Liu *et al.* 2008) as well as a smaller average clutch size (Olsen *et al.* 2008a). These considerations suggest that the CPSS is on a separate evolutionary trajectory from the inland subspecies and is worthy of conservation as a distinct evolutionary unit.

Coastal plain swamp sparrow habitat and range

Preferred marshes for the CPSS have salinities well below 10 parts per thousand (ppt) and support a relatively diverse flora of *Spartina* grasses, rushes (*Schoenoplectus* spp.), reeds (*Phragmites australis*), and shrubs (*Iva frutesens* and *Baccharis halimifolia*). The subspecies winters in a similar shrub–grass marsh community, which lies between the maritime loblolly pine (*Pinus taeda*) forests and extensive tracts of black needle rush (*Juncus romerainus*) in the southeastern

USA. The CPSS migrates north to breed despite the presence of very similar habitat in the wintering range that lacks breeding sparrows in the summer months. Breeding populations are known in coastal marshes from northern New Jersey (40.9° N) to the southern Chesapeake shore of the Delmarva Peninsula in Maryland (38.3° N) and the upper Rappahannock River in Northern Virginia (37.9° N) (Greenberg and Droege 1990; Beadell *et al.* 2003; Watts *et al.* 2008). The subspecies completely leaves its breeding range in the fall (Greenberg *et al.* 2010) and has been found wintering from extreme south-coastal Virginia (36.7° N) to Charleston, SC (32.7° N) (Greenberg *et al.* 2007, 2010), but appears to be most common in the northern part of this coastal zone. Thus the migration, while complete, is relatively short, ranging from about 200 to 600 km. It is rather curious, then, that swamp sparrows would complete an energetically costly and risky northward migration, when the environment they leave appears structurally identical to the niche they occupy further north and possesses no obvious competitors. This seasonal within-habitat shift provides an excellent system for exploring the effects of gradients within key environmental factors on demographic processes that determine the breeding niche.

Seasonal fecundity model

To estimate the number of young produced per female during a breeding season (seasonal productivity) for female CPSS we adapted the regular Markov chain model developed for dickcissel (*Spiza americana*) and eastern meadowlark (*Sturnella magna*) by Etterson *et al.* (2009). Thus our breeding season model is a generalization of Markov chain methods for estimating avian nest survival (Etterson and Bennett 2005), which in turn are a generalization of standard methods for estimating nest survival (Mayfield 1975; Johnson 1979; Bart and Robson 1982). The form of the model we use here employs six parameters, which are

- m_p the daily probability that a nest is destroyed through predation
- m_f the daily probability that a nest is destroyed through flooding
- a the age (in days since the first egg was laid) at which fledglings leave a successful nest
- w_s the expected time required between fledging and the first egg in a subsequent nesting attempt
- w_f the expected time required between failure and the first egg in a subsequent nest
- T the length of the breeding season, defined as the amount of time separating the first egg of the first nest (among all females) and the first egg of the last nest of the season (again among all females).

Analysis of the resulting Markov chain model gives the expected number of successful broods per female (b) in a typical breeding season (Etterson *et al.* 2009; see Eq. (13.A1.2) in the Appendix to this chapter). To estimate seasonal fecundity (β) we estimated mean clutch size per nest (c) and assumed negligible hatching failure and equal ratio of male to female offspring so that $\beta \approx \frac{bc}{2}$.

We parameterized our seasonal fecundity model using nest data from a breeding population of CPSS at Woodland Beach, Delaware, that has been intensively monitored for the last 7 years (Greenberg *et al.* 2006, 2007; Etterson *et al.* 2007; Olsen 2007; Liu *et al.* 2008; Olsen *et al.* 2008a, 2008b). Greenberg *et al.* (2006) had already reported values for CPSS at Woodland Beach for clutch size, predation rates, flooding rates, and breeding season length (c, m_p, m_f, and T). These were used in our present model with transformations described in the next section (equations are provided in the Appendix to this chapter). Because nests were monitored daily and females were color-banded, we were able to estimate the remaining three parameters (a, w_s, and w_f) as the simple mean observed values, under the assumption that females did not disperse off the study site between nesting attempts and intermediate nests were not missed. The behavior of female swamp sparrows allows assumptions regarding our nest-locating ability that are much safer than for typical grassland-nesting songbirds, because females utter a distinctive series of call notes whenever they leave the nest during nest building, egg incubation, and nestling brooding (McDonald and Greenberg 1991). Over all years we located nests 3.6 ± 0.3 days (mean ± SE) after the start of incubation ($n = 275$), for those nests for which clutch completion date could be calculated (by laying or hatching date). Thus we are confident that we did not miss many nesting attempts.

The fecundity curve

Of the seven parameters required for our seasonal fecundity model (m_p, m_f, a, w_f, w_s, T, and c), Greenberg *et al.* (2006) published estimates of four (m_p, m_f, T, and c) derived from 12 breeding populations of three species of tidal-marsh sparrows distributed from 29.8° to 43.6° N latitude (Gulf County, Florida, to Scarborough Marsh, Maine; see Table1 in Greenberg *et al.* 2006). We used their data to estimate gradients in m_p, m_f, T, and c by first logit transforming (m_p, m_f) or log transforming (T, c) the observed values to express them on the continuum, and then estimating linear regressions of the transformed data on latitude. Full details of the transformations and regressions are provided in the Appendix.

The niche model

To expand our fecundity model to predict the population growth rate of CPSS across its range, we embedded our estimated fecundity curve (above) within a simple, two-stage population projection model assuming constant survival among habitats (Pulliam 1988).

$$\lambda = P_A + P_J\beta \tag{13.1}$$

where P_A and P_J are the annual adult and juvenile survival probabilities, respectively, and β is the annual per-female rate of production of female offspring (from the fecundity model above).

Survival estimates for the niche model

To estimate P_A we used 6 years of mark–recapture data (2002–2007, $n = 385$) for breeding adult CPSS at Woodland Beach, Delaware. We estimated separate apparent survival rates (φ) and recapture rates (p) for males ($n_m = 234$) and females ($n_f = 151$). Parameter estimates were generated using the "recapture-only" model in the program MARK (White and Burnham 1999). Preliminary demographic modeling of our Woodland Beach population suggested that the resulting female apparent survival rate (φ_f), in combination with other model parameters, would project a declining population at Woodland Beach. We suspected that this result was due to greater dispersal bias in the apparent survival estimate for females than for males, rather than a true difference in survival, and so we analyzed the population model (Eq. 13.1) twice, using each survival value (φ_f and φ_m) in turn for P_A.

We do not know juvenile survival rates in CPSS. Of 540 nestlings banded over the course of the study, we relocated only 34 after their hatch year. Therefore we solved for potential values of juvenile survival by fitting the niche model to the known northern range boundary of CPSS. In other words, we used the niche model to determine what value (if any) of P_J would result in a predicted value of $\lambda = 1$ at the northern range boundary (40.87° N). If this problem has a solution, then there will be a corresponding point further south at which, again, $\lambda = 1$, which we define to be the niche model's prediction of the location of the southern boundary. The precision analysis described below is a hypothetical exercise conducted under the assumption that the input parameters and model predictions, derived as described above, are correct.

Model precision

We used the Delta method (Daley 1979; Houllier *et al.* 1989) to discover how precisely the output metric (the location of the southern distributional limit, where $\lambda = 1$) would be estimated, conditional on hypothesized values for the estimated coefficients of variation of input parameters. The same question can be stated in a form akin to a power analysis: how precisely must the input parameters be estimated to predict the southern distributional limit to a given level of precision? We analyzed the model with the coefficients of variation in all input parameters set to three hypothetical levels of precision (0.01, 0.001, 0.0001). Equations are provided in the Appendix. Our precision analyses incorporate some variation due to fixed environmental effects (the latitudinal gradients), but we do not consider random environmental effects (environmental stochasticity).

Results

Annual apparent survival for male swamp sparrows was 0.52 (SE, CV: 0.04, 0.08) and 0.45 (0.07, 0.16) for females. Recapture probability for males was 0.67 (0.06, 0.09) and 0.43 (0.10, 0.23) for females. Mean age at fledging (a) was 22.3 (0.13, 0.006) days after the first egg was laid. The mean time between failure and the first egg of a replacement clutch (w_f) was 10.6 (0.88, 0.08) days, and between fledging and the first egg of the next nest (w_s) was 11.9 (1.4, 0.12) days. The estimated gradients in clutch size (c), breeding season length (T), flooding (m_f), and predation (m_p) are depicted graphically in Figure 13.1. As expected, predicted values of clutch size and flooding risk increased with latitude, whereas breeding season length and predation rate decreased with latitude (Fig. 13.1).

Maximum estimated reproductive success occurred at 39.68° N latitude, at which point the fitted estimate was approximately 1.67 female offspring per female (Fig. 13.2a). Thus, maximum fecundity was predicted to occur within CPSS current range, 37 km north of our Woodland Beach site (39.35° N). With this fecundity curve (Fig. 13.2a) and assuming $P_A = \varphi_m$, a juvenile survival value of $P_J = 0.288$ results in a niche curve for which $\lambda = 1$ at 40.87° N (Fig. 13.2b). On the latter curve, the southern latitude at which $\lambda = 1$ occurs at 38.53° N, close to the most southern (38.07° N) known breeding population of coastal plain swamp sparrows (Fig. 13.2b). For $P_A = \varphi_f$ and $P_J = 0.288$, the niche curve is entirely below unity (Fig. 13.2b). However, assuming $P_A = \varphi_f$, a juvenile survival value of $P_J = 0.328$ results in a niche curve for which $\lambda = 1$ at 40.87° N (not pictured). On the latter curve, the southern latitude at which $\lambda = 1$ occurs at 38.52° N, again very close to the most southern (38.07° N) known breeding population.

The difference between the niche curve (Fig. 13.2b) with $P_J = 0.288$ and $P_A = \varphi_m$ versus $P_A = \varphi_f$ can serve as a natural perturbation analysis. Thus, a 13% reduction in adult survival is enough to drive the entire growth rate curve well below sustainability. A similar perturbation in the opposite direction would

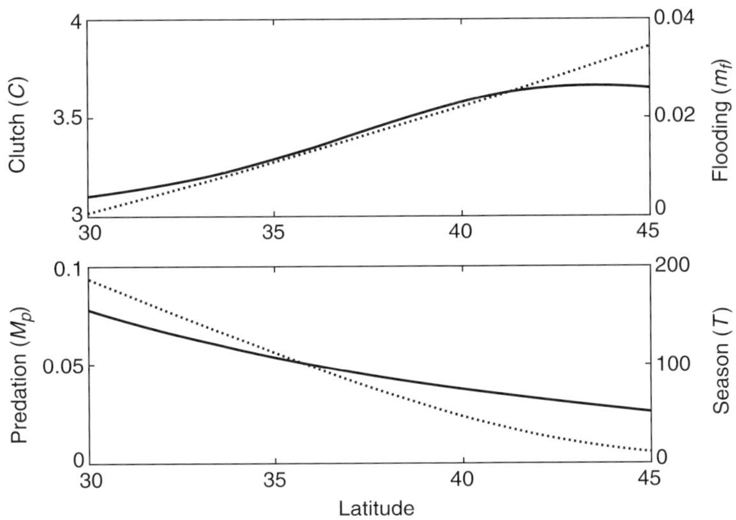

FIGURE 13.1.
Estimated latitudinal gradients in clutch size (top, solid line), daily probability of inundation (top, dotted line), daily probability of predation (bottom, dotted line), and breeding season length (bottom, solid line).

TABLE 13.1. *Estimated sampling variance, confidence intervals (degrees latitude) and width (km) of the southern distributional limit of coastal plain swamp sparrows (CPSS) for assumed levels of precision (coefficient of variation, CV) of estimated input parameters.*

	Southern limit		
CV	Variance[1]	95% CI (° lat.)	Width (km)
0.01	44.57	24.52–50.70	2,914
0.001	2.76	34.36–40.86	723
0.0001	0.26	36.61–38.61	222

[1] Estimated sampling variance around the predicted location of the southern range limit of the coastal plain swamp sparrow.

result in a distribution that is far too large (not pictured). Similarly, the results of the power analysis suggest that the southern boundary may be very difficult to locate with confidence, even with very precisely estimated input parameters (Table 13.1). Even with presumed coefficients of variation (CV) of 10^{-5}, this model can only predict the location of the southern range limit to within 222 km (Table 13.1).

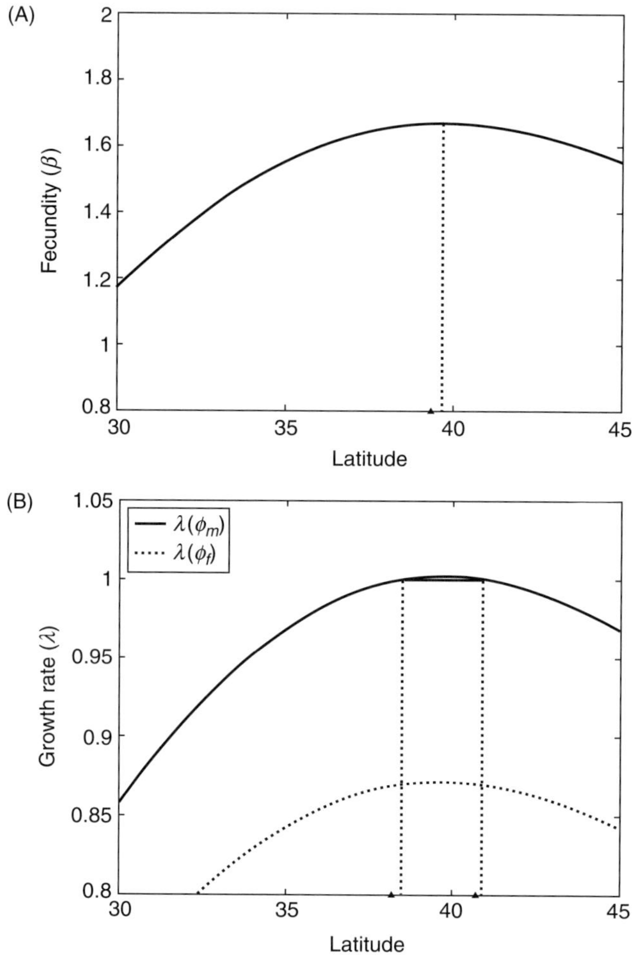

FIGURE 13.2.
(a) Estimated number of female offspring fledged per female as a function of latitude. Vertical dotted line shows the optimal breeding latitude (maximum predicted fecundity). The triangle on the horizontal axis is placed at the latitude of our Woodland Beach, DE, study site. (b) Estimated population growth rate for different values of annual adult survival predicted as a function of latitude. Vertical dotted lines, topped by horizontal line at $\lambda = 1$, indicate points of intersection of the growth rate curve with the line $\lambda = 1$. The triangles on the latitude axis indicate northernmost and southernmost known breeding populations of CPSS.

Discussion

Our model results show the promise of applying source–sink theory to predicting species' distributions. The shape of the niche curve (Fig. 13.2a) confirms that the latitudinal gradients in fecundity (nest predation, nest inundation, clutch size, and breeding season length) are a coherent and sufficient explanation for the breeding range limits of CPSS (Fig. 13.2b). The maximum of the fecundity curve occurs within the heart of the breeding range of the subspecies. The value(s) of P_J required to fit the niche model to the northern boundary are plausible and the resulting estimates of the location of the southern range limit are about 60 km from the location of the southernmost known breeding population. Thus our model demonstrates the kind of insights that can be gained through the unification of source–sink theory with the Hutchinsonian niche concept (Hutchinson 1957) to predict species distributional limits, as suggested by Pulliam (2000).

However, our model is subject to considerable uncertainty at many scales. Large sensitivities result in large variance around the projected location of the southern boundary. Relatively small perturbations in survival result in large shifts in the growth rate curve (Fig 13.2b), and power analyses suggest that enormous effort would be required to estimate the range limit to within 200 km. The former result is a good example of the relative importance of the sensitivities. Our growth rate model (Eq. 13.1) is more sensitive to changes in survival than fecundity. This also highlights the importance of careful management of this subspecies because small anthropogenic effects on survival (e.g., changes in predator communities near human habitation, increased risk during migration near human structures, decreased wintering habitat quality, or increased frequency or intensity of winter storms) could have a large impact on the species' extinction probability. This exercise, then, directs our future research attention toward understanding the causes of mortality throughout the year, including potential gradients in survival.

Fortunately, there are good reasons to believe that our estimates of uncertainty are conservative. First, our application of the Delta method did not incorporate covariances among model parameters, many of which are likely to be negative if there are life-history tradeoffs, say between survival and reproduction. Negative covariances among estimated parameters will necessarily reduce the estimated model variance because the squared sensitivities (see Appendix) are necessarily positive. Second, our model included no population regulation, whether in the form of density dependence, habitat saturation, or dispersal limitation, any of which would create negative feedbacks that would also serve to reduce the total variance (assuming a projection model with a stable equilibrium).

A useful generalization to our conceptual model would be to include mechanistic explanations for the observed latitudinal gradients. For example, the clutch-size gradient we report is not unique to tidal-marsh sparrows. It is a well-known gradient, which applies quite generally across avian taxa (Skutch 1949), that has been hypothesized to result from diverse causes, including global gradients in ambient temperatures during laying (Stoleson and Beissinger 1999; Cooper *et al.* 2005; Olsen *et al.* 2008a), nest predation pressure (Eggers *et al.* 2006; Greenberg *et al.* 2006, Olsen *et al.* 2008a), and concentrated resource pulses at higher latitudes (Rabenold 1979; Ricklefs 1980). In contrast to clutch size, nest predation rates vary negatively with latitude (Ricklefs 1969, but see Martin 1996). While most latitudinal comparisons of nest predation rates involve temperate–tropical comparisons, decreased predation rates on artificial nests have been found along latitudinal gradients within the temperate zone (Andrén *et al.* 1985; Berg *et al.* 1992) and the latter authors hypothesized that these changes were due to changes in the nest-predator community with latitude. The gradient in breeding season length (T) reported by Greenberg *et al.* (2006) also generalizes beyond tidal-marsh sparrows, especially for re-nesting and/or multiple-brooded birds (e.g., James and Shugart 1974) and is probably ultimately caused by a combination of temperature and food availability (Immelmann 1973), although its proximate control is strongly influenced by endocrine response to changes in photoperiod (Hahn *et al.* 2004).

Another useful model generalization would be to incorporate annual variation in demographic parameters, in particular in response to the projected effects of climate change (changes in sea level and resulting effects on nest flooding rates, and increased temperature effects on clutch size). Other mechanisms generating our gradients and, in turn, our latitude-specific fecundity rates are also likely to vary between years. In years with abundant food resources, the breeding season will be longer, and birds will attempt more nests and experience greater fecundity (Nagy and Holmes 2004). In bad years they will attempt fewer nests and experience lower fecundity. Similarly, nest predation rates in this system vary from year to year (Etterson *et al.* 2007; Olsen *et al.* 2008a). These factors are also likely to vary locally around the large-scale latitudinal trends. In comparison with the models of Holt and colleagues (Holt and Keitt 2000; Holt *et al.* 2005), the productivity gradients we have described are best considered as gradients in extinction probability (or its complement), but other processes such as colonization rates probably also influence habitat occupancy near the range boundary. Taken together, these observations suggest that the distribution of CPSS sources and sinks is probably quite dynamic and characterized by shifting source–sink dynamics near the range boundary. Repeat survey data from near the southern range limit also support this hypothesis (Beadell *et al.* 2003).

The latter considerations are a good reminder that our model, while geographically explicit, was not fully spatially explicit because it did not incorporate dispersal probabilities. Doing so would likely increase the variance (except where dispersal limitation is involved in population regulation, as described above). We know very little about dispersal in CPSS, except that juvenile return rates are very low and males are more philopatric than females (as evidenced by the differences in recapture rate, $\hat{p}_m >> \hat{p}_f$), both of which are common passerine patterns (Greenwood 1980; Clarke *et al.* 1997). Uncertainty surrounding the incorporation of dispersal into spatially explicit models has been a contentious topic (Mooij and DeAngelis 2003; Etterson and Nagy 2008) and a thorough exploration is outside the scope of this chapter. Further, estimation of dispersal rates is very difficult in birds, even with banded populations (Clobert and Lebreton 1991; Robinson and Hoover, Chapter 20, this volume). However, the configuration of habitat in CPSS may be particularly conducive to dispersal estimation because tidal salt-marsh is found in a narrow curvilinear band along the eastern North American seaboard. Thus the problem of relocating marked individuals may be relatively easier with CPSS.

As we continue to modify species' environments on continental and global scales, we need spatial demographic models to guide management and remediation efforts. Our model of coastal plain swamp sparrows is a promising example of how conceptual models of source–sink dynamics, when linked to empirical factors that determine vital rates, can be useful in planning and managing habitat at large spatial scales. However, our results also suggest that much more work should be done to reconcile the promise of spatial demographic models within the often large uncertainties surrounding their outputs. We believe this question should be at the forefront of the study of spatial demography.

References

Andrén, H., P. Angelstam, E. Lindstrom and P. Widen (1985). Differences in predation pressure in relation to habitat fragmentation: an experiment. *Oikos* **45**: 273–277.

Ballentine, B. and R. Greenberg (2010). Common garden experiment reveals genetic control of phenotypic divergence between swamp sparrow subspecies that lack divergence in neutral genotypes. *Plos One* **5**: e10229.

Bart, J. and D. S. Robson (1982). Estimating survivorship when the subjects are visited periodically. *Ecology* **63**: 1078–1090.

Beadell, J., R. Greenberg, S. Droege and J. A. Royle (2003). Distribution, abundance, and habitat affinities of the coastal plain swamp sparrow. *Wilson Bulletin* **115**: 38–44.

Berg, A., S. G. Nilsson and U. Bostrom (1992). Predation on artificial wader nests on large and small bogs along a south-north gradient. *Ornis Scandinavica* **23**: 13–16.

Caswell, H. (2001). *Matrix Population Models: Construction, Analysis, and Interpretation*, 2nd edition. Sinauer Associates, Sunderland, MA.

Clarke A. L., B. E. Saether and E. Roskaft (1997). Sex biases in avian dispersal: a reappraisal. *Oikos* **79**: 429–438.

Clobert, J. and J.-D. Lebreton (1991). Estimation of demographic parameters in bird populations. In *Bird Population Studies* (C. M. Perrins, J.-D. Lebreton and G. J. M. Hirons, eds.). Oxford University Press, Oxford, UK: 75–104.

Cooper, C. B., W. M. Hochachka, G. Butcher and A. A. Dhondt (2005). Seasonal and latitudinal trends in clutch size: thermal constraints during laying and incubation. *Ecology* **86**: 2018–2031.

Daley, D. J. (1979). Bias in estimating the Malthusian parameter for Leslie matrices. *Theoretical Population Biology* **15**: 257–263.

Eggers, S., M. Griesser, M. Nystrand and J. Ekman (2006). Predation risk induces changes in nest-site selection and clutch size in the Siberian jay. *Proceedings of the Royal Society B – Biological Sciences* **273**: 701–706.

Etterson, M. and R. Bennett (2005). Including transition probabilities in nest-survival estimation: a Mayfield–Markov chain. *Ecology* **86**: 1414–1421.

Etterson, M. A and L. R. Nagy (2008). Is mean squared error a consistent indicator of accuracy for spatially structured demographic models? *Ecological Modelling* **211**: 202–208.

Etterson, M. A. and T. Stanley (2008). Incorporating classification uncertainty in competing risks nest failure modeling. *Auk* **125**: 687–699.

Etterson, M. A., B. J. Olsen and R. Greenberg (2007). The analysis of covariates in multi-fate Markov chain nest failure models. *Studies in Avian Biology* **34**: 55–64.

Etterson, M. A., R. S. Bennett, E. L. Kershner and J. W. Walk (2009). Markov chain estimation of avian seasonal fecundity. *Ecological Applications* **19**: 622–630.

Gaona, P., P. Ferraras and M. Delibes (1998). Dynamics and viability of a metapopulation of the endangered Iberian lynx (*Lynx pardinus*). *Ecological Monographs* **68**: 349–370.

Greenberg, R. and S. Droege (1990). Adaptations to tidal marshes in breeding populations of the swamp sparrow. *Condor* **92**: 393–404.

Greenberg, R., P. J. Cordero, S. Droege and R. C. Fleischer (1998). Morphological adaptation with no mitochondrial DNA differentiation in the coastal plain swamp sparrow. *Auk* **115**: 706–712.

Greenberg, R., C. Elphick, J. C. Nordby, C. Gjerdrum, H. Spautz, G. Shriver, B. Schmeling, P. Marra, N. Nur, B. J. Olsen and M. Winter (2006). Flooding and predation: trade-offs in the nesting ecology of tidal-marsh sparrows. *Studies in Avian Biology* **32**: 96–109.

Greenberg, R., P. P. Marra and M. J. Wooller (2007). Stable-isotope (C, N, H) analyses help locate the winter range of the coastal plain swamp sparrow (*Melospiza georgiana nigrescens*). *Auk* **124**: 1137–1148.

Greenberg, R., B. J. Olsen and M. A. Etterson (2010). Patterns of seasonal abundance and social segregation in inland and coastal plain swamp sparrows in a Delaware tidal marsh. *Condor* **112**: 159–167.

Greenwood, P. J. (1980). Mating systems, philopatry and dispersal in birds and mammals. *Animal Behavior* **28**: 1140–1162.

Grenier, J. L. and Greenberg, R. (2005). A biographic pattern in sparrow bill morphology: parallel adaption to tidal marshes. *Evolution* **59**: 1588–1595.

Grzybowski, J. A. and C. M. Pease (2005). Renesting determines seasonal fecundity in songbirds: What do we know? What should we assume? *Auk* **122**: 280–292.

Hahn, T. P., M. E. Pereyra, S. M. Sharbaugh and G. E. Bentley (2004). Physiological responses to photoperiod in three cardueline finch species. *General and Comparative Endocrinology* **137**: 99–108.

Holt, R. D. and T. H. Keitt (2000). Alternative causes for range limits: a metapopulation perspective. *Ecology Letters* **3**: 41–47.

Holt, R. D., T. H. Keitt, M. A. Lewis, B. A. Maurer and M. L. Taper (2005). Theoretical models of species' borders: single species approaches. *Oikos* **108**: 18–27.

Houllier, F., J. D. Lebreton and D. Pontier (1989). Sampling properties of the asymptotic behaviour of age- or stage-grouped population models. *Mathematical Biosciences* **95**: 161–177.

Hutchinson, G. E. (1957). Concluding remarks. *Cold Spring Harbor Symposia on Quantitative Biology* **22**: 415–427.

Immelmann, K. (1973). Role of the environment in reproduction as source of "predictive" information. In *Breeding Biology of Birds* (D. S. Farner, ed.). National Academy of Sciences, Washington, DC: 121–147.

James, F. C. and H. H. Shugart (1974). The phenology of the nesting season of the American robin (*Turdus migratorius*) in the United States. *Condor* **76**: 159–168.

Johnson, D. H. (1979). Estimating nest success: the Mayfield method and an alternative. *Auk* **96**: 651–661.

Kemeny, J. L. and J. G. Snell (1983). *Finite Markov Chains*. Springer-Verlag, New York.

Liu, I. A., B. Lohr, B. J. Olsen and R. Greenberg (2008). Macrogeographic vocal variation in subpecies of swamp sparrow (*Melospiza georgiana*). *Condor* **110**: 102–109.

Martin, T. E. (1995). Avian life-history evolution in relation to nest sites, nest predation, and food. *Ecological Monographs* **65**: 101–127.

Mayfield, H. F. (1975). Suggestions for calculating nest success. *Wilson Bulletin* **87**: 456–466.
McDonald, M. V. and R. Greenberg (1991). Nest departure calls in female songbirds. *Condor* **93**: 365–373.
Mooij, W. M. and D. L. DeAngelis (2003). Uncertainty in spatially explicit animal dispersal models. *Ecological Applications* **13**: 794–805.
Nagy, L. R. and R. T. Holmes (2004). Factors influencing fecundity in migratory songbirds: is nest predation the most important? *Journal of Avian Biology* **35**: 487–491.
Olsen, B. J. (2007). Life history divergence and tidal salt marsh adaptations of the coastal plain swamp sparrow. PhD dissertation, Virginia Polytechnic Institute and State University, Blacksburg, VA.
Olsen, B. J., J. M. Felch, R. Greenberg and J. R. Walters (2008a). Causes of reduced clutch size in a tidal marsh endemic. *Oecologia* **15**: 421–435.
Olsen, B. J., R. Greenberg, R. C. Fleischer and J. R. Walters (2008b). Extrapair paternity in the swamp sparrow, *Melospiza georgiana*: male access or female preference? *Behavioral Ecology and Sociobiology* **63**: 285–294.
Pulliam, H. R. (1988). Sources, sinks, and population regulation. *American Naturalist* **132**: 652–661.
Pulliam, H. R. (1995). Sources and sinks: empirical evidence and population consequences. In *Population Dynamics in Ecological Space and Time* (O. E. Rhodes, R. K. Chesser and M. H. Smith, eds.). University of Chicago Press, Chicago, IL: 45–71.
Pulliam, H. R. (2000). On the relationship between niche and distribution. *Ecology Letters* **3**: 349–361.
Rabenold, K. N. (1979). Reversed latitudinal diversity gradient in avian communities of eastern deciduous forests. *American Naturalist* **114**: 275–286.
Remes, V. (2000). How can maladaptive habitat choice generate source–sink population dynamics? *Oikos* **91**: 579–582.
Ricklefs, R. E. (1969). An analysis of nesting mortality in birds. *Smithsonian Contributions to Zoology* **9**: 1–48.
Ricklefs, R. E. (1980). Geographical variation in clutch size among passerine birds: Ashmole's hypothesis. *Auk* **97**: 38–49.
Rowe, C. L. and W. A. Hopkins (2003). Anthropogenic activities producing sink habitats for amphibians in the local landscape: a case study of lethal and sublethal effects of coal combustion residues in the aquatic environment. In *Amphibian Decline: An Integrated Analysis of Multiple Stressor Effects* (G. Linder, S. K. Krest and D. W. Sparling, eds.). Society of Environmental Toxicology and Chemistry (SETAC), Pensacola, FL: 271–282.
Skutch, A. F. (1949). Do tropical birds rear as many young as they can nourish? *Ibis* **91**: 430–455.
Stoleson, S. H. and S. R. Beissinger (1999). Egg viability as a constraint on hatching synchrony at high ambient temperatures. *Journal of Animal Ecology* **68**: 951–962.
Watts, B. D., M. D. Wilson, F. M. Smith, B. J. Paxton and J. B. Williams (2008). Breeding range extension of the coastal plain swamp sparrow. *Wilson Journal of Ornithology* **120**: 393–395.
White, G. C. and K. P. Burnham (1999). Program MARK: survival estimation from populations of marked animals. *Bird Study* **46**(Suppl.): 120–138.

Appendix

Our fecundity model treats seasonal productivity as a regular homogeneous Markov chain with transition matrix (**M**) in the following form:

$$\mathbf{M}=\begin{bmatrix} 0 & \mathbf{0} & 0 & 0 & \mathbf{0} & 0 & \mathbf{0} & 0 & 1_1 \\ s_a & \mathbf{0} & 0 & 0 & \mathbf{0} & 1-s_a & \mathbf{0} & 0 & 0 \\ \mathbf{0} & \ddots & \mathbf{0} & \mathbf{0} & \mathbf{0} & \vdots & \mathbf{0} & \mathbf{0} & \mathbf{0} \\ 0 & \mathbf{0} & s_1 & 0 & \mathbf{0} & 1-s_1 & \mathbf{0} & 0 & 0 \\ 0 & \mathbf{0} & 0 & 1_{w_s} & \mathbf{0} & 0 & \mathbf{0} & 0 & 0 \\ \mathbf{0} & \mathbf{0} & \mathbf{0} & \mathbf{0} & \ddots & \mathbf{0} & \mathbf{0} & \mathbf{0} & \mathbf{0} \\ 0 & \mathbf{0} & 0 & 0 & \mathbf{0} & 1_{w_s-w_f} & \mathbf{0} & 0 & 0 \\ \mathbf{0} & \mathbf{0} & \mathbf{0} & \mathbf{0} & \mathbf{0} & \mathbf{0} & \ddots & \mathbf{0} & \mathbf{0} \\ 0 & \mathbf{0} & 0 & 0 & \mathbf{0} & 0 & \mathbf{0} & 1_2 & 0 \end{bmatrix} \tag{13.A1.1}$$

In Eq. (13.A1.1) each row or column corresponds to a single day. Thus each transition probability in the matrix describes the probability of being in a given state (column) on a given day conditional on the state (row) the female was in the previous day. The daily survival parameters (s_i) are the complement of the sum of all daily failure probabilities (m_p, m_f, m_o; see below) and the subscript on s indicates the daily ageing of the nest up to a, at which age it fledges. The subdiagonal of ones (i.e., $[1_{ws}, \ldots, 1_{ws-wf}, \ldots\ 1_2]$) incorporates the periods of time required between a failed nest (w_f) versus a successful nest (w_s) and the first egg in the next nesting attempt. Bold zeros in the matrix (Eq. 13.A1.1) represent portions of the matrix that, when expanded to the same dimension as the corresponding ellipses, are vectors or submatrices of zeros of the appropriate dimension (Etterson *et al.* 2009).

Note that the parameter T does not appear in the transition matrix. It is used to impose a plausible biological limit on the number of nests a female can attempt in a breeding season, without assuming a fixed number of nesting attempts per female per season (e.g., Grzybowski and Pease 2005). The expected number of successful broods (b) a female will raise during a breeding season lasting T days is the expected number of times the female passes through the state of having just fledged a nest (row 1) conditional on having started the season with the laying of her first egg (row $a+1$) in her first nest. This quantity can be obtained from the fundamental matrix, Z, of $\mathbf{M}$, where $Z=[I-(\mathbf{M}-A)]^{-1}$, $A=\lim_{t\to\infty}(\mathbf{M}^t)$, and I is the identity matrix of the same dimension as $\mathbf{M}$ (Kemeny and Snell 1983). Thus:

$$E(b)\to z_{a+1,1}+\alpha_1(T-1) \tag{13.A1.2}$$

The arrow in Eq. (13.A1.2) indicates that this formula is asymptotically valid.

The data provided by Greenberg *et al.* (2006) required transformation before they could be used in the above fecundity model. First, we converted the reported overall nest survival rates (S, which were already corrected for discovery bias) to site-specific daily nest survival rates, s, using the equation $s \approx \sqrt[a]{S}$. Next we estimated the probability that an unsuccessful nest failed due to predation versus other causes (P_p) as the simple proportion of depredated nests among all failed nests (after Etterson and Stanley 2008). Then the maximum likelihood estimates of the daily probabilities of failure due to predation (m_p) versus flooding and other causes ($m_f + m_0$) are given by $m_p = P_p(1-s)$ and $m_f + m_0 = (1-P_p)(1-s)$ (Etterson and Stanley 2008). After estimating the above site-specific probabilities, we transformed them to the real number scale using the logit transformation and performed linear regression on the logits. For example, to estimate a nest survival gradient, we fitted the equation:

$$\log\left(\frac{s}{1-s}\right) = \alpha_0 + \alpha_1{}^{*}\text{latitude}. \qquad (13.\text{A}1.3)$$

The resulting regression coefficients (α_i) describe the gradients, and the latitude-specific probabilities can be obtained by exponentiating. A similar equation was fitted to the predation gradient (m_p). The flooding gradient (including some failures due to other causes) was estimated as the complement of the sum of the daily survival probability and the daily probability of nest predation (i.e., $m_f + m_0 \approx 1 - s - m_p$). Clutch size ($c$) and breeding season length (T) were log-transformed to the real scale and then regressed on latitude, as above.

For the analysis of Pulliam's model (Eq. 13.1), we make use of the sensitivities and elasticities of λ to changes in the demographic parameters. The sensitivities are:

$$\text{a. } \frac{\partial\lambda}{\partial\beta} = P_J, \text{ b. } \frac{\partial\lambda}{\partial P_A} = 1, \text{ and c. } \frac{\partial\lambda}{\partial P_J} = \beta. \qquad (13.\text{A}1.4)$$

The elasticities are:

$$\text{a. } e_{P_J} = \frac{P_J\beta}{\lambda}, \text{ b. } e_{P_A} = \frac{P_A}{\lambda}, \text{ and c. } e_\beta = \frac{P_J\beta}{\lambda}. \qquad (13.\text{A}1.5)$$

To estimate sampling variance in the population growth rate, we used the Delta method (Daley 1979; Houllier *et al.* 1989):

$$\text{var}(\lambda) \approx \left(\frac{\partial\lambda}{\partial\theta}\right)^T \text{cov}(\theta)\left(\frac{\partial\lambda}{\partial\theta}\right). \qquad (13.\text{A}1.6)$$

In Eq. (13.A1.6), θ is a column-vector of model parameters. For the survival parameters, the partial derivatives of λ with respect to θ are the sensitivities (Eq. 13.A1.4a,c). However, for fecundity we used the second-order sensitivities of λ to the regression coefficients (the α_i from Eq. (13.A1.3), which were obtained using the product rule; Caswell 2001). The matrix cov(θ) is the variance–covariance matrix of estimated parameters. When the covariances are zero, Eq. (13.A1.6) can be reformulated using the elasticities (Eq. 13.A1.5):

$$\mathrm{var}(\lambda) \approx \lambda^2 E_\theta^2 \mathrm{CV}_\theta^2. \qquad (13.\mathrm{A}1.7)$$

In Eq. (13.A1.7), E_θ^2 is a row-vector of squared elasticities and CV_θ^2 is a column-vector of squared coefficients of variation. We derived Eq. (13.A1.7) to allow the sampling variance of the estimated population growth rate to be expressed conditionally on similarly scaled hypothetical levels of precision in the estimated input parameters.

JEFFREY M. DIEZ AND ITAMAR GILADI

14

Scale-dependence of habitat sources and sinks

Summary

Studies of population dynamics are necessarily contingent on scale, both spatial and temporal extent and grain of study. Observed population dynamics may vary across scales, and different processes may drive these patterns at different scales. Habitat sources and sinks are driven by variation in demographic vital rates such as survival, growth, and reproduction, which often vary widely across spatial and temporal scales. The knowledge that patterns may vary across scales, and different driving variables may be relevant at different scales, is intuitive to ecologists. Merging this awareness of scale with quantitative studies of population dynamics has proven difficult, however. The overall aims of this chapter are to show how scale has influenced studies of source–sink dynamics, and to highlight an emerging statistical approach for better quantifying population dynamics at different scales. After a brief review of how issues of scale are central to understanding source–sink dynamics, we show how hierarchical models can help quantify demographic variation across different scales and make predictions for sparsely populated sites. We use a brief case study of the demography of a forest herb, *Hexastylis arifolia* ("little brown jug"), to highlight how demographic rates and predicted population growth rates may be quantified at different scales. We conclude with a discussion of important extensions to this work, including the incorporation of dispersal, and the possible implications of scale for assessments of source–sink dynamics.

Sources, Sinks and Sustainability, ed. Jianguo Liu, Vanessa Hull, Anita T. Morzillo and John A. Wiens.
Published by Cambridge University Press.

Background

Shortly after the publication of source–sink ideas (Van Horne 1983; Holt 1985; Pulliam 1988), another influential paper outlined the importance of scale in ecology (Wiens 1989). The formalization of the source–sink concept helped to change the way in which ecologists view the spatial structure of populations, and the notion of scale has infused all studies of ecological patterns and processes. Nonetheless, significant work remains to be done in order to integrate these ideas by incorporating an explicit analysis of scale into source–sink studies. Because scale terminology has been used in different ways, it is worth pointing out that we follow the common definitions of scale as consisting of both "extent" – the overall spatial area or temporal duration of a study, and "grain" – the spatial or temporal unit of observation (O'Neill *et al.* 1986). We use the word "scale" when referring to both of these together.

Source–sink dynamics may be "scale-dependent" in at least three general ways. First, observed dynamics will vary with the grain and extent at which they are viewed. This is true for population dynamics generally, of which source–sink dynamics are a special case. For example, a fluctuating population may appear to be increasing or decreasing depending on the temporal extent of observations, and the degree to which the population is seen as fluctuating will depend also on the grain of sampling. Thus, the answers to questions about population persistence and population trends over time are necessarily contingent on temporal extent and grain. Observed dynamics across space instead of time are analogous. These effects of spatial and temporal extent and grain of study – i.e., the "lens" used to view a system – influence our interpretation of a wide range of ecological processes (Wiens 1989), and source–sink dynamics are no exception.

The second general way in which source–sink dynamics depend on scale is that underlying processes influencing demographic performance (e.g., density dependence, resource supply, predation, disturbances) operate at different scales. Thus, not only do apparent dynamics change as a result of increasing or decreasing grain size of sampling, but the basic processes that influence how well a species performs actually vary over different spatial and temporal scales. Thirdly, species mobility is a vital determinant of the relevant spatial scales at which source–sink dynamics occur. In particular, the degree of mobility relative to the grain of habitat heterogeneity will, in large part, determine the relevant scales for measuring source–sink dynamics.

We explore these relationships between scale and source–sink dynamics in two ways in this chapter. First, we briefly review how scale is inherent to studies of source–sink dynamics, and second, we use a case study to show how hierarchical models may help fill a gap in this research by quantifying demographic variation across scales. In the review we first highlight some of the reasons that different "lenses" of scale have been applied in different cases, reflecting both underlying species–habitat relationships and researchers' perceptions of the system. Then we examine more closely the processes underlying scale-dependence of source–sink dynamics. In particular, we discuss (1) how abiotic and biotic heterogeneity creates variation in habitat suitability at different scales, and (2) how the mobility of species determines the extent to which habitat heterogeneity is important for source–sink dynamics.

Given this summary, we then introduce the basics of hierarchical models and use a case study to highlight how they can help elucidate scale-dependent demographic rates. We conclude with a discussion about ways to expand such analysis to better explore the mechanisms behind observed patterns of demographic variability, approaches to integrate dispersal into this framework, and implications for understanding source–sink dynamics. Throughout this chapter we primarily address how source–sink dynamics are influenced by spatial scale, but briefly discuss extensions to address temporal scale at the end.

Scale in the source–sink literature

Studies of habitat sources and sinks have explored a wide range of spatial extents and grains, ranging from submetric for annual plants (Kadmon and Tielborger 1999; Thomson 2007), a few meters for a palm species (Berry *et al.* 2008), tens of meters for terrestrial insects (Boughton 2000; Cronin 2007), hundreds of meters for small mammals (Kreuzer and Huntly 2003; Schooley and Branch 2007) and aquatic insects (Caudill 2003), kilometers for non-migratory birds (Nystrand *et al.* 2010), hundreds of kilometers for marine fish (Roberts 1997; but see Cowen *et al.* 2006), and continental for migratory birds (Lloyd *et al.* 2005). The choice of spatial scale(s) in any given study may be based on the apparent grain of habitat heterogeneity, and/or on putative dispersal distances of the study organism. The smallest grain at which habitat heterogeneity is detected sets a lower limit to the grain relevant for studying source–sink dynamics. The upper limit to spatial extent may be more likely set by species mobility, dictating the distances over which populations may be demographically connected. Many studies use an *a priori* working definition of habitats as source or sink that is based on general appearance including, for example, topography, soil type, vegetation type, and vegetation cover (Kadmon and Shmida 1990; Dias *et al.* 1996; Murphy 2001), or based on spatial heterogeneity of factors that are assumed to affect suitability, such as disturbance regime, nutrient supply, predation pressure, or parasite abundance (Kunin 1998; Caudill 2005). The distribution of habitat patches that obey this working definition (which is frequently verified or modified *post hoc*) dictates the scale of investigation.

The spatial arrangement of habitat patches in a study system may also play a major role in determining the scale of investigation. These arrangements may take several forms:

- source and sink patches are juxtaposed, often when one patch type forms the matrix in which the other patch type is embedded as isolated patches;
- both source and sink patches form distinct islands separated by completely inhospitable environment;
- habitat quality varies gradually within an apparently uniform environment.

When patches are adjacent, studies of source–sink dynamics often have quite small spatial extents, focusing on the "contact zone" between two habitat types (Kunin 1998; Thomson 2007; Berry *et al.* 2008). For organisms that occupy distinct patches in an otherwise inhospitable matrix, the spatial extent of investigation often matches or slightly exceeds the distances separating neighboring patches (Blondel *et al.* 1999; Kreuzer and Huntly 2003). The case of a gradual change in habitat quality is rarer in the source–sink literature, and may be more open to alternative definitions of the relevant scales. This situation is discussed in more depth below in the context of our case study.

Habitat quality, demographic performance and scale

A primary reason why observed source–sink dynamics depend on scale is that habitat suitability varies at different scales. The classification of a habitat patch as a source or sink may change as it is either aggregated with other patches (increased grain size) or dissected into smaller patches (Figure 14.1). This may result both from the averaging effects of aggregating heterogeneous patches, and from additional processes that may be occurring at larger scales. Thus, because species niches are multidimensional, it is typically difficult to identify a unique scale at which key environmental factors vary for a species. Heterogeneity of different environmental variables, biotic interactions, and disturbances creates mosaics of habitat suitability at different scales. These overlapping scales of variability may drive patterns of species' performance and population growth in complex ways and, as a result, often no single "correct" scale will exist at which to study the population dynamics of any given species (Thomas and Kunin 1999).

A long history of demographic studies has documented variation in vital rates (e.g., survival, growth, reproduction) and population growth rates within and among populations (e.g., Moloney 1988; Oostermeijer *et al.* 1996; Vavrek *et al.* 1996; Damman and Cain 1998). This demographic variation reflects, in

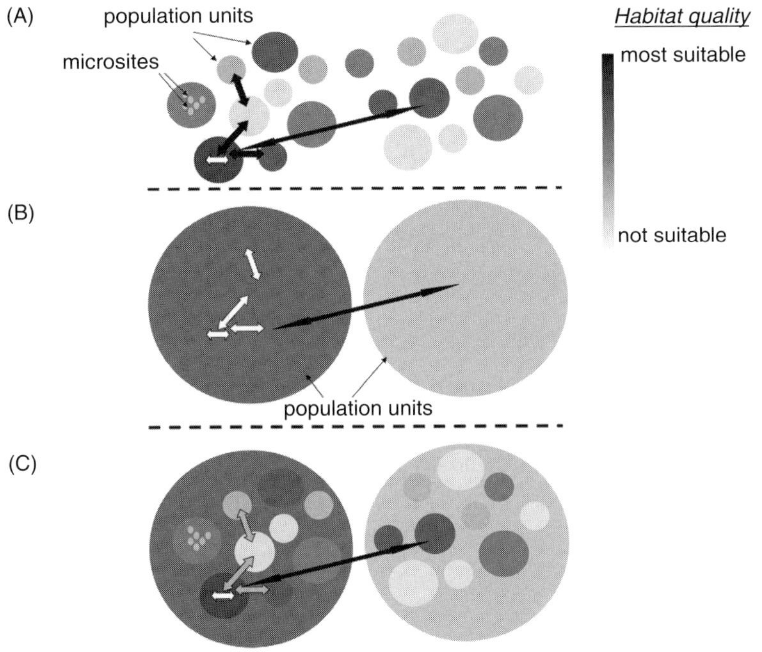

FIGURE 14.1.
Conceptual diagram of the interplay between grain size and movement as a function of how "population units" (*sensu* Thomas and Kunin 1999) are defined. Shading of population units represents a gradient of habitat quality, and arrows show movements of individuals. Arrows are colored white when movements are within population units, as defined by one's study design, and black when moving between units. As the grain size of study changes from finer (a) to more coarse (b) the importance of between-population movement decreases relative to within-population processes. Most species, however, are subject to varying habitat quality at multiple scales simultaneously (c), shown here as a multiscale combination of (a) and (b). These differences across scales may derive from different underlying environmental processes. Gray arrows in (c) represent intermediate dispersal between fine-scale units but within larger units. In all but the most discretely patchy systems, there is likely to be a substantial "gray area" to this match-up between dispersal and habitat suitability.

part, the heterogeneity of habitat suitability. A growing number of examples also document the mechanisms behind differing performance at multiple scales. For example, the demography of forest birds can be affected both by local nest-site characteristics and by regional patterns of forest fragmentation (Robinson *et al.* 1995; Reid *et al.* 2006; With *et al.* 2006; Robinson and Hoover, Chapter 20, this volume). Regional levels of fragmentation influence the abundance of predators and nest parasites associated with open habitats, which affects nest productivity of birds nesting in forest fragments. At a more local scale, nest predation and parasitism may both increase with proximity to forest edges, and at a still more local scale some nest sites are likely to be better protected than others. The same process, in this case habitat fragmentation, may thus affect productivity through two different mechanisms at two different spatial scales (Lloyd *et al.* 2005). In desert plants, water availability is the main factor explaining source–sink dynamics (Kadmon and Tielborger 1999), and this varies with both microtopography and across larger gradients. For a galling sawfly, McGeoch and Price (2005) demonstrated that both local host-plant characteristics and larger-scale moisture gradients can affect performance. It was only through an explicit attention to scale that these different processes were discovered.

Therefore, the demographic performance of species is likely to vary across spatial scales, and mechanisms can often be identified to help explain these patterns. A growing number of studies frame these patterns using a distinction between "local" and "landscape" effects on productivity. Because such factors at different scales may interact, examining them together is critical.

Mobility and scale

In addition to variation in habitat suitability at different scales, species mobility is the other major factor that clearly changes the relevant spatial extent and grain for any given study. The pairing of scale and taxon in the examples listed above probably reflects an intuitive matching of the scale of a study with what is perceived as a "feasible" dispersal distance for the organism in question. Nevertheless, empirical support for this quantitative matching (in the form of measures of dispersal) is absent from most source–sink studies (Runge *et al.* 2006). Migration in source–sink models is usually treated as a spatially implicit process occurring among habitat compartments, while assuming no mortality during migration (Pulliam 1988; McPeek and Holt 1992; Doak 1995; Watkinson and Sutherland 1995). In accordance with this view, most empirical studies of source–sink dynamics focus on estimating migration rates between patch types while ignoring the spatial arrangement of patches and the potential effects of the matrix that separates the patches (Gundersen *et al.* 2001).

This "closed system" and compartmentalized view of the real world is inherently scale-invariant, although under certain circumstances it may provide a reasonable approximation of reality (for example, when source and sink habitats are juxtaposed, or when the study organism(s) can recognize and move at a negligible cost among well-defined habitat patches). However, for most organisms, migration rate decreases with the distance separating patches and increases with species' dispersal capabilities. This scale-dependent migration success may significantly affect the dynamics of the whole system (With and King 1999; Grear and Burns 2007).

Extending the definitions of Pulliam (1988) for source and sink habitats, Runge *et al.* (2006) proposed a new metric for distinguishing between sources and sinks that focuses on the contribution of a patch to the whole metapopulation (implemented by Pasinelli *et al.*, Chapter 10, this volume). This contribution metric combines within-patch growth rate with successful emigration. Runge *et al.* (2006) argued that the estimation of emigration success of any given patch (and consequently its definition as a source or a sink) may be scale-dependent because it is expected to increase with the number of "receiving" patches, which in turn is determined by the study extent. However, even within a given extent, emigration success may be scale-dependent as it is influenced by dispersal abilities, the spatial arrangement of potential receiving patches, and the effects of the intervening matrix on dispersal behavior and dispersal-related mortality (Cronin 2007). Thus, the contribution of a patch to the whole metapopulation and the very definition of a patch as a source or a sink (*sensu* Runge *et al.* 2006) are scale-dependent.

Despite the importance of scale-dependent migration for source–sink dynamics, the estimation of migration rates in source–sink studies is uncommon (Bowne and Bowers 2004; Runge *et al.* 2006), and the estimation of scale-dependent migration rates even rarer (Thompson *et al.* 2002). The study of the source–sink dynamics of Florida scrub jays by Breininger and coworkers is an exception (Breininger and Oddy 2004). In this well-studied system the excellent record of dispersal (Stith *et al.* 1996, cited by Breininger and Carter 2003), which is expressed as the probability of migration from natal to site of first breeding as a function of the number of territories separating these locations, is coupled with a very detailed knowledge of both spatial and temporal heterogeneity in habitat quality. Lacking such a detailed knowledge of dispersal, most studies of source–sink dynamics have to rely on ancillary data from various sources, if they exist, for estimating migration rates.

Recently, the ability to quantify long-distance events has dramatically improved (Nathan *et al.* 2003; Tackenberg 2003) and the importance of long-distance dispersal for many ecological and evolutionary processes has been highlighted (Bohrer *et al.* 2005; Trakhtenbrot *et al.* 2005; Nathan 2006). Nevertheless, the relevance of long-distance dispersal for source–sink dynamics should be treated cautiously. With the notable exception of some migratory birds (Hobson *et al.* 2004; Tittler *et al.* 2006), long-distance dispersal is successfully accomplished by an extremely small proportion of emigrants. The small number of individuals involved may be insufficient for sustaining source–sink systems, especially when population growth rate in the sink is significantly below replacement level (With *et al.* 2006). This may be true even in systems which were previously assumed to be open to substantial between-patch (long-distance) dispersal, such as marine systems (Cowen *et al.* 2000, 2006). Therefore, evidence in support of the existence of between-patch dispersal (for example, by using molecular techniques) may not be a sufficient indication that populations are demographically linked, as implied by source–sink dynamics (Dias 1996; Milot *et al.* 2008). The demonstration of demographic coupling within many source–sink systems requires a much more rigorous treatment of scale-dependent between-patch dispersal than what typically appears in the current literature.

Synthesis

We draw from the above discussion three primary points relevant to studies of population dynamics in heterogeneous landscapes.

1. Species often experience variation in habitat quality, and consequently in population dynamics, at different scales.
2. Different abiotic and biotic processes may cause these differences across scales.
3. Species' mobility relative to habitat heterogeneity largely determines the relevant spatial scales.

These ideas suggest a few key challenges for future source–sink studies: how do we identify the scales at which key population processes are occurring, how do we quantify both the population processes at these different scales and the factors influencing them, and how do we incorporate species dispersal? As reviewed above, different approaches have been taken to choosing the scale(s) of study based on species mobility and/or habitat heterogeneity. When discrete scales of variability are not evident in a system, one might start with individual-level demographic data and begin to scale up in a variety of ways. Other statistical methods have been used recently for detecting the scale(s) at which heterogeneity in habitat quality is most pronounced (Schooley and Branch 2007), or at which source–sink pairs are demographically connected (Tittler *et al.* 2006). Recognizing that ecological phenomena are best understood by using a multiscale approach (Kotliar and Wiens 1990), some studies of source–sink dynamics have been designed to cover more than one scale (e.g., With *et al.* 2006). Here we show how hierarchical models may help to fill a gap in this research by quantifying population processes at different scales and identifying factors driving them.

Hierarchical models for incorporating scale

Hierarchical models are increasingly used in ecological research, and have been explained in depth elsewhere (Clark 2007; Gelman and Hill 2007; McMahon and Diez 2007). After a brief review of their basic structure, we focus our discussion on the key attributes that make them useful for quantifying sources and sinks at different scales: hierarchical models (1) permit "data sharing," such that predictions for unoccupied and sparsely populated sites are informed by sites with data; (2) help maintain reasonable estimates of uncertainty where there is significant within-population variability; and (3) allow simultaneous evaluation of abiotic covariates at multiple levels.

The defining characteristic of a hierarchical model is the use of additional parameters to describe distributions of parameters. The core idea is that individual parameters of any kind of model (e.g., intercept or slope terms in a regression) are themselves considered as random variables drawn from a "higher-level" distribution. This "higher-level" distribution can itself be described by additional parameters and so on. The way in which parameters are nested describes the "structure" of the model. This is the key point of their application to identifying scale-dependent sources and sinks: this parameter structure can be designed to use individual-level demographic data to predict performance at different scales. The choice of scales, as discussed above, will reflect both study design and hypotheses about relevant population dynamic processes.

We first illustrate the principles of hierarchical modeling with a simple example: assume that we have information on the survival of 1,000 individuals distributed across different microsites and across ten populations. Each individual survived or did not, indicated by data $Y_i = \{0,1\}$. With an interest in scale, we want to understand how the probability of surviving varies across three spatial scales: in different microsites, across different populations, and the overall metapopulation. This is achieved in the structure of parameters, beginning with the data at the individual level. Survival of individuals is a Bernoulli process (event either occurred or did not) such that $Y_i \sim \text{Bernoulli}(p_i)$ where p_i is each individual's probability of surviving. These individual-level probabilities are drawn from microsite-level probabilities as $\text{logit}(p_i) \sim \mathit{Normal}(\phi_{mp}, \sigma_m^2)$. As in a generalized linear model, the logit function transforms the probability of survival onto a continuous scale, and we assume that these values are normally distributed with a mean ϕ_{mp} for each microsite m and population p, with estimated variance σ_m^2. These microsite-level values are themselves drawn from population-level values as $\phi_{mp} \sim \mathit{Normal}(\gamma_p, \sigma_p^2)$, where γ_p are the population means and σ_p^2 is the variance at the population level. Finally, the population-level means are drawn from an overall distribution $\gamma_p \sim \mathit{Normal}(\psi_0, \sigma_0^2)$, where ψ_0 is the overall mean and σ_0^2 is the variance across populations. An assumption of hierarchical models is that subunits are *exchangeable*; that is, the subunits must be considered related enough to have been drawn from a common distribution (Gelman *et al.* 2004). This does not mean that subunit values must be equivalent, and may be explained using covariates. In such cases, subunits are considered to be *conditionally exchangeable* within the model.

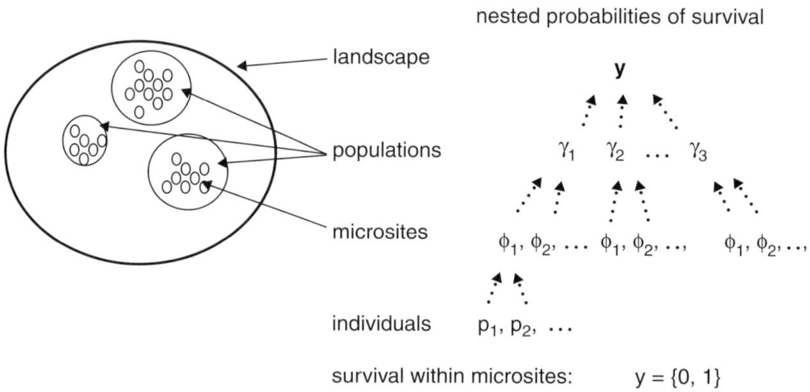

FIGURE 14.2.
Conceptual diagram of how hierarchical parameters map onto different spatial scales. Survival is measured for individuals ($y = \{0,1\}$), and probabilities of survival are estimated for each subsequent level above: microsites, populations, and the landscape. The nested structure of parameters may reflect study design, hypotheses about the structure of processes influencing survival, or preferably both.

With parameters structured in this way, the probability of survival can be described at different spatial scales (Fig. 14.2). This hierarchical structure is significant for quantifying population dynamics for at least two reasons. First, predicted microsite-level values will be different than simply taking means of the microsite data. Instead, they will be more akin to weighted averages that are influenced by values of other microsites within the same populations. This is enabled via the nestedness of parameter values. The degree to which particular microsite values are influenced by others will depend on how many individuals are in the microsite (i.e., how many data points are there?) and how extreme that microsite is compared with others (i.e., is it an outlier?). The more individuals a microsite contains, the more confident we are of its estimate and the less it will rely on "data sharing" from other sites. This "data sharing" is a critical advantage when we are interested in predicting the suitability of sites for which there are limited data (Clark *et al.* 2005): often the case in ecological studies.

The second useful aspect of the hierarchical structure is that it maintains a realistic measure of uncertainty in predictions when there is within-population demographic variability. A traditional approach might calculate mean survival rates for each population, ignoring microsite variation, and our confidence in that mean would increase as the number of individuals increases. If the variability between microsites is *real* variability, due to microsite environmental conditions, genetic differences, etc., instead of measurement *error*, then we want to preserve that variability in our population-level estimates (Clark 2003; Clark *et al.* 2004). This simple hierarchical structure is the key to the utility of demographic models described below.

Case study

Study system

For a brief case study to illustrate how these principles are relevant for predicting sources and sinks, we use demographic data collected as part of a larger study investigating controls on the distribution and demography of forest herb species in southeastern North America. The data used in this example describe the growth, survival, and reproduction for one of these species, *Hexastylis arifolia* ("little brown jug"), in eight populations, from the 2003–2004 census period.

Hexastylis arifolia is an evergreen herbaceous perennial belonging to the family Aristolochiaceae, ranging from southeastern Virginia to northwestern Florida and west to eastern Tennessee and southern Louisiana. In this range, it is commonly found in mature deciduous forests and occasionally in mixed pine–hardwood forests and on the edges of swamp forests (Gonzalez 1972). Vegetative reproduction is limited, and populations are maintained and propagated mainly by seed production (Gonzalez 1972). Age of first reproduction is estimated at 7–10 years, and the life span may exceed 20 years. The flower and subsequently the fruit (typically one per reproductive individual) are carried on a short peduncle (1–4 cm long) and are located at ground level under or barely above the leaf litter.

A mature fruit may have as many as 50 seeds, but usually the number is lower (mean ~20). The seed (weighing 8–12 mg) bears an elaiosome, a lipid-rich appendage that attracts several species of ants, the main dispersal agents of the plant. Most (>90%) of the seeds that are dispersed by ants are taken to distances of less than 1 m (mean = 0.81 m, SD = 0.86), but a few seeds may travel up to 16 m (I. Giladi, 2004, unpublished data). In the sites where this study was conducted, seed dispersal occurs in late May to early June. In the spring of the following year (10 months after dispersal) seedlings emerge with two cotyledons and no true leaf. A true, heart-shaped leaf appears only in the second year. Populations of *H. arifolia* are usually dense, continuous, and their boundaries are well defined. While surveying large tracts of forest, only once was a single plant found further than 50 m from any continuous population.

Plants within the eight populations were individually marked and revisited yearly to measure growth, survival, and reproduction. Each of the eight populations was sampled using a study grid (20 × 24 m or 10 × 12 m) that was divided into 2 × 2 m cells, corresponding to a reasonable "microsite" level of study for forest herbs, reflecting the microtopography of the forest floor. Microsites vary in light availability and soil moisture, which influence plant performance (Warren 2007). The eight populations were distributed across two landscapes: four populations in the "North" landscape in the US Forest Service Lake Russell wildlife management area in northern Georgia (34°30′ N latitude; 240–480 m elevation), and four populations in the "South" landscape in Whitehall Forest near Athens, Georgia (33°52′ N latitude, 150–240 m elevation) (see Giladi 2004 for additional study details). Although all plants within study grids were marked, additional unmarked individuals were present outside each grid, and additional unmarked populations existed in each landscape.

Vital rates and demographic models

Identifying habitat sources and sinks entails quantifying spatial variation in productivity, which is not straightforward to calculate for long-lived species. However, vital rates, including growth, survival and reproduction, are commonly assembled within demographic models to predict population growth rates, which give a measure of productivity. Population growth rates do not clearly identify sources and sinks without the incorporation of dispersal and consideration of density dependence (Watkinson and Sutherland 1995), but are a useful tool for identifying potential sources and sinks. We begin by focusing on how individual vital rates are modeled hierarchically, and later address how these are compiled for predicting population growth rates.

Our modeling approach largely follows Clark (2003), with modifications for the life history of the species and the structure of data collection (individuals within microsites, within populations, within landscapes). We do not revisit the detail of the methods, but focus on how the hierarchical model structure (mirroring that of the above survival example) allows prediction of each vital rate and population growth rates across three spatial scales.

Each vital rate, represented as a transition on a life-cycle diagram (Fig. 14.3), can be modeled using a sampling distribution appropriate to the rate. For example, probabilities of transitions between the seedling, juvenile, and adult classes (distiguished based on leaf sizes, which in this species provide a good predictor of survival and fruiting probabilites; Giladi 2004) are estimated as a multinomial process:

$$Y_{mplij} \sim Multinomial(\phi_{mplij}, n_{mpli})$$

where Y_{mplij} is the number of plants observed to make the transition from stage class i to class j in microsite m, population p, and landscape l out of the n_{mpli} plants in the previous year; ϕ_{mplij} is the probability of making that transition. Individuals are allowed to progress or regress between juvenile and adult classes, remain in their class, or die. The transition probabilities are given Dirichlet prior distributions, which is a multivariate generalization of the beta distribution, and is therefore a natural way to constrain probabilities to the 0–1

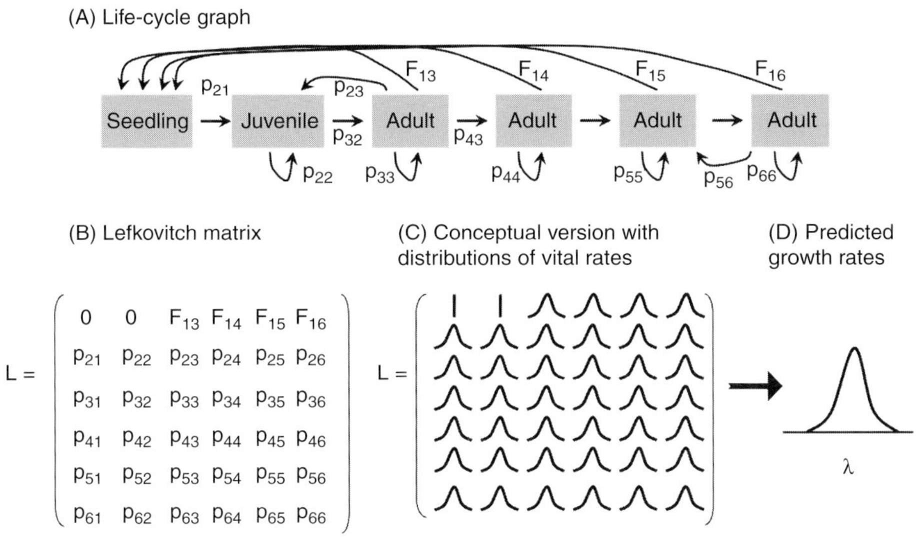

FIGURE 14.3.
Overview of demographic models. (a) Life-cycle diagram for *Hexastylis arifolia* ("little brown jug"). The adult stages were chosen based on leaf size. Transitions between life stages form the basis for a stage-structured demographic matrix model (b) used to predict population growth rates. Not all transitions are shown for clarity. (c) In the hierarchical Bayesian framework, one may conceptually think of individual transition probabilities as distributions, which when used for eigenanalysis at each Markov chain Monte Carlo (MCMC) iteration (see text) results in a distribution of predicted population growth rates (d).

interval (see Gelman *et al.* 2004: 83). These prior distributions have parameters described by additional hyperparameters, such as those that characterize hierarchical models. As described above for the simple survival model, transition probabilities at the microsite level are modeled as drawn from population-level distributions, which are in turn drawn from landscape-level distributions. Thus, the posterior distributions of parameters at these different levels may be used to describe demographic performance at the different scales.

The fecundity model is similar, but uses different distributions. Fecundity, the number of seedlings produced per adult, is a combination of the probability of fruiting, the number of seeds produced per fruit (assuming one fruit per reproductive adult, see description of the study plant), and the probability that a seed will germinate to become a seedling in the following year. The probability of fruiting is modeled as a binomial process $F_{smpl} \sim Binomial(\phi_{smpl}, n_{smpl})$, where F_{smpl} is the number of fruiting individuals in microsite m, population p, and landscape l produced by the n_{smpl} plants of size class s. The parameter ϕ_{smpl} represents the probabilities of fruiting in each microsite.

The number of seeds produced by each adult plant was estimated from dissected fruits and was assumed not to vary spatially. The probability of seed germination was modeled using a binomial distribution for number of seedlings found in 2004, given the estimated number of seeds produced in 2003. It was assumed that the observed seedlings were derived from seeds produced within the same population. This latter assumption is based on the extremely short dispersal distances of this ant-dispersed plant and on the observation of virtually zero recruitment of seedlings in any population that completely failed to reproduce in a previous year (Giladi 2004). Nonetheless, our estimates of fecundity are likely to be biased due to the effects of dispersal, and methods to account for such effects should be used (see Pulliam *et al.*, Chapter 9, this volume).

These models were fit in a Bayesian framework, although hierarchical models need not be Bayesian (Raudenbush and Bryk 2002; Lele *et al.* 2007). As Bayesian models, parameters are given prior distributions, which combine with likelihood models to yield posterior probability distributions for each parameter (such as transition probabilities or fecundity). In this case, prior distributions that assumed no prior knowledge of the parameter estimates (so-called "non-informative" priors) were used for all parameters. However, the hierarchical structure of the model (through "data sharing," as described earlier) in effect helps to inform all parameter estimates. The more data entering through the likelihood, the less a site's estimate will be influenced by other sites.

The resulting posterior distributions were derived iteratively using a Markov chain Monte Carlo (MCMC) procedure. At each iteration of the MCMC procedure, all parameters were estimated, conditional on values of all other parameters; a Lefkovitch matrix was assembled; dominant eigenvalues were calculated; and distributions of population growth rates were obtained. This was performed using parameters at the microsite, population, and landscape levels. Thus, this approach uses standard matrix model techniques (Caswell 2001), but within an iterative framework in which variability in individual parameters propagates to influence the distribution of predicted population growth rates. Models for all transition probabilities were fit using the MCMC procedure in OpenBUGS v 2.10 (Thomas *et al.* 2006), using the BRugs package in R 2.8.0 (R Development Core Team 2008).

The concepts and utility of hierarchical modeling will apply to a wide range of demographic modeling beyond this case study. Although we used stage-structured demographic models in this case study, the principles of hierarchical modeling would be similar when using other demographic frameworks (e.g., integral projection models or individual-based models). Stage-structured demographic modeling can be useful for teasing apart how stage-specific vital rates may vary across scales, but such models will not always be appropriate. Many other analyses for characterizing specific aspects of species demography, such as elasticity analysis or life-table response analyses, may also be interesting to explore at different scales using the same hierarchical framework. We focus here on population growth rates as the key element relevant to identifying sources and sinks.

Results

Models yielded posterior probability distributions for all vital rates across the three spatial scales of interest in this study: microsites, populations, and landscapes. These were then used to predict population growth rates at each of these scales (Fig. 14.4). Population growth rates are summarized by posterior probability distributions at each level.

Several overall patterns of population growth rates are noteworthy. First, most population growth rates are substantially less than 1, the level at which population growth is sufficient for replacement over time. Also, although still largely below 1, the growth rates of populations from the North landscape tend to be higher than from the South landscape. There are a number of possible reasons for these low growth rates, which we discuss below. A second pattern evident in the population growth rates is that microsites have substantially greater uncertainty than populations, as seen by the wider posterior distributions. This reflects the limited data at the microsite level and effects of "sharing" inference across microsites. Third, even population- and landscape-level estimates of population growth, for which there are many plants available for estimation, have considerable associated uncertainty. This uncertainty reflects both estimation error and the significant variability of demographic rates at finer spatial scales, maintained in these estimates through the hierarchical model structure.

Posterior distributions for population growth rates were used to calculate the probability that growth rates meet replacement levels ($\lambda > 1$), calculated as the proportion of MCMC draws that are greater than 1 (equivalent to the integral of the probability distribution from 1 to ∞). These probabilities were then mapped out in space for each of the microsites and populations (Fig. 14.5).

Discussion

There are a number of possible explanations for the low population growth rates observed here, related both to the ecology of the species and limitations of our models. First, it is plausible that the negative growth rates accurately reflect poor conditions in the region for the species. Although our current analysis was restricted to the 2003–2004 growing period, low growth rates of *H. arifolia* were consistently estimated in our study site in 1999–2004 (Giladi 2004; Warren 2007).

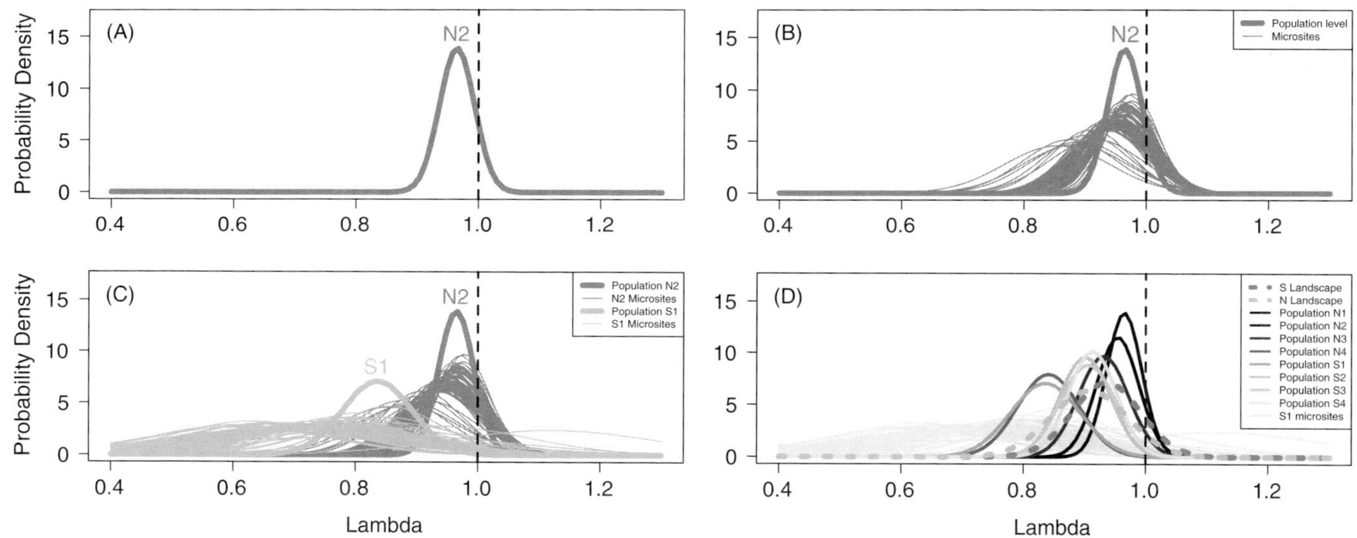

FIGURE 14.4.
Posterior probability distributions for population growth rates (λ). Distributions of population growth rates were obtained for each microsite, population, and landscape. Shown are the distributions for population N2 (a); population N2 and its microsites (b); two populations – one from the "North" landscape and one from the "South" landscape – and their microsites (c); and both of the landscapes, all populations, and the same two populations' microsites (d). As for probability distributions, population growth rates can be summarized in a number of ways. For example, integrating over the portion of each curve that lies above 1 directly yields predicted probabilities that $\lambda > 1$.

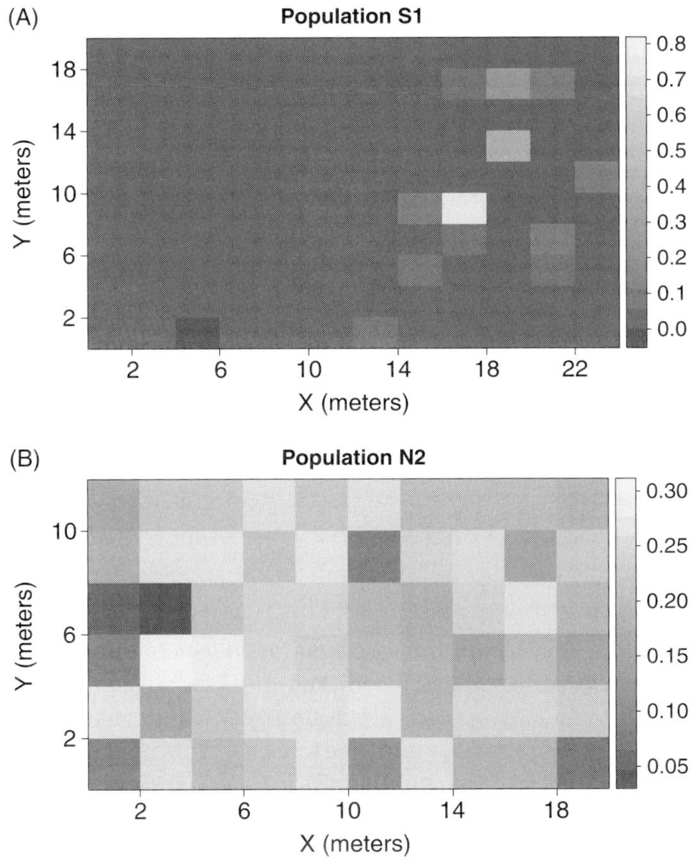

FIGURE 14.5.
Probabilities of positive population growth rates ($\lambda > 1$) in a southern landscape population (a) and a northern landscape population (b). The probability that microsites have positive population growth is a useful step toward identifying habitat sources and sinks. The cells of these grids (2 × 2 m) capture some of the fine-scale variability in microsite suitability for this understory herb. While there are substantial differences in overall population-level estimates, some microsites are clearly predicted to be acting as local demographic "hot spots" and "cold spots."

The southeastern USA had been under persistent drought conditions for several years, a fact that may explain the low growth rates for all the populations. This hypothesis is further supported by lower growth rates in the southern landscape, which is consistently hotter and drier than the northern landscape.

Second, the negative growth rates may be accurate in this year yet not reflect long-term dynamics. Although low population growth rates were observed for several years surrounding the year reported here, our estimates may not capture the longer-term temporal axis necessary to properly understand source–sink dynamics for long-lived species. There are many reasons why species exhibit temporal variations in performance, even within particular habitats.

For example, environmental conditions change over time due to disturbance or shifting climatic regimes (Boughton 1999; Virgl and Messier 2000; Johnson 2004), limiting the relevance of short-term demographic studies for predicting population trends. Infrequent disturbances may be critical to some stage transitions but difficult to predict. For example, canopy gaps created by tree-falls or other disturbances may trigger reproductive bursts and other demographic changes in understory plants (Collins *et al.* 1985; Whigham 2004).

Quantifying the effects of temporal scale on population dynamics will always be a daunting task, but other modeling tools may help. Stochastic demographic models, for example, may help to project the effects of temporal environmental variation on longer-term growth rates (Fieberg and Ellner 2001). Hierarchical models may again prove useful, as predictions of future dynamics are analogous to predicting at locations with little information. To the degree that information can be "shared" across years and sites, hierarchical models will aid in making forward projections. A final potential cause of the low observed population growth rates may have been limitations of the models. Most notably, the current models lack estimates of dormancy because we only examined one transition. If dormancy is prevalent, we may have overestimated mortality, leading to an underestimation of population growth rates. Dormancy of adult plants is known from other long-lived perennial herbs (Shefferson *et al.* 2001; Kery *et al.* 2005) and occurs in *H. arifolia* as well (Giladi 2004).

Explaining patterns

This case study has shown how hierarchical models may help fill a key need, suggested by the literature review, to quantify population growth rates at different scales. This is an important step toward understanding how source–sink dynamics may change across scales. When scales of variability are evident for a species, e.g., from discrete suitable versus unsuitable habitat delineations, then models may be structured to match these scales. When relevant scales of variability are not so evident, demographic data at the individual level may still be scaled up using hierarchical models to explore patterns of variability. In that case, different pattern recognition methods may help link the individuals to the relevant levels of variability.

An important but more difficult subsequent step is to understand the causes of observed patterns: why do species exhibit demographic variation at the scales they do? Indeed, much of the theoretical and empirical work describing source–sink dynamics (as well as metapopulations) has focused on describing patterns and their implications for population dynamics. Much less is understood about the underlying causes, possibly because of difficulties both with study design and with analysis.

We suggest two primary extensions to the type of analysis presented here that may help explain patterns of variability. First, abiotic or biotic variables may be evaluated for their effects on demography at different scales. To test the hypothesis about the importance of drought, in our case study for example, we could incorporate soil moisture into the models directly, in order to evaluate how well it explains variation in vital rates and overall population growth rates. This is a straightforward extension of multilevel models (Gelman and Hill 2007). Moreover, different abiotic variables may be relevant at microsite versus population and landscape scales, which may be explicitly explored. This approach for testing relationships at different scales is becoming more widely used in ecology (Clark and Gelfand 2006; Diez 2007; McMahon and Diez 2007; Byers *et al.* 2008). If demographic rates can be related to environmental variables within this framework, these relationships may be used to predict population performance and potential source and sink habitat across landscapes.

A second useful extension to help explain patterns across different scales will be comparisons across species. Multispecies comparative studies may help elucidate life-history characteristics (e.g., dispersal syndrome, growth form) that are important for generating observed patterns at different scales. For example, habitat configuration will vary in importance for species with different life-history traits (Dupré and Ehrlén 2002). For plant populations, differences in dispersal characteristics, clonality, and niche characteristics may influence observed patterns. Pearson and Fraterrigo (Chapter 6, this volume) show, using a simulation study, that species' niche width and dispersal can influence patterns of within- and between-population source–sink dynamics. In animals, behavioral characteristics may be critical in shaping habitat selection and mobility. A number of studies have shown how habitat selection by animals may be scale-dependent (Orians and Wittenberger 1991; Sodhi *et al.* 1999; Apps *et al.* 2001; Chalfoun and Martin 2007; Ciarniello *et al.* 2007). As more examples in which patterns are quantified at multiple scales accumulate, comparative studies between different taxa may test the generality of these relationships.

Integrating dispersal

Because source–sink dynamics ultimately depend on both demography and dispersal, identifying scales of demographic variability should also help evaluate whether dispersal is sufficient to maintain a source–sink system. Maps of potential source and sink patches at different grain sizes (e.g., Fig. 14.5) can be used to detect the relevant scale(s) of autocorrelation in habitat quality (Meisel and Turner 1998). The joint distribution of habitat autocorrelation and dispersal distances (i.e., dispersal kernel) is then the most informative about relevant scales for understanding population dynamics. In this case study, the typical dispersal distances of the ant-dispersed *H. arifolia* are on the order of a few meters (Giladi 2004; Zelikova *et al.* 2008). Therefore, it is likely that source–sink dynamics in this species may primarily occur within populations. Ant-dispersed species may also exhibit occasional long-distance dispersal (hundreds to thousands of meters) (Vellend *et al.* 2003), but such long-distance events appear too rarely in *H. arifolia* to sustain large-scale source–sink systems (Giladi 2004).

Maps of demographic habitat suitability may also be used to design a network structure of interconnected patches, where the pairwise connectivity is based on distance- and potentially habitat-specific dispersal. The dynamics of such a network can then be analyzed using a variety of possible methods. For example, With *et al.* (2006) used a simulation study to explore the conditions under which migration from source habitats (continuous landscapes) can rescue populations of songbirds in sinks (fragmented landscapes). They calculated the migration rates necessary for sustaining a sink habitat for various combinations of attributes that affect the sink reproductive deficit. The viability of sink habitat (and thus the persistence of source–sink systems) in the real world can be evaluated by comparing the calculated threshold migration rates with independent estimates of migration rates, preferably at comparable distances to those separating potential sources and sinks. A similar approach was advocated by Runge *et al.* (2006) for a proper identification of source and sink habitats. Their contribution metric (see the section "Mobility and scale") includes a term for successful emigration, which expresses the probability that an individual of a certain age, in a given patch, is alive in a different patch in the following season. Acknowledging that estimating these probabilities of successful migration is hard, Runge *et al.* (2006) proposed to numerically solve for the contribution metric over a wide range of values, and provided a simple example that included two patches and two age classes. However, with the exception of a reference to the effect of study extent on estimating these probabilities, they stop short of pointing to the effect of scale on these probabilities.

An interesting approach for constructing spatial matrix population models has recently been presented by Hunter and Caswell (2005). Using a special permutation matrix (vec-permutation matrix) they formulated a model that allows the estimation of growth rates of each subpopulation and of the whole metapopulation while retaining the structure of individual populations' matrices and migration matrices. This method allows for an easy analysis of the sensitivity and elasticity of these growth rates to various stage- and patch-specific demographic and dispersal parameters. Using estimates of habitat-specific demography and migration rates among several juxtaposed habitat types, Grear and Burns (2007) applied the vec-permutation approach for analyzing source–sink dynamics of the white-footed mouse in two landscapes.

They considered three migration scenarios, where emigrants either stay within their natal patch, emigrate only (and then disappear or die), or emigrate and arrive safely as immigrants to other patches. The latter two scenarios represent two extremes of a continuum of emigration success. Intermediate rates of emigration success, which result from the fact that emigration success is scale-dependent, can be used for constructing more realistic migration matrices for that type of model. In addition, the method can be used to test the sensitivity of the population growth rates (at the subpopulation and at the metapopulation scales) to the dispersal parameter, either "locally" by perturbation analysis or "globally" by simulating a wide range of values.

Implications and conclusions

There are various ways in which scale-dependent population dynamics may influence predictions of source–sink theory or alter assessments of population viability. At the very least, changes in scale can alter the relative distribution of individuals in source and sink habitats, and change the relative importance of within-population dynamics compared with transfers between populations (Fig. 14.1). These shifts in the perceived structure of populations may have numerous implications for their predicted dynamics. For example, local demographic "hot spots" and "cold spots" may help explain phenomena at larger scales. If species have local sites in which individuals perform substantially better (e.g., Hatchwell *et al.* 1996; also see Fig. 14.5), these "hot spots" may help explain perceived sinks that become sources at low densities (so-called "pseudo-sinks"). Moreover, if species are susceptible to within-population extinctions (Pearson and Fraterrigo, Chapter 6, this volume), then profiles of how habitat suitability varies within large populations will help assess risk.

Pulliam (1988) also highlighted the conservation implications of correctly identifying sources and sinks, warning that using the wrong proxies for habitat quality (e.g., presence and/or abundance patterns) may concentrate conservation efforts toward sinks rather than sources. The case study we presented demonstrates the importance of scale for the very delineation of patches as sources or sinks. The sustainability of a source–sink metapopulation depends on the reproductive surplus in sources relative to the reproductive deficit in sinks, on the percentage of populations residing in sink habitats, and on the probability that emigrants from source habitat will become immigrants in sink habitat. This transition from emigrants to immigrants is clearly scale-dependent (how many sink habitats surround the source, how far away they are, etc.). Although many conservation efforts have focused on sink habitats in efforts to increase landscape connectivity, sink habitats may also negatively affect the dynamics of source habitat (Gundersen *et al.* 2001). Further, if sinks act as ecological traps (Delibes *et al.* 2001; Battin 2004), increasing connectivity may have negative effects on source habitat viability and the sustainability of the metapopulation. Despite the importance of source–sink connectivity, an inherently scale-dependent property, and our knowledge of scale-dependent habitat selection (Orians and Wittenberger 1991; Sodhi *et al.* 1999; Apps *et al.* 2001; Ciarniello *et al.* 2007), models analyzing ecological traps are still mostly spatially implicit (Battin 2004 and references therein), and extensions to a spatially realistic scale-dependent approach are needed, especially if we wish to use such models in practice.

Finally, although the Bayesian framework was primarily used here as a convenient way to fit the hierarchical models, the output of posterior probabilities that sites have positive population growth may also be a useful application to conservation planning. By further including probabilities of exporting or importing propagules from other sites, this approach may be used to estimate probabilities that a site is a source or sink. This would provide a useful quantitative link to conceiving of sources and sinks as points along a continuum of population dynamic behavior (Thomas and Kunin 1999), and be a more realistic measure of habitat quality for land managers.

Although the scale-dependence discussed in this chapter is perhaps intuitive to ecologists, it is not widely explored within the source–sink literature. Our challenge lies in using this awareness to shape our questions, study designs, analyses, and interpretations in a way that that better captures the complexity of these processes.

Acknowledgments

This work would have not been possible without the mentoring and financial support from H. Ronald Pulliam during the PhDs of both J. M. Diez and I. Giladi. Further financial support came from NSF grants DEB-0235371 to H. Ronald Pulliam and DEB-9632854 (the Coweeta LTER) to the University of Georgia Research Foundation. J. M. Diez also thanks Richard Duncan, Philip Hulme, and their lab group for helpful discussions of ideas, and appreciates the useful comments of three reviewers. The authors are very grateful to the editors for the invitation to participate in the symposium.

References

Apps, C. D., B. N. McLellan, T. A. Kinley and J. P. Flaa (2001). Scale-dependent habitat selection by mountain caribou, Columbia Mountains, British Columbia. *Journal of Wildlife Management* **65**: 65–77.

Battin, J. (2004). When good animals love bad habitats: ecological traps and the conservation of animal populations. *Conservation Biology* **18**: 1482–1491.

Berry, E. J., D. L. Gorchov, B. A. Endress and M. H. H. Stevens (2008). Source–sink dynamics within a plant population: the impact of substrate and herbivory on palm demography. *Population Ecology* **50**: 63–77.
Blondel, J., P. C. Dias, P. Ferret, M. Maistre and M. M. Lambrechts (1999). Selection-based biodiversity at a small spatial scale in a low-dispersing insular bird. *Science* **285**: 1399–1402.
Bohrer, G., R. Nathan and S. Volis (2005). Effects of long-distance dispersal for metapopulation survival and genetic structure at ecological time and spatial scales. *Journal of Ecology* **93**: 1029–1040.
Boughton, D. A. (1999). Empirical evidence for complex source–sink dynamics with alternative states in a butterfly metapopulation. *Ecology* **80**: 2727–2739.
Boughton, D. A. (2000). The dispersal system of a butterfly: a test of source–sink theory suggests the intermediate-scale hypothesis. *American Naturalist* **156**: 131–144.
Bowne, D. R. and M. A. Bowers (2004). Interpatch movements in spatially structured populations: a literature review. *Landscape Ecology* **19**: 1–20.
Breininger, D. R. and G. M. Carter (2003). Territory quality transitions and source–sink dynamics in a Florida scrub-jay population. *Ecological Applications* **13**: 516–529.
Breininger, D. R. and D. M. Oddy (2004). Do habitat potential, population density, and fires influence scrub-jay source–sink dynamics? *Ecological Applications* **14**: 1079–1089.
Byers, J. E., A. M. H. Blakeslee, E. Linder, A. B. Cooper and T. J. Maguire (2008). Controls of spatial variation in the prevalence of trematode parasites infecting a marine snail. *Ecology* **89**: 439–451.
Caswell, H. (2001). *Matrix Population Models: Construction, Analysis, and Interpretation*. Sinauer Associates, Sunderland, MA.
Caudill, C. C. (2003). Empirical evidence for nonselective recruitment and a source–sink dynamic in a mayfly metapopulation. *Ecology* **84**: 2119–2132.
Caudill, C. C. (2005). Trout predators and demographic sources and sinks in a mayfly metapopulation. *Ecology* **86**: 935–946.
Chalfoun, A. D. and T. E. Martin (2007). Assessments of habitat preferences and quality depend on spatial scale and metrics of fitness. *Journal of Applied Ecology* **44**: 983–992.
Ciarniello, L. M., M. S. Boyce, D. R. Seip and D. C. Heard (2007). Grizzly bear habitat selection is scale dependent. *Ecological Applications* **17**: 1424–1440.
Clark, J. S. (2003). Uncertainty and variability in demography and population growth: a hierarchical approach. *Ecology* **84**: 1370–1381.
Clark, J. S. (2007). *Models for Ecological Data: An Introduction*. Princeton University Press, Princeton, NJ.
Clark, J. S. and A. E. Gelfand (eds.) (2006). *Hierarchical Modelling for the Environmental Sciences*. Oxford University Press, Oxford, UK.
Clark, J. S., S. LaDeau and I. Ibanez (2004). Fecundity of trees and the colonization–competition hypothesis. *Ecological Monographs* **74**: 415–442.
Clark, J. S., G. A. Ferraz, N. Oguge, H. Hays and J. DiCostanzo (2005). Hierarchical Bayes for structured, variable populations: from recapture data to life-history prediction. *Ecology* **86**: 2232–2244.
Collins, B. S., K. P. Dunne and S. T. A. Pickett (1985). Responses of forest herbs to canopy gaps. In *The Ecology of Natural Disturbance and Patch Dynamics* (S. T. A. Pickett and P. S. White, eds.). Academic Press, London.
Cowen, R. K., K. M. M. Lwiza, S. Sponaugle, C. B. Paris and D. B. Olson (2000). Connectivity of marine populations: open or closed? *Science* **287**: 857–859.
Cowen, R. K., C. B. Paris and A. Srinivasan (2006). Scaling of connectivity in marine populations. *Science* **311**: 522–527.
Cronin, J. T. (2007). From population sources to sieves: the matrix alters host-parasitoid source–sink structure. *Ecology* **88**: 2966–2976.
Damman, H. and M. L. Cain (1998). Population growth and viability analyses of the clonal woodland herb, *Asarum canadense*. *Journal of Ecology* **86**: 13–26.

Delibes, M., P. Ferreras and P. Gaona (2001). Attractive sinks, or how individual behavioural decisions determine source–sink dynamics. *Ecology Letters* **4**: 401–403.

Dias, P. C. (1996). Sources and sinks in population biology. *Trends in Ecology and Evolution* **11**: 326–330.

Dias, P. C., G. R. Verheyen and M. Raymond (1996). Source–sink populations in Mediterranean blue tits: evidence using single-locus minisatellite probes. *Journal of Evolutionary Biology* **9**: 965–978.

Diez, J. M. (2007). Hierarchical patterns of symbiotic orchid germination linked to adult proximity and environmental gradients. *Journal of Ecology* **95**: 159–170.

Doak, D. F. (1995). Source–sink models and the problem of habitat degradation: general models and applications to the Yellowstone grizzly. *Conservation Biology* **9**: 1370–1379.

Dupré, C. and J. Ehrlén (2002). Habitat configuration, species traits and plant distributions. *Journal of Ecology* **90**: 796–805.

Fieberg, J. and S. P. Ellner (2001). Stochastic matrix models for conservation and management: a comparative review of methods. *Ecology Letters* **4**: 244–266.

Gelman, A. and J. Hill (2007). *Data Analysis Using Regression and Multilevel/Hierarchical Models.* Cambridge University Press, Cambridge, UK.

Gelman, A., J. B. Carlin and H. S. S. B. Rubin (2004). *Bayesian Data Analysis*, 2nd edition. Chapman and Hall/CRC, New York.

Giladi, I. (2004). The role of habitat-specific demography, habitat-specific dispersal, and the evolution of dispersal distances in determining current and future distributions of the ant-dispersed forest herb, *Hexastylis arifolia*. PhD dissertation, Institute of Ecology, University of Georgia, Athens, GA.

Gonzalez, V. C. (1972). The ecology of *Hexastylis arifolia*, an evergreen herb in the North Carolina deciduous forest. PhD dissertation, Department of Botany, Duke University, Durham, NC.

Grear, J. S. and C. E. Burns (2007). Evaluating effects of low-quality habitats on regional population growth in *Peromyscus leucopus*: insights from field-parameterized spatial matrix models. *Landscape Ecology* **22**: 45–60.

Gundersen, G., E. Johannesen, H. P. Andreassen and R. A. Ims (2001). Source–sink dynamics: how sinks affect demography of sources. *Ecology Letters* **4**: 14–21.

Hatchwell, B. J., D. E. Chamberlain and C. M. Perrins (1996). The demography of blackbirds *Turdus merula* in rural habitats: is farmland a sub-optimal habitat? *Journal of Applied Ecology* **33**: 1114–1124.

Hobson, K. A., L. I. Wassenaar and E. Bayne (2004). Using isotopic variance to detect long-distance dispersal and philopatry in birds: an example with ovenbirds and American redstarts. *Condor* **106**: 732–743.

Holt, R. D. (1985). Population dynamics in two-patch environments: some anomalous consequences of an optimal habitat distribution. *Theoretical Population Biology* **28**: 181–208.

Hunter, C. M. and H. Caswell (2005). The use of the vec-permutation matrix in spatial matrix population models. *Ecological Modelling* **188**: 15–21.

Johnson, D. M. (2004). Source–sink dynamics in a temporally heterogeneous environment. *Ecology* **85**: 2037–2045.

Kadmon, R. and A. Shmida (1990). Spatiotemporal demographic processes in plant populations: an approach and a case-study. *American Naturalist* **135**: 382–397.

Kadmon, R. and K. Tielborger (1999). Testing for source–sink population dynamics: an experimental approach exemplified with desert annuals. *Oikos* **86**: 417–429.

Kery, M., K. B. Gregg and M. Schaub (2005). Demographic estimation methods for plants with unobservable life-states. *Oikos* **108**: 307–320.

Kotliar, N. B. and J. A. Wiens (1990). Multiple scales of patchiness and patch structure: a hierarchical framework for the study of heterogeneity. *Oikos* **59**: 253–260.

Kreuzer, M. P. and N. J. Huntly (2003). Habitat-specific demography: evidence for source–sink population structure in a mammal, the pika. *Oecologia* **134**: 343–349.

Kunin, W. E. (1998). Biodiversity at the edge: a test of the importance of spatial "mass effects" in the Rothamsted Park Grass experiments. *Proceedings of the National Academy of Sciences of the USA* **95**: 207–212.
Lele, S. R., B. Dennis and F. Lutscher (2007). Data cloning: easy maximum likelihood estimation for complex ecological models using Bayesian Markov chain Monte Carlo methods. *Ecology Letters* **10**: 551–563.
Lloyd, P., T. E. Martin, R. L. Redmond, U. Langner and M. M. Hart (2005). Linking demographic effects of habitat fragmentation across landscapes to continental source–sink dynamics. *Ecological Applications* **15**: 1504–1514.
McGeoch, M. A. and P. W. Price (2005). Scale-dependent mechanisms in the population dynamics of an insect herbivore. *Oecologia* **144**: 278–288.
McMahon, S. M. and J. M. Diez (2007). Scales of association: hierarchical linear models and the measurement of ecological systems. *Ecology Letters* **10**: 437–452.
McPeek, M. A. and R. D. Holt (1992). The evolution of dispersal in spatially and temporally varying environments. *American Naturalist* **140**: 1010–1027.
Meisel, J. E. and M. G. Turner (1998). Scale detection in real and artificial landscapes using semivariance analysis. *Landscape Ecology* **13**: 347–362.
Milot, E., H. Weimerskirch and L. Bernatchez (2008). The seabird paradox: dispersal, genetic structure and population dynamics in a highly mobile, but philopatric albatross species. *Molecular Ecology* **17**: 1658–1673.
Moloney, K. A. (1988). Fine-scale spatial and temporal variation in the demography of a perennial bunchgrass. *Ecology* **69**: 1588–1598.
Murphy, M. T. (2001). Habitat-specific demography of a long-distance, neotropical migrant bird, the eastern kingbird. *Ecology* **82**: 1304–1318.
Nathan, R. (2006). Long-distance dispersal of plants. *Science* **313**: 786–788.
Nathan, R., G. Perry, J. T. Cronin, A. E. Strand and M. L. Cain (2003). Methods for estimating long-distance dispersal. *Oikos* **103**: 261–273.
Nystrand, M., M. Griesser, S. Eggers and J. Ekman (2010). Habitat-specific demography and source–sink dynamics in a population of Siberian jays. *Journal of Animal Ecology* **79**: 266–274.
O'Neill, R. V., B. T. Milne, M. G. Turner and R. H. Gardner (1986). *A Hierarchical Concept of Ecosystems*. Princeton University Press, Princeton, NJ.
Oostermeijer, J. G. B., M. L. Brugman, E. R. DeBoer and H. C. M. DenNijs (1996). Temporal and spatial variation in the demography of *Gentiana pneumonanthe*, a rare perennial herb. *Journal of Ecology* **84**: 153–166.
Orians, G. H. and J. F. Wittenberger (1991). Spatial and temporal scales in habitat selection. *American Naturalist* **137**: S29–S49.
Pulliam, H. R. (1988). Sources, sinks, and population regulation. *American Naturalist* **132**: 652–661.
R Development Core Team (2008). *R: A Language and Environment for Statistical Computing*. R Foundation for Statistical Computing, Vienna, Austria [available at www.R-project.org]
Raudenbush, S. W. and A. S. Bryk (2002). *Hierarchical Linear Models: Applications and Data Analysis Methods*. Sage Publications, London.
Reid, J. M., E. M. Bignal, S. Bignal, D. I. McCracken and P. Monaghan (2006). Spatial variation in demography and population growth rate: the importance of natal location. *Journal of Animal Ecology* **75**: 1201–1211.
Roberts, C. M. (1997). Connectivity and management of Caribbean coral reefs. *Science* **278**: 1454–1457.
Robinson, S. K., F. R. Thompson, T. M. Donovan, D. R. Whitehead and J. Faaborg (1995). Regional forest fragmentation and the nesting success of migratory birds. *Science* **267**: 1987–1990.
Runge, J. P., M. C. Runge and J. D. Nichols (2006). The role of local populations within a landscape context: defining and classifying sources and sinks. *American Naturalist* **167**: 925–938.
Schooley, R. L. and L. C. Branch (2007). Spatial heterogeneity in habitat quality and cross-scale interactions in metapopulations. *Ecosystems* **10**: 846–853.

Shefferson, R. P., B. K. Sandercock, J. Proper and S. R. Beissinger (2001). Estimating dormancy and survival of a rare herbaceous perennial using mark–recapture models. *Ecology* **82**: 145–156.
Sodhi, N. S., C. A. Paszkowski and S. Keehn (1999). Scale-dependent habitat selection by American redstarts in aspen-dominated forest fragments. *Wilson Bulletin* **111**: 70–75.
Tackenberg, O. (2003). Modeling long-distance dispersal of plant diaspores by wind. *Ecological Monographs* **73**: 173–189.
Thomas, A., R. B. O'Hara, U. Ligges and S. Sturtz (2006). Making BUGS open. *R News* **6**: 12–17.
Thomas, C. D. and W. E. Kunin (1999). The spatial structure of populations. *Journal of Animal Ecology* **68**: 647–657.
Thompson, F. R., T. M. Donovan, R. M. DeGraaf, J. Faaborg and S. K. Robinson (2002). A multi-scale perspective of the effects of forest fragmentation on birds in eastern forests. In *Effects of Habitat Fragmentation on Birds in Western Landscapes: Contrasts With Paradigms from the Eastern United States* (T. L. George and D. S. Dobkin, eds.). Studies in Avian Biology 25, Cooper Ornithological Society, Norman, OK.
Thomson, D. M. (2007). Do source–sink dynamics promote the spread of an invasive grass into a novel habitat? *Ecology* **88**: 3126–3134.
Tittler, R., L. Fahrig and M. A. Villard (2006). Evidence of large-scale source–sink dynamics and long-distance dispersal among wood thrush populations. *Ecology* **87**: 3029–3036.
Trakhtenbrot, A., R. Nathan, G. Perry and D. M. Richardson (2005). The importance of long-distance dispersal in biodiversity conservation. *Diversity and Distributions* **11**: 173–181.
Van Horne, B. (1983). Density as a misleading indicator of habitat quality. *Journal of Wildlife Management* **47**: 893–901.
Vavrek, M. C., J. B. McGraw and H. S. Yang (1996). Within-population variation in demography of *Taraxacum officinale*: maintenance of genetic diversity. *Ecology* **77**: 2098–2107.
Vellend, M., J. A. Myers, S. Gardescu and P. L. Marks (2003). Dispersal of *Trillium* seeds by deer: implications for long-distance migration of forest herbs. *Ecology* **84**: 1067–1072.
Virgl, J. A. and F. Messier (2000). Assessment of source–sink theory for predicting demographic rates among habitats that exhibit temporal changes in quality. *Canadian Journal of Zoology – Revue Canadienne de Zoologie* **78**: 1483–1493.
Warren, R. J. (2007). Linking understory evergreen herbaceous distributions and niche differentiation using habitat-specific demography and experimental common gardens. PhD dissertation, University of Georgia, Athens, GA.
Watkinson, A. R. and W. J. Sutherland (1995). Sources, sinks and pseudo-sinks. *Journal of Animal Ecology* **64**: 126–130.
Whigham, D. E. (2004). Ecology of woodland herbs in temperate deciduous forests. *Annual Review of Ecology Evolution and Systematics* **35**: 583–621.
Wiens, J. A. (1989). Spatial scaling in ecology. *Functional Ecology* **3**: 385–397.
With, K. A. and A. W. King (1999). Dispersal success on fractal landscapes: a consequence of lacunarity thresholds. *Landscape Ecology* **14**: 73–82.
With, K. A., G. R. Schrott and A. W. King (2006). The implications of metalandscape connectivity for population viability in migratory songbirds. *Landscape Ecology* **21**: 157–167.
Zelikova, T. J., R. R. Dunn and N. J. Sanders (2008). Variation in seed dispersal along an elevational gradient in Great Smoky Mountains National Park. *Acta Oecologica – International Journal of Ecology* **34**: 155–162.

15

Effects of experimental population removal for the spatial population ecology of the alpine butterfly, *Parnassius smintheus*

Summary

For spatially segregated populations, the dispersal of organisms among local populations is a fundamental process affecting population dynamics and persistence. We present preliminary results from a long-term, large-scale experiment examining the effects of population removal for surrounding populations. During 2001–2006 we removed adult butterflies from two large populations within a system of 17 subpopulations of the Rocky Mountain Apollo butterfly, *Parnassius smintheus*. Surrounding populations were monitored using individual mark–recapture methods. We found that population removal reduced immigration into surrounding populations. Correspondingly, within-generation abundance of these populations was reduced. There was little effect of the loss of immigration for local population persistence. Only one confirmed local extinction occurred during the removals, but it was in a population expected to be highly impacted by the removals. Populations experiencing apparent local extinctions (observation of no adults within a flight season), and thus evidence of extremely low abundance, were much less connected than populations that did not experience such low abundances. Taken together, these results point to a system where immigration is too infrequent to impact the persistence of most populations, but may be frequent enough to recolonize or rescue the smallest populations from extinction. Given the fragmentation of these meadows, due to forest encroachment, the results imply that the persistence of this system will be more greatly impacted by the loss of habitat than by the isolation of populations, provided that some large populations remain.

Sources, Sinks and Sustainability, ed. Jianguo Liu, Vanessa Hull, Anita T. Morzillo and John A. Wiens.
Published by Cambridge University Press.

Introduction

The study of spatial population ecology has grown dramatically since being dubbed the "final frontier" for ecological theory (Kareiva 1994). Empirical study has provided information about the dispersal and spatial population structure of numerous species (Matter 1996; Lewis *et al.* 1997; Sutcliffe *et al.* 1997; Lele *et al.* 1998; Boughton 1999; Brommer and Fred 1999; Marsh *et al.* 1999; Baguette *et al.* 2000; Kean and Barlow 2000; Crone *et al.* 2001; Wahlberg *et al.* 2002). These studies have demonstrated a variety of spatial population structures such as metapopulation, source–sink, and patchy population dynamics (Harrison 1994). Theoretical models have often presaged the observed dynamics, exploring the implications of many of these spatial structures (Levins 1970; Pulliam 1988; Gotelli 1991; Hanski 1999; Wilson *et al.* 1999; Hastings 2001; Kaitala *et al.* 2001; Matter 2001).

For all spatially segregated populations, the dispersal of organisms among local populations is a fundamental process. The dynamics of local populations are linked via dispersal, producing emergent properties not necessarily apparent at the local population level. In a metapopulation context, dispersal results in the recolonization of unoccupied habitat. If local populations are at risk of extinction, a colonization–extinction dynamic may result where the system persists, as long as the rate of recolonization equals or exceeds extinction (Levins 1970). As highlighted in this volume, immigration may maintain populations with negative intrinsic growth (sinks) (Pulliam 1988; Thomas *et al.* 1996) and, in extreme cases, rescue populations that would otherwise become extinct (Brown and Kodric-Brown 1977; Gonzalez *et al.* 1998).

Despite advances in theory and empirical knowledge, many of the ascribed effects of dispersal for spatial population ecology remain untested experimentally. A few long-term datasets have focused on metapopulation dynamics (Thomas *et al.* 1996; Moilanen *et al.* 1998; Boughton 1999; Hanski 1999), showing results consistent with theory. As an example, data amassed by Hanski and coworkers for the butterfly *Melitaea cinxia* have shown distance-dependent colonization and extinction (Hanski 1994). Mark–recapture work (Kuussaari *et al.* 1996) has been used to parameterize models and predict dynamics for *M. cinxia* (Hanski *et al.* 1995; Kuussaari *et al.* 1998) and other species (Wahlberg *et al.* 1996). One of the clearest demonstrations of the effects of dispersal in a spatial system comes from a natural experiment involving the butterfly *Euphydryas editha* (Thomas *et al.* 1996). These butterflies live on two host plants within rocky outcrops and on a novel host plant within forest clearcuts. Population growth rates of the butterfly were substantially higher in the clearcuts than in outcrops, and migration was asymmetrical from clearcuts (source) to outcrops (sink). A rare summer frost killed all host plants within the clearcuts but not in the outcrops, and consequently population size in the outcrops fell by two-thirds (Thomas *et al.* 1996). Immigrants from the clearcuts made up a significant fraction of the outcrop populations, but populations in the outcrops were not extinction-prone. Thus, dispersal merely added to population sizes in the outcrops rather than rescuing them from any potential extinction. As such Thomas *et al.* (1996) dubbed these "pseudo-sinks."

As compelling as these observational studies are, most work detailing the effects of dispersal for spatial population dynamics has been short-term and non-experimental, leaving open possible effects of other factors such as spatially correlated extinction or colonization events. Experimental and non-experimental studies of habitat fragmentation are numerous (see Holt and Debinski 2003 for reviews); however, these studies are almost always confounded by having both population and habitat loss. We build on our previous work and incorporate an experimental approach to examine the role of dispersal for spatial population dynamics. Here we present initial results from an experimental study investigating the role of population removal for the spatial population dynamics of the Rocky Mountain Apollo butterfly, *Parnassius smintheus* Doubleday. Specifically, we assess how the removal of two large populations affects immigration, population size, and persistence of surrounding populations. We were particularly interested in determining whether abundance and extinction risk in small populations that had previously experienced extinctions are affected by the loss of nearby populations.

Methods

Study species

Parnassius smintheus is an abundant butterfly in alpine and sub-alpine meadows in the Rocky Mountains of North America, although congeners are threatened in Europe and elsewhere (Väisänen and Somerma 1985). The butterflies' host plants, lance-leaved stonecrop (*Sedum lanceolatum*) and, to a lesser degree, ledge stonecrop (*Rhodiola integrifolia*) occur in gravelly sites above the tree-line (Fownes and Roland 2002; Roslin *et al.* 2008). At our site *P. smintheus* is the only herbivore of *S. lanceolatum*, although it is fed on by other species elsewhere in its range. *Parnassius smintheus* is univoltine, with a flight period from July to September in our study area. Males fly more frequently and are thus more apparent than the sedentary females. Nonetheless, estimated dispersal distances are similar between the sexes (Roland *et al.* 2000).

Our previous research with this species shows that their dispersal among meadow habitats is limited by intervening forest habitat, which imposes an edge effect (Ross *et al.* 2005) and decreases movement distances (Matter *et al.* 2004), thus affecting immigration rates (Roland *et al.* 2000), genetic structure (Keyghobadi *et al.* 2005), and local population dynamics (Roland and Matter 2007).

Study site and population structure

Experiments were conducted in a network of 17 sub-alpine meadows along Jumpingpound Ridge, Kananaskis, Alberta, Canada (Fig. 15.1). Meadows are located above the tree-line (~2,100 m) and are bordered on their lower slopes by forest consisting of lodgepole pine, *Pinus contorta*; sub-alpine fir, *Abies lasiocarpa*; and Engelmann spruce, *Picea engelmannii*. Vegetation in the meadows is composed of grasses, sedges, mountain avens, and numerous species of wildflowers (Ezzeddine and Matter 2008).

Study of this system began in 1995. Initially, meadows and their butterfly subpopulations were somewhat arbitrarily assigned using forest habitat and ridgelines as meadow boundaries, then using the centroids of butterfly capture to determine subpopulation locations along the ridge (Roland *et al.* 2000; Matter *et al.* 2003). These intuitive assumptions regarding population boundaries have proved to be valid. When defined in this manner, each meadow contains a semi-independent population with differing population growth rates (Roland and Matter 2007) and genetic structure (Keyghobadi *et al.* 1999). The strength of the between-population correlation for these factors is largely determined by the distance between subpopulations that comprises forested habitat. Currently, all meadows and their subpopulations are separated by forest borders, but in some cases the borders are thin or incomplete (e.g., the border between meadows L and M is a patch of forest with openings to the northeast and southwest). The Jumpingpound system is isolated from other populations. The closest single population is across a forested valley and is located 6 km to the west. Larger networks of populations are located on Lusk Ridge – 6 km to the northwest, and Powderface Ridge – 10 km to the southeast. We have conducted simultaneous mark–recapture on these three ridges over several years and have never seen any movement. Genetic analyses also indicate little to no gene flow between the Ridges (Keyghobadi *et al.* 2005).

The Jumpingpound system functions along the continuum between a metapopulation and a patchy population. Between-population dispersal is moderate, particularly for well-connected populations; however, small, isolated populations (e.g., N, R, and Y) experience repeated local extinction and recolonization (S. F. Matter, personal observation).

Experiments and general hypotheses

To investigate the effects of neighboring populations for the spatial population dynamics of *Parnassius smintheus* we used an experimental

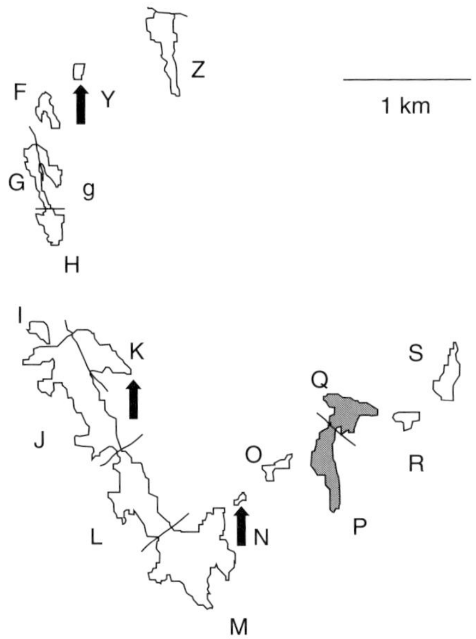

FIGURE 15.1.
Depiction of the study area. Meadows along Jumpingpound Ridge, Kananaskis Country, Alberta, Canada are outlined, based on aerial photographs from 1993. Bars indicate boundaries between meadows where the tree-line is thin or incomplete. The area between meadows is primarily forested, but open area that does not contain the butterfly's host plants is not shown. Note that meadow Z is isolated from a large meadow (Cox Hill, not shown) that occasionally contains a low density of butterflies; however, the butterfly's host plants do not occur there. Butterflies were removed from populations in meadows P and Q (shaded). Populations experiencing apparent extinction (2001–2006) are indicated with arrows.

approach. Beginning in 2001, all butterflies that we captured in meadows P and Q were removed (Fig. 15.1). Meadows were visited every 3–4 days during the entire flight season and butterflies were captured by hand-netting. The purpose of the removals was not necessarily to eliminate these populations, but to substantially decrease the potential immigrant pool for other meadows.

Correspondingly, if immigration is an important component of population size, we expected the abundance of butterflies in meadows affected by the removals to decrease. Populations in the nearby meadows N, O, R, and S are smaller than in P and Q, and a reduction in immigration was predicted to be greatest for them; thus we hypothesized that abundance would be substantially reduced in these meadows through the loss of immigrants.

In this system small meadows have low population abundance and are periodically observed to have no adult butterflies. If these populations are true sink populations, we expected the number of extinctions to increase for populations affected by the removals – particularly the populations in meadows N and R, which are small and have previously experienced apparent local extinctions.

Mark–recapture and population estimates

To evaluate the effects of the removals we used mark–recapture techniques. Butterflies in all meadows were censused three to six times during the adult flight periods during the summers of 1995–1996 and from 2001 to 2006. We captured butterflies using hand nets and each was given a unique three-letter mark on its hind wings. For all captures, we recorded the date, location based on *x*, *y* coordinates derived from aerial photographs, sex, and identity mark. Location is accurate to ~20 m (Roland *et al.* 2000). Population size in each meadow during each census period was estimated using Craig's method (Craig 1953; Southwood 1994). The frequency of butterflies captured once, twice, thrice, etc. was assumed to follow a Poisson distribution. The zero term of the distribution, the number not caught, was estimated and added to the number caught, to arrive at a point estimate of population size. In general, we continued capturing until about 75% of captures were recaptures, which produced high accuracy for this estimate (Craig 1953). We have shown that this method provides a highly accurate estimate of population size, even more so than capture–recapture estimators (Matter and Roland 2004). Note that we have not tried to separate immigrants, thus we include the effect of migration in our estimate. Our estimate of abundance in each meadow in each year was the maximal abundance observed. This estimate was used because of differing phenologies among meadows and differences in the timing of censuses.

During 1997–2000 (and to 2002 for meadows I, J, and K), population size in each meadow was estimated using transect counts (Pollard 1977). Here, each observer walked a path through the middle (along the longest axis) and around the circumference of a meadow, tallying the number of *P. smintheus* observed at any distance in front of them. *P. smintheus* fly more frequently when it is sunny (Ross *et al.* 2005), therefore the observations were conducted during full sun. As a guide, we stopped walking and counting if we could no longer see our shadow. For each survey there were between two and four observers, and meadows were surveyed two or three times each year. To arrive at a population estimate for each survey, we calculated the mean of the number of butterflies reported by the observers. Transect counts were converted to a "common currency" of population size, as estimated by Craig's method via a regression equation relating the metrics (Matter and Roland 2004).

The number of immigrants into populations in each meadow was enumerated for each year. An immigrant was defined as a marked butterfly subsequently arriving in a different meadow. This method underestimates the actual number of immigrants, but should be unbiased after controlling in the statistical analysis for the number of censuses in each meadow and the population sizes of source populations in the connectivity metric.

A local extinction was defined as the observation of no adult butterflies in a meadow during normal mark–recapture followed by observation of no larval feeding scars on host plants. For the latter, we sampled between 15 and 200 randomly placed 1.69 m^2 quadrats per meadow. The number of quadrats was scaled to meadow area. Within each quadrat we determined the number of host plants and any evidence of larval feeding damage.

Analyses

Based on previous investigation of the dispersal of this species (Matter *et al.* 2004), we could make specific predictions regarding the effects of the removals for immigration into surrounding populations. To do so, we calculated the connectivity (*S*) for each meadow *j* at time *t* using the best-fit connectivity equation based on data from 1995–1996 (S. F. Matter, unpublished):

$$S_{j(t)} = (A_{j(t)}^{im} \sum_{k \neq j} \exp(-\alpha_f d_{fjk} - \alpha_m d_{mjk}) \mu A_{k(t)}^{em}) N_{k(t)}$$

where d_f and d_m are the distances between meadows *j* and *k* comprising forested and open or "meadow" habitat, *A* is meadow area, and *N* is the population size at time *t*. The parameters α_f and α_m describe the inverse of mean dispersal distance through forest and meadow habitat, respectively. The parameters μ and *em* describe the rate and scaling of emigration with population size, respectively. We note that connectivity changes are dependent upon population size (*N*) in each surrounding population *k*.

Parameters were estimated for combined data from 1995 and 1996 using the virtual migration model (Hanski *et al.* 2000; Matter *et al.* 2004). Distances composed of forest and meadow habitat were determined from aerial photographs taken in 1993. Because these butterflies rarely move across valleys (Roland *et al.* 2000), distances were measured along the ridge. The estimated parameter values were used to calculate connectivity: $\mu = 0.04$, $em = -0.23$, $im = 0.93$, $\alpha_f = 4.47$, and $\alpha_m = 1.83$. It should be noted that connectivity as calculated in the current study differs from Matter *et al.* (2004), who showed that a simpler metric of connectivity was a better predictor of immigration. In Matter *et al.* (2004), to facilitate comparison with another system, we used total geometric distance between populations rather than separate distances through forest and meadow along the ridge. We examined the effects of population removal on the number of immigrants by comparing a statistical model using the connectivity metric including meadows P and Q with a model using the connectivity metric where the effects of these meadows were excluded – giving the predicted effect of the removals. When including P and Q in the connectivity metric, we used the maximum number of butterflies removed in a day as its estimate of population size. We included the methodological covariate of the number of times each meadow was censused in a year. This method was used because more traditional statistical methods (e.g., ANOVA) assume that the treatment is applied equally to all subjects, at least within treatment levels. That is clearly not the case here, because the effect of population removal will decrease with distance. Additionally, prior to the removals we only have data from 1995 and 1996 to compare with respect to immigration.

We expected that if removals affect immigration, models not including P and Q would show a better fit to the number of immigrants than models including P and Q. To avoid bias due to butterflies not being marked in meadows I, J, and K in 2001 and 2002, we restricted our analysis of immigration to 2003–2006. Because the number of immigrants consists of count data, we used generalized linear models assuming a Poisson error distribution and a log link function (McCullagh and Nelder 1989). We corrected for overdispersion by using a quasi-Poisson distribution with estimated dispersion for significance tests. To compare the statistical models, because there were the same number of parameters for each, we simply examined the deviance explained, which is analogous to variance explained in linear regression.

To evaluate the effects of population removals on abundance we used a similar procedure. We compared the fit of connectivity models either including or not including meadows P and Q for the maximum abundance observed in each meadow from 2001 to 2006. For the abundance models we assumed a Gaussian error distribution and an identity link function (McCullagh and Nelder 1989). Abundance was square root transformed to improve linearity.

To determine the effects of population removal on the extinction rate we tallied the number of apparent (no butterflies observed) and confirmed extinctions (no butterflies and no larval feeding damage). Here the expectation is that both apparent and confirmed extinctions should increase following population removals.

Results

From 2001 to 2006 we captured 4,681 individual butterflies along Jumpingpound Ridge a total of 9,993 times, giving a mean of 2.1 captures per individual. Over this time an additional 3,687 butterflies were removed from meadows P and Q (Fig. 15.2). We observed 322 between-meadow dispersal events from 2003 to 2006. In 2003 there was a ridge-wide population crash; only 93 individuals were observed in 2003, excluding removals. The

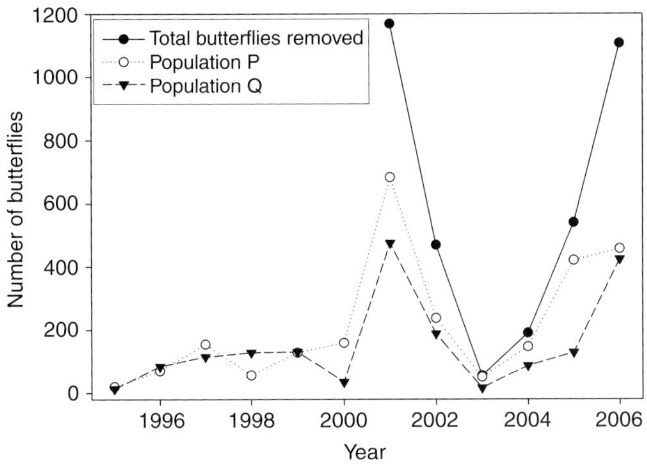

FIGURE 15.2.
The number of butterflies in meadows P (empty circles) and Q (filled inverted triangles) prior to and during population removal. The dashed and dotted lines indicate maximal population size estimated each year (see text for details). The solid line indicates the combined total number of butterflies removed from both meadows in each year.

crash appears to have resulted from early larval or overwintering egg mortality; mating success was normal in 2002 but the number of larval feeding scars decreased substantially in 2003.

Connectivity either including or excluding meadows P and Q was related to the number of immigrants (Fig. 15.3). Removal of butterflies from meadows P and Q reduced the number of immigrants to surrounding meadows. The statistical model using the connectivity term not including meadows P and Q as a source of immigrants was a better predictor of the number of immigrants to each meadow than was the model including the nominal abundance in P and Q (Table 15.1). Similarly, local abundance was reduced in populations close to the removals (Fig. 15.4). The connectivity model excluding meadows P and Q was a better predictor of abundance than was the connectivity model including their effect, indicating that the removal of populations was responsible for the reduction in abundance (Table 15.1).

Despite the reductions in immigration and abundance resulting from removals, there was surprisingly little effect on local extinction. From 2001 to 2006 there were five observations of no adult butterflies over an adult flight season among all populations. No adult butterflies were seen or captured in meadows Y and N in 2003 and 2004, and in meadow K in 2003. The mean connectivity of these populations in the years when adults were observed (0.16 ± 0.04 SE) was significantly lower (separate variance $t = 6.88$, df = 64.8, $P < 0.01$)

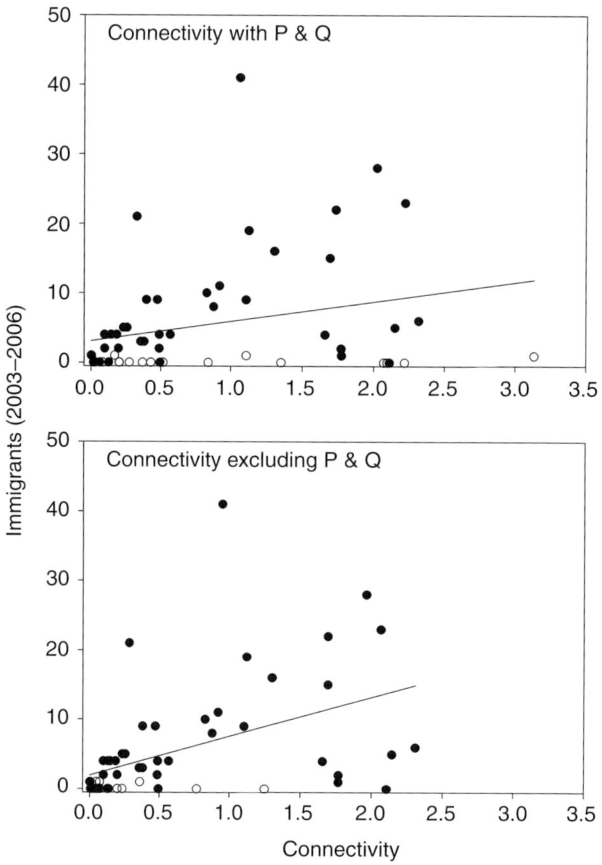

FIGURE 15.3.
Effects of the removal of populations in meadows P and Q on immigration into other populations. Both panels show the relationship between connectivity and the number of immigrants into a meadow. For the top panel, connectivity was calculated including meadows P and Q, while the bottom panel shows connectivity without meadows P and Q. If the removal of butterflies from meadows P and Q resulted in a reduction in immigration, we would expect the connectivity metric not including their effect to be a better predictor of the number of immigrants, which is the case. Note that the removal has little effect on many of the subpopulations. Fitted lines are for a simple linear regression, while the full model included the number of times each meadow was censused. Open circles show population for which removals would have the largest impact on connectivity. Filled circles show populations for which removals were predicted to have little effect. Details of the full statistical models are provided in Table 15.1.

than all others (0.86 ± 0.09). Only for the population in meadow N in 2004 could we confirm this as a local extinction based on larval feeding damage. Damage was seen in meadow Y in both years and in meadows N and K in 2003. The connectivity for the population in meadow N in 2004 assuming an effect of P and Q was 0.43, while the connectivity assuming P and Q do not supply

TABLE 15.1. *Analysis of deviance table for the effects of population removal on immigration (a) and abundance (b). Significance tests for immigration are based on quasi-Poisson distributions and estimated dispersion (ϕ). Abundance used a Gaussian distribution and was square root transformed.*

	df	Deviance explained	Beta	F	P	ϕ
(a) Immigration						
Null	56	553.5				
Model 1						8.4
Intercept			−0.19 ± 0.59			
Censuses	1	64.5	0.35 ± 0.12	7.7	0.01	
Connectivity with P & Q	1	57.4	0.51 ± 0.19	16.4	0.01	
Model 2						6.7
Intercept			−0.51 ± 0.48			
Censuses	1	64.5	0.36 ± 0.10	9.6	<0.01	
Connectivity without P & Q	1	167.7	0.93 ± 0.18	25.0	<0.01	
(b) Abundance						
Null	89	2042.4				
Model 1						18.2
Intercept			4.12 ± 0.68			
Connectivity with P & Q	1	441.4	2.30 ± 0.47	24.3	<0.01	
Model 2						16.4
Intercept			4.10 ± 0.60			
Connectivity without P & Q	1	601.2	3.07 ± 0.51	36.6	<0.01	

immigrants to meadow N was 0.20. This result supports our hypothesis that population removal would increase the risk of extinction; however, because it is based on only one event, it is tenuous.

Discussion

The removal of local populations reduced the number of immigrants to surrounding populations, resulting in reduced population abundance – results consistent with both general theoretical expectations and predictions for this system. Despite the reduced immigration and lower abundance, extinction rates were largely unaffected. This result is somewhat surprising given the ridge-wide population crash in 2003. Even though populations were

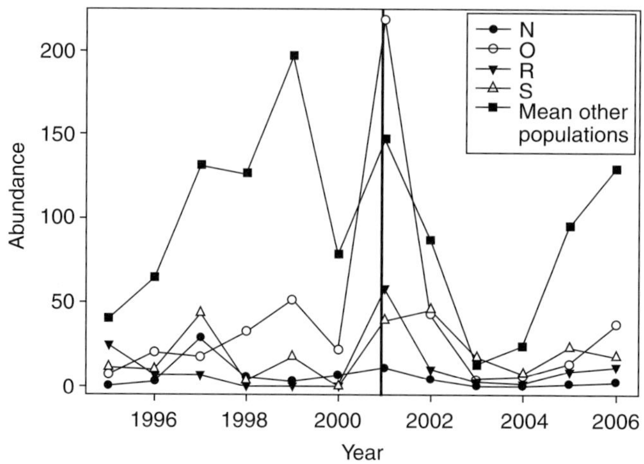

FIGURE 15.4.
The abundance of populations most greatly affected by removal of butterflies in meadows P and Q (N, O, R, S) and the mean abundance of other populations along Jumpingpound Ridge. The bar indicates the year in which removals began. Note that the predicted effect on abundance is not equal among populations, but depends on the distance from the removals, i.e., populations in meadows O or R should experience the greatest impact, dependent on local abundance in the surrounding populations.

extremely low, we saw only apparent extinctions for populations in meadows N and Y that year. For N the apparent extinction was followed by a confirmed extinction in 2004, while for Y there was evidence of feeding damage in 2004. The extinction in N was consistent with population removal lowering persistence, but Y is distant from the removals and should have been unaffected. Additionally, both of these populations experienced one apparent extinction in the six years prior to removal. Thus, it is difficult to characterize the extinction in N as anything but a "normal" occurrence. If immigration is important for local persistence, we would have expected extinctions for populations close to the removals, in particular in meadow R, which had three consecutive years of apparent extinction prior to removals.

A second somewhat surprising result involved our population removals. Despite removing a substantial fraction of these populations each year, their abundance was not substantially reduced, generally reflecting the pattern of abundance seen for the rest of the ridge. This result potentially indicates that we failed to remove a sufficient number of females, which have a lower capture rate, to impact abundance. Alternatively, density-dependent effects (Roslin *et al.* 2008) may have been reduced, such that the reduced number of larvae had a greater probability of survival, resulting in little change in abundance. Given that we removed far more individuals from P and Q than would ever immigrate into any other population, the loss of immigrants from local populations is not likely to have strong between-generation effects on abundance or population growth.

The results, taken as a whole, point to a system where the few immigrants arriving in most populations appear to be virtually inconsequential for their long-term persistence. Somewhat contrastingly, those populations that experienced local extinctions were less connected than those that did not, indicating that immigration may be important for the persistence of these populations (Hanski *et al.* 1995; Morris, Chapter 3, this volume). This result may be more due to small populations having a higher extinction rate (Richter-Dyn and Goel 1972) than small populations attracting fewer immigrants.

These differences illustrate subtleties between source–sink dynamics and the rescue effect (Brown and Kodric-Brown 1977). In source–sink dynamics we expect immigrants to bolster the size of local populations with inherently negative growth rates; whereas, for the rescue effect, an immigrant or immigrants may prevent immediate extinction. Although the rescue effect is more often considered in studies of metapopulation dynamics because of the focus on extinction and colonization, in some ways the rescue effect is the endpoint of a continuum of source–sink dynamics set by the effects of immigration on local population size and dynamics.

Our results have important implications for conservation and sustainability in alpine systems. The elevation of the tree-line is rising in many alpine areas throughout the world (Dyer and Moffett 1999; Walther *et al.* 2002; Millar *et al.* 2004). At our site, forest encroachment over the last 70 years has reduced meadow area by over 70% (Roland *et al.* 2000). For species such as *P. smintheus*, which are restricted to habitats above the tree-line, the rising tree-line results in habitat fragmentation. Analogous to traditional notions of habitat fragmentation (Debinski and Holt 2000), the rising tree-line reduces habitat area, albeit from the perimeter rather than by bisection, and isolates populations by increasing the distance between habitat fragments. If forest is a barrier to dispersal, as it is for *P. smintheus*, the effective isolation between fragments will also increase. The results of our study indicate some positives and some negatives with regard to the fate of *P. smintheus* in the face of continued tree-line rise. We view it as a positive that the persistence of most populations is not dependent upon immigration; we do not have a few sources and several sinks. Most populations appear to be able to maintain their abundance through local reproduction, even under the ridge-wide population crash seen in 2003. Conversely, although quite resilient, our smallest populations (Y, R, N) may not be large enough to maintain themselves for long without immigration or recolonization. For these populations there likely is not enough habitat for their long-term persistence.

What is the future for these populations if the tree-line continues to rise? We know that some populations will be lost completely, such as those in meadows N and R, which have conditions throughout that seem to be suitable for tree seedling establishment and growth. Other meadows contain a mixture of microhabitats, some of which may not be suitable for forest and may provide refugia from the advancing trees for host plants, nectar flowers, and the butterflies. We can be certain that the continued rise in tree-line will have adverse effects for *P. smintheus*. This continued rise in tree-line will reduce meadow area and thus butterfly population sizes as well as increase their isolation. Both factors will increase the risk of local extinction. However, if enough habitat area remains as refugia, the risk of extinction for individual populations may still be low. Because of the strong effect of forest, *P. smintheus* may be an early indicator of things to come for other alpine species. Many species, particularly plants and the invertebrates that feed on them, are restricted to alpine habitats and will be facing similar conditions as the tree-line rises. Additionally, many alpine species are less abundant than *P. smintheus* and may already be experiencing negative impacts.

In summary, our experiments have revealed a spatial population structure for *P. smintheus* where many populations exchange migrants at a fairly low rate. For most populations this immigration affects short-term abundance but has little effect on long-term persistence. However, infrequent immigration may fend off extinction for the smallest populations via a rescue effect.

Acknowledgments

We thank Norine Ambrose, Mark Caldwell, Renée Cormier, Sue Cotterill, Dave Dennewitz, Liz Duermit, Maya Ezzeddine, Anna Fiskin, Marian Forrester, Sheri Fownes, Matt Frantz, Chris Garrett, Mike Gaydos, Rebecca Hamilton, Fiona Johnson, Kurt Illerbrun, Nusha Keyghobadi, Kenny Kim, Tiffany Lucas, Jeff Mashburn, Doug Meldrum, Evelyn Robinson, Andy Ross, Dave Roth, Kris Sabourin, Chris Schmidt, Lynnette Scott, Dana Sjoström, Kristy Ward, Anne Wick, Arliss Winship, and Alyson Winkelaar for help in the field. We thank the Ohio Supercomputing Network for parallel processing time. This work was supported by NSF DEB-0326957 to S. F. Matter, and NSERC Operating and Discovery Grants to J. Roland.

References

Baguette, M., S. Petit and F. Queva (2000). Population spatial structure and migration of three butterfly species within the same habitat network: consequences for conservation. *Journal of Applied Ecology* **37**: 100–108.

Boughton, D. A. (1999). Empirical evidence for complex source–sink dynamics with alternative states in a butterfly metapopulation. *Ecology* **80**: 2727–2739.
Brommer, J. E. and M. S. Fred (1999). Movement of the Apollo butterfly *Parnassius apollo* related to host plant and nectar plant patches. *Ecological Entomology* **24**: 125–131.
Brown, J. H. and A. Kodric-Brown (1977). Turnover rates in insular biogeography: effect of immigration on extinction. *Ecology* **58**: 445–449.
Craig, C. C. (1953). On the utilization of marked specimens in estimating populations of flying insects. *Biometrika* **40**: 170–176.
Crone, E. E., D. Doak and J. Pokki (2001). Ecological influences on the dynamics of a field vole metapopulation. *Ecology* **82**: 831–843.
Debinski, D. M. and R. D. Holt (2000). A survey and overview of habitat fragmentation experiments. *Conservation Biology* **14**: 342–355.
Dyer, J. M. and K. E. Moffett (1999). Meadow invasion from high-elevation spruce–fir forest in south-central New Mexico. *Southwestern Naturalist* **44**: 444–456.
Ezzeddine, M. and S. F. Matter (2008). Nectar flower use and electivity by butterflies in sub-alpine meadows. *Journal of the Lepidopterists' Society* **62**: 138–142.
Fownes, S. and J. Roland (2002). Effects of meadow suitability on female movement and oviposition behaviour in the alpine butterfly, *Parnassius smintheus* Doubleday. *Ecological Entomology* **27**: 457–466.
Gonzalez, A., J. H. Lawton, F. S. Gilbert, T. M. Blackburn and I. Evans-Freke (1998). Metapopulation dynamics, abundance, and distribution in a microecosystem. *Science* **281**: 2045–2047.
Gotelli, N. (1991). Metapopulation models: the rescue effect, propagule rain, and the core-satellite hypothesis. *American Naturalist* **138**: 768–776.
Hanski, I. (1994). A practical model of metapopulation dynamics. *Journal of Animal Ecology* **63**: 151–162.
Hanski, I. (1999). *Metapopulation Ecology*. Oxford University Press, Oxford, UK.
Hanski, I., J. Alho and A. Moilanen (2000). Estimating the parameters of survival and migration in metapopulations. *Ecology* **81**: 239–251.
Hanski, I., T. Pakkala, M. Kuussaari and G. Lei (1995). Metapopulation persistence of an endangered butterfly in a fragmented landscape. *Oikos* **72**: 21–28.
Harrison, S. (1994). Metapopulations and conservation. In *Large Scale Ecology and Conservation Biology* (P. J. Edwards, R. M. May and N. Webb, eds.). Blackwell, Oxford, UK: 111–128.
Hastings, A. (2001). Transient dynamics and persistence of ecological systems. *Ecology Letters* **4**: 215–220.
Holt, R. D. and D. M. Debinski (2003). Reflections on landscape experiments and ecological theory: tools for the study of habitat fragmentation. In *How Landscapes Change* (G. A. Bradshaw and P. A. Marquet, eds.). Springer, Berlin, Heidelberg, New York: 201–223.
Kaitala, V., E. Ranta and P. Lundberg (2001). Self-organized dynamics in spatially structured populations. *Proceedings of the Royal Society – Biological Sciences Series B* **268**: 1655–1660.
Kareiva, P. (1994). Space: the final frontier for ecological theory. *Ecology* **75**: 1.
Kean, J. M. and N. D. Barlow (2000). Effects of dispersal on local population increase. *Ecology Letters* **3**: 479–482.
Keyghobadi, N., J. Roland and C. Strobeck (1999). Influence of landscape on the population genetic structure of the alpine butterfly *Parnassius smintheus* (Papilionidae). *Molecular Ecology* **8**: 1481–1495.
Keyghobadi, N., J. Roland and C. Strobeck (2005). Genetic differentiation and gene flow among populations of the alpine butterfly, *Parnassius smintheus*, vary with landscape connectivity. *Molecular Ecology* **14**: 1897–1909.
Kuussaari, M., M. Nieminen and I. Hanski (1996). An experimental study of migration in the Glanville fritillary butterfly, *Melitaea cinxia*. *Journal of Animal Ecology* **65**: 791–801.

Kuussaari, M., I. Saccheri, M. Camara and I. Hanski (1998). Allee effect and population dynamics in the Glanville fritillary butterfly. *Oikos* **82:** 384–392.

Lele, S., M. L. Taper and S. Gage (1998). Statistical analysis of population dynamics in space and time using estimating functions. *Ecology* **79:** 1489–1502.

Levins, R. (1970). Extinction. In *Some Mathematical Problems in Biology* (M. Gesternhaber, ed.). American Mathematical Society, Providence, RI: 75–107.

Lewis, O. T., C. D. Thomas, J. K. Hill, M. I. Brookes, T. P. R. Crane, Y. A. Graneau, J. L. B. Mallet and O. Rose (1997). Three ways of assessing metapopulation structure in the butterfly *Plebejus argus*. *Ecological Entomology* **22:** 283–293.

Marsh, D. M., E. H. Fegraus and S. Harrison (1999). Effects of breeding pond isolation on the spatial and temporal dynamics of pond use by the tungara frog, *Physalaemus pustulosus*. *Journal of Animal Ecology* **68:** 804–814.

Matter, S. F. (1996). Interpatch movement of the red milkweed beetle, *Tetraopes tetraophthalmus*: individual responses to patch size and isolation. *Oecologia* **105:** 447–453.

Matter, S. F. (2001). Synchrony, extinction and dynamics of spatially segregated heterogeneous populations. *Ecological Modelling* **141:** 217–226.

Matter, S. F. and J. Roland (2004). Relationships among population estimation techniques: an examination for *Parnassius smintheus*. *Journal of the Lepidopterists' Society* **58:** 189–195.

Matter, S. F., J. Roland, N. Keyghobadi and K. Sabourin (2003). The effects of isolation, habitat area and resources on the abundance, density and movement of the butterfly, *Parnassius smintheus*. *American Midland Naturalist* **150:** 26–36.

Matter, S. F., J. Roland, A. Moilanen and I. Hanski (2004). Migration and survival of *Parnassius smintheus*: detecting effects of habitat for individual butterflies. *Ecological Applications* **14:** 1526–1534.

McCullagh, P. and J. A. Nelder (1989). *Generalized Linear Models*, 2nd edition. Chapman and Hall, London.

Millar, C. I., R. D. Westfall, D. L. Delany, J. C. King and L. J. Graumlich (2004). Response of subalpine conifers in the Sierra Nevada, California, USA to 20th-century warming and decadal climate variability. *Arctic, Antarctic, and Alpine Research* **36:** 181–200.

Moilanen, A., A. T. Smith and I. Hanski (1998). Long-term dynamics in a metapopulation of the American pika. *American Naturalist* **152:** 530–542.

Pollard, E. (1977). A method for assessing changes in the abundance of butterflies. *Biological Conservation* **12:** 115–134.

Pulliam, H. R. (1988). Sources, sinks, and population regulation. *American Naturalist* **132:** 652–661.

Richter-Dyn, N. and N. S. Goel (1972). On the extinction of a colonizing species. *Theoretical Population Biology* **3:** 406–433.

Roland, J. and S. F. Matter (2007). Encroaching forests decouple alpine butterfly population dynamics. *Proceedings of the National Academy of Sciences* **104:** 13702–13704.

Roland, J., N. Keyghobadi and S. Fownes (2000). Alpine *Parnassius* butterfly dispersal: effects of landscape and population size. *Ecology* **81:** 1642–1653.

Roslin, T., H. Syrjälä, J. Roland, P. Harrison, S. Fownes and S. F. Matter (2008). Caterpillars on the run: induced defences create spatial patterns in host plant damage. *Ecography* **31:** 335–347.

Ross, J. A., S. F. Matter and J. Roland (2005). Edge avoidance and movement of the butterfly *Parnassius smintheus* in matrix and non-matrix habitat. *Landscape Ecology* **20:** 127–135.

Southwood, T. R. E. (1994). *Ecological Methods*, 2nd edition. Chapman and Hall, London.

Sutcliffe, O. L., C. D. Thomas and D. Peggie (1997). Area-dependent migration by ringlet butterflies generates a mixture of patchy population and metapopulation attributes. *Oecologia* **109**:229–234.

Thomas, C. D., M. C. Singer and D. A. Boughton (1996). Catastrophic extinction of population sources in a butterfly metapopulation. *American Naturalist* **148:** 957–975.

Väisänen, R. and P. Somerma (1985). The status of *Parnassius mnemosyne* (Lepidoptera, Papilionidae) in Finland. *Notulae Entomologicae* **65:** 109–118.

Wahlberg, N., T. Klemetti, V. Selonen and I. Hanski (2002). Metapopulation structure and movements in five species of checkerspot butterflies. *Oecologia* **130**: 33–43.
Wahlberg, N., A. Moilanen and I. Hanski (1996). Predicting the occurrence of endangered species in fragmented landscapes. *Science* **273**: 1536–1538.
Walther, G., E. Post, P. Convey, A. Menzel, C. Parmesan, T. J. C. Beebee, J. Fromentin, O. Hoegh-Guldberg and F. Bairlaein (2002). Ecological responses to recent climate change. *Nature* **416**: 389–395.
Wilson W. G., S. P. Harrison, A. Hastings and K. McCann (1999). Exploring stable pattern formation in models of tussock moth populations. *Journal of Animal Ecology* **68**: 94–107.

Part IV

Improvement of source–sink management

To sustain species, it is essential to manage both source and sink populations and their associated habitats, as well as factors affecting populations and habitats. Effective management requires a solid understanding of how populations and habitats respond to various management regimes. Source–sink theory has greatly assisted in the process of developing management plans and setting management priorities. It has enriched biodiversity conservation and offers considerable promise for the future. The chapters in this section all show the importance of the source–sink concept in improving management of populations and their associated habitats. They demonstrate that the source–sink concept helps capture spatial variation in population demographics, which, in turn, can assist in identifying management targets that allow for adequate connectivity between sources and sinks.

The first two chapters of this section synthesize source–sink research and management implications in protected areas as well as marine and estuarine systems. Protected areas are perceived as cornerstones of biodiversity conservation. Around the world there are more than 100,000 protected areas. In Chapter 16, after reviewing the literature published during the past two decades, Hansen classifies protected areas into three types: sinks that may be vulnerable to degradation of source populations in unprotected surrounding lands due to land use intensification, sources that lose individuals to attractive sinks in the surrounding unprotected areas because of the failure of these individuals (including the fittest ones) to recognize the threats in the attractive sinks (mainly due to human activities), and sources that provide surplus individuals to subsidize harvested populations in surrounding areas. He systematically explores these different types of protected areas with examples from around the world, graphically illustrating gradients in biophysical conditions and land use intensity relative to protected areas. To achieve sustainable populations, Hansen calls on natural resource managers to consider source–sink dynamics when defining greater ecosystems, to create and expand protected areas, to modify human behavior in attractive sinks, and to monitor population demography in and around protected areas.

Most of the focus on source–sink dynamics has been on terrestrial systems, but there are also increased research and management efforts in aquatic systems. By reviewing the existing literature, Lipcius and Ralph (Chapter 17) find some strong evidence of source–sink dynamics in marine and estuarine species. Furthermore, they report that the source–sink dynamics of aquatic species are affected by habitat quality, dispersal, fishing, predation, and interactions between these factors. In order to conserve or restore marine and estuarine species, Lipcius and Ralph suggest that it is important to identify the connectivity between populations and to conserve both source populations and interconnected networks of sources and sinks. Because of uncertainties and complex interactions between various factors, there is no single optimal conservation strategy. The authors therefore suggest a bet-hedging conservation strategy coupled with additional research on metapopulations across sources and sinks in marine and estuarine systems.

The remaining chapters of this section provide in-depth discussions of the application of source–sink theory to management in terrestrial and aquatic systems. In Chapter 18, Wing analyzes the spatial patterns of sea-urchin demographics across the New Zealand coastline. The sea urchin *Evechinus chloroticus*, endemic to New Zealand, is a critical species in coastal kelp forests. Through a fine-scale survey of *E. chloroticus* population structure, Wing reveals that at many inner-fjord sites there were strong recruitment events with a large frequency of recently emergent recruits. In fact, at some sites, populations were made up of only a single juvenile cohort with no adults present. These patterns indicate a source–sink population structure across the productivity gradients. The results have important implications for the effectiveness and sustainability of marine reserves and commercial exclusion zones. For example, the fact that many of the source populations for sea urchins exist outside of current marine protected areas suggests a need to revise current management plans.

Many management plans have been developed around the world, but their effectiveness has not been widely evaluated in the context of source–sink dynamics. In Chapter 19, Smith and colleagues assess the effectiveness of the Land Management Plan of the US Forest Service for the Tongass National Forest in southeastern Alaska, which includes an integrated system of large, medium and small old-growth forest reserves to sustain viable plant and wildlife populations in the heavily harvested forest. Specifically, they use a population-growth model and dispersal parameters to determine effective distances that would allow northern flying squirrels (*Glaucomys sabrinus*, an indicator species) to colonize and persist in sink habitat in small reserves. While the sustainability of flying squirrels in small reserves relies on dispersal from larger reserves, small reserves can nonetheless play an important role in sustaining a viable overall population by providing breeding habitat for females and by serving as stepping-stones for dispersal among subpopulations. The results suggest that most small reserves are not functionally connected. Active management is therefore needed to hasten the succession of second-growth stands toward old-growth status, increase primary habitat in small reserves, add stepping-stones, and improve the permeability of habitat comprising the matrix to enhance the likelihood of viable northern flying squirrel populations.

It has long been hypothesized that forest fragmentation and habitat loss on the breeding grounds of migratory songbirds affect source–sink dynamics through mechanisms such as increasing populations of nest predators and promoting brood parasitism. Despite these threats, it has been hypothesized that source populations in mainly forested landscapes can also rescue sink populations in mainly agricultural regions. In Chapter 20, Robinson and Hoover present a variety of evidence to support these hypotheses. They find that habitat fragmentation and loss create sink habitats, which can turn into ecological traps due to high rates of nest predation and nest parasitism. Furthermore, the impacts of fragmentation on reducing nesting success appear to vary across regions, and it is not clear whether large habitat patches always serve as sources. However, maintaining and restoring large forest tracts in mainly forested regions would likely increase source habitat, decrease sink habitat, and sustain bird populations.

Source–sink dynamics are present not only in the forest overstory, but also in forest understory species. Understory species such as palms (*Chamaedorea radicalis*) provide essential food for wildlife and livestock, materials for floral displays in international floral and horticultural markets, and non-timber forest products for millions of people around the globe. In Chapter 21, Berry and colleagues evaluate the impact of leaf harvesting of the understory palm using stage-structured transition matrices. To understand the sustainability of palms, the authors incorporate sources, sinks, and demographic effects of harvest practices and livestock browsing into population models. They show that, while it is possible for the understory palm to grow above the replacement rate under either leaf harvesting or livestock browsing, simultaneous leaf harvesting and livestock browsing lead to unsustainable populations. The authors provide recommendations for sustainable leaf-harvest strategies, including removing livestock from harvested populations, implementing certification programs, increasing the economic value of palm leaves, and enhancing recruitment in harvested populations through seed and seedling enrichment plantings.

Corridors are important mechanisms for promoting biodiversity conservation through connecting sources and sinks across landscapes. Much theoretical and empirical work has revealed positive roles of corridors in creating or maintaining population sources, but negative effects of corridors have not been studied as widely. Haddad and colleagues (Chapter 22) review the effectiveness of wildlife corridors for protecting species in a source–sink context. Through synthesizing findings on seed mass and number, bird-dispersed seed rain, and bird nest predation and density from a large and long-running corridor experiment (the Savannah River Site Corridor Experiment), they illustrate how corridors may promote the development of sources in certain circumstances and create sinks in others by changing the dynamics of predators (natural enemies) and competitors. The authors also discuss whether the benefits of corridors in allowing for dispersal outweigh the potential negative consequences of edge effects. They argue that corridors can serve as important tools, enabling landscape managers to generate, connect or sustain sources, provided that concentrated efforts are made toward mitigating the negative effects of corridors while restoring connectivity to fragmented landscapes.

The reality of source–sink population structures and dynamics, documented so clearly in the chapters of this and previous sections, creates a challenge for resource managers. Management plans that treat all populations in the same way may backfire depending on the source–sink status of the populations, and conservation efforts focused on source (or sink) populations (perhaps inadvertently) may fail if the relevant sink (or source) populations are not also conserved. There is no single, simple solution to this challenge, but these chapters provide ample evidence showing management that incorporates the spatial dynamics of populations is worthwhile.

ANDREW HANSEN

16

Contribution of source–sink theory to protected area science

Summary

The concept of source–sink population dynamics may be especially relevant to protected areas. Places set aside as nature reserves often have steep gradients in climate, topography, and other abiotic factors that result in spatially explicit population dynamics occurring within them. Protected areas are also frequently placed in relatively extreme parts of the landscape with regard to climate, soils, elevation, and water. Consequently, spatially explicit population dynamics may occur between protected areas and the more moderate surrounding landscape. The goal of this chapter is to evaluate the contribution that source–sink theory has made to understanding population viability in and around protected areas. A review of the literature for the past 20 years indicates that the source–sink concept has been applied to protected areas primarily in three ways.

1. Protected areas may be sinks for some species, due to the more extreme biophysical conditions within them. These sink populations may be vulnerable to loss of source areas in unprotected surrounding lands. Land use intensification around reserves may drive the degradation of these sources and reduce viability of the species in the protected area.
2. The areas surrounding protected areas may become "attractive" sinks due to human activities and lead to loss of viability of the source population. Large carnivores appear to be especially vulnerable to this dynamic.
3. Protected areas may serve as population source areas that supplement hunted or fished populations in surrounding areas. Many

Sources, Sinks and Sustainability, ed. Jianguo Liu, Vanessa Hull, Anita T. Morzillo and John A. Wiens. Published by Cambridge University Press.

marine protected areas have been designated as a means of allowing more sustainable fisheries in surrounding waters.

I summarize the conceptual basis of each of these scenarios, provide examples, and draw implications for conservation and management.

Introduction

In the 20 years since the publication of Pulliam (1988) on source–sink population dynamics, the theory has been the basis for advances in ecological theory (as shown by the chapters in this book). It has also contributed to strategies for conservation of species. One type of conservation application has involved the viability of species in and around protected areas. Protected areas are places where human activities are minimized in order to provide for the maintenance of natural ecological processes and native species (Possingham *et al.* 2006). They serve as places for the protection of nature, research sites into the functioning of intact ecosystems, benchmarks for understanding change in human-altered systems, and areas for human enjoyment and renewal (Arcese and Sinclair 1997). It is increasingly apparent, however, that many protected areas are not functioning as envisioned. Ecological processes such as disturbance regimes have been altered and native species have been lost in many protected areas (Newmark 1987, 1995, 1996; Rivard *et al.* 2000; Parks and Harcourt 2002). Source–sink theory offers a basis for understanding some of the factors that influence the function of protected areas.

The spatially explicit population dynamics that are the basis of source–sink theory may be particularly likely in and around protected areas. Protected areas may be subject to source–sink dynamics more than areas selected at random for three reasons. First, protected areas are often portions of larger ecosystems, and native species move out of the protected area and across the larger ecosystem to obtain needed resources over the annual cycle (Wright and Thompson 1935; Craighead 1979; Newmark 1985; Hansen and DeFries 2007). Second, protected areas are often surrounded by areas of more intense land use (Wittemyer *et al.* 2008). The resulting gradient in land use intensity and human interaction with native species can influence birth and death rates and create source–sink dynamics (McKinney 2002; Hansen *et al.* 2005). Hence, protected areas may be sources that maintain subpopulations in sinks in the surrounding more intensively used lands. Third, protected areas tend to be located in landscapes with strong gradients in topography, climate, soils, and other biophysical factors (Pressey 1994), which can lead to differential habitat quality and spatially explicit population dynamics (as demonstrated by Wing,

Chapter 18, this volume). In fact, protected areas are typically in the harsher parts of these biophysical gradients (e.g., high mountains, deserts, low-productivity soils) (Scott *et al.* 2001) and may have relatively low habitat quality, resulting in populations moving outside of protected areas to better habitats seasonally or for some life-history phases. The lower-quality habitats in some protected areas may result in them being population sinks for some species (Hansen and DeFries 2007).

Over the past 20 years of theoretical development, two refinements have been added to the source–sink model that are highly relevant to protected areas. "Refuge" habitats are places where survival of a population is relatively high but reproduction is relatively low, so the subpopulation is a weak source (Naves *et al.* 2003). This typically occurs in "suboptimal refuge areas with scarce nutritional resources but a lower risk of human-induced mortality that may allow for population persistence" (Naves *et al.* 2003: 1277). Such populations are often endangered species that have been displaced to the edges of their former ranges. "Attractive sinks" are places where habitat quality is good, allowing potentially high reproduction and high survival, but where either reproduction or survival is reduced by forces that are not detected by the organism (Delibes *et al.* 2001a). Hence, the organism selects habitat based on perceived quality, but suffers either high mortality or low reproduction there due to hazards that are not detected, such as hunting or the presence of pesticides (Gundersen *et al.* 2001).

A key issue for protected areas is how the presence of source–sink dynamics influences the viability of the metapopulation (Fig. 16.1). The original source–sink theory focused on how a subpopulation in a sink could be sustained by dispersal from a source and maintain a viable metapopulation. Loss of viability from the source due to loss of individuals to the sink was not expected because of the assumption that intraspecific density-dependent competition for the good habitat would lead to the fittest individuals occupying the source and would force subdominant individuals into the sinks (Pulliam 1988; Delibes *et al.* 2001a). This is highly relevant to protected areas because populations in protected areas in harsh biophysical settings may be sinks that are maintained by sources outside the protected area (Fig. 16.1a). However, populations in protected areas are vulnerable if land use change or other factors cause the external source to become a sink.

The assumption of density-dependent habitat selection does not hold, however, in the case of attractive sinks. Because of the inability of the organism to perceive the threats in the attractive sink, even the fittest individuals may move from sources to attractive sinks, reducing population size in the source and reducing the viability of the metapopulation (Delibes *et al.* 2001a, 2001b; Gundersen *et al.* 2001). Thus, sources or refuges in protected areas could

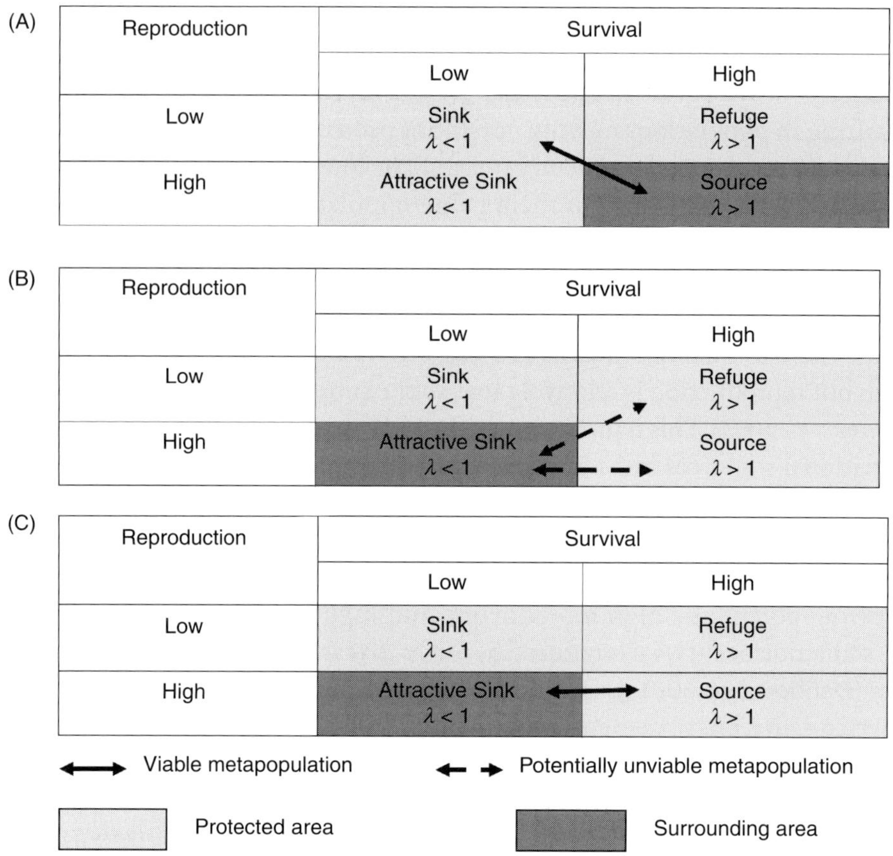

FIGURE 16.1.
Types of source and sink populations based on levels of survival and reproduction. Subpopulations in protected areas are denoted by the light gray shading and those outside of protected areas by dark gray shading. Interactions between sources or sinks in protected areas and those outside of protected areas are indicated by the lines. (a) A sink subpopulation in the protected area is maintained by a viable source subpopulation outside the protected area. (b) A source or refuge subpopulation in a protected area loses viability due to emigration to an attractive sink. (c) A source subpopulation in a protected area provides for a sustainable harvest in attractive sinks in surrounding areas.

become unviable if too many individuals from the source are lost to attractive sinks (Fig. 16.1b).

A third scenario may allow the sustained harvesting of individuals in sink habitats. Under the density-dependent model, protected areas can be created or maintained as sources which supply surplus individuals to surrounding areas that are attractive sinks through fishing or hunting pressure (McCullough 1996). This scenario is being used as the basis for creating marine reserves to sustain fisheries in surrounding areas (Lubchenco *et al.* 2003).

This chapter explores and illustrates each of these three source–sink dynamics with examples from various protected areas around the world. The examples were derived from a literature review of studies of source–sink dynamics in protected areas for the period 1988–2008. These scenarios are graphically illustrated by depicting gradients in biophysical conditions, and thus habitat conditions and land use intensity, relative to protected areas. Following these examples, the implications for management are presented.

Protected area as a sink

Because protected areas were often established in places not productive for agriculture or other natural resource extraction, they are often in harsher biophysical conditions than the surrounding areas (Scott *et al.* 2001). These conditions may constrain organisms directly or reduce primary productivity and food availability for consumers, resulting in lower habitat quality in protected areas. If habitat quality is sufficiently low, the protected area may be a population sink for a species that is maintained by dispersal from sources in the surrounding areas that have higher habitat quality (Fig. 16.2). In this case, the presence of the species in the protected area is dependent upon the maintenance of the source areas in the unprotected portion of the landscape (Dias 1996; Delibes *et al.* 2001a). If land use intensification leads to conversion of the source areas to sinks, the metapopulation is at risk of local extinction.

The implications for protected areas of source–sink dynamics where density-dependent habitat selection in the source regulates movement from sources to sinks were summarized by Delibes *et al.* (2001a: 283) as follows:

> The results of the scenario corresponding to an "avoided" sink (one occupied by overflow individuals from the preferred source) are well known (Pulliam, 1988). The source is extinction-resistant but the sink is not. Thus, reserves must be established in sources since preserving only sink habitats will probably lead to population extinction.

Example: Song bird population dynamics in Greater Yellowstone

Yellowstone and Grand Teton National Parks in the Northern Rocky Mountains in the western USA are centered on the Yellowstone Plateau and surrounding mountains (Fig. 16.3a). Climate is harsh at these higher elevations, with a snow-free growing season as short as 2 months. The volcanically derived soils are also poor in nutrients and water-holding capacity. Consequently, net primary productivity is relatively low in the protected areas (Fig. 16.3b). In contrast, the valley bottoms along rivers flowing from the Yellowstone Plateau

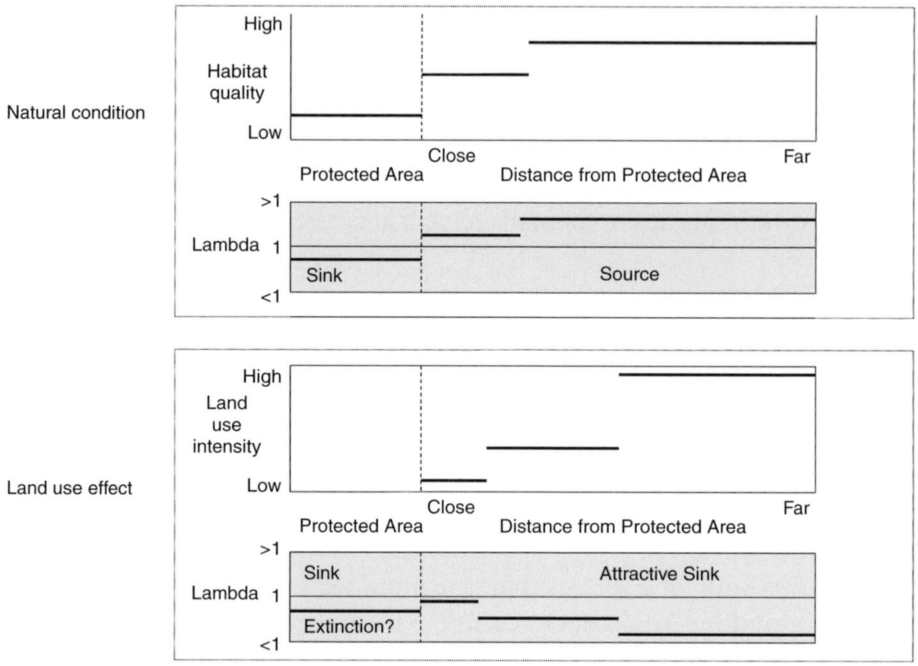

FIGURE 16.2.
Model depicting subpopulations in a protected area being a sink maintained by an external source under natural conditions (top). This results from the protected area being placed at the harsher end of the habitat quality gradient across the landscape (in the absence of human impacts). Human presence in the landscape leads to conversion of the unprotected source to a sink, reducing the viability of the subpopulation in the protected area (bottom).

have substantially longer growing seasons, more fertile soils, and relatively higher primary productivity (Hansen *et al.* 2000).

Hansen *et al.* (1999) found that bird species were not distributed randomly relative to these abiotic gradients. Rather, they were associated with landscape settings that had warmer June temperatures, were lower in elevation, were situated on alluvial parent materials, and/or had higher primary productivity. These sites were dominated by the deciduous forest cover types and were relatively rare in the study area. "Hot spots," where bird species were numerous and abundant, covered less than 3% of the study area and were mostly on or near private lands (Fig. 16.3c). Only about 7% of the hot spots were within protected areas. Within the private lands, land use was more intense near hot spots (Hansen *et al.* 2002). Rural homes were placed closer to biodiversity hot spots than would be expected from a random distribution.

The primary means by which rural home development influenced birds in hot spots was by favoring avian brood parasites and predators (Hansen and Rotella 2002). These were especially abundant near rural residences. Consequently, those bird species susceptible to nest parasitism and predation suffered very

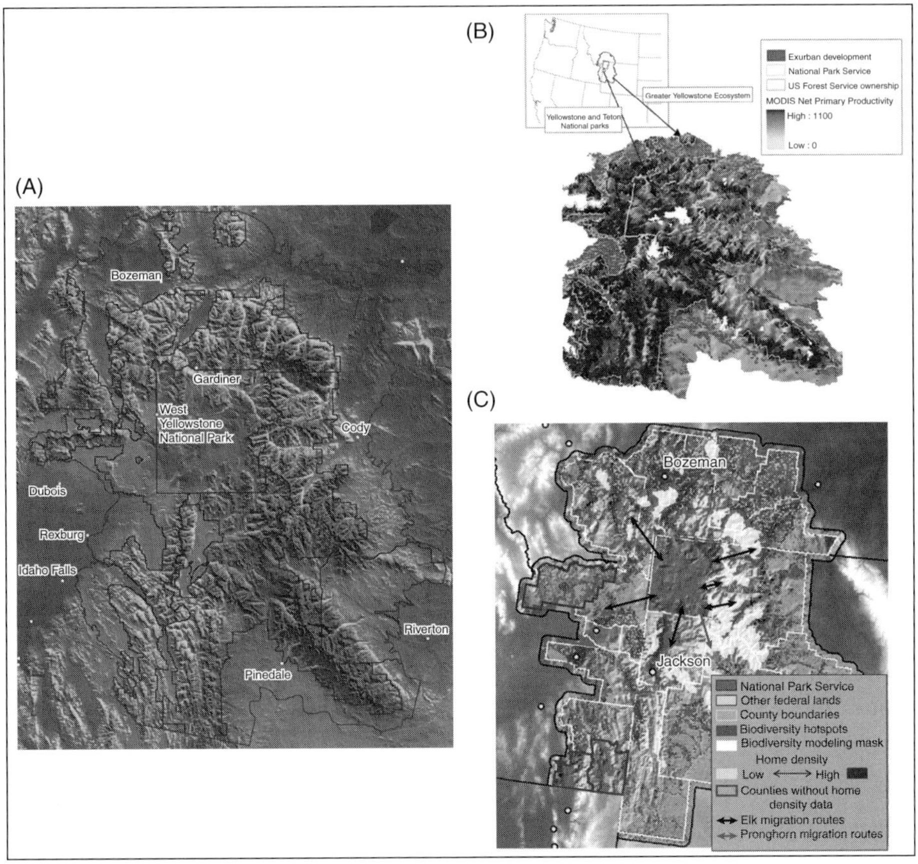

FIGURE 16.3.
(a) Shaded relief map of the Greater Yellowstone Ecosystem (GYE) in the northwestern USA. Black lines enclose public lands, with Yellowstone National Park (YNP) in the center. (b) Distribution of primary productivity across GYE as derived from the MODIS satellite product. (c) Locations within GYE of native ungulate migration routes (arrows), high bird species richness, and rural residential home development. Color version available online at: www.cambridge.org/9780521199476.

low rates of reproduction. In the case of the yellow warbler (*Dendroica petechia*), for example, nearly half of the nests in hot spots near homes suffered cowbird parasitism, and only about 20% of the nests successfully fledged young. The American robin (*Turdus migratorius*), in contrast, is better able to defend its nests against cowbirds and predators. Nest success in hot spots near homes for this species was nearly 50% higher than that for the yellow warbler. An important measure of the consequences of these differences in reproduction is the net population growth. Estimated population growth of the American robin was positive in these hot spots. The yellow warbler population, in contrast, was projected to be declining in these hot spots, likely due to the increased nest parasitism and predation associated with rural homes.

TABLE 16.1. *Simulated net population growth of yellow warblers across the Greater Yellowstone Ecosystem across land allocations with and without the effect of land use. Modified from Hansen and Rotella (2002).*

Land type	Area (ha)	Current population size	Net population change without homes	Net population change with homes
Private	808	2,942	309	−85
National Forest	4,251	2,003	41	6
National Park	984	804	−28	−35

Projections of the population dynamics of yellow warblers across the northwest portion of the Greater Yellowstone Ecosystem revealed that rural home development near hot spots in the lowlands has likely increased the probability of extinction of these species even in protected areas such as Yellowstone National Park (YNP) (Hansen and Rotella 2002). Due to the harsh climate at higher elevations, yellow warblers in the hot spots in YNP have relatively low reproduction, and these habitats are probably population sinks (Table 16.1). In pre-settlement times, populations in these high-elevation sinks were likely maintained by immigration from the low-elevation hot spots that were population source areas. Expansion of rural home development and other intense land uses in the lowlands has likely converted biodiversity hot spots that were source areas for species like yellow warbler to population sinks, thereby reducing the viability of subpopulations in the nature reserves. When the viability of yellow warbler populations in the nature reserves in the study area were modeled assuming no immigrants from outside the reserves, the probability of extinction was projected to be high. Nearly half of the simulated populations went extinct within 50 years. This example illustrates that changes in biotic interactions in small portions of the private lands can theoretically lead to an increased risk of extinction tens of kilometers away in nature reserves.

Other examples

Sinclair (1995) suggested that lion (*Panthera leo*) subpopulations in woodland habitats outside of Serengeti National Park in East Africa are sources for subpopulations occupying plains habitats within the park. Both subpopulations are regulated by resident prey. Resident prey are relatively stable in the woodland habitats, allowing this habitat to be a net exporter of lions. The plains habitats are subject to the vagaries of weather and migratory prey movements, and lions cannot maintain themselves on the low resident prey numbers. Hence, they are sink habitats. The metapopulation may not be viable, however, because hunting and poaching in the source habitats has led to increased lion mortality. Careful management of mortality in the source areas is probably needed to maintain the sink population in the protected area.

While the two examples cited indicate that protected areas may be sinks dependent upon surrounding sources, it is unclear how commonly this scenario occurs. These examples may stimulate additional research in protected areas that have the characteristics that could lead to this scenario.

Protected area as source at risk from an attractive sink

The first scenario represents a conservation concern because the protected area is a sink for the species, and persistence in the protected area is dependent upon the fate of the source areas in the unprotected portion of the landscape. In the second scenario described here, in contrast, the protected area represents a source habitat. Under the assumption of density-dependent habitat selection in the source, the subpopulation in the protected area is predicted to remain viable regardless of the fate of the subpopulation in the adjacent unprotected lands. With the typical pattern of habitat quality increasing outside the protected area due to more favorable biophysical conditions (Fig. 16.4), these surrounding areas are also expected to be sources. Recent conceptual developments identifying "attractive sinks," however, have led to concern that human impacts on the unprotected lands may put at risk the subpopulation in the source habitats in the protected area.

Gundersen *et al.* (2001) demonstrated experimentally that high rates of mortality in a dispersal habitat patch (sink) could lead to reduced population growth in a source habitat. Delibes *et al.* (2001a) used a simulation modeling approach to show that high enough rates of mortality in sink habitats can lead to extinction of subpopulations in source areas, depending upon several factors including demography in sources and sinks, rates of dispersal, proportion of habitat that is a sink, etc. Delibes *et al.* (2001b) provide an explanation for these results that deals with the ability of organisms to perceive habitat quality. Most source–sink models assume that animals dispersing actively are able to recognize and, if possible, avoid substandard (sink) habitats. Thus, they should preferentially select the source, and only the poorer competitors would occupy sinks. However, if individuals fail to select habitat properly the metapopulation may be at risk (as demonstrated by Robinson and Hoover, Chapter 20, this volume). The subpopulation in the source may become too small to remain viable if too many individuals disperse to a sink where demographic performance is poor.

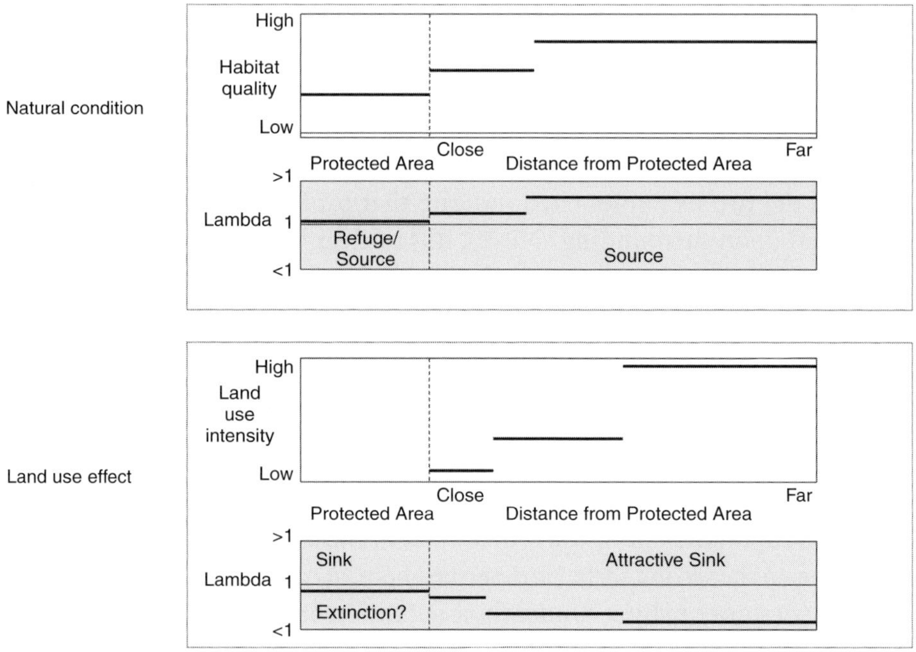

FIGURE 16.4.
Model depicting subpopulations in a protected area being sources that are vulnerable to loss to attractive sinks. Under natural conditions, habitat quality is sufficiently high over the landscape such that the entire population is a source (top). Human-induced mortality reduces reproduction in unprotected parts of the landscape, creating an attractive sink, and loss of individuals to this sink puts at risk the source in the protected area (bottom).

Attractive sinks are thought to typically arise in association with human activity:

> the actual anthropogenically caused poor habitat conditions may be difficult to detect by individual animals, as the causes of mortality or reduced breeding would be different from those in their evolutionary history. Hence maladaptive behaviors may be expected because most of these attractive sinks probably will be perceived by animals as sources. (Delibes *et al.* 2001a: 278)

These human activities include hunting and poaching, which increase mortality rates, and the application of pesticides, which can reduce reproductive rates (Delibes *et al.* 2001b).

Example: Yellowstone grizzly bears

When the grizzly bear (*Ursus arctos*) was declared an endangered species in the USA in 1973, the population in the contiguous 48 states was

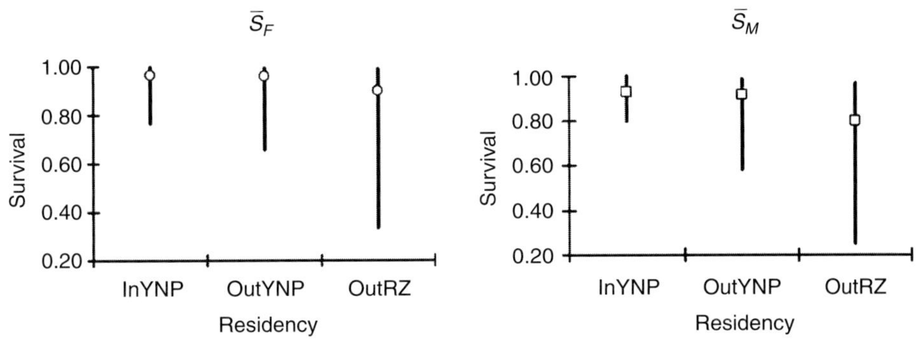

FIGURE 16.5.
Effect of location on estimates and 95% confidence intervals of annual survival (*S*) for female (F) and male (M) study sample grizzly bears in the Greater Yellowstone Ecosystem, 1983–2001. Locations were inside Yellowstone National Park (InYNP), inside the grizzly bear recovery zone (RZ) but outside YNP (OutYNP), and outside the RZ (OutRZ) (from Schwartz *et al.* 2007).

approximately 250 individuals, with most located in Yellowstone National Park (YNP) (Craighead 1979). Recovery efforts led to a population expansion to the approximately 400–600 individuals alive today, and a range expansion to the public and private lands surrounding YNP (Schwartz *et al.* 2006).

Extensive demographic analysis has been conducted on this population by Schwartz *et al.* (2007) for the period 1983–2002. Based on reproduction levels, they suggest that habitat quality is higher outside of YNP, due to higher primary productivity and the factors that drive it. They found that the reproductive rate across the population was 0.32 cubs per female per year, an adequate level to support population growth. This reproductive rate was found to correlate with population density and measures of habitat quality (winter severity and whitebark [*Pinus albicaulis*] pinecone production). With regard to mortality, they found that 85% of mortality was caused by humans. Adult survival rates varied across land allocations. Survival was highest inside YNP, slightly lower on the federal lands within the grizzly bear recovery zone surrounding YNP (which has been managed to minimize bear mortality), and lowest on public and private lands surrounding the recovery zone (Fig. 16.5). The primary correlates with survival rates were the density of developments, roads, and homes found within a bear's home range, plus the amount of time a bear spent in areas open to ungulate hunting during autumn (Schwartz *et al.* 2010). Using these demographic estimates to model population growth across the Greater Yellowstone Ecosystem, Schwartz *et al.* (2010) found that YNP and the surrounding federal lands were population source areas. However, the surrounding private lands were population sinks (Fig. 16.6). The strength and size of the source areas are currently sufficient to more than compensate for the sinks, and the population has grown over the study period.

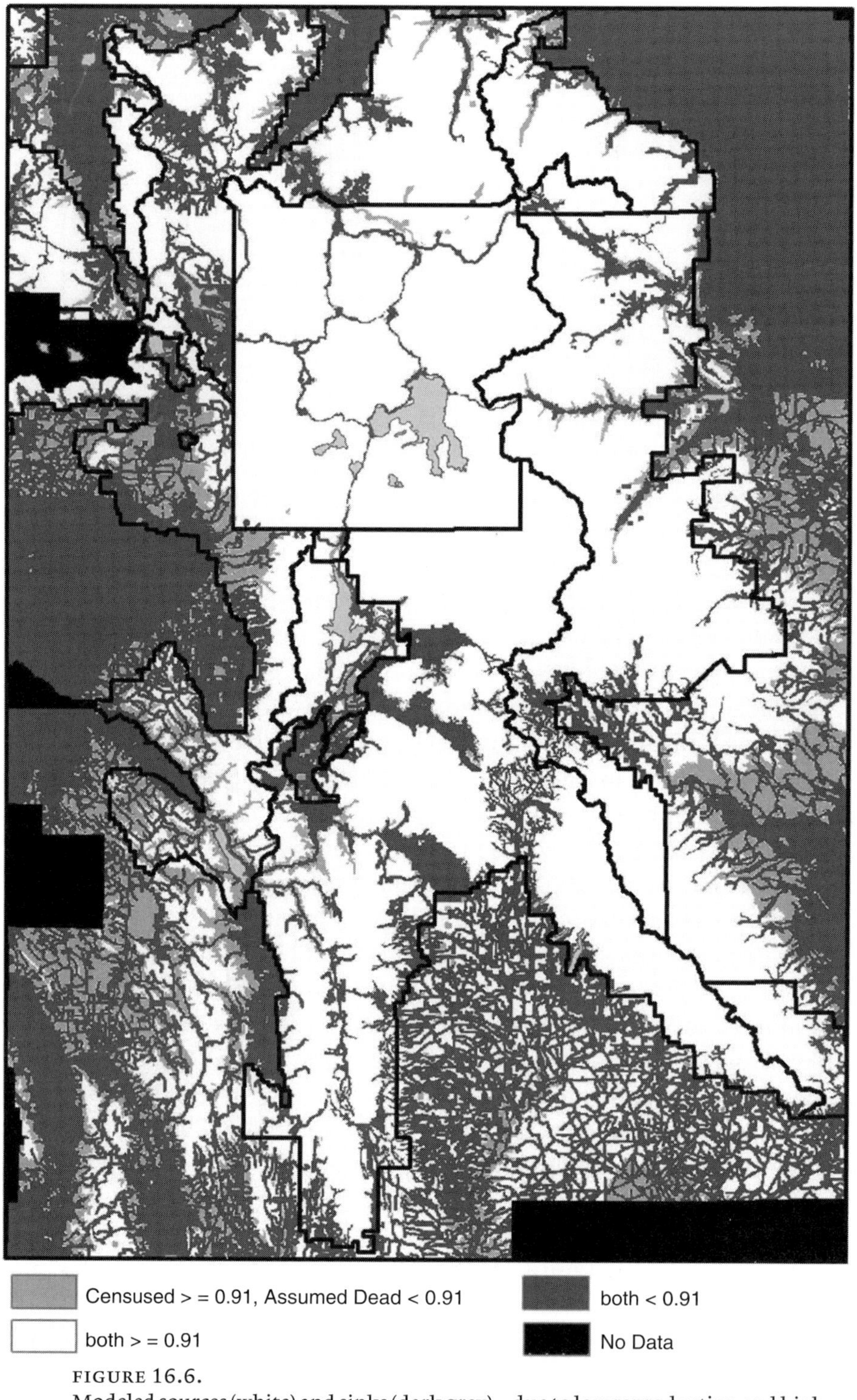

FIGURE 16.6.
Modeled sources (white) and sinks (dark grey) – due to low reproduction and high mortality; and light grey – due to low birth rates of grizzly bears in the Greater Yellowstone Ecosystem based on a female survival cutoff of 0.9 (from Schwartz *et al.* 2010). Color version available online at: www.cambridge.org/9780521199476.

A key question is whether future human development outside YNP could lead to mortality rates that result in a loss of viability of the population in YNP. Simulation models indicate an increased chance of population decline of the entire population if average female survival drops below 0.91 (Schwartz *et al.* 2007). Current estimates of female survival (0.95) exceed this level. However, the human population size, rural home density, and backcountry recreation are growing rapidly in the GYE (Greater Yellowstone Ecosystem; Gude *et al.* 2007). Regulatory agencies have concluded that future risks are unlikely to be sufficient to cause population extinction and the species was delisted in the USA in April 2007. The potential for loss of habitat quality under climate change (e.g., loss of the key food species, whitebark pine) and increasing human-caused mortality associated with human population expansion has led conservation organizations to challenge this delisting of the species.

In summary, these studies suggest that the private lands surrounding YNP represent an attractive sink for Yellowstone grizzly bears. This sink does not currently appear to put the metapopulation at risk. However, careful monitoring of human development and bear demography across the system is clearly warranted in order to anticipate and prevent loss of viability of the bear source areas in Yellowstone National Park. As part of the delisting process, the Conservation Strategy restricts all forms of development within Yellowstone National Park and the recovery zone to 1998 levels, and requires state and federal agencies to monitor grizzly bear demographics, and document changes in human development to ensure continued viability of the bear source areas within Yellowstone National Park and the Recovery Zone.

Other examples

Brown bears (*Ursus arctos*) in the Cantabrian Mountains of Spain are an endangered relict from a distribution that once covered the entire Iberian Peninsula. Human-induced mortality is thought to be the primary driver of the range constriction. Naves *et al.* (2003) introduced the concept of population refuges to describe weak source areas for brown bears, where reproduction is low due to poor habitat and mortality is low due to little human influence. They found that the brown bear population in the eastern part of the study area mainly occupies areas of suboptimal natural habitat and relatively low human impact. Bear demographic data suggest that 38% of this area may function as a source and 41% as a refuge due to low habitat quality but also low human influence. The western population was located mainly in areas with high human impact but otherwise good natural quality. Some 41% was considered to be attractive sink and 16% as sink. Both reproduction and mortality rates were higher in the western population and it experienced a mean annual decrease of 4–5% during 1982–1995.

Although this example does not involve protected areas, it does illustrate how habitat quality and human activity are often correlated, with humans being concentrated in the highest-quality habitats, and that this can have large effects on population viability. The authors concluded that the main management goal in the western population should be the reduction of human-induced mortality and, in the eastern population, extensive reforestation to increase habitat quality.

The pattern of elevated mortality for large carnivores outside of protected areas was found for several mammals in sub-Saharan Africa (Woodroffe and Ginsberg 1998), the lynx (*Lynx pardinus*) in Spain (Gaona *et al.* 1998), grizzly bears in the US and Canadian Rockies (Carroll *et al.* 2004; Neilsen *et al.* 2006), and wolves (*Canis lupus*) in the midwestern USA (Haight *et al.* 1998). Lynx in the Iberian Peninsula show a similar demography across habitat quality and land use gradients to the brown bears studied by Naves *et al.* (2003). Gaona *et al.* (1998) concluded that Donana National Park was a likely population source and surrounding areas a likely sink due to human-induced mortality. The estimated probability of extinction of the population within 100 years is 22–34%, depending on the assumptions made. In the Rocky Mountains of the USA and Canada, Carroll *et al.* (2004) simulated spatially explicit demography of grizzly bears and concluded that increasing the size and connectivity of protected areas increased population persistence.

The examples above suggest that it is not uncommon for some wildlife species to migrate to what appear to be high-quality habitats outside of protected areas and then suffer high levels of human-induced mortality. Species with particular life-history strategies are probably the most susceptible to this situation. Specifically, this means those animals that are more likely to come into contact with humans, such as species with large home ranges that extend outside protected areas (Woodroffe and Ginsberg 1998) or large predators that are dangerous to livestock and thus are less likely to be tolerated by people living in regions surrounding protected areas. Other examples include species that are likely to have a strong demographic response to human-induced mortality, such as those that are long-lived, have high survival rates, and exhibit population growth primarily set by female mortality (Naves *et al.* 2003).

I am unaware, however, of any documented cases of mortality in attractive sinks outside of protected areas being sufficient to cause the extinction of a source population in the protected area. The high rates of extinction of large carnivores in many African protected areas (Woodroffe and Ginsberg 1998) are consistent with this hypothesis. It is to be hoped that development of the concept of attractive sinks will stimulate additional research on potentially vulnerable species in protected areas.

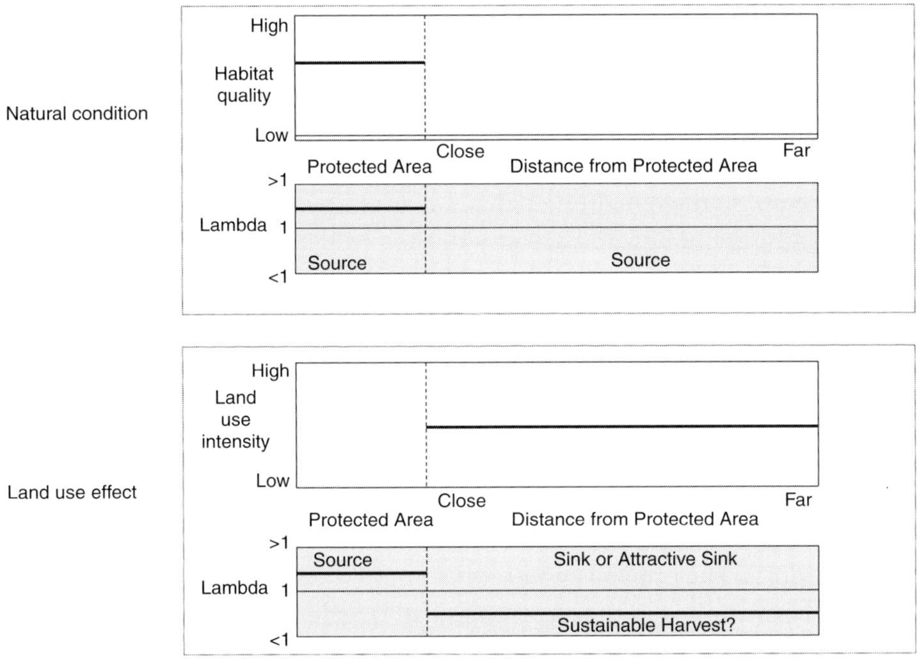

FIGURE 16.7.
Model depicting the protected area as a source that contributes emigrants to surrounding areas where sustainable harvesting creates an attractive sink.

Protected area as source allowing sustainable harvest in surrounding areas

The third scenario is a case where population sources in protected areas are used to support human harvesting in the surroundings. Under this model, habitat quality is sufficiently high in the protected area for it to be a population source area (Fig. 16.7). No assumptions need be made about habitat quality outside the protected area. Individuals dispersing outside of the protected area are harvested, creating a sink. However, the level of harvest is controlled to maintain the protected area as a source. This approach has been advocated to support hunting systems surrounding terrestrial protected areas and for the creation of marine protected areas to allow a more sustainable fishery than traditional methods. Detailed examples of the results of this approach are few and many questions remain.

Terrestrial applications

The concept that the productivity of game in hunted areas can be supplemented by movement from source areas has received widespread attention only in recent decades.

McCullough (1996) examined harvest theory in the context of spatially structured populations. He described a harvest system where the landscape is divided into patches with a minimum patch size large enough to contain a viable subpopulation of the target species. A subset of these patches is withdrawn from harvest in order to ensure a viable population and to act as source areas. Harvest on the remaining patches may remove all or some of the population within them. Future harvest will then be dependent upon either dispersal from the protected patches or additional production within the hunted patches. The proportion of the landscape open to harvest may be increased up to the point where total harvest is maximized, which will be the maximum sustainable yield. By strictly enforcing no hunting in an adequate proportion of source areas, this system is protected from the accelerating overharvesting that is characteristic of quota harvest systems (McCullough 1996). A similar approach was presented by Joshi and Gadgil (1991).

Novaro *et al.* (2000) examined the application of spatial control of harvest in the Neotropics. The previous non-spatial approaches for sustainable hunting that had been applied in the Neotropics had largely ignored the potential for dispersal into hunting areas and assumed that animal production was derived from within the hunted patch. Novaro *et al.* (2000) suggested that dispersal into hunted areas from surrounding areas can be significant. They reviewed the literature on subsistence hunting in the Neotropics and found that hunting is often conducted in areas adjacent to relatively undisturbed habitat that may act as sources of animals for the hunted sites. They also identified species that may show such dispersal dynamics. Many of the potential sources for the game populations they identified occur in protected areas. They conclude that recognition of protected areas as vital sources of game is important in that it would help change the negative attitude which many local people hold about these areas.

Strong empirical tests of the role of protected areas in supporting source–sink hunting systems have apparently not yet been done in terrestrial systems. Novaro *et al.* (2005) used empirical data and simulation models of culpeo foxes (*Pseudalopex culpaeus*) in a study area in Argentina and concluded that they exhibit source–sink dynamics between cattle ranches (no hunting) and sheep ranches (hunted). Circumstantial evidence was provided by Naranjo and Bodmer (2007), indicating that the Lacandon Forest Reserve in Mexico acts as a source for ungulates hunted on surrounding lands. Source–sink theory was used as the basis for a management plan for sustainable harvest of puma (*Puma concolor*) in the western USA based on protecting animals in source areas. Laundré and Clark (2003) concluded that closing 63% of puma habitat to hunting would ensure long-term puma population viability while permitting traditional hunting levels in other areas.

Marine applications

Source–sink theory has also stimulated considerable thinking on the benefits of marine protected areas (MPA) for sustainable fisheries. Crowder *et al.* (2000) explained that the use of MPA is attractive because they can provide insurance against management uncertainty and can simplify the management of fisheries. The latter is true because an area completely closed to fishing is easier to police than one in which multiple gear regulations must be enforced. Like terrestrial protected areas, emigration of fish from sources to attractive sinks is expected to be driven by density-dependent habitat selection. An additional mechanism of dispersal in marine systems is water-borne larval export. Eggs and larvae produced in larger numbers by the greater spawning stock biomass within reserves are transported to habitat outside of the protected area by water currents, leading to an increased recruitment in the fished areas. However, these mechanisms lead to substantial challenges in developing source–sink fisheries. Defining habitat quality and positioning MPA in population source areas are difficult in oceanic ecosystems (Crowder *et al.* 2000). The extreme complexity and unpredictability of marine currents make it very difficult to backtrack dispersed larvae to their physical sources. Thus, designing the spatial distribution of MPA and harvest areas is very challenging.

The most widely cited study demonstrating a sustained-yield fishery using MPA is that of Russ *et al.* (2004). During the period 1983–2001 they monitored changes in biomass of surgeonfish (Acanthuridae) and jacks (Carangidae) in a no-take reserve occupying 10% of the coral-reef fishing area at Apo Island, Philippines, and at a site open to fishing. Underwater visual censuses were used to assess fish densities inside and outside the MPA. They found that fish biomass in the MPA increased threefold during the study period; fish biomass did not change significantly across the entire harvest areas, although density did increase in proximity to the MPA; and fishery catch in the harvest zone was higher during 1985–2001 than before the MPA was established. The authors concluded:

> The benefits of the reserve to local fisheries at the island were higher catch, increased catch rate, and a reduction in fishing effort. The fishery and tourism benefits generated by the reserve have enhanced the living standard of the fishing community. (Russ *et al.* 2004: 597)

In reviewing the state of MPA as a basis for sustainable fisheries, Sale *et al.* (2005) concluded that the approach is widely advocated. However, rigorous empirical tests of the effect of MPA designation on fish density within the reserves are few, and well-studied effects outside of reserve borders are even rarer. The current network of MPA is small and often not placed in population source areas due to socio-economic factors and lack of knowledge. Gaps in knowledge include:

- distance and direction of dispersal, especially of marine larvae, but also juvenile and adult phases, which are needed to determine MPA size and location;
- water movements near coastlines, and their effects on dispersal, and on siting of MPA;
- spatial patterns of habitat quality and population performance;
- the effects of release from fishing on ecosystem processes such as trophic cascades, which may inhibit recovery of target species density in MPA;
- well-studied cases of MPA establishment leading to a sustained-yield fishery.

Acknowledging and remedying these gaps in knowledge should lay the foundations for a more effective application of no-take fishery reserves.

In summary, the use of terrestrial and marine protected areas as sources for sustainable harvest in the surrounding lands is supported by theory and has widespread appeal. Successful application of the approach will require greater knowledge of spatial patterns of habitat quality, target species demography and movement, ecosystem processes, and landscape design in order to achieve sustainable harvests. This use of protected areas provides additional justification for the creation and maintenance of protected areas.

Conclusions and management implications

During the two decades since the publication of Pulliam (1988), source–sink theory has strongly influenced theory and practice involving conservation in and around protected areas. The available studies on this topic largely depict three categories of applications: protected areas as sinks dependent upon outside sources; protected areas as sources vulnerable to extinction due to human activity, creating attractive sinks in the surroundings; and protected areas as sources supporting sustainable harvest in surrounding areas. Current knowledge on these relationships points to various management strategies to increase species viability within protected areas and to increase their contribution to sustainable harvest in the surrounding areas.

Define greater ecosystems. The connectivity of subpopulations within protected areas with those outside is one example of the ecological processes and flows that often link protected areas to some larger surrounding ecosystem (Hansen and DeFries 2007). Defining such "greater ecosystems" (Grumbine 1990) is essential for developing a management approach that places the protected area in the context of the surrounding area that influences its functioning. Criteria for defining such greater ecosystems are presented in DeFries *et al.* (2010) and Hansen *et al.* (2011). Among these criteria are population source and sink areas.

Create or expand protected areas. Existing protected areas were often established without the benefit of knowledge of source–sink relationships or other spatially explicit processes. Identifying and maintaining source areas that are outside protected area boundaries may be necessary in order to ensure population viability within protected areas. Innovative approaches for functionally expanding the sizes of protected areas, such as the use of conservation easements, are being widely employed in many places globally (Theobald *et al.* 2005). Newly created reserves can be placed and designed based on source–sink and other current theory to better maintain the natural ecosystem and possibly support sustainable harvest in surrounding landscapes or seascapes (Margules and Pressey 2000).

Modify human behavior in attractive sinks. In many cases, the negative effects of human activities on the demographics of at-risk species are unintended and/or avoidable. Examples of unintended impacts include road kill, death or displacement of wildlife by pets, and spread of disease by livestock or by feeding stations. Intended but avoidable impacts include the killing of dangerous animals to protect human lives or property. Both unintended and avoidable impacts can often be greatly reduced through the education of local residents. Effective strategies are often low in cost and effort, and may be widely embraced and supported by local citizens who appreciate the benefits of living more sustainably (Rosenzweig 2003; Hansen *et al.* 2005).

Monitor population demography. The functioning of source–sink systems may be very sensitive to the proportions of source and sink habitats, population vital rates, lags in population response, and many other factors (Doak 1995; Delibes *et al.* 2001b; Gundersen *et al.* 2001). Incomplete demographic information may be misleading on the status of populations in sources and sinks. For example, animal censuses in places that function as attractive sinks could lead to misinterpretation, as the apparently stable or growing sink population could be misidentified as a source (Delibes *et al.* 2001a). Furthermore, harvesting or other human-induced mortality in the sink could be temporarily sustained by the sources, before resulting in a rapid collapse of the whole system. Careful demographic monitoring is essential to generate knowledge of management strategies that will maintain the function of source–sink systems. Progress is being made on diagnostic tools in this regard. Jonzen *et al.* (2005) offer a statistical method for estimating the monitoring efforts in source or sink habitats needed to detect declining reproductive success in source habitats.

This review leads to the conclusion that much research, monitoring, and creative management is still required in order to sustain source–sink populations in and around protected areas. However, the source–sink concept has thus far greatly enriched conservation biology and has considerable promise for future contributions.

References

Arcese, P. and A. R. E. Sinclair (1997). The role of protected areas as ecological baselines. *Journal of Wildlife Management* **61**: 587–602.

Carroll, C., R. F. Noss, P. C. Pacquet and N. H. Schumaker (2004). Extinction debt of protected areas in developing landscapes. *Conservation Biology* **18**: 1110–1120.

Craighead, F. (1979). *Track of the Grizzly*. Sierra Club Books, San Francisco, CA.

Crowder, L. B., S. J. Lyman, W. F. Figueira and J. Priddy (2000). Source–sink population dynamics and the problem of siting marine reserves. *Bulletin of Marine Science* **66**: 799–820.

DeFries, R., F. Rovero, P. Wright, J. Ahumada, S. Andelman, K. Brandon, J. Dempewolf, A. Hansen, J. Hewson and J. Liu (2010). From plot to landscape scale: linking tropical biodiversity measurements across spatial scales. *Frontiers in Ecology and the Environment* **8**: 153–160.

Delibes, M., P. Gaona and P. Ferreras (2001a). Effects of an attractive sink leading into maladaptive habitat selection. *American Naturalist* **3**: 277–285.

Delibes, M., P. Ferreras and P. Gaona (2001b). Attractive sinks, or how individual behavioural decisions determine source–sink dynamics. *Ecology Letters* **4**: 401–403.

Dias, P. C. (1996). Sources and sinks in population biology. *Trends in Ecology and Evolution* **11**: 326–330.

Doak, D. F. (1995). Source–sink models and the problem of habitat degradation: general models and applications to the Yellowstone grizzly. *Conservation Biology* **9**: 1370–1379.

Gaona, P., P. Ferreras and M. Delibes (1998). Dynamics and viability of a metapopulation of the endangered *Iberian lynx* (*Lynx pardinus*). *Ecological Monographs* **68**: 349–370.

Grumbine, E. (1990). Protecting biological diversity through the greater ecosystem concept. *Natural Areas Journal* **10**: 114–120.

Gude, P. H., A. J. Hansen and D. A. Jones (2007). Biodiversity consequences of alternative future land use scenarios in Greater Yellowstone. *Ecological Applications* **17**: 1004–1018.

Gundersen, G., E. Johannesen, H. P. Andreassen and R. A. Ims (2001). Source–sink dynamics: how sinks affect demography of sources. *Ecology Letters* **4**: 14–21.

Haight, R. G., D. J. Mladenoff and A. P. Wudenven (1998). Modeling disjunct gray wolf populations in semi-wild landscapes. *Conservation Biology* **12**: 879–888.

Hansen, A. J. and R. DeFries (2007). Ecological mechanisms linking protected areas to surrounding lands. *Ecological Applications* **17**: 974–988.

Hansen, A. J. and J. J. Rotella (2002). Biophysical factors, land use, and species viability in and around nature reserves. *Conservation Biology* **16**: 1–12.

Hansen, A. J., J. J. Rotella and M. L. Kraska (1999). Dynamic habitat and population analysis: a filtering approach to resolve the biodiversity manager's dilemma. *Ecological Applications* **9**: 1459–1476.

Hansen, A. J., J. J. Rotella, M. L. Kraska and D. Brown (2000). Spatial patterns of primary productivity in the Greater Yellowstone ecosystem. *Landscape Ecology* **15**: 505–522.

Hansen, A. J., R. Raske, B. Maxwell, J. J. Rotella, A. Wright, U. Langner, W. Cohen, R. Lawrence and J. Johnson (2002). Ecology and socioeconomics in the New West: a case study from Greater Yellowstone. *BioScience* **52**: 151–168.

Hansen, A. J., R. Knight, J. Marzluff, S. Powell, K. Brown, P. Hernandez and K. Jones (2005). Effects of exurban development on biodiversity: patterns, mechanisms, research needs. *Ecological Applications* **15**: 1893–1905.

Hansen, A. J., C. Davis, N. B. Piekielek, J. Gross, D. M. Theobald, S. Goetz, F. Melton, R. DeFries (2011). Delineating the ecosystems containing protected areas for monitoring and management, Bio Science. In press.

Jonzen, N., J. R. Rhodes and H. P. Possingham (2005). Trend detection in source–sink systems: when should sink habitats be monitored? *Ecological Applications* **15**: 326–334.
Joshi, N. V. and M. Gadgil (1991). On the role of refugia in promoting prudent use of biological resources. *Theoretical Population Biology* **40**: 211–229.
Laundré, J. and T. W. Clark (2003). Managing puma hunting in the western United States: through a metapopulation approach. *Animal Conservation* **6**: 159–170.
Lubchenco, J., S. R. Palumbi, S. D. Gaines and S. Andelman (2003). Plugging a hole in the ocean: the emerging science of marine reserves. *Ecological Applications* **13**: S3–S7.
Margules, C. R. and R. L. Pressey (2000). Systematic conservation planning. *Nature* **405**: 243–253.
McCullough, D. R. (1996). Spatially structured populations and harvest theory. *Journal of Wildlife Management* **60**: 1–9.
McKinney, M. L. (2002). Urbanization, biodiversity, and conservation. *BioScience* **52**: 883–890.
Naranjo, E. J. and R. E. Bodmer (2007). Source–sink systems and conservation of hunted ungulates in the Lacandon Forest, Mexico. *Biological Conservation* **138**: 412–420.
Naves, J., T. Wiegand, E. Revilla and M. Delibes (2003). Endangered species constrained by natural and human factors: the case of brown bears in northern Spain. *Conservation Biology* **17**: 1276–1289.
Newmark, W. D. (1985). Legal boundaries of western North American National Parks: a problem of congruence. *Biological Conservation* **33**: 197–208.
Newmark, W. D. (1987). A land-bridge island perspective on mammalian extinctions in western North American parks. *Nature* **325**: 430–432.
Newmark, W. D. (1995). Extinction of mammal populations in western North American national parks. *Conservation Biology* **9**: 512–526.
Newmark, W. D. (1996). Insularization of Tanzanian parks and the local extinction of large mammals. *Conservation Biology* **10**: 1549–1556.
Nielsen, S. E., G. B. Stenhouse and M. S. Boyce (2006). A habitat-based framework for grizzly bear conservation in Alberta. *Biological Conservation* **130**: 217–229.
Novaro, A. J., K. H. Redford and R. E. Bodmer (2000). Effect of hunting in source–sink systems in the Neotropics. *Conservation Biology* **14**: 713–721.
Novaro, A. J., M. C. Funes and R. S. Walker (2005). An empirical test of source–sink dynamics induced by hunting. *Journal of Applied Ecology* **42**: 910–920.
Parks, S. A. and A. H. Harcourt (2002). Reserve size, local human density, and mammalian extinctions in the US protected areas. *Conservation Biology* **16**: 800–808.
Possingham, H. P., K. A. Wilson, S. J. Aldeman and C. H. Vynne (2006). Protected areas: goals, limitations, and design. In *Principles of Conservation Biology* (M. J. Groom, G. K. Meffe and C. R. Carroll, eds.). Sinauer Associates, Sunderland, MA: 509–551.
Pressey, R. L. (1994). Ad hoc reservations: forward or backward steps in developing representative reserve systems? *Conservation Biology* **8**: 662–668.
Pulliam, H. R. (1988). Sources, sinks, and population regulation. *American Naturalist* **132**: 652–661.
Rivard, D. H., J. Poitevin, D. Plasse, M. Carleton and D. J. Currie (2000). Changing species richness and composition in Canadian National Parks. *Conservation Biology* **14**: 1099–1109.
Rosenzweig, M. L. (2003). *Win–Win Ecology: How the Earth's Species can Survive in the Midst of Human Enterprise*. Oxford University Press, New York.
Russ, G. R., A. C. Alcala, A. P. Maypa, H. P. Calumpong and A. T White (2004). Marine reserve benefits local fisheries. *Ecological Applications* **14**: 597–606.
Sale, P. F., R. K. Cowen, B. S. Danilowicz, G. P. Jones, J. P. Kritzer, K. Lindeman, S. Planes, N. V. C. Polunin, G. R. Russ, Y. J. Sadovy and R. S. Steneck (2005). Critical science gaps impede use of no-take fishery reserves. *Trends in Ecology and Evolution* **20**: 74–80.
Schwartz, C. C., M. A. Haroldson, K. A. Gunther and D. Moody (2006). Distribution of grizzly bears in the Greater Yellowstone Ecosystem in 2004. *Ursus* **17**: 63–66.
Schwartz, C. C., M. A. Haroldson, G. C. White, R. B. Harris, S. Cherry, K. A. Keating, D. Moody and C. Servheen (2007). Temporal, spatial, and environmental influences on the demographics of grizzly bears in the Greater Yellowstone Ecosystem. *Wildlife Monographs* **161**: 1–68.

Schwartz, C. C., M. A. Haroldson and G. C. White (2010). Hazards affecting grizzly bear survival in the Greater Yellowstone Ecosystem. *Journal of Wildlife Management* **74:** 654–667.

Scott, J. M., F. W. Davis, R. G. McGhie, R. G. Wright, C. Groves and J. Estes (2001). Nature reserves: do they capture the full range of America's biological diversity? *Ecological Applications* **11:** 999–1007.

Sinclair, A. R. E. (1995). Serengeti past and present. In *Serengeti II: Dynamics, Management and Conservation of an Ecosystem* (A. R. E. Sinclair and P. Arcese, eds.). University of Chicago Press, Chicago, IL: 3–30.

Theobald, D. M., T. Spies, J. Kline, B. Maxwell, N. T. Hobbs and V. H. Dale (2005). Ecological support for rural land use planning. *Ecological Applications* **5:** 1906–1914.

Wittemyer, G., P. Elsen, W. T. Bean, A. Coleman, O. Burton and J. S. Brashares (2008). Accelerated human population growth at protected area edges. *Science* **321:** 123–126.

Woodroffe, R. and J. R. Ginsberg (1998). Edge effects and the extinction of populations inside protected areas. *Science* **280:** 2126–2128.

Wright, G. M. and B. Thompson (1935). *Fauna of the National Parks of the US* USDA Department of Interior, Washington, DC.

17

Evidence of source–sink dynamics in marine and estuarine species

Summary

We review the evidence for source–sink dynamics in marine and estuarine species ranging from algae and seagrasses to invertebrates and vertebrates. There are only a few species with strong evidence for source–sink dynamics, primarily due to the logistical difficulties inherent in demonstrating source–sink dynamics convincingly, but there is extensive circumstantial evidence for the existence of source–sink dynamics, indicating that the issue requires serious consideration and further examination. The most prevalent mechanisms underlying source–sink dynamics include variation in habitat quality (natural or anthropogenic), dispersal, predation, and fishery exploitation, as well as interactions between these factors. In efforts to conserve or restore marine and estuarine metapopulations, optimal results are most likely to be achieved by identifying the connectivity between populations and preserving source populations or interconnected networks of sources and sinks. Further investigation of source–sink dynamics is critically needed to promote the effective conservation and restoration of marine and estuarine species.

Introduction

Source–sink structure can greatly affect the dynamics of marine and estuarine metapopulations (Lipcius *et al.* 1997, 2005, 2008; Roberts 1997, 1998; Crowder *et al.* 2000), but the evidence is scattered and lacking generalizations regarding the patterns and consequences for conservation and restoration. In

Sources, Sinks and Sustainability, ed. Jianguo Liu, Vanessa Hull, Anita T. Morzillo and John A. Wiens. Published by Cambridge University Press.

this chapter, we review the evidence for source–sink dynamics in marine and estuarine species, we determine whether there are general patterns, we identify likely mechanisms underlying the patterns, and we discuss the implications for conservation and restoration strategies.

To conduct the review, we used our personal libraries and searched Science Direct, JSTOR, Cambridge Scientific Abstracts, and Google Scholar using a combination of keywords: (1) source–sink dynamics, dispersal or metapopulation, and (2) marine, estuarine or coastal. Our review may not include all examples of source–sink dynamics because many earlier studies were conducted without the conceptual framework of source–sink dynamics. For instance, Katz (1985) clearly demonstrates the existence of source–sink dynamics in intertidal barnacle populations, yet that study made no mention of source–sink dynamics. Consequently, there are likely to be many more unrecognized examples of source–sink dynamics awaiting discovery.

Criteria for determination of source–sink dynamics

Attempts to identify the criteria necessary to define source–sink dynamics range from the classic works of Pulliam (1988) and Holt (1985) to more recent efforts that integrate the role of dispersal (Figueira and Crowder 2006; Runge *et al.* 2006; White 2008; Krkosek and Lewis 2010). We have adopted the demographic criteria originally formulated by Pulliam (1988) to categorize populations as sources or sinks by their demographic rates, whereby for source populations, births > deaths and emigration > immigration, resulting in net export of individuals to the metapopulation. For sinks, births < deaths and emigration < immigration, such that sinks require subsidies from source populations for long-term persistence. We employ this established population-dynamics definition of source–sink dynamics, which emphasizes demographic rates (Pulliam 1988), rather than that whereby "sources" and "sinks" pertain to the origins and destinations, respectively, of dispersive stages (Roberts 1997, 1998; Cowen *et al.* 2000), and categorize the evidence for source–sink dynamics as either "definitive," "probable," or "feasible." Within each of these sections we discuss the evidence under subheadings defined by the likely mechanisms driving source–sink dynamics (i.e., habitat quality, predation, dispersal, and fishery exploitation). These mechanisms are not mutually exclusive and may act in concert; when other mechanisms are also active in addition to the dominant mechanism, we discuss them as well. When possible, we distinguish between sources, sinks, pseudo-sinks, black-hole sinks, and ecological traps (Robertson and Hutto 2006). Our focus is upon ecological criteria, recognizing that metapopulations may be ecologically heterogeneous while being genetically panmictic.

At the time of this review, there have been no documented cases where birth (b), death (d), emigration (e) and immigration (i) rates have been measured directly in the field for an interconnected set of source and sink populations in marine and estuarine species (Table 17.1). There are, however, field demonstrations of sink populations that go locally extinct and are subsequently re-established annually from unidentified source populations; this situation defines examples of "*definitive*" source–sink dynamics. Use of this criterion is in alignment with the demographic rate approach of Pulliam (1988) as well as with more recent efforts that emphasize dispersal (Figueira and Crowder 2006; Runge *et al.* 2006; White 2008; Krkosek and Lewis 2010). In no case has a specific source population been identified for any species. Consequently, we rely on inferential evidence to identify the likely source populations, such as through modeling of dispersal pathways from sources to sinks. Moreover, there are instances where sinks have been identified, but where the sinks may not occur over the long term. In this case, we define the sinks as "*transient*" at the present time, although future research may deem these to be permanent sinks.

The examples of "*probable*" source–sink dynamics are evident in field estimates of demographic rates, in hydrodynamic models of larval dispersal among interconnected populations, and in field experiments of survival whereby specific populations have very low survival rates. In these instances there is reasonable inferential or circumstantial evidence for source–sink dynamics, but the existence of sinks has not been demonstrated conclusively.

For examples of "*feasible*" source–sink dynamics, we used information indicating that there are interconnected populations, and that some of the populations have the potential to go locally extinct. These cases are questionable and require extensive investigation to prove or disprove the existence of sources and sinks.

Evidence of source–sink dynamics

Definitive sources and sinks

Habitat quality

The few direct field demonstrations of population sinks include a suite of marine invertebrates, from barnacles to bivalves. A metapopulation of the Peruvian bay scallop, *Argopecten purpuratus*, undergoes extinctions and recolonization events in seagrass beds off the coast of Peru and Chile caused by El Niño; apparently there is a source population in Independence Bay, Peru (Wolff and Mendo 2000).

The harpacticioid copepod, *Tigriopus brevicornis*, occupies intertidal rock pools in the North Atlantic linked by larval dispersal (Harris 1973). Populations

TABLE 17.1. *Listing of species examined for evidence of source–sink dynamics.*

Class	Species	Habitat	Dispersal mode	Methods	Evidence	Mechanisms	Sources
Phaeophyceae (brown algae, giant kelp, sea palm)	*Macrocystis pyrifera*	Marine	Passive dispersal	Modeling, Empirical	Definitive	Dispersal (passive)	Gaylord *et al.* 2006; Reed *et al.* 2006
	Postelsia palmaeformis	Marine, Intertidal	Passive dispersal	Empirical	Probable	Habitat quality	Paine 1988
Liliopsida (seagrass, eelgrass)	*Zostera marina*	Marine, Estuarine	Passive dispersal, Vegetative propagation	Modeling, Empirical	Feasible	Dispersal (passive)	Harwell and Orth 2002
				Empirical, Genetics			Williams and Orth 1998
Polychaeta (marine worms)	*Pectinaria koreni*	Marine, Benthic	Active larval and postlarval dispersal	Genetics, Empirical	Probable	Dispersal	Jolly *et al.* 2003
				Empirical, Modeling	Probable	Habitat quality, Dispersal	Thiébaut *et al.* 1998; Ellien *et al.* 2000
Oligochaeta (worms)	*Paranais litoralis*	Marine, Benthic	Budding, Emigration	Empirical	Feasible	Habitat quality	Junkins *et al.* 2006; Nilsson *et al.* 1997, 2000
Gastropoda (snails, abalone)	*Haliotis discus discus, H. madaka, H. gigantean*	Marine, Bays	Active larval dispersal	Modeling, Empirical	Feasible	Dispersal, Exploitation	Miyake *et al.* 2009
	Haliotis sorenseni, H. corrugate, H. fulgens	Marine	Active larval dispersal	Empirical	Feasible	Exploitation	Rogers-Bennett *et al.* 2002
	Nucella emarginata	Marine	Crawling larvae	Genetics, Empirical	Definitive, Transient	Dispersal	Wares *et al.* 2001

Bivalvia (scallops, clams, oysters, mussels)	*Argopecten purpuratus*	Marine, Bays	Active larval dispersal	Modeling	Definitive	Habitat quality, Exploitation, Dispersal	Wolff and Mendo 2000
	Argopecten irradians concentricus	Marine, Bays	Active larval dispersal	Empirical	Definitive	Predation	Peterson *et al.* 2001
				Genetics, Empirical		Dispersal	Marko and Barr 2007
	Dreissena polymorpha	Riverine, Estuarine	Active larval and postlarval dispersal	Empirical	Definitive, Transient	Dispersal	Stoeckel *et al.* 1997
				Modeling		Dispersal	Akçakaya and Baker 1998
				Empirical		Dispersal	Horvath *et al.* 1996
	Mytilus edulis	Marine, Intertidal	Active larval dispersal	Empirical, Modeling	Feasible	Predation, Dispersal	Katz 1985
	Mytilus californianus, M. galloprovincialis	Marine, Intertidal	Active larval dispersal	Empirical	Feasible	Dispersal	Becker *et al.* 2007; Phillips 2007
	Perna perna	Marine, Intertidal	Active larval dispersal	Empirical	Feasible	Dispersal	Erlandsson and McQuaid 2004; Porri *et al.* 2006
	Crassostrea virginica	Estuarine	Active larval dispersal	Modeling	Probable	Dispersal	Lipcius *et al.* 2008; North *et al.* 2008
Maxillopoda (barnacles, copepods)	*Balanus glandula, Chthamalus fissus*	Marine	Active larval dispersal	Genetics	Probable	Dispersal	Wares *et al.* 2001
	Semibalanus balanoides	Marine, Intertidal	Active larval dispersal	Empirical, Modeling	Definitive	Predation, Dispersal	Katz 1985
				Empirical	Probable	Habitat quality	Bertness *et al.* 1991; Leslie *et al.* 2005

TABLE 17.1. *(cont.)*

Class	Species	Habitat	Dispersal mode	Methods	Evidence	Mechanisms	Sources
	Tigriopus brevicornis	Intertidal pools	Active larval dispersal	Empirical	Definitive	Habitat quality	Johnson 2001
Malacostraca (shrimp, isopods, amphipods, lobsters, crabs)	*Neomegamphopus hiatus, Melita nitida, Bemlos unicornis*	Live in pen shells	Brooding, Dispersal as adults or large juveniles	Empirical	Probable	Habitat quality	Munguia *et al.* 2007
	Stenopus hispidus	Marine, Coral reefs	Active larval dispersal	Empirical, Modeling	Feasible	Habitat quality, Dispersal	Chockley *et al.* 2008
	Panulirus argus	Marine, Coral reefs	Active larval dispersal	Empirical, Modeling	Probable	Dispersal, Habitat quality	Lipcius *et al.* 1997, 2001
	Homarus americanus	Marine	Active larval dispersal	Modeling	Probable	Dispersal, Exploitation	Fogarty and Botsford 2006; Xue *et al.* 2008
				Modeling, Empirical	Probable	Dispersal, Habitat quality	Incze *et al.* 2010
Echinoidea (sea urchins)	*Evenchinus chloroticus*	Marine, Estuarine	Active larval dispersal	Empirical, Modeling	Probable	Habitat quality, Dispersal	Wing *et al.* 2003, 2008; Wing 2009
	Strongylocentrotus purpuratus	Marine	Active larval dispersal	Genetics, Empirical	Feasible	Dispersal	Wares *et al.* 2001
	Strongylocentrotus franciscanus	Marine	Active larval dispersal	Modeling	Feasible	Exploitation	Quinn *et al.* 1993

	Strongylocentrotus purpuratus, S. franciscanus	Marine	Active larval dispersal	Empirical	Feasible	Dispersal	Ebert *et al.* 1994
Asteroidea (sea stars)	*Pisaster ochraceus*	Marine	Active larval dispersal	Empirical	Feasible	Habitat quality	Sanford and Menge 2007
Osteichthyes (bony fishes)	*Theragra chalcogramma*	Marine	Active larval dispersal	Genetics	Feasible	Dispersal	Olsen *et al.* 2002
	Clinocottus embryum, C. globiceps, Oligocottus maculosus	Marine, Tidepools	Active larval dispersal	Modeling	Feasible	Habitat quality	Pfister 1998
	Eucyclogobius newberryi	Estuarine	Benthic	Empirical	Probable	Habitat quality	Lafferty *et al.* 1999
	Thalassoma bifasciatum	Coral reef	Active larval dispersal	Empirical	Probable	Dispersal	Swearer *et al.* 1999
	Reef fish	Marine	Active larval dispersal	Modeling	Probable	Dispersal	Armsworth 2002; Bode *et al.* 2006
	Sebastes spp.	Marine	Active larval dispersal	Empirical	Feasible	Habitat quality, Exploitation	Paddack & Estes 2000
	Oncorhynchus spp.	Estuarine	Active larval dispersal	Modeling	Feasible	Habitat quality	Hill *et al.* 2003
	Salmo salar	Marine, Anadromous	Active larval dispersal	Modeling, Genetics	Feasible	Habitat quality	Hindar *et al.* 2004
	Gadus spp.	Marine	Active larval dispersal	Empirical, Modeling	Feasible	Dispersal	Brickman *et al.* 2007

in intertidal rock pools exhibit a metapopulation structure due to their sporadic local extinction and recolonization (Johnson 2001). Copepodites and adults migrate to pools further upshore during spring tides, potentially increasing the persistence of upshore populations. Higher pools can act as refuges from predation and washout during high tides and storms, while lower pools can act as refuges from desiccation during calm, dry times. This suggests that the pools are not permanent sources or sinks, but that the ability of each pool to either sustain itself or be colonized from sources depends on environmental conditions.

Certain species of kelp, including *Macrocystis* spp., reproduce via passive dispersal of propagules and form productive kelp forests separated by sand flats (Gaylord *et al.* 2006). The flats, when expansive, inhibit vegetative or sexual reproduction between kelp forests and therefore facilitate a metapopulation structure comprised of multiple forests. Expansive patches and forests of a *Macrocystis* metapopulation along the Northeast Pacific coast had gone extinct due to biological or physical disturbances, and were recolonized through passive dispersal of propagules, indicating source–sink dynamics. When considering the kelp metapopulation, the scale of connectivity was estimated at 1 km, which would require additional dispersal through active mechanisms, possibly by fish or invertebrates that might transport propagules. Connectivity of *Macrocystis* spp. is likely strongly influenced by oceanographic conditions, with most populations connected to one to three others, such that stepping-stone exchange between populations would be the most important exchange pathway (Reed *et al.* 2006).

Predation

Within the Cape Lookout lagoonal system in North Carolina, cownose rays prey upon a bay scallop (*Argopecten irradians*) population in a productive seagrass bed, causing the population to crash annually before fall spawning (Peterson *et al.* 2001). Each year this population is replenished by larvae from surrounding populations, clearly demonstrating a source–sink metapopulation. Bay scallops are hermaphroditic broadcast spawners with a short pelagic larval phase, which allows for transport of larvae between seagrass beds in a given basin, but also tends to isolate populations between adjacent basins (Peterson and Summerson 1992; Peterson *et al.* 2001). Genetic analysis of *A. irradians concentricus* in North Carolina suggested that populations exchange fewer than four migrants per generation (Marko and Barr 2007), indicating that source–sink dynamics are operating in ecological time but not necessarily over an evolutionary time scale.

In the rocky intertidal of southern New England, the predatory snail, *Urosalpinx cinerea*, consumes the barnacle *Semibalanus balanoides* in the lower intertidal to extinction (Katz 1985). The higher intertidal zone acts as a refuge from predation due to severe physical stress on consumers, and the adults surviving there produce larvae and juveniles that recolonize the lower intertidal (Katz 1985), which is direct evidence of source–sink dynamics.

Dispersal

Nucella emarginata is a predatory snail that feeds on mussels in the rocky intertidal of the Pacific Northwest. It has crawling larvae with a correspondingly low propensity for dispersal. The species has recently expanded its range southward, which is indicative of transient source–sink dynamics (Wares *et al.* 2001). Whether the new populations will continue to depend on source–sink dynamics is unclear, as there is little evidence of further southward migration or demographic information on the newly established population. Similarly, the invasive zebra mussel, *Dreissena polymorpha*, was introduced into Lake St. Clair and then spread throughout the Great Lakes and nearby estuarine systems such as the Hudson River (Horvath *et al.* 1996; Stoeckel *et al.* 1997). Zebra mussels have a planktonic larval stage lasting 1–4 weeks, and postlarval dispersal occurs when juveniles or adults are detached from their settlement sites such as on the bottoms of recreational fishing boats. The lake populations apparently served as sources for estuarine populations (Horvath *et al.* 1996; Stoeckel *et al.* 1997; Akçakaya and Baker 1998). If, as reported by Horvath *et al.* (1996), the downstream populations could not be sustained without additional recruitment from upstream populations, then a source–sink metapopulation model would be appropriate (Akçakaya and Baker 1998).

Probable sources and sinks

Habitat quality

The amphipods *Neomegamphopus hiatus*, *Melita nitida* and *Bemlos unicornis* live in pen shells in the Gulf of Mexico (Munguia *et al.* 2007). *Neomegamphopus* and *Melita* disperse as adults, and *Bemlos* as juveniles. Pen shells are an ephemeral hard substrate that persists about one year after the death of the individual. Empty pen shells are colonized by amphipods from existing shells, whereas amphipods in deteriorating shells must colonize newly deposited pen shells. Consequently, there is a progression of pen shell quality such that recently deposited pen shells are sources for deteriorated pen shells acting as sinks, depending on the local distribution of pen shells and their amphipod inhabitants.

The sea palm, *Postelsia palmaeformis*, is an intertidal kelp that competes with plant and animal species for the limiting resource, space (Paine 1979). There is higher persistence of local populations established on bare rock than those established on either algal or faunal substrate. Populations with fewer than 30 individuals tend toward extinction, but one mature or drifting individual from a source is capable of maintaining or founding a local cluster in a sink (Paine 1988).

Spatial structure in growth and reproduction driven by food availability has been postulated to produce reproductive sources and sinks in the sea urchin *Evenchinus chloroticus* (Wing *et al.* 2003; Wing, Chapter 18, this volume). In New Zealand fjords, growth was lowest in inner-fjord sites (nominal sinks), apparently caused by poor food quality, whereas abundant algae near the fjord entrance provided sufficient nutrition for urchin growth and reproductive output at these nominal sources (Wing *et al.* 2008). Rates of larval supply and recruitment were lower at the fjord entrances than the inner fjord, likely because of the differences in the physical environments and hydrodynamics (Wing 2009). This situation represents a system generated by an interaction between habitat quality (i.e., food availability) and hydrodynamics, resulting in a source–sink structure.

In the endangered tidewater goby (*Eucyclogobius newberryi*), males dig burrows in which they care for clutches of eggs, which take about 10 days to develop. Populations occur in shallow, brackish environments that tend to be small in area. Dispersal among estuaries appears unlikely as there is no marine dispersal phase, and there is some genetic evidence that movement is extremely limited (Lafferty *et al.* 1999). Analysis of historic presence–absence data showed high rates of extinction and recolonization of local populations. In this case the core-satellite metapopulation designation may be appropriate, as populations in large wetlands are relatively stable when compared with those in small wetlands.

Dispersal

The tubicolous polychaete *Pectinaria koreni* is a dominant worm in muddy sediments in the English Channel, with a pelagic larval stage lasting 15 days, which can be transported more than 70 km from the natal habitat (Jolly *et al.* 2003). Transient (sink – Baie des Veys) and stable (source – Baie de Seine) local populations of *P. koreni* were discovered in France. An analysis of four microsatellite loci, however, did not find significant differences in gene diversity or allelic richness between the Baie des Veys and the eastern Baie de Seine populations (Jolly *et al.* 2003). Ellien *et al.* (2000) used a 2D hydrodynamic model to explore the impact of larval dispersal on populations of *P. koreni* and found that under most wind conditions, larvae are generally retained within the natal habitat. Only during extreme wind events would larvae be exchanged between bays or populations, such that source–sink dynamics would be dependent on acute environmental events.

Populations of the eastern oyster, *Crassostrea virginica*, exist as discrete oyster reefs (= populations) separated by unstructured bottom. Gametes are released directly into the water column, and fertilization success decreases at low population densities (Levitan and Petersen 1995). The larval stage is pelagic and lasts 2–3 weeks, allowing for ample exchange between reefs. Modeling studies with virtual metapopulations based on actual field distributions in Chesapeake Bay suggest that source–sink dynamics are in play both between individual oyster reefs (Lipcius *et al.* 2008) and between populations in different tributaries (North *et al.* 2008), with both patterns being driven by the features of dispersal pathways.

The Caribbean spiny lobster (*Panulirus argus*) has a pelagic larval stage that lasts weeks to months. Evidence for source–sink dynamics in *P. argus* comes from field and modeling studies of abundance and larval advection in Exuma Sound, Bahamas, where adult density and postlarval supply varied significantly across four sites (Lipcius *et al.* 1997). Adult densities were highest at the one site protected from exploitation, as compared with three exploited sites, and the exploited site with lowest adult densities received more postlarvae than all other sites. Hydrodynamic modeling demonstrated that the disparity in larval settlement was due to currents that advected larvae from all sites to the site with highest postlarval abundance and low adult abundance. Two of the sites had sufficient self replenishment to be designated as sources, whereas a third site contributed very few larvae to the metapopulation and was designated a sink. The site with the lowest adult density and high postlarval settlement from all other sites was designated a pseudo-sink.

Coral reefs support some of the most diverse communities in the world (Graham *et al.* 2006). Individual reefs support local populations that are interconnected through the pelagic larval stages. Many species have some degree of homing, such that larvae may return to natal reefs even with pelagic larvae durations longer than 30 days (Swearer *et al.* 1999). Within populations of the blue-headed wrasse, *Thalassoma bifasciatum*, local retention can be very important, supplying 15–60% of the local recruits. However, the pattern may differ on the leeward and windward shores, such that recruits on the leeward side are mostly retained from local areas and on the windward side from up-current source populations (Swearer *et al.* 1999). Hydrodynamic modeling of the Great Barrier Reef suggested that there must be a larval dispersal pathway for reef fish that allows feedback from a local population to itself (Armsworth 2002). The connectivity patterns of larvae depicted two potential subregions, with transport primarily from north to south. A few local populations, called “gateway reefs,” were capable of transporting larvae from sink to source regions (Bode *et al.* 2006), probably producing source–sink dynamics.

Fishery exploitation

Heavily exploited inshore stocks of the American lobster, *Homarus americanus*, are apparently sustained through a subsidy of about 12% from offshore stocks, which are relatively unexploited, as a proportion of the postlarvae are brought inshore by prevailing currents and directed swimming (Fogarty and Botsford 2006; Incze *et al.* 2010). Local retention of *H. americanus* in the Gulf of Maine was predicted to be 20–40%, with counter-clockwise currents transporting larvae between populations. Northwest zones receive larvae from eastern zones, and southeastern zones from the coast of Maine (Xue *et al.* 2008). An individual-based, coupled biophysical model predicted that two of the 15 populations within the Gulf of Maine are sources that produce most of the competent larvae and therefore subsidize the entire metapopulation (Incze *et al.* 2010).

Feasible sources and sinks

Habitat quality

Various populations of eelgrass, *Zostera marina*, have become locally extinct and subsequently re-established, possibly through viable seeds within floating, flowering shoots (Orth *et al.* 1984), which can disperse more than 1 km (Harwell and Orth 2002). Although there may be little gene flow between populations (Williams and Orth 1998), interpopulation connectivity can be high even at low levels of gene flow, suggesting a source–sink structure.

The naid oligochaete *Paranais litoralis* is a dominant benthic worm in muddy sediments of the northeastern USA that reproduces primarily via budding (Nilsson *et al.* 1997, 2000). Although immigration and emigration affect population sizes and growth rates of different populations, emigration is triggered by resource depletion, which is indicative of a continuously shifting mosaic of sources and sinks depending on local resource availability (Nilsson *et al.* 1997; Junkins *et al.* 2006).

On the Oregon coast, higher per individual and per unit area larval production of *Balanus glandula* at particular intertidal sites was linked to higher primary productivity nearshore (Leslie *et al.* 2005). These areas of high productivity may be sources for other sites in the area. Similarly, variation in growth and reproduction of *S. balanoides* between coastal and estuarine habitats is mostly caused by differences in primary productivity and current velocities. If survival is constant, estuarine habitats may be regional sources, exporting larvae to coastal sink habitats (Bertness *et al.* 1991).

Spatially explicit fisheries data were used in southern California to identify the most productive areas for abalone, and fishery-independent data were used to assess the impact of marine protected areas on population persistence

(Rogers-Bennett *et al.* 2002). Elevated densities of juveniles and adults with high gonad indices in shallow, food-rich habitats suggested that these areas acted as sources for nearby depleted populations in sinks.

The marine shrimp *Stenopus hispidus* lives in crevices and overhangs of coral reefs. Dispersal occurs primarily during the larval phase, which can last 17–30 weeks. In a demographic study, offshore reefs had high settlement and abundance, but few large mature adults (Chockley *et al.* 2008). Inshore reefs had much lower settlement and abundance, but the shrimp were almost twice the size as those collected offshore. This disparity resulted in tenfold higher reproductive output from the inshore reefs than from offshore reefs, and thus the potential for source–sink dynamics. The offshore population was a probable sink, because so few of the individuals mature that, on average, each settler did not replace itself. The inshore population, in contrast, produced more than one larva per settler and could be considered a source.

Pisaster ochraceus, a large sea star in the northeastern Pacific Ocean, is a broadcast spawner with a long-distance planktotrophic larval stage (Todd and Doyle 1981). Contrary to expectation, higher prey quality did not lead to higher per capita reproductive output for two *P. ochraceus* populations on the coast of Oregon (Sanford and Menge 2007). Higher prey quality led to faster growth, so individuals of the same size were younger at the site with high-quality prey than at the site with poor-quality prey, and probably put less energy into reproduction. In contrast, the sea stars at the site with low-quality prey had consistently high reproductive output and may have been a reproductive source.

The mostly anadromous salmoniformes generally return to natal streams for reproduction, and straying – returning to a non-natal stream – is low. If homing were perfect, the populations in each stream would be completely closed. However, straying means that homing is not perfect, and thus salmonid populations may be modeled using source–sink dynamics. For example, populations of Atlantic salmon (*Salmo salar*) in the Sognefjorden district of Norway are dominated by one river in terms of size and productivity. In demographic models, total effective metapopulation size increased with population abundance in that single river (Hindar *et al.* 2004), indicating that this river may act as a source. In a two-patch metapopulation model of Pacific salmon, *Oncorhynchus* spp., populations in each patch continuously rescued the other from extinction, suggesting source–sink dynamics (Hill *et al.* 2003).

Scorpaeniformes, including sculpins and rockfish, demonstrate high site fidelity and dispersal occurs primarily during the larval phase (Paddack and Estes 2000). For two of the three species of tidepool sculpin (*Clinocuttus embryum*, *C. globiceps*, and *Oligocottus maculosus*) analyzed in one study, extinction probabilities decreased as tidepool size and population increased (Pfister

1998). Tidepools likely contributed to a common larval pool, so if some pools produced significantly more larvae than others, source–sink dynamics could occur.

Dispersal

Hydrodynamic processes may produce source–sink dynamics for various species. Populations of the mussel *Mytilus californianus* vary spatially in reproductive output north and south of Point Conception, such that the southern population contributes substantially more to the metapopulation larval pool (Phillips 2007), suggesting that the southern population is a source for the northern one. Similarly, *M. californianus* and *M. galloprovincialis* demonstrated high self-seeding in northern populations and high import of larvae in southern populations (Becker *et al.* 2007). Genetic analysis of two barnacle species (*Balanus glandula* and *Chthamalus fissus*) along the northwest Pacific coast indicated that dispersing larvae were predominantly advected southward, driven by the southward flow of ocean currents across Point Conception (Wares *et al.* 2001). Genetic analysis of the purple sea urchin (*Strongylocentrotus purpuratus*) on the northwest Pacific coast implied southward gene flow across Point Conception (Wares *et al.* 2001), suggesting a source–sink structure. Recruitment was higher and less variable just south of Point Conception than either to the north or south (Ebert *et al.* 1994).

Perna perna along the South African coastline exhibits strong spatial variation in adult abundance, and a high correlation of recruits with adult densities (Erlandsson and McQuaid 2004; Porri *et al.* 2006). This is likely due to differential delivery of larvae caused by the effects of small-scale topography on local hydrodynamics, which drives spatial differences in adult mussel abundance, and potentially source–sink dynamics.

Larvae of the blue crab, *Callinectes sapidus*, develop in a common larval pool on the continental shelf, after which a postlarval stage re-invades estuaries. When wind conditions are not favorable for local retention, larval exchange is probably north to south along the Atlantic coast and east to west in the Gulf of Mexico (Fogarty and Botsford 2006), resulting in irregular levels of exchange and source–sink dynamics. Sinks may form in bays, lagoons and estuaries where local population depletion or extinction can occur due to stressful environmental conditions, disease or overexploitation (Lipcius and Stockhausen 2002).

The gadiformes, which include cod and pollock, are broadcast spawners that spend several months as planktonic eggs and larvae (Begg and Marteinsdottir 2000). The discovery of pollock in Prince William Sound, where the population had been absent, as the population in nearby Shelikof Strait increased, supported a source–sink origin (Olsen *et al.* 2002). Hydrodynamic modeling of larval drift of Icelandic cod (*Gadus morhua*) determined that larvae released from the northeastern spawning grounds primarily drift out of Icelandic waters (Brickman *et al*. 2007). The southeastern spawning grounds were predicted to contribute decreasing quantities of larvae with clockwise distance from the area. Of the 13 spawning grounds, one was predicted to produce about a quarter of the larvae, seven to produce less than about 10%, and the remaining five to produce the rest, about equally.

Fishery exploitation

The fisheries for abalone (*Haliotis* spp.) have characteristically declined, with the supporting populations experiencing serial or spatial depletion (Rogers-Bennett *et al*. 2002). Abalone are broadcast spawners and fertilization success is reduced at low adult densities (Babcock and Keesing 1999). Sampling of large abalone species in Japan found relatively high larval densities in the fishing grounds, where there were few adults, and a high correlation between the larval densities in a refugium from exploitation and the fishing grounds. Hydrodynamic modeling showed that the refugium could be a reproductive source for the fishing grounds (Miyake *et al*. 2009).

For sea urchin populations, sinks may result from two types of Allee effects at low population densities, either reduced fertilization success in areas of low adult densities or higher per capita mortality rates caused by enhanced predation on juveniles when they settle in areas of low adult density (Quinn *et al*. 1993). These effects suggest that a mosaic of high and low densities of sea urchins could result in source and sink populations. Areas of low density can result from high fishery exploitation, such that harvest refugia, areas of high urchin density, may act as sources for low-density fishing grounds where Allee effects prevail.

Rockfish (*Sebastes* spp.) are fished heavily on the west coast of the USA and may be driven to local extinction. In a comparison of 10 species of kelp forest rockfish between marine reserves and exploited areas, density and size of rockfish were greater in the reserves than in exploited areas, such that the reproductive potential per unit area was higher in the reserves (Paddack and Estes 2000). These results suggest that the reserves (sources) may help to sustain exploited populations (sinks), both through adult emigration and larval dispersal.

Characteristics and mechanisms of source–sink dynamics

While there are no global generalizations regarding the prevalence and mechanisms of source and sink populations, certain patterns emerged from the literature.

- Biotic or abiotic factors can serve as the mechanism producing source–sink dynamics.
- Source–sink dynamics can occur at various spatial and temporal scales.
- Source–sink dynamics can arise from natural or anthropogenic causes, such as variation in food availability (natural) and overexploitation (anthropogenic), or through an interaction between natural and anthropogenic causes.

The original basis for modeling source–sink dynamics was to understand the role of varying habitat quality on demographic rates (Pulliam 1988). In marine and estuarine species inhabiting shallower habitats, habitat quality remains a key mechanism driving source–sink dynamics, but dispersal, predation, and fishery exploitation also play influential roles.

Future investigations may determine that sources and sinks exist under diverse environmental conditions. The most recently explored marine environment is the deep ocean, extending from the bathyal zone at 4,000 m to the abyssal zone at 6,000 m in depth (Rex *et al.* 2005). The transition from the bathyal to abyssal zones is gradual, in both physical and biological characteristics, without obvious distinct boundaries or barriers. The abyss, covering over half of the sea floor, is characterized by low biomass, density, and diversity of macrofauna (Rex *et al.* 2005), and by species existing at low densities but with pelagic larvae. Many abyssal species (e.g., mollusks) appear to be range extensions of bathyal species (Rex *et al.* 2005), such that some abyssal populations may act as sinks, receiving larvae from bathyal populations but incapable of self-maintenance.

There are likely to be additional anthropogenic processes mediating source–sink dynamics. Eutrophication is the natural process through which a water body ages, as the accumulation of nutrients leads to excess primary productivity and, eventually, low concentrations of dissolved oxygen. However, high anthropogenic inputs of nitrogen to shallow coastal and estuarine systems have led to an increased rate of eutrophication, and an increased occurrence of hypoxia and anoxia (= low dissolved oxygen; Diaz and Rosenberg 2008). Annual summertime hypoxia, the most common form of hypoxia, accounts for more than half of the known occurrences, including those in Chesapeake Bay, the Gulf of Mexico, Adriatic Sea, and Black Sea (Diaz 2001). Mass mortalities are often associated with severe annual hypoxia and sublethal effects on reproduction occur in copepods (Marcus *et al.* 2004), gastropods (Cheung *et al.* 2008), and amphipods (Hoback and Barnhart 1996). Death of the organisms living in hypoxic regions, particularly if the hypoxia occurs prior to the time of first reproduction, can result in hypoxic areas acting as population sinks (Long 2007). Individuals disperse to hypoxic areas during recruitment in the spring and early summer, but are unable to contribute significantly to the larval pool.

Consequently, the seasonally hypoxic and anoxic habitats attract larval or post-larval settlers from source habitats, but the juveniles and adults perish annually prior to reproduction, which is suggestive of an ecological trap (Robertson and Hutto 2006). The worldwide dead zones (Diaz and Rosenberg 2008) could be a leading mechanism for anthropogenic production of source–sink dynamics.

Implications of source–sink dynamics for conservation

The conservation and restoration of marine and estuarine species displaying metapopulation source–sink dynamics requires careful attention to the interplay between habitat quality and hydrodynamic processes driving interpopulation connectivity. Source populations, which are optimal for conservation and restoration efforts, can be distinct geographically and may be a small percentage of a metapopulation. Sink areas, which contribute little to metapopulation persistence, can nonetheless benefit from habitat restoration because larvae from source reefs recruiting to these areas can enhance habitat quality in the case of ecosystem engineers or fishery yield in exploited species. Optimal metapopulation conservation may be attained by preserving source populations while allowing the exploitation of sink populations linked to the sources via larval dispersal. However, the connectivity and network structure of sources and sinks are complex (Lipcius *et al.* 2008), such that a strategy of preserving sources may be too simplistic and lead to conservation or restoration failure (see also Wiens and Van Horne, Chapter 23, this volume). Unfortunately, the structure of metapopulations and potential sources and sinks are often unknown. In these cases, we recommend a bet-hedging conservation strategy complemented by further research into metapopulation structure.

In the case of marine reserves, these protected habitats and populations have the potential to act as sources of both adults and larvae to nearby fished areas (Lipcius *et al.* 2001, 2005, 2008). The extent to which a reserve will supplement nearby areas is influenced by the dispersal ability of the species (Gerber *et al.* 2004), the degree of poaching (Byers and Noonburg 2007), and the source–sink dynamics of the species (Crowder *et al.* 2000). Consequently, the interaction between fishing pressure, habitat quality, and hydrodynamic connectivity could produce relatively complex patterns of metapopulation connectivity and source–sink dynamics, which will make it difficult to arrive at a single optimal conservation strategy.

References

Akçakaya, R. and P. Baker (1998). *Zebra Mussel Demography and Modeling: Preliminary Analysis of Population Data from Upper Midwest Rivers*. USACE Contract Report EL-98-1. US Army Corps of Engineers, Washington, DC.

Armsworth, P. (2002). Recruitment limitation, population regulation, and larval connectivity in reef fish metapopulations. *Ecology* **83**: 1092–1104.
Babcock, R. and J. Keesing (1999). Fertilization biology of the abalone *Haliotis laevigata*: laboratory and field studies. *Canadian Journal of Fisheries and Aquatic Sciences* **56**: 1658–1678.
Becker, B., L. Levin, F. Fodrie and P. McMillan (2007). Complex larval connectivity patterns among marine invertebrate populations. *Proceedings of the National Academy of Sciences of the USA* **104**: 3267–3272.
Begg, G. and G. Marteinsdottir (2000). Spawning origins of pelagic juvenile cod *Gadus morhua* inferred from spatially explicit age distributions: potential influences on year-class strength and recruitment. *Marine Ecology Progress Series* **202**: 193–217.
Bertness, M., S. Gaines, D. Bermudez and E. Sanford (1991). Extreme spatial variation in the growth and reproductive output of the acorn barnacle *Semibalanus balanoides*. *Marine Ecology Progress Series* **75**: 91–100.
Bode, M., L. Bode and P. Armsworth (2006). Larval dispersal reveals regional sources and sinks in the Great Barrier Reef. *Marine Ecology Progress Series* **308**: 17–25.
Brickman, D., G. Marteinsdottir, K. Logemann and I. Harms (2007). Drift probabilities for Icelandic cod larvae. *ICES Journal of Marine Science* **64**: 49–59.
Byers, J. and E. Noonberg (2007). Poaching, enforcement, and the efficacy of marine reserves. *Ecological Applications* **17**: 1851–1856.
Cheung, S., H. Chan, C. Liu and P. Shin (2008). Effect of prolonged hypoxia on food consumption, respiration, growth, and reproduction in marine scavenging gastropod *Nassarius festivus*. *Marine Pollution Bulletin* **57**: 280–286.
Chockley, B., C. St. Mary and C. Osenberg (2008). Population sinks in the Upper Florida Keys: the importance of demographic variation in population dynamics of the marine shrimp *Stenopus hispidus*. *Marine Ecology Progress Series* **360**: 135–145.
Cowen, R., K. Lwiza, S. Sponaugle, C. Paris and D. Olson (2000). Connectivity of marine populations: open or closed? *Science* **287**: 857–859.
Crowder, L., S. Lyman, W. Figueira and J. Priddy (2000). Source–sink population dynamics and the problem of siting marine reserves. *Bulletin of Marine Science* **66**: 799–820.
Diaz, R. (2001). Overview of hypoxia around the world. *Journal of Environmental Quality* **30**: 275–281.
Diaz, R. and R. Rosenberg (2008). Spreading dead zones and consequences for marine ecosystems. *Science* **321**: 926–929.
Ebert, T., S. Schroeter, J. Dixon and P. Kalvass (1994). Settlement patterns of red and purple sea urchins (*Strongylocentrotus franciscanus* and *S. purpuratus*) in California, USA. *Marine Ecology Progress Series* **111**: 41–52.
Ellien, C., E. Thiébaut, A.-S. Barnay, J.-C. Dauvin, F. Gentil and J.-C. Salomon (2000). The influence of variability in larval dispersal on the dynamics of a marine metapopulation in the eastern Channel. *Oceanologica Acta* **23**: 423–442.
Erlandsson, J. and C. McQuaid (2004). Spatial structure of recruitment in the mussel *Perna perna* at local scales: effects of adults, algae, and recruit size. *Marine Ecology Progress Series* **267**: 173–185.
Figueira, W. and L. Crowder (2006). Defining patch contribution in source–sink metapopulations: the importance of including dispersal and its relevance to marine systems. *Population Ecology* **48**: 215–224.
Fogarty, M. J. and L. W. Botsford (2006). Metapopulation dynamics of coastal decapods. In *Marine Metapopulations* (J. P. Kritzer and P. F. Sale, eds.). Academic Press/Elsevier, Amsterdam: 271–320.
Gaylord, B., D. Reed, P. Raimondi and L. Washburn (2006). Macroalgal spore dispersal in coastal environments: mechanistic insights revealed by theory and experiment. *Ecological Monographs* **76**: 481–502.
Gerber, L., S. Heppell, F. Ballantyne and E. Sala (2004). The role of dispersal and demography in determining the efficacy of marine reserves. *Canadian Journal of Fisheries and Aquatic Sciences* **62**: 863–871.

Graham, N., S. Wilson, S. Jennings, N. Polunin, J. Bijoux and J. Robinson (2006). Dynamic fragility of oceanic coral reef ecosystems. *Proceedings of the National Academy of Sciences of the USA* **103**: 8425–8429.
Harris, R. (1973). Feeding, growth, reproduction, and nitrogen utilization by the harpacticoid copepod *Tigriopus brevicornis*. *Journal of the Marine Biological Association of the UK* **35**: 785–800.
Harwell, M. and R. Orth (2002). Long-distance dispersal potential in a marine macrophyte. *Ecology* **83**: 3319–3330.
Hill, M., A. Hastings and L. Botsford (2003). The effect of small dispersal rates on extinction times in structured metapopulation models. *American Naturalist* **160**: 389–402.
Hindar, K., J. Tufto, L. Saettem and T. Balstad (2004). Conservation of genetic variation in harvested salmon populations. *ICES Journal of Marine Science* **61**: 1389–1397.
Hoback, W. and M. Barnhart (1996). Lethal limits and sublethal effects of hypoxia on the amphipod *Gammarus pseudolimnaeus*. *Journal of the North American Benthological Society* **15**: 117–126.
Holt, R. (1985). Population dynamics in two-patch environments: some anomalous consequences of an optimal habitat distribution. *Theoretical Population Biology* **28**: 181–208.
Horvath, T., G. Lamberti, D. Lodge and W. Perry (1996). Zebra mussel dispersal in lake-stream systems: source–sink dynamics? *Journal of the North American Benthological Society* **15**: 564–575.
Incze, L., H. Xue, N. Wolff, D. Xu, C. Wilson, R. Steneck, R. Wahle, P. Lawton, N. Pettigrew and Y. Chen (2010). Connectivity of lobster (*Homarus americanus*) populations in the coastal Gulf of Maine. Part II. Coupled biophysical dynamics. *Fisheries Oceanography* **19**: 1–20.
Johnson, M. (2001). Metapopulation dynamics of *Tigriopus brevicornis* (Harpacticoida) in intertidal rock pools. *Marine Ecology Progress Series* **211**: 215–224.
Jolly, M., F. Viard, G. Weinmayr, F. Gentil, E. Thiébaut and D. Jollivet (2003). Does the genetic structure of *Pectinaria koreni* (Polychaeta: Pectinariidae) conform to a source–sink metapopulation model at the scale of the Baie de Seine? *Helgoland Marine Research* **56**: 238–246.
Junkins, R., B. Kelaher and J. Levinton (2006). Contributions of adult oligochaete emigration and immigration in a dynamic soft-sediment community. *Journal of Experimental Marine Biology and Ecology* **330**: 208–220.
Katz, C. (1985). A nonequilibrium marine predator–prey interaction. *Ecology* **66**: 1426–1438.
Krkosek, M. and M. Lewis (2010). An R_0 theory for source–sink dynamics with application to *Dreissena* competition. *Theoretical Ecology* **3**: 25–43.
Lafferty, K., C. Swift and R. Ambrose (1999). Extirpation and recolonization in a metapopulation of an endangered fish, the tidewater goby. *Conservation Biology* **13**: 1447–1453.
Leslie, H., E. Breck, F. Chan, J. Lubchenco and B. Menge (2005). Barnacle reproductive hotspots linked to nearshore ocean conditions. *Proceedings of the National Academy of Sciences of the USA* **102**: 10534–10539.
Levitan, D. and C. Petersen (1995). Sperm limitation in the sea. *Trends in Ecology and Evolution* **10**: 228–231.
Lipcius, R. and W. Stockhausen (2002). Concurrent decline of the spawning stock, recruitment, larval abundance, and size of the blue crab *Callinectes sapidus* in Chesapeake Bay. *Marine Ecology Progress Series* **226**: 45–61.
Lipcius, R., W. Stockhausen, D. Eggleston, L. Marshall and B. Hickey (1997). Hydrodynamic decoupling of recruitment, habitat quality, and adult abundance in the Caribbean spiny lobster: source–sink dynamics? *Marine and Freshwater Research* **48**: 807–815.
Lipcius, R., W. Stockhausen and D. Eggleston (2001). Marine reserves for Caribbean spiny lobster: empirical evaluation and theoretical metapopulation dynamics. *Marine and Freshwater Research* **52**: 1589–1598.
Lipcius, R., L. Crowder and L. Morgan (2005). Metapopulation structure and marine reserves. In *Marine Conservation Biology* (E. Norse and L. B. Crowder, eds.). Island Press, Washington, DC: 328–345.

Lipcius, R., D. Eggleston, S. Schreiber, R. Seitz, J. Shen, M. Sisson, W. Stockhausen and H. Wang (2008). Importance of metapopulation connectivity to restocking and restoration of marine species. *Reviews in Fisheries Science* **16**: 101–110.

Long, W. (2007). Hypoxia and *Macoma balthica*: ecological effects on a key infaunal benthic species. PhD dissertation, College of William & Mary, Williamsburg, VA.

Marcus, N. H., C. Richmond, C. Sedlacek, G. A. Miller and C. Oppert (2004). Impact of hypoxia on the survival, egg production, and population dynamics of *Acartia tonsa* Dana. *Journal of Experimental Marine Biology and Ecology* **301**: 111–128.

Marko, P. and K. Barr (2007). Basin-scale patterns of mtDNA differentiation and gene flow in the bay scallop *Argopecten irradians concentricus*. *Marine Ecology Progress Series* **349**: 139–150.

Miyake, Y., S. Kimura, T. Kawamura, T. Horii, H. Kurogi and T. Kitagawa (2009). Simulating larval dispersal processes for abalone using a coupled particle-tracking and hydrodynamic model: implications for refugium design. *Marine Ecology Progress Series* **387**: 205–222.

Munguia, P., C. Mackie and D. Levitan (2007). The influence of stage-dependent dispersal on the population dynamics of three amphipod species. *Oecologia* **153**: 533–541.

Nilsson, P., J. Kurdziel and J. Levinton (1997). Heterogeneous population growth, parental effects, and genotype–environment interactions of a marine oligochaete. *Marine Biology* **130**: 181–191.

Nilsson, P., J. Levinton and J. Kurdziel (2000). Migration of a marine oligochaete: induction of dispersal and microhabitat choice. *Marine Ecology Progress Series* **207**: 89–96.

North, E., Z. Schlag, R. Hood, M. Li, L. Zhong, T. Gross and V. Kennedy (2008). Vertical swimming behavior influences the dispersal of simulated oyster larvae in a coupled particle-tracking and hydrodynamic model of Chesapeake Bay. *Marine Ecology Progress Series* **359**: 99–115.

Olsen, J., S. Merkouris and J. Seeb (2002). An examination of spatial and temporal genetic variation in walleye pollock (*Theragra chalcogramma*) using allozyme, mitochondrial DNA, and microsatellite data. *Fisheries Bulletin* **100**: 752–764.

Orth, R., K. Heck Jr. and J. van Montfrans (1984). Faunal communities in seagrass beds: a review of the influence of plant structure and prey characteristics on predator–prey relationships. *Estuaries* **7**: 339–350.

Paddack, M. and J. Estes (2000). Kelp forest fish populations in marine reserves and adjacent exploited areas of central California. *Ecological Applications* **10**: 855–870.

Paine, R. (1979). Disaster, catastrophe, and local persistence of the sea palm *Postelsia palmaeformis*. *Science* **205**: 685–687.

Paine, R. (1988). Habitat suitability and local population persistence of the sea palm *Postelsia palmaeformis*. *Ecology* **69**: 1787–1794.

Peterson, C. and H. Summerson (1992). Basin-scale coherence of population dynamics of an exploited marine invertebrate, the bay scallop: implications of recruitment limitation. *Marine Ecology Progress Series* **90**: 257–272.

Peterson, C., F. Fodrie, H. Summerson and S. Powers (2001). Site-specific and density-dependent extinction of prey by schooling rays: generation of a population sink in top-quality habitat for bay scallops. *Oecologia* **129**: 349–356.

Pfister, C. A. (1998). Extinction, colonization, and species occupancy in tidepool fishes. *Oecologia* **114**: 118–126.

Phillips, N. (2007). A spatial gradient in the potential reproductive output of the sea mussel *Mytilus californianus*. *Marine Biology* **151**: 1543–1550.

Porri, F., C. McQuaid and S. Radloff (2006). Spatio-temporal variability of larval abundance and settlement of *Perna perna*: differential delivery of mussels. *Marine Ecology Progress Series* **315**: 141–150.

Pulliam, H. (1988). Sources, sinks, and population regulation. *American Naturalist* **132**: 652–661.

Quinn, J. F., S. R. Wing and L. W. Botsford (1993). Harvest refugia in marine invertebrate fisheries: models and applications to the red sea urchin, *Strongylocentrotus franciscanus*. *American Zoologist* **33**: 537–550.

Reed, D., B. Kilan, P. Raimondi, L. Washburn, B. Gaylord, and P. Drake (2006). A metapopulation perspective on the patch dynamics of giant kelp in southern California. In *Marine Metapopulations* (J. P. Kritzer and P. F. Sale, eds.). Academic Press/Elsevier, Amsterdam: 353–386.

Rex, M., C. McClain, N. Johnson, R. Etter, J. Allen, P. Bouchet and A. Waren (2005). A source–sink hypothesis for abyssal biodiversity. *American Naturalist* **165**: 163–178.

Roberts, C. (1997). Connectivity and management of Caribbean coral reefs. *Science* **278**: 1454–1457.

Roberts, C. (1998). Sources, sinks, and the design of marine reserve networks. *Fisheries* **23**: 16–19.

Robertson, B. and R. Hutto (2006). A framework for understanding ecological traps and an evaluation of existing evidence. *Ecology* **87**: 1075–1085.

Rogers-Bennett, L., P. Hakker, K. Karpov and D. Kushners (2002). Using spatially explicit data to evaluate marine protected areas for abalone in southern California. *Conservation Biology* **16**: 1308–1317.

Runge, J., M. Runge and J. Nichols (2006). The role of local populations within a landscape context: defining and classifying sources and sinks. *American Naturalist* **167**: 925–938.

Sanford, E. and B. Menge (2007). Reproductive output and consistency of source populations in the sea star *Pisaster ochraceus*. *Marine Ecology Progress Series* **349**: 1–12.

Stoeckel, J., D. Schneider, L. Soeken, K. Blodgett and R. Sparks (1997). Larval dynamics of a riverine metapopulation: implications for zebra mussel recruitment, dispersal, and control in a large-river system. *Journal of the North American Benthological Society* **16**: 586–601.

Swearer, S., J. Caselle, D. Lea and R. Warner (1999). Larval retention and recruitment in an island population of a coral-reef fish. *Nature* **402**: 799–802.

Thiébaut, E., Y. Lagadeuc, F. Olivier, J. Dauvin and C. Retière (1998). Do hydrodynamic factors affect the recruitment of marine invertebrates in a macrotidal area? The case study of *Pectinaria koreni* (Polychaeta) in the Bay of Seine (English Channel). *Hydrobiologia* **375/376**: 165–176.

Todd, C. and R. Doyle (1981). Reproductive strategies of marine benthic invertebrates: a settlement-timing hypothesis. *Marine Ecology Progress Series* **4**: 75–83.

Wares, J., S. Gaines and C. Cunningham (2001). A comparative study of asymmetric migration events across a marine biogeographic boundary. *Evolution* **55**: 295–306.

White, J. (2008). Spatially coupled larval supply of marine predators and their prey alters the predictions of metapopulation models. *American Naturalist* **171**: 179–194.

Williams, S. and R. Orth (1998). Genetic diversity and structure of natural and transplanted eelgrass populations in the Chesapeake and Chincoteague Bays. *Estuaries* **21**: 118–128.

Wing, S. (2009). Decadal-scale dynamics of sea urchin population networks in Fiordland, New Zealand are driven by juxtaposition of larval transport against benthic productivity gradients. *Marine Ecology Progress Series* **378**: 125–134.

Wing, S., M. Gibbs and M. Lamare (2003). Reproductive sources and sinks within a sea urchin, *Evechinus chloroticus*, population of a New Zealand fjord. *Marine Ecology Progress Series* **248**: 109–123.

Wing, S., R. McLeod, K. Clark and R. Frew (2008). Plasticity in the diet of two echinoderm species across an ecotone: microbial recycling of forest litter and bottom-up forcing of population structure. *Marine Ecology Progress Series* **360**: 115–123.

Wolff, M. and J. Mendo (2000). Management of the Peruvian bay scallop (*Argopecten purpuratus*) metapopulation with regard to environmental change. *Aquatic Conservation: Marine and Freshwater Ecosystems* **10**: 117–126.

Xue, J., L. Incze, D. Xu, N. Wolff and N. Pettigrew (2008). Connectivity of lobster populations in the coastal Gulf of Maine. Part I. Circulation and larval transport potential. *Ecological Modelling* **210**: 193–211.

STEPHEN R. WING

18

Population networks with sources and sinks along productivity gradients in the Fiordland Marine Area, New Zealand: a case study on the sea urchin *Evechinus chloroticus*

Summary

Many coastal marine populations are made up of networks of discrete subpopulations of relatively sedentary adults linked by larval dispersal at the mesoscale (10–100 km). Patterns in abundance, and structure of populations at this scale are strongly influenced by the interaction of hydrodynamic forcing on larval dispersal with patterns in adult productivity and larval production. This issue is particularly important in the 14 fjords that indent the southwest coast of New Zealand, which present a highly fragmented and diverse array of marine habitats with strong gradients in benthic productivity, and with larval transport in each fjord dominated by estuarine circulation. Population structure of sea urchins (*Evechinus chloroticus*) was illustrated across this region by examining trends in size frequency distributions from 53 sites sampled in 2002. A consistent pattern was observed, with sea urchin populations from the kelp-dominated fjord entrances consistently displaying a large adult mode (mean test diameter 110–130 mm) with evidence for gradual recruitment of individuals into the population, while populations from the inner fjords showed two distinct patterns. At many inner-fjord sites a large frequency of recently emergent recruits 50–60 mm indicated strong recruitment events and high demographic variability, while at some sites populations had undergone complete mortality of the adult mode and at sampling were made up of only a single juvenile cohort. These patterns likely reflect a source–sink population structure across the benthic productivity gradients within each fjord. This has important

Sources, Sinks and Sustainability, ed. Jianguo Liu, Vanessa Hull, Anita T. Morzillo and John A. Wiens. Published by Cambridge University Press.

implications for the efficacy of a network of eight new marine reserves (10,241 ha) and 14 commercial exclusion zones (46,002 ha) established within the inner fjords under the Fiordland Marine Management Act 2005, and for the sustainability of this fragmented marine system.

Introduction

Classical single-species management of fisheries relies on the assumption that stocks are well mixed throughout the management area so that increases in mortality from fishing are evenly spread throughout the population (Hart and Reynolds 2002). This "dynamic pool" assumption is challenged by the fact that many coastal stocks of finfish and invertebrates are highly subdivided (Orensanz and Jamieson 1998; Thorrold *et al.* 2001), and many marine populations persist as source–sink networks (Wing *et al.* 2003; Sanford and Menge 2007) or metapopulations (Kritzer and Sale 2006; Fogarty and Botsford 2007). Spatial management, in particular the establishment of marine reserves, is becoming a frequently implemented alternative tool for conserving marine populations and communities (Dugan and Davis 1993; Lauck *et al.* 1998). This shift in management strategy has been supported by theoretical advances in understanding the regulation of population networks, and application of these principles to marine systems (Pulliam 1988; Hastings and Botsford 1999; Hixon *et al.* 2002). The resulting investment in reserve systems by managers follows a large and growing scientific literature that demonstrates how spatial management applies to spatially structured populations, and in many cases records dramatic responses of fished species to local exclusion of fishing pressure (NRC 2001; Halpern 2003).

The life history of individual species and physical transport mechanisms influence connectivity and the ability of some populations to self-seed within reserves, or to provide propagules to surrounding areas (Botsford *et al.* 2001; Gaines *et al.* 2003; Almany *et al.* 2007). Consequently, success in the application of marine protected areas can be enhanced by an understanding of how populations are connected by dispersal across fragmented landscapes (Fogarty 1998; Jones *et al.* 1999; Swearer *et al.* 1999; Thorrold *et al.* 2001), and how heterogeneity of adult habitat quality affects reproductive output (Crowder *et al.* 2000; Sanford and Menge 2007). In this context, positioning marine reserves with reference to source–sink dynamics and essential habitat of focal species is recognized as an important issue (Fogarty 1999; Crowder *et al.* 2000), although this presents challenges for the management of diverse communities (Roberts 1998).

Here a case study on population structure of the New Zealand sea urchin (*Evechinus chloroticus*) was considered relative to the subdivided nature of habitats and strong productivity gradients within the New Zealand fjords. The implications of physical forcing on dispersal and bottom-up control on population structure across the Fiordland landscape were considered in the context of a newly implemented network of marine protected areas designed to protect marine biodiversity in the region (Wing *et al.* 2004). The sea urchin *Evechinus chloroticus* (Echinodermata: Echinoidea) is endemic to New Zealand and a common member of rocky reef communities. Its distribution includes the North and South Islands, Stewart Island, the Snares, Chatham Island, and Three Kings Islands (Pawson 1965; Dix 1970a). Adults are generally found subtidally in shallow waters above 10 m. They possess an annual reproductive cycle and produce typical planktotrophic echinopluteus larvae with a pelagic existence of 1–2 months (Dix 1969, 1970b; Walker 1984). Because adults are benthic grazers, dispersal is limited to planktonic drifting and weak swimming of *E. chloroticus* larvae. Juveniles less than 40 mm test diameter are usually cryptic, living in rock crevices and under boulders until they emerge as new recruits into the adult population at 30–40 mm test diameter (Shears and Babcock 2002).

Evechinus chloroticus is a critical species in the ecology of coastal kelp forests in New Zealand (Choat and Andrew 1986; Shears and Babcock 2002), influencing both the composition and productivity of algal reefs (Choat and Schiel 1982). Because recruitment is typically sporadic and potentially driven by larval supply from distant populations along the coast, the spatial distribution of populations is a critical feature of their regional dynamics.

Previous studies of the population structure of *E. chloroticus* within Doubtful Sound revealed a strong source–sink structure in the population, with high per capita larval production linked to populations inhabiting kelp beds at the entrances of the fjords (Wing *et al.* 2003). These wave-exposed regions support high-quality macroalgal food sources, resulting in rapid growth rates, large adult sizes, and high gamete production (Lamare and Mladenov 2000; Lamare *et al.* 2002). Larvae produced at the fjord entrance are likely retained by estuarine circulation in the fjord, where they settle out in high numbers in the inner-fjord habitats (Lamare 1998; Wing *et al.* 2003). In these areas estuarine algae and microbial recycling of forest litter make up a poor-quality diet for *E. chloroticus*, and both growth and per capita gamete production is reduced (Lamare and Wing 2001; Wing *et al.* 2008).

In the present study a fine-scale survey of *E. chloroticus* population structure, and abundance of the common kelp *Ecklonia radiata*, provide the basis for examining population structure throughout the 14 fjords within the Fiordland landscape. Specifically, the data are used to examine differences in average test diameter of sea urchin populations with distance from the outer coast, and with density of *E. radiata*. Variability in these measures serves as an effective proxy for relative differences in growth and per capita gamete production for *E. chloroticus* in this system (Wing *et al.* 2003). Relationships between fraction of newly emergent recruits in the population with distance from the outer coast provide a proxy for recruitment patterns and demographic variability among populations, while assessment of population structure relative to management zones provides an analysis of the position of likely reproductive sources relative to a newly established network of marine protected areas.

Materials and methods

Sea urchin size structure

For the 53 survey sites, sea urchins (*Evechinus chloroticus*) were collected using scuba, test diameter was measured with Vernier calipers, and the sea urchins were returned to the seabed. Care was taken at each site to collect all sea urchins in a swath 0–20 m deep along the shore for 50–200 m, depending on density, and to search for cryptic juveniles in rubble fields and crevices. Sample sizes ranged from 34 to 264 (average 128) and reflected a haphazard accumulation of individuals from the population.

Sea urchin and kelp abundance survey

At a subset of sites, a stratified random survey was conducted using paired 2 m^2 quadrats at depths of 5, 10 and 15 m. At each stratum a series of six paired quadrats was quantified for sea urchins *Evechinus chloroticus* and the common kelp *Ecklonia radiata* and other conspicuous species of brown algae. In the case of both kelps and sea urchins, if an organism was located on the border of the quadrat, it was counted. At each quadrat site the first lay of the quadrat was quantified and then the quadrat was flipped horizontally at the same depth stratum for a second count. This resulted in a series of six replicate 4 m^2 quadrats at each depth stratum. I consider that 4 m^2 is an appropriate spatial scale for this type of sampling because it averages across a large enough area to resolve local density patterns (Wing *et al.* 2003).

Distance to fjord entrance

For each fjord, the distance from study sites to the outer coast was calculated using a geographic information system (GIS) with 50 m horizontal resolution in the inner fjords. In this case a mean coastline raster line served as a zero line for the distance algorithm "r.cost" in GRASS 5.3 (Geographical Resources Analysis Support System; ITC-irst, Trento, Italy). The r.cost algorithm uses a "knight's move" distance calculation on a square grid to accurately provide distances around complex coastlines (Neteler and Mitasova 2002).

Data analysis

Average test diameter of sea urchins among study sites was analyzed using Ward's hierarchical clustering to form two distinct clusters (Legendre and Legendre 1998). Linear regression was then used to test for the relationship between average test diameter and distance from the fjord entrance across all sites.

Average density of *Ecklonia radiata* was calculated for the depth-stratified survey for a subset of 30 sites where information was available for both size structure of *Evechinus chloroticus* populations and density of *E. radiata*. Type II linear regression was used to test the relationship between average test diameter among sites and density of *E. radiata*. Density of *E. radiata* was then considered relative to the two distinct clusters based on average size. In the absence of a transformation that achieved normality, a Wilcoxon/Kruskal–Wallis test was used in lieu of ANOVA to compare the density of *E. radiata* between the two groups of sites defined by the average test diameter clusters.

Sea urchin size distributions from all sites were visually examined for evidence of distinct cohorts of new recruits and evidence for absence of the adult mode. The fraction of individuals in the newly emergent cohorts was then calculated as the proportion of individuals below 70 mm for each size distribution. A second calculation was made of the proportion of individuals below 50 mm test diameter for each size distribution, to reflect approximate size at emergence. A Wilcoxon/Kruskal–Wallis test was then applied to compare the fraction of new recruits in the two clusters of sites.

Sites were stratified according to management zone within the Fiordland Marine Management Act 2005. There are 14 commercial exclusion zones within Fiordland that extend from areas with kelp forest habitat toward the entrances of the fjords, to the head of each fjord, giving a total area of 46,002 ha, or 59% of inner-fjord habitat (Wing *et al.* 2004). Ten marine reserves are distributed through the fjords with a total area of 10,241 ha, or 13% of inner-fjord habitat, but little representation of kelp forest habitat (Wing *et al.* 2004). Areas outside of the commercial exclusion zones are designated "open to fishing" and are managed under the individual transferable quota system (ITQ) and special regulations on recreational take within the Fiordland Marine Area. In the case of sea urchin fishery, a total annual quota of 480 tonnes is allocated by the Ministry of Fisheries for the Fiordland and surrounding area as one management unit (SUR5). Fractions of sites with sea urchin populations that were classified in the cluster with relatively large mean test diameters were calculated for each management zone (open to fishing, commercial exclusion, marine reserve).

Results

The 53 study sites were distributed throughout Fiordland, with replicate sites along the axis of each of the 14 fjords (Fig. 18.1). Sites were not situated on the fully wave-exposed outer coast but each fjord had sites located at the entrance sill in semi wave-exposed habitat.

Analysis of the mean test diameter of *Evechinus* among sites with a Ward's hierarchical clustering algorithm defined two distinct clusters of populations. Populations from sites in cluster 1 had a mean test diameter of 120.9 mm (SE 1.47), while populations from sites in cluster 2 had a mean test diameter of 78.8 mm (SE 2.5) (Fig. 18.2). The location of sites within these two clusters were plotted on a map of Fiordland showing their distribution relative to management zones and positions within each fjord (Fig. 18.1).

Linear regression of average test diameter among sites with distance from the outer coast demonstrated a significant relationship (Test diameter (mm) = 115 – 1.8 Distance (km); $r^2 = 0.45$, $P < 0.0001$, $n = 53$). A distinct group of sites within 5 km of the outer coast had mean test diameters between 110 and 130 mm, while mean test diameter declined and variance increased among populations with distance from the outer coast (Fig. 18.2).

Type II linear regression demonstrated a strong relationship between average test diameter of sea urchins among sites and density of the important food resource *Ecklonia radiata* (Test diameter (mm) = 69.2 + 19.9 Density (2 m^{-2}); $R(x) = 0.82$, $P < 0.0001$, $n = 30$). Results of the Wilcoxon/Kruskal–Wallis test indicated that beds of *E. radiata* were significantly more dense at sites within cluster 1, where the average test diameter of sea urchins was relatively large ($\chi^2 = 16.72$, df = 1, $P < 0.0001$) (Fig. 18.3).

Examination of size distributions revealed three distinct patterns, which were distinctly separated along the axis of some fjords, such as the example shown from George Sound (Figs. 18.1 and 18.4). *E. chloroticus* populations at sites within cluster 1 typically had a single large adult mode with only small numbers of newly emergent recruits in the population (Fig. 18.4a, b). *Evechinus* populations at sites within cluster 2 typically had an adult mode accompanied by a large cohort of newly emergent juveniles of 30–50 mm test diameter (Fig. 18.4c), or had lost the adult mode and were made up solely of a newly emergent juvenile cohort (Fig. 18.4d).

Results of the Wilcoxon/Kruskal–Wallis test on fractions of new recruits among sites between the two clusters demonstrated a significant difference in this proxy for recruitment and demographic variability between the entrance sites and those in the inner-fjord habitats ($\chi^2 = 22.51$, $P < 0.0002$) (Fig. 18.5).

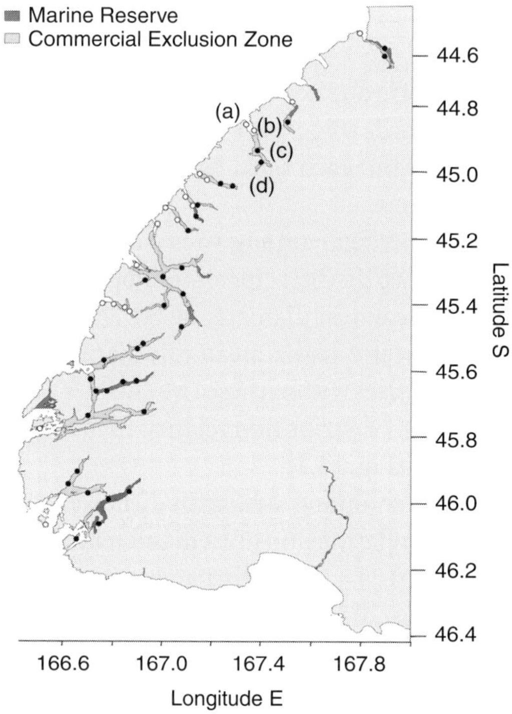

FIGURE 18.1.
Distribution of study sites and management zones across the Fiordland Marine Area. Open circles indicate populations of sea urchins in cluster 1, while closed circles indicate populations in cluster 2. Lettered sites (a)–(d) correspond to the positions of sites in Figure 18.4.

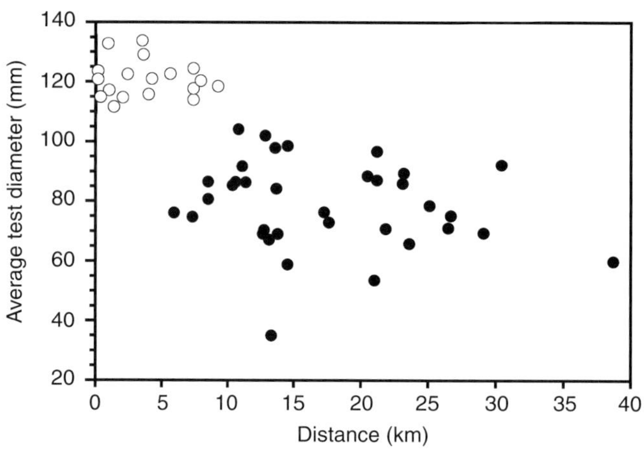

FIGURE 18.2.
Relationship between average test diameter (mm) of sea urchins from each site and distance from the fjord entrances (km). Open circles indicate populations of sea urchins in cluster 1, while closed circles indicate populations in cluster 2.

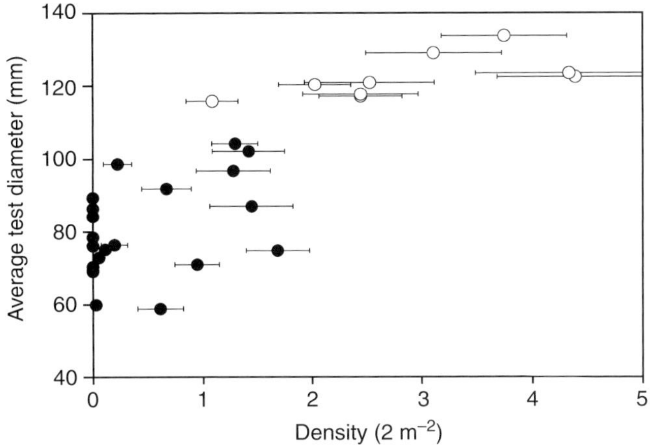

FIGURE 18.3.
Relationship between density (2 m^{-2}) of *Ecklonia radiata* and the average test diameter (mm) of sea urchins. Error bars indicate one standard error (SE). Open circles indicate populations of sea urchins in cluster 1, while closed circles indicate populations in cluster 2.

Analysis of the distribution of sites within cluster 1 relative to the management units indicated that the majority of populations in this group were found in areas of the Fiordland Marine Area that are open to commercial fishing (Fig. 18.6). In this case, 12 of 13 sites sampled in areas open to commercial fishing, 6 of 30 sites sampled in the commercial exclusion zones, and none of the nine sites sampled in marine reserves, were in cluster 1.

Discussion

One consequence of the fragmented habitat structure and circulation patterns within Fiordland is that a variety of marine populations may be less well mixed and more likely to persist as population networks within and among fjords, in some cases with a strong reproductive source–sink structure (Wing *et al.* 2003, 2008). The isolated basins of the inner fjords and predominant estuarine circulation play an important role in limiting larval dispersal and gene flow between fjords for marine organisms with a larval dispersal phase. Evidence for this comes from several sources including genetic studies of the eleven-armed sea star, *Coscinasterias muricata* (Sköld *et al.* 2003; Perrin *et al.* 2004), and the brachiopods *Liothyrella neozelanica* and *Terebratella sanguinea* (Ostrow *et al.* 2001). Similarly, a highly divided stock structure and low rates of movement between regions has been shown for the blue cod (*Parapercis colias*), indicating these populations may be vulnerable to localized depletion from overfishing (Carbines and McKenzie 2004; Rodgers and Wing 2008).

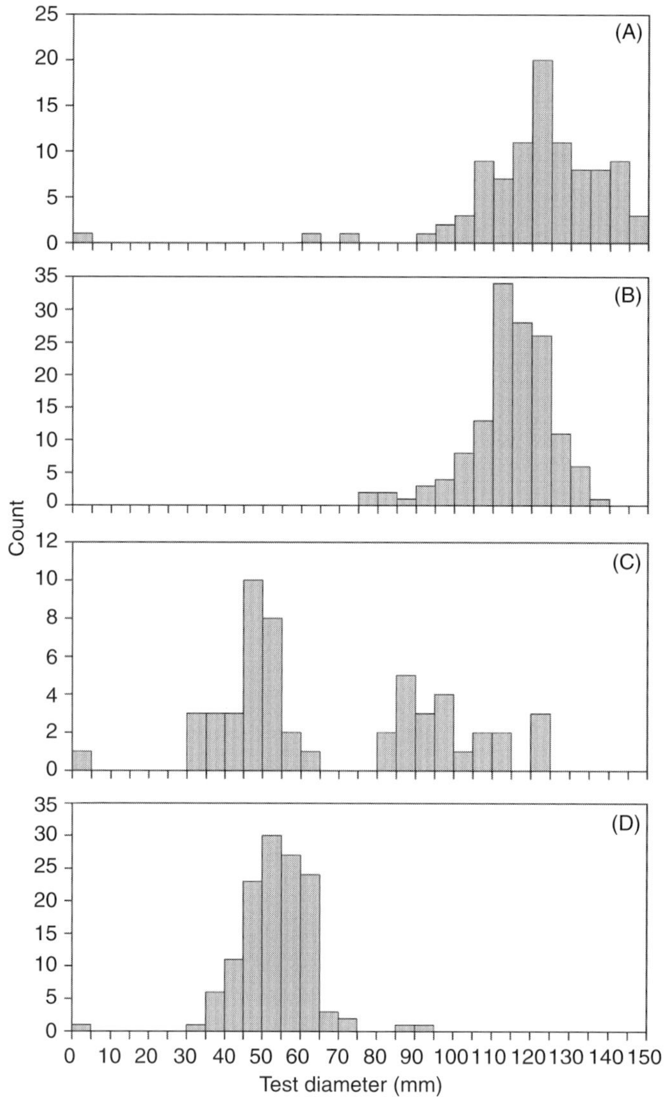

FIGURE 18.4.
Example of variability in size distribution of sea urchins along the axis of George Sound from (a) entrance of fjord to (d) head of fjord. Site positions for each panel are indicated on Figure 18.1.

The results of the present study demonstrate that productivity gradients can have a strong influence on population structure across this coastal landscape. In this case the existence of highly productive and dense kelp forests at the entrance of each of the New Zealand fjords (Wing *et al.* 2007) results in sea urchin populations with a large average test diameter and a consistent adult mode in the size distributions. These sites have relatively few newly emergent juveniles, and rates of larval supply and recruitment are likely low in the

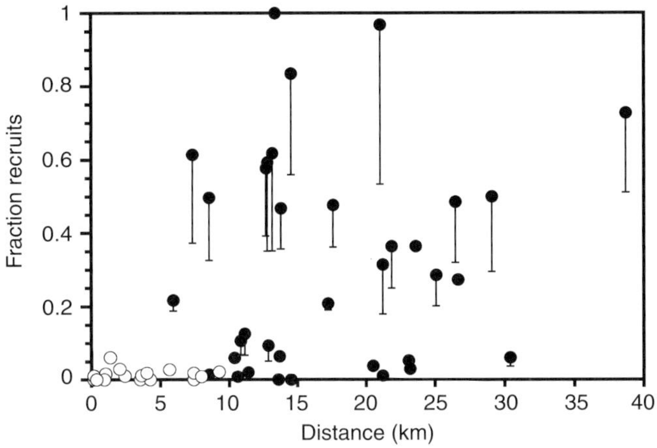

FIGURE 18.5.
Relationship between fraction of newly emergent recruits (<70 mm test diameter) in each population and distance from the fjord entrances. Open circles indicate populations of sea urchins in cluster 1, while closed circles indicate populations in cluster 2. Error bars indicate fraction of population made up of juveniles <50 mm test diameter.

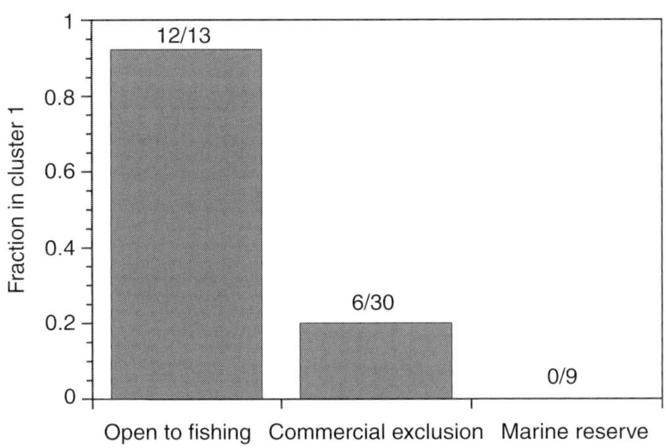

FIGURE 18.6.
Fraction of sites surveyed within areas open to fishing, commercial exclusion zones, and marine reserves that comprise cluster 1: a single large adult mode (115–130 mm test diameter).

high-advection environments at the fjord entrances. In contrast, those populations located in the inner fjords had adult modes with relatively small test diameters and abundant cohorts of newly emergent juveniles making up the size distribution. In some cases the adult mode was absent and populations were made up entirely of newly emergent juveniles, indicating a recent colonization event following a population crash. These populations occupy habitat with poor quality and low density of macroalgal food, but have high rates of larval supply, aided by retention of larvae in the inner fjords by estuarine circulation.

Estuarine circulation is a dominant process in the inner basins of the New Zealand fjords, where freshwater inputs from abundant rainfall (up to 7 m per year) result in a distinct low-salinity layer (Stanton and Pickard 1981). This low-salinity water flows seaward and entrains seawater from below. In order to maintain salt balance along this hydroclinic gradient, the underlying seawater is slowly advected up-fjord, resulting in retention of larvae and phytoplankton in the inner basin. This process can be arrested by strong up-fjord wind events that may last up to 2–4 days (Gibbs *et al.* 2000); however, the average circulation over the time scale of a month is dominated by gravitational or estuarine circulation. Variability in the average volume of freshwater inputs among the 14 fjords, and differences in topography, particularly at the entrances of the fjords, likely result in a range of strengths of estuarine circulation and consequently different retention times for larvae between basins.

The physical gradients within Fiordland, including strong gradients in salinity, wave exposure, and irradiance, result in productivity gradients along the axes of the fjords (Goebel *et al.* 2005; Cornelisen *et al.* 2007; Wing *et al.* 2007). Kelp forest habitat at the entrances of the fjords and abundant growth of phytoplankton in the mid-fjord region are typical patterns of new production within the basins. In the inner-fjord habitats less labile resources, such as riverine inputs of forest litter and microbial recycling of detritus, dominate carbon inputs to the system (McLeod and Wing 2007). These gradients in carbon flux to populations that span the physical gradients in the fjords result in strong bottom-up forcing on population structure (Wing *et al.* 2008).

For example, in Doubtful Sound the average test diameter of *E. chloroticus* has been shown to co-vary between sites with the asymptotic size L_∞ from a Richards growth model, and with per capita gamete production (Lamare and Mladenov 2000; Lamare *et al.* 2002; Wing *et al.* 2003). Wing *et al.* (2003) used a three-dimensional hydrodynamic model to test the likely dispersal pathways of larvae from these entrance sites and their sensitivity to environmental forcing. Model runs were carried out under a range of wind and rain forcing conditions to test the effects of estuarine circulation on loss of larvae to the low-salinity layer, where they perish from osmotic stress, and on the distribution of larvae throughout the fjord. Larval trajectories were tested from different regions of the fjord that harbored populations with high and low per capita gamete production. Correlations between observed settlement patterns and the results of these model runs indicated that the most likely source of larvae into the inner-fjord habitats was from the productive kelp forest regions at the entrance of the fjord. These patterns indicated that the population was most likely made up of source regions at the fjord entrance with high rates of adult growth and gamete production but low rates of larval supply, and sink regions in the inner-fjord habitats with low rates of growth and gamete production but abundant larval supply from the entrainment of larvae due to the estuarine circulation. These source populations shared the characteristics of cluster 1 in the present study, having large average test diameter, access to abundant macroalgal food resources, and low rates of recruitment. Sink populations shared the characteristics of cluster 2 in the present study, having small average test diameter, poor-quality food resources, and a high proportion of newly emergent recruits.

Wing *et al.* (2008) used stable-isotope analysis to examine the carbon inputs to sea urchin populations along the nutritional gradient between the entrance of Doubtful Sound and the habitats at the head of the fjord, where riverine inputs of forest litter dominate the carbon pool. Abundant kelp forests at the entrances of the fjords corresponded to high rates of assimilation of diet, particularly *Ecklonia radiata*, high growth rates, and large average size of individuals. In the inner-fjord habitat these high-quality food sources were replaced with a community of estuarine algae, benthic diatoms, and forest litter. Isotope analysis indicated that low rates of assimilation of diet in these regions corresponded with low growth rates and small average sizes. Interestingly, in the areas where forest litter was common, isotope analysis of sea urchin diet samples indicated the presence of microbial recycling of cellulose material.

The patterns observed in the present study were consistent with these more detailed studies within Doubtful Sound. The apparent source–sink structure of populations in Fiordland presents some important consequences for effective management via marine protected areas. Nevertheless some caveats apply to this indirect evidence for source–sink patterns from population structure, and alternative hypotheses for the observed patterns must be considered.

The observed patterns in population structure provide only indirect evidence for the processes that connect the source–sink structure of sea urchin populations across Fiordland. Evidence on nutrition and population vital rates from Doubtful Sound is provided as being consistent with the hypothesis that regions of the wider Fiordland landscape harbor population sources and sinks; however, direct evidence for larval entrainment is limited to observations by Lamare (1998). Furthermore, Perrin *et al.* (2003), using microsatellite markers to detect genetic variability among sea urchin populations throughout Fiordland, have presented evidence that populations within some fjords may be reproductively isolated from the populations on the outer coast and in adjacent fjords. This suggests the hypothesis that the population in Long Sound was probably reproductively isolated from the regional population network. It follows that source and supply of larvae within Long Sound would be influenced by the effect of the productivity gradient within the fjord on the reproductive output of sea urchins. Lamare *et al.* (2002) report that sea urchins in the inner-fjord habitats do produce viable gametes, though at a much lower per capita output than those in more productive regions. This indicates that some of the larval pool within each fjord could be locally sourced, and this likely varies between fjords according to environmental forcing conditions.

Spatially explicit models of sea urchin populations have been used to address the interaction of adult population structure, larval dispersal, and management through the use of marine protected areas. For example Quinn *et al.* (1993) used a spatially explicit stage-structured model, including Allee effects, to demonstrate that in a population of sea urchins occupying a homogeneous coastline, the application of marine protected areas results in the development of a strong source–sink structure. Furthermore, they observed population persistence even under high rates of harvest pressure outside of the reserves. In this case the source populations contribute to outside areas via a common larval pool. In addition, Morgan and Botsford (2001) demonstrate that deviations from the assumption of homogeneous contributions to the larval pool result in risks to population persistence within protected areas when knowledge of source habitat for recruits or dispersal pathways is lacking. They highlight the importance of directional dispersal as a structuring mechanism in the persistence of the population network. In this case, inclusion of source areas for larvae in the protected area network are vital for regional persistence. In the case study presented here, a highly divided regional population with potential source–sink structure within each fjord presents distinct challenges to the effective management of populations and has indirect effects on communities.

Marine protected areas with exclusion of commercial fishing, as implemented in the Fiordland Marine Management Act 2005, were designed to include regions in each of the fjords to account for the probable subdivided nature of populations among fjords across the landscape. In addition, information on the distribution of kelps and the structure of sea urchin populations was used to help insure that the region of the highly productive kelp forest habitat found toward the entrances of the fjords was included in each of the commercial exclusion zones (Wing *et al.* 2004). Here, within the constraints of legislation, marine protected areas were situated to encompass as much of the gradient in benthic productivity, and therefore the highest diversity of habitats, along the axis of the fjord as possible (Wing *et al.* 2005). In these regions, recreational fishing was also limited so that catch allowances were one-third of those in the open fishing regions. For those species that are reliant on kelp forests for optimal growth and reproduction, these regions were considered critical habitat for population persistence.

In the present analysis, a majority of the sites with sea urchin populations in cluster 1, which are likely productive source populations, fall outside of the commercial exclusion zones (12 of 18), and none are found within the new marine reserves. Although our distribution of sites does not resolve population structure at the extreme seaward boundaries of the protected areas, the spatial distribution of sites in cluster 1 indicates that extensive kelp forest habitats outside of the marine protected areas may harbor the most productive sea urchin populations. This pattern likely results in a large subsidy of larvae and new recruits to the inner-fjord areas from the populations inhabiting the entrances and seaward habitats of each fjord. The magnitude of this subsidy may be influenced by variability in estuarine circulation between the fjords due to the properties of the low-salinity layer (Stanton and Pickard 1981) and entrance topography. Because sea urchins play a critical role in the structure of rocky reef communities and are important grazers on both macroalgae and benthic invertebrates, subsidies of recruits from outside of the marine protected areas will likely have an important influence on the rocky reef communities under spatial management. A corollary is that exploitation or variability within these likely source populations would have potentially large indirect influences on populations and communities within the reserve network (as implicated in Hansen, Chapter 16, this volume). This result highlights the potential influence of population source areas outside of marine protected areas on the internal dynamics of reserves, and the importance of bottom-up forcing as well as physical influences on larval dispersal on population network structure (Wing 2009). The direct implication for conservation in this system is that sea urchin populations, and other populations that rely on productive kelp bed areas as sources (Rodgers and Wing 2008; Jack *et al.* 2009), are probably not adequately protected by a marine reserve network that does not include extensive areas of these critical kelp forest habitats. This highlights the requirement to include areas harboring reproductive source populations in the judicious application of marine reserve networks and adaptive management of existing networks.

Acknowledgments

This research was made possible by contributions from Sara Rutger, Kim Clark, Kirsten Rodgers, Hamish Bowman, Franz Smith, Rebecca McLeod, Lucy Jack, and James Leichter. Support was provided from the University of Otago Department of Conservation and the Marsden Fund.

References

Almany, G., M. Berumen, S. Thorrold, S. Planes and G. Jones (2007). Local replenishment of coral reef fish populations in a marine reserve. *Science* **316**: 742–744.

Botsford, L., A. Hastings and S. Gaines (2001). Dependence of sustainability on the configuration of marine reserves and larval dispersal distance. *Ecology Letters* **4**: 144–150.
Carbines, G. and J. McKenzie (2004). *Movement Patterns and Stock Mixing of Blue Cod in Dusky Sound in 2002*. Ministry of Fisheries, Wellington, New Zealand.
Choat, J. and N. Andrew (1986). Interactions amongst species in a guild of subtidal benthic herbivores. *Oecologia* **68**: 387–394.
Choat, J., and D. Schiel (1982). Patterns of distribution and abundance of large brown algae and invertebrate herbivores in subtidal regions of northern New Zealand. *Journal of Experimental Marine Biology and Ecology* **60**: 129–162.
Cornelisen, C., S. Wing, K. Clark, M. Bowman, R. Frew and C. Hurd (2007). Patterns of macroalgal stable carbon and nitrogen isotope signatures: interaction between physical gradients and nutrient source pools. *Limnology and Oceanography* **52**: 820–832.
Crowder, L., S. Lyman, W. Figuerina and J. Priddy (2000). Source–sink population dynamics and the problem of siting marine reserves. *Bulletin of Marine Science* **66**: 799–820.
Dix, T. (1969). Larval life span of the echinoid *Evechinus chloroticus* (Val.). *New Zealand Journal of Marine and Freshwater Research* **3**: 13–16.
Dix, T. (*1970a*). Biology of *Evechinus chloroticus* (Echinoidea: Echinometridae) from different localities. 1. General. *New Zealand Journal of Marine and Freshwater Research* **4**: 91–116.
Dix, T. (*1970b*). Biology of *Evechinus chloroticus* (Echinoidea: Echinmetridae) from different localities. 3. Reproduction. *New Zealand Journal of Marine and Freshwater Research* **4**: 385–405.
Dugan, J. and G. Davis (1993). Applications of marine refugia to coastal fisheries management. *Canadian Journal of Fisheries and Aquatic Sciences* **50**: 2029–2042.
Fogarty, M. (1998). Implications of migration and larval interchange in American lobster (*Homarus americanus*) stocks: spatial structure and resilience. In *North Pacific Symposium on Invertebrate Stock Assessment and Management* (G. Jamieson and A. Campbell, eds.). *Canadian Journal of Fisheries and Aquatic Science*, Naniamo, BC, Canada: 273–283.
Fogarty, M. (1999). Essential habitat, marine reserves and fishery management. *Trends in Ecology and Evolution* **14**: 133–134.
Fogarty, M. and L. Botsford (2007). Population connectivity and spatial management of marine fisheries. *Oceanography* **20**: 112–123.
Gaines, S., B. Gaylord and J. Largier (2003). Avoiding current oversights in marine reserve design. *Ecological Applications* **13**: S32–S46.
Gibbs, M., M. Bowman and D. Dietrich (2000). Maintenance of near surface stratification in Doubtful Sound, a New Zealand fjord. *Estuarine Coastal and Shelf Science* **51**: 683–704.
Goebel, N., S. Wing and P. Boyd (2005). A mechanism for onset of diatom blooms in a fjord with persistent salinity stratification. *Estuarine Coastal and Shelf Science* **64**: 546–560.
Halpern, B. (2003). The impact of marine reserves: do they work and does size matter? *Ecological Applications* **13**: S117–S137.
Hart, P. and J. Reynolds (2002). *Handbook of Fish Biology and Fisheries: Vol. 2*. Blackwell, Oxford, UK.
Hastings, A. and L. W. Botsford (1999). Equivalence in yield from marine reserves and traditional fisheries management. *Science* **284**: 1537–1538.
Hixon, M., S. Pacala and S. Sandin (2002). Population regulation: historical context and contemporary challenges of open vs. closed systems. *Ecology* **83**: 1490–1508.
Jack, L., S. R. Wing and R. J. McLeod (2009). Prey base shifts in red rock lobster *Jasus edwardsii* in response to habitat conversion in Fiordland marine reserves: implications for effective spatial management. *Marine Ecology Progress Series* **381**: 213–222.
Jones, G., M. Milicich, M. Emslie and C. Lunow (1999). Self-recruitment in a coral reef fish population. *Nature* **402**: 802–804.
Kritzer, J. P. and P. F. Sale (2006). *Marine Metapopulations*. Academic Press/Elsevier, London.
Lamare, M. (1998). Origin and transport of larvae of the sea urchin *Evechinus chloroticus* (Echinodermata: Echinoidea) in a New Zealand fjord. *Marine Ecology Progress Series* **174**: 107–121.

Lamare, M. and P. Mladenov (2000). Modelling somatic growth in the sea urchin *Evechinus chloroticus* (Echinoidea: Echinometridea). *Journal of Experimental Marine Biology and Ecology* **243**: 17–43.

Lamare, M. and S. Wing (2001). Calorific content of New Zealand marine macrophytes. *New Zealand Journal of Marine and Freshwater Research* **35**: 335–341.

Lamare, M., P. Brewin, M. Barker and S. Wing (2002). Reproduction of the sea urchin *Evechinus chloroticus* (Echinodermata: Echinodea) in a New Zealand fjord. *New Zealand Journal of Marine and Freshwater Research* **36**: 219–232.

Lauck, T., C. Clark, M. Mangel and G. Munro (1998). Implementing the precautionary principle in fisheries management through marine reserves. *Ecological Applications* **8**: S72–S78.

Legendre, P. and L. Legendre (1998). *Numerical Ecology*. Elsevier, Amsterdam.

McLeod, R. and S. Wing (2007). Hagfish in the New Zealand fjords are supported by chemoautotrophy of forest carbon. *Ecology* **88**: 809–816.

Morgan, L. and L. Botsford (2001). Managing with reserves: modelling uncertainty in larval dispersal for a sea urchin fishery. In *Spatial Processes and Management of Marine Populations*. Alaska Sea Grant College Program, Fairbanks, AK: 667–684.

Neteler, M. and H. Mitasova (2002). *Open Source GIS: A GRASS GIS Approach*. Kluwer Academic Publishers, Boston, MA.

NRC (National Research Council) (2001). *Marine Protected Areas: Tools for Sustaining Ocean Ecosystems*. National Academies Press, Washington, DC.

Orensanz, J. M. and G. S. Jamieson (1998). The assessment and management of spatially structured stocks: an overview of the North Pacific Symposium on Invertebrate Stock Assessment and Management. In *Proceedings of the North Pacific Symposium on Invertebrate Stock Assessment and Management* (G. Jamieson and A. Campbell, eds.). *Canadian Journal of Fisheries and Aquatic Science* **125**: 441–459.

Ostrow, D., S. Wing, P. Mladenov and M. Roy (2001). Genetic differentiation of *Terebratella sanguinea* in the New Zealand fjords: a dispersal barrier in the marine environment? In *Brachiopods Past and Present* (C. Howard, C. Brunton, L. Robin, M. Cocks and S. Long, eds.). The Natural History Museum, London: 159.

Pawson, D. (1965). The distribution of echinoderms along the east coast of New Zealand. *Transactions of the Royal Society of New Zealand* **6**: 245–252.

Perrin, C., M. Roy and S. Wing (2003). Genetic differentiation amongst populations of the sea urchin *Evechinus chloroticus* and the sea star *Coscinasterias muricata* in New Zealand's fjords. In *Echinoderm Research 2001* (J. Féral and B. David, eds.). Balkema, Rotterdam, The Netherlands.

Perrin, C., S. Wing and M. Roy (2004). Population genetic structure amongst populations of the sea star *Coscinasterias muricata* in the New Zealand fjords. *Molecular Ecology* **13**: 2183–2195.

Pulliam, H. (1988). Sources, sinks, and population regulation. *American Naturalist* **132**: 652–661.

Quinn, J., S. Wing and L. Botsford (1993). Harvest refugia in marine invertebrate fisheries: models and applications to the red sea urchin, *Strongylocentrotus franciscanus*. *American Zoologist* **33**: 537–550.

Roberts, C. (1998). Sources, sinks, and the design of marine reserve networks. *Fisheries* **23**: 16–19.

Rodgers, K. and S. Wing (2008). Spatial structure and movement of blue cod (*Parapercis colias*) in Doubtful Sound, New Zealand, inferred from δ13C and δ15N. *Marine Ecology Progress Series* **359**: 239–248.

Sanford, E. and B. Menge (2007). Reproductive output and consistency of source populations in the sea star *Pisaster ochraceus*. *Marine Ecology Progress Series* **349**: 1–12.

Shears, N. and R. Babcock (2002). Marine reserves demonstrate top-down control of community structure on temperate reefs. *Oecologia* **132**: 131–142.

Sköld, M., S. Wing and P. Mladenov (2003). Genetic subdivision of the sea star *Coscinasterias muricata* in the fjords of New Zealand. *Marine Ecology Progress Series* **250**: 163–174.

Stanton, B. and G. Pickard (1981). Physical oceanography of the New Zealand fiords. *New Zealand Oceanographic Institute Memoir* **88**: 3–37.

Swearer, S., J. Caselle, D. Lea and R. Warner (1999). Larval retention and recruitment in an island population of a coral-reef fish. *Nature* **402**: 799–802.

Thorrold, S., C. Latkoczy, P. Swart and C. Jones (2001). Natal homing in a marine fish metapopulation. *Science* **291**: 297–299.

Walker, M. (1984). Larval lifespan, larval settlement, and early growth of *Evechinus chloroticus* (Val.). *New Zealand Journal of Marine and Freshwater Research* **18**: 393–397.

Wing, S. (2009). Decadal scale dynamics of sea urchin population networks in Fiordland, New Zealand are driven by juxtaposition of larval transport against benthic productivity gradients. *Marine Ecology Progress Series* **378**: 125–134.

Wing, S., M. Gibbs and M. Lamare (2003). Reproductive sources and sinks within a sea urchin, *Evechinus chloroticus*, population of a New Zealand fjord. *Marine Ecology Progress Series* **248**: 109–123.

Wing, S., M. Bowman, F. Smith and S. Rutger (2004). *Analysis of Biodiversity Patterns and Management Decision Making Processes to Support Stewardship of Marine Resources and Biodiversity in Fiordland: A Case Study*. Report 2 of 3, Ministry for the Environment, Wellington, New Zealand.

Wing, S., M. Bowman, F. Smith and S. Rutger (2005). *Analysis of Biodiversity Patterns and Management Decision Making Processes to Support Stewardship of Marine Resources and Biodiversity in Fiordland: A Case Study*. Report 3 of 3, Ministry for the Environment, Wellington, New Zealand.

Wing, S., J. Leichter, C. Perrin, S. Rutger, M. Bowman and C. Cornelisen (2007). Topographic shading and wave exposure influence morphology and ecophysiology of *Ecklonia radiata* (C. Agardh 1817) in Fiordland, New Zealand. *Limnology and Oceanography* **52**: 1853–1864.

Wing, S., R. McLeod, K. Clark and R. Frew (2008). Plasticity in diet of two echinoderm species across an ecotone: microbial recycling of forest litter and bottom-up forcing of population structure. *Marine Ecology Progress Series* **360**: 115–123.

19

Source–sinks, metapopulations, and forest reserves: conserving northern flying squirrels in the temperate rainforests of Southeast Alaska

Summary

Reserves are a common strategy used to ensure the viability of wildlife populations, but their effectiveness is rarely empirically evaluated. The Tongass National Forest implemented a conservation plan (TLMP) in 1997 to maintain biological diversity across Southeast Alaska, the cornerstone of which was an integrated system of large, medium, and small old-growth reserves (OGRs). Small OGRs were intended to facilitate functional connectivity between larger reserves and ensure well-distributed populations of forest-dependent wildlife. The northern flying squirrel (*Glaucomys sabrinus*) was selected as an indicator of wildlife communities that operate at small spatial scales because its abundance has been correlated with old-growth forest structure and processes and because of specific habitat requirements for efficient locomotion. Previous research predicted that small OGRs were unlikely to support flying squirrels over a 100-year time horizon. Consequently, the presence and persistence of flying squirrels in small OGRs depended on dispersal from larger reserves. Using data from telemetry experiments, we determined effective distances immigrants could move through landscapes composed of old-growth and managed forests. Effective distance accounted for the resistance of habitats such as clearcuts that are difficult for flying squirrels to traverse. We used findings of previous studies to parameterize a logistic population growth model incorporating dispersal to determine the number of dispersers necessary to enable a flying squirrel population in a small OGR to persist for 25 and 100 years. We combined that information with a function relating the probability

Sources, Sinks and Sustainability, ed. Jianguo Liu, Vanessa Hull, Anita T. Morzillo and John A. Wiens. Published by Cambridge University Press.

of successful dispersal with effective distance to estimate the maximum effective distances between OGRs that would ensure flying squirrels colonize and persist in small OGRs for 25 and 100 years. Our findings underscore the essential role of immigration in sustaining sinks and facilitating metapopulation viability among unsustainable fragmented populations (i.e., sinks). They also demonstrate the extent to which permeability of landscape elements can influence the probability of dispersal and functional connectivity of subpopulations in a managed matrix.

Introduction

Habitat loss and fragmentation are widely regarded as major threats to the viability of wildlife populations (Wilcove *et al.* 1986; Rolstad 1991; Fahrig and Merriam 1994; Wiens 1995; Reed 2004). Along the northwestern Pacific coast, for example, habitat loss and fragmentation from timber harvests have threatened wildlife species that rely on late-seral forests (Richards *et al.* 2002). Of particular concern is maintaining the minimum population size (i.e., "ecologically effective" density, *sensu* Soulé *et al.* 2003: 1239) of essential species to fulfill their critical roles in ecosystem processes (Carey 2000). For many species, fragmentation can cause population declines beyond what would be expected solely from the total amount of habitat loss (Schumaker 1996) because of critical thresholds in their responses to landscape structure (With and Crist 1995).

A consequence of habitat fragmentation can be the fracturing of wildlife populations into subpopulations distributed among residual habitat patches. Disjunct population segments may take on the characteristics of patchily distributed populations, metapopulations (i.e., groups of separate but functionally connected subpopulations), source–sinks, or insular populations depending on the size and quality of patches or the dispersal and connectivity between patches. Empirical evidence suggests that the population dynamics of a wide range of wildlife species in fragmented landscapes resembles a pattern of stochastic local extinctions and recolonizations (Opdam *et al.* 1985; Opdam 1991; Hanski 1994). Thus, for wildlife populations to persist in heterogeneous landscapes, individual habitat patches must be large enough to provide for viable subpopulations (e.g., Hanski 1994; Smith and Person 2007), or the juxtaposition of suitable habitat must allow for interpatch migration to colonize patches that become vacant from localized extinctions (Burkey 1989; Bender *et al.* 1998). Nonetheless, habitat fragmentation may create an uneven distribution with respect to size and quality of patches and source–sink dynamics may arise in which unsustainable subpopulations in small or poor-quality habitat fragments (sinks) are maintained by frequent dispersal from larger source populations (Pulliam 1988). Under those circumstances, landscape connectivity (the degree to which landscape facilitates or impedes movement between resource patches; see Taylor *et al.* 1993) is a critical feature largely determining species' distributions (Verbeylen *et al.* 2003) and persistence (Fahrig and Merriam 1985; Beier 1993; Ferreras 2001; Richards *et al.* 2002).

Establishing habitat reserves is a common strategy used to increase the probability that wildlife populations remain viable and well distributed in developed and fragmented landscapes (FEMAT 1993; USDA Forest Service 1997). Depending on size and habitat quality, some reserves sustain viable populations that produce a surplus of offspring (i.e., source populations), many of which disperse to other habitat patches. Smaller or poorer-quality reserves are less likely to sustain viable populations (Smith and Person 2007), in which case they function as sinks that are subsidized by the surplus of emigrants from source populations (Pulliam 1988). Nevertheless, sinks can play a key role in sustaining viable metapopulations by serving as breeding habitat for females (thereby contributing to the overall population) and by functioning as stepping-stones that facilitate dispersal among subpopulations (Pulliam 1988; Pearson and Fraterrigo, Chapter 6, this volume). Unfortunately, the efficacy of reserves to sustain wildlife populations is rarely empirically evaluated, even for individual species (Ferreras 2001; Richards *et al.* 2002; Carroll *et al.* 2003). Because metapopulation or source–sink dynamics may occur in populations distributed among habitat reserves (Hansen, Chapter 16, this volume), conservation strategies that rely on reserves must consider not only the amounts of habitat to be retained, but also the spatial configurations of habitat across landscapes (Schumaker 1996), the characteristics of the intervening matrix (Wiens *et al.* 1993; Ferreras 2001), and temporal dynamics (Leroux *et al.* 2007).

A provocative example of land use planning incorporating habitat reserves at a large scale is the conservation strategy for the Tongass National Forest in southeast Alaska (USDA Forest Service 1997). Southeast Alaska, which comprises the Alexander Archipelago and a narrow mainland strip, is almost entirely national forest, encompassing ~6.8 million hectares. The island topography divides much of the forest among thousands of islands that range in area from 0.5 ha to 6,700 km^2 (USDA Forest Service 1997). Approximately 160,000 ha of the most productive forest stands have been clearcut logged (USDA Forest Service 2003: 3–43), prompting concerns about the cumulative impacts of timber harvests on forest plant and animal communities and the sustainability of ecosystem functions and services (Everest *et al.* 1997; Shaw 1999). To address those concerns, the US Forest Service revised the Tongass Land Management Plan (TLMP) in 1997 to include a conservation strategy designed to sustain terrestrial wildlife communities. The 1997 TLMP was recently amended (USDA Forest Service 2008), but the underlying framework for the conservation strategy remains fundamentally unchanged.

A cornerstone of the TLMP strategy was an integrated system of large (≥16,200 ha), medium (≥4,050 ha and <16,200 ha), and small (≥650 ha and <4,050 ha) old-growth forest reserves (OGR; USDA Forest Service 1997: K-1), which – together with other non-development lands – was expected to sustain viable populations without demographic contributions from the matrix. That is, the conservation strategy explicitly assumed that wildlife species that were intended to be conserved through the reserve system occurred exclusively within old-growth forests. Medium and large reserves were established across landscapes at intervals of ≤12.9 km and ≤32.2 km, respectively. The old-growth tracts were expected to support viable populations of forest-dependent species either within individual reserves (e.g., small mammals, mustelids) or among a collection of reserves (e.g., wolves, bears) that sustain a metapopulation structure (*sensu* Hanski and Gilpin 1991; USDA Forest Service 1997: N-16; Flynn *et al.* 2004). A small OGR comprising ≥16% of national forest lands was established in each Value Comparison Unit (i.e., management unit) and was intended to provide functional habitat for animals dispersing between large and medium reserves and to ensure that the species of concern have a relatively high likelihood of occurring in each watershed >4,000 ha (USDA Forest Service 1997: N-21). This implies that small OGRs must support populations of focal species of sufficient size and persistence (low risk of extinction) to ensure a high probability of dispersal from reserves as well as a high probability that populations will be present within individual watersheds (i.e., well distributed). Consequently, small OGRs must sustain populations for long enough to produce a sufficient number of dispersers to ensure successful emigration to neighboring reserves, or the distances separating small OGRs from larger OGRs must be sufficiently small to ensure a steady flow of immigrants and emigrants to and from small OGRs. Resistance of the landscape to movement by focal species and size and quality of small OGRs will likely determine the success of either strategy.

To monitor the effectiveness of the OGR strategy, the northern flying squirrel (*Glaucomys sabrinus*) was proposed as one of several management indicator species (Suring 1993) because its abundance may be correlated with old-growth forest structure and processes (Carey 2000) and because of specific habitat requirements for efficient locomotion (Scheibe *et al.* 2006). Its small size and home range (<5 ha; Smith 2007) make it particularly well suited for evaluating the effectiveness of small OGRs (Smith *et al.* 2005). Previously, Smith and Person (2007) modeled the population dynamics of northern flying squirrels to determine the minimum patch size and habitat composition to sustain flying squirrel populations over the 100-year planning horizon. Their simulations indicated that flying squirrels had a low probability (0.66–0.73) of persisting in isolated small OGRs; furthermore, the minimum patch size to sustain populations with a high probability (≥0.95) of persistence was an order of magnitude larger than the "preferred" prescription of 650 ha reserves with a composition of ≥50% productive old-growth forest (USDA Forest Service 1997: Appendix K). Consequently, small OGRs specified by TLMP may function as sinks or metapopulation segments dependent on recolonization from larger reserves. Because dispersal by flying squirrels may be critical for the viability of flying squirrel populations within small – and possibly larger – OGRs, S. Pyare and W. P. Smith, in related studies, used field behavioral trials, radiotelemetry, and landscape analysis to quantify landscape connectivity in highly modified landscapes (Pyare and Smith 2005), and Flaherty *et al.* (2008) studied behavioral and energetic aspects of dispersal in managed landscapes. We combine results from Pyare and Smith (2005) and results from population modeling described by Smith and Person (2007) to evaluate the efficacy of small OGRs as a functionally connected network of reserves that provide temporary functional habitat for animals dispersing between large and medium OGRs. Our specific objectives were

1. to determine the number of immigrants required for flying squirrel populations in small OGRs to persist for 25 and 100 years;
2. to determine the maximum landscape resistance that could exist between small and larger reserves and still ensure the required number of immigrants.

We used those results to assess the value of small OGRs specified by TLMP for the conservation of northern flying squirrels within the Tongass National Forest. Because the conservation strategy assumed that viability of northern flying squirrels depended solely on old-growth forest reserves, our analysis excluded any contribution of the matrix to population viability except in considering landscape permeability and the probability of dispersal among old-growth reserves. The results of our analyses are useful in establishing general guidelines for designing a network of functionally connected reserves. We offer specific recommendations for the distribution of reserves to improve the likelihood of viable northern flying squirrel populations in managed landscapes of Southeast Alaska.

Study area

Southeast Alaska is unique because of many naturally fragmented landscapes, a dynamic geological history (MacDonald and Cook 1996), and its expansive coastal temperate rainforest (Harris and Farr 1974; Alaback 1982). It has a cool, wet (200–600 cm precipitation) maritime climate: mean monthly temperatures range from 13°C in July to 1°C in January (Searby 1968).

Coniferous rainforests occur at lower elevations (<600 m) throughout the region. Fragmentation of old-growth forest habitats has increased substantially since the mid-twentieth century because of extensive clearcut logging throughout the region (USDA Forest Service 1997, 2003).

Smith and Nichols (2003) provide more detail about the region and the two major forest types that represent the ends of a continuum of forest canopy and a range of soil conditions in this region. High-volume old-growth stands (Julin and Caouette 1997) are primarily Sitka spruce (*Picea sitchensis*)–western hemlock (*Tsuga heterophylla*) forests (DeMeo *et al.* 1992) with a dense canopy of tall (>60 m), large (≤2.5 m diameter) trees (Harris and Farr 1974; Alaback 1982) interspersed with small openings created by windthrow. Old-growth forests have not experienced any logging or anthropogenic disturbance and can vary in age from 300 to more than 500 years old, depending on exposure and natural disturbance. Secondary forests typically do not develop old-forest characteristics until 350 years following catastrophic disturbance (Nowacki and Kramer 1998). This is the primary habitat of the Prince of Wales flying squirrel, *Glaucomys sabrinus griseifrons* (Smith and Nichols 2003; Smith *et al.* 2004) and supports the highest mean densities recorded for the species (Smith *et al.* 2003; Smith 2007). In contrast, peatland-scrub/mixed-conifer forests typically have a sparse canopy, and understory vegetation varies from thick patches of shrubs to open peatland heaths. Flying squirrels can be abundant in this habitat (Smith and Nichols 2003; Smith *et al.* 2003; Smith 2007), but sex ratios in those populations are skewed toward males and the habitat probably does not sustain populations without immigration (Smith and Person 2007).

We used data from study sites located on Prince of Wales Island (55.9° N, 133.2° W; Fig. 19.1) to conduct our evaluation of small OGRs. Prince of Wales is the largest island in the Alexander Archipelago, a group of more than 20,000 islands that – together with a narrow mainland strip – comprise Southeast Alaska. Four decades of intensive logging in the study area has created an ideal setting for studying functional connectivity and the population dynamics of northern flying squirrels in managed landscapes. The landscape surrounding our study sites were composed of about 1% clearcut, 9% young (<25-year-old) second-growth, 38% older (25–50-year-old) second-growth, and 49% old-growth forests, including lower-volume peatland-scrub/mixed-conifer forests. The remaining 3% was non-forested, which included open peatland heaths.

In clearcuts, residual shrubs and tree seedlings grow rapidly within 5 years after canopy removal (Alaback 1982). A shrub community dominates the vegetation initially, but Sitka spruce and western hemlock seedlings, which become established at about the same time, overtop the shrub layer within 8–10 years after logging (Harris and Farr 1974). Shrubs and herbs are effectively eliminated after the forest canopies close, between 25 and 35 years post-harvest and

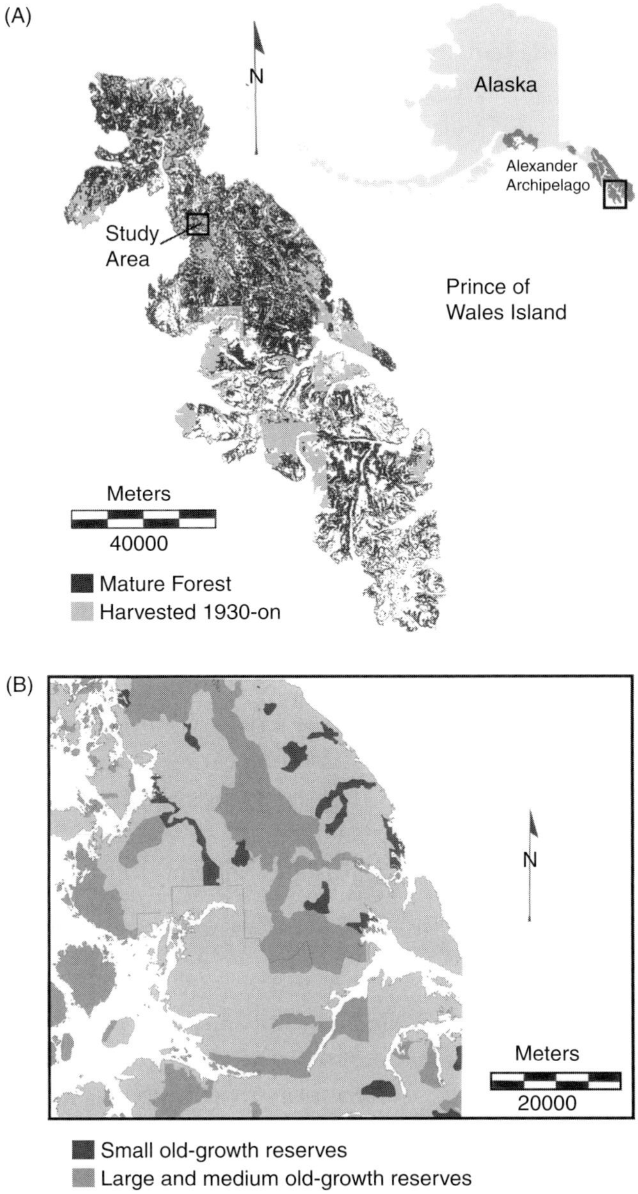

FIGURE 19.1.
(a) Study area, an extensively managed landscape of coastal temperate rainforest, Prince of Wales Island, Alexander Archipelago, Southeast Alaska. (b) Boundaries of old-growth reserves prescribed in the 1997 Tongass Land Management Plan (USDA Forest Service 1997) for northern Prince of Wales Island. Additional information regarding land use designations for the Tongass National Forest, including size and placement of old-growth reserves, can be found at http://steffenrasile.com/staging/tetratech/tongass/implementation-maps.php.

usually do not become re-established until the stand age approaches 150 years (Alaback 1982). Secondary forests remain even-aged for up to 300 years before gradually transitioning into an uneven-aged condition (Harris and Farr 1974). Mean stem density in unthinned young ($\bar{x}$ = 19 years) second-growth stands ranges around 3,300–5,300 trees/ha, with the composition of spruce being 31–52%. Older ($\bar{x}$ = 57–60 years) second growth averages 3,450–5,300 trees/ha and an overstory composition that is 29–42% Sitka spruce (Deal and Farr 1994). Average canopy height of nearby 19-year-old stands was 9–10 m, whereas the height of 58-year-old stands averaged 35–36 m (DeMars 2000). The sparse understory and lower abundance of preferred food items (Flaherty *et al.* 2008) render older second-growth stands unsuitable as habitat for northern flying squirrels; the structure of unthinned second-growth stands likely precludes their use as travel corridors. Indeed, small old-growth fragments (<50 ha) surrounded by second growth can become isolated and the local flying squirrel population extirpated (E. A. Flaherty and M. Ben-David, unpublished data).

The ability of northern flying squirrels to move through managed landscapes is significantly influenced by forest structure, and factors to take into consideration include perceptual range and fine-scale movements (Flaherty *et al.* 2008), cost of transport (Scheibe *et al.* 2006), and effective movement rates (Pyare and Smith 2005). Generally, the average length of straight-line movement increases and cost of transport decreases as forest canopy increases in height and stem density decreases. Friction (i.e., resistance to movement), and presumably risk of predation, is higher in clearcuts than mature forests; unthinned second-growth stands obstruct visibility and impede gliding.

Methods

We tested the functionality of small OGRs in two ways. First, we estimated the number of juvenile immigrants required for a flying squirrel population that had been reduced to a breeding pair within a small OGR to persist for short (25-year) and long (100-year) time periods. We excluded adult dispersal because our focus was evaluating the role of immigration in facilitating demographic rescue. We assumed that persistence over 25 years would represent the role of small OGRs as stepping-stones between larger reserves or as population sinks; whereas persistence for 100 years, which was the planning horizon in the 1997 TLMP, would also imply greater population stability in small OGRs contributing to metapopulation dynamics. We related the achievement of those persistence goals to dispersal rates from source populations and to the probability of successful dispersal as a function of habitat composition and distance between reserves. We assumed that source populations were supported solely by primary habitat because peatland-scrub/mixed-conifer forests depended on immigration to sustain flying squirrel populations and likely produced few emigrants (Smith and Person 2007), and second growth was unsuitable (Smith and Nichols 2003; Flaherty *et al.* 2008).

The second way in which we examined functionality of small OGRs was to predict the number of immigrants necessary to ensure that a vacant reserve would be colonized by a breeding pair (i.e., dispersers included ≥1 member of each sex) given distance from a source population and the habitat composition of the landscape separating the reserves.

Emigration rate and dispersal probability

We used the proportion of radio-marked juvenile northern flying squirrels that left their natal area during autumn to estimate emigration rate. To distinguish juvenile movements as emigration, Pyare *et al.* (2010) computed the mean ± 95% confidence interval (CI) of Euclidean distances between natal dens and all subsequent den locations (67 dens among 21 juveniles). Permanent relocations with a displacement distance greater than the upper 95% CI were classified as emigrations.

Pyare and Smith (2005) estimated the probability of successful juvenile dispersal as a function of distance and habitat type. They used experimental translocations to measure the elapsed time and distance moved by flying squirrels coursing through different habitats: the ratio of distance traveled to elapsed time (effective movement rate) became a surrogate for landscape resistance. Specifically, they derived habitat-specific "effective distance" estimates to represent the effective travel distance of moving through a habitat patch after explicitly taking into account its landscape resistance. For example, because travel times were about 16 times greater in recent clearcuts than in intact old-growth forest, the effective distance of a clearcut crossing was relatively 16× greater than moving a similar distance through uninterrupted old-growth forest. Effective distance estimates were derived experimentally for a subset of common habitats in the study area and subsequently modeled as a function of vertical forest height (Table 19.1). To estimate cumulative landscape resistance, the total effective distance was calculated by adding the effective distance of all habitats encountered during a specific dispersal event. Thus, dispersing through 1 km of continuous old-growth forest was equivalent to dispersing through 0.1 km of a recent clearcut; intermediate effective distances result from flying squirrels moving across a combination of habitats (Table 19.1). Because the exact pathway of dispersal events was not known, Pyare and Smith (2005) used minimum effective distances between points of origin and dispersal in all subsequent analyses. To estimate the dispersal probability of juveniles, the total effective distance of each juvenile that dispersed from the natal

TABLE 19.1. *Habitat-specific estimates of effective distance, which represents the distance that animals can effectively move through a patch after considering its resistance, i.e., the degree to which the habitat facilitates or impedes movement. Estimates were initially derived from a series of translocation experiments in a related study (Pyare and Smith 2005) and then subsequently modeled as a function of the vertical height of vegetation.*

Habitat type	Effective distance (m-equivalent)
Late successional conifer	1.00
Mesic conifer–peatland	1.00
Regenerating conifer stands (6–80 years)	1.00–15.58
Open peatland	5.34
Brush	5.34
Recent clearcuts (≤ 5 years)	15.58
Open other (alpine/development)	15.58
Major roads	15.58
Minor roads	15.58
Minor streams	15.58
Minor freshwater bodies	15.58
Saltwater crossings	999.000
Major freshwater bodies	999.000
Major streams/rivers	999.000

den to a new den was calculated, and the resulting proportions of the juvenile sample that dispersed over a range of observed effective distances was then used to derive the following dispersal probability function (Fig. 19.2):

$$P = -0.4559 * \text{Ln}\left[\text{effective distance, ft}\right] + 3.7644. \quad (19.1)$$

Probability of persistence

We followed the procedures of Smith and Person (2007) to compute probability of persistence (P_t) from predictions of time to extinction (T_e) using the following equation:

$$P_t = e^{-t/T_e} \quad (19.2)$$

where t is the time period for which P_t is calculated. Rearranging Eq. (19.1), we computed T_e for $P_t = 0.95$ and $t = 25$ years (comparable to the time horizon of the revised TLMP conservation strategy; USDA Forest Service 2008) and $t = 100$ years (time horizon of 1997 TLMP and typical of viability analyses of vertebrate populations). Time to extinction for 95% probability of persistence over a period of 25 years was 490 years; for a time horizon of 100 years, T_e was

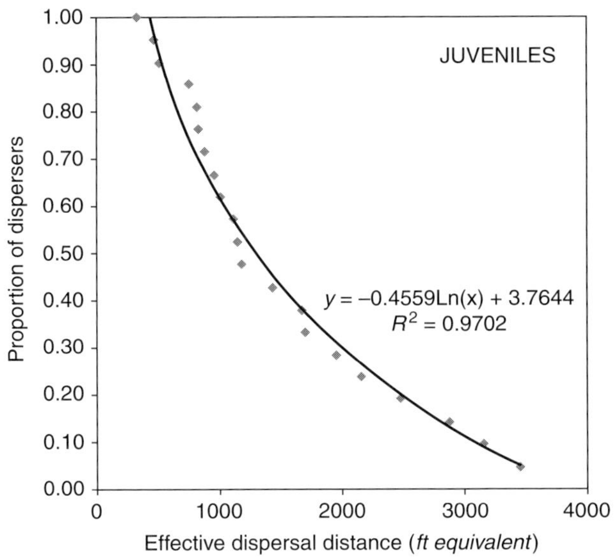

FIGURE 19.2.
Proportion of juvenile northern flying squirrels (*Glaucomys sabrinus*) that dispersed in managed landscapes of Southeast Alaska relative to effective distance (ft-equivalent), which was calculated by adding the effective distance (i.e., resistance or permeability) of different landscape elements, from Pyare and Smith (2005).

1,950 years. Smith and Person (2007) used a model described by Foley (1994) to estimate time to extinction (T_e) for northern flying squirrels in isolated small (650 ha) old-growth reserves:

$$T_e = \frac{k^2}{v}\left[1 + \frac{2}{3}\left(\frac{kr}{v}\right)\right]; \tag{19.3}$$

where k is the natural log of the population carrying capacity (K) of the reserve, r is the per capita rate of increase when the population is at low density relative to K, and v is the variance of r. The model incorporates density-dependent population growth, which dampens large fluctuations in population growth and increases the predictions of T_e compared with results from models that ignore density-dependent population growth. We rearranged terms in Eq. (19.3) to compute r, given T_e, K and v:

$$r = \left(\frac{3v}{2k}\right)\left[\left(\frac{T_e v}{k^2}\right) - 1\right]. \tag{19.4}$$

We chose $v = 0.39$ because it was based on the variance in r reported for a 43-year population study of northern flying squirrels (Fryxell *et al.* 1998); the longest time series of population data in the literature for this species. Carrying capacity (K) was estimated for the two prescriptions of habitat composition for small old-growth reserves specified in TLMP (USDA Forest Service 1997) and for a third scenario in which the entire reserve was composed of primary habitat for flying squirrels (Smith and Person 2007). For the "minimum" prescription in which ≥25% of the habitat within the reserve was primary habitat, $K =$ 454 squirrels. For the "preferred" prescription in which ≥50% of a small OGR was primary habitat, $K = 910$ squirrels. If the entire reserve was primary habitat, $K = 1{,}820$ squirrels. We computed r for each value of K and for T_e equal to 490 and 1,950 years. This procedure generated estimates of r (for each value of K) necessary at low densities of flying squirrels to achieve a 95% probability of populations in small OGRs persisting for 25 and 100 years.

Smith and Person (2007) estimated an intrinsic value of $r = 0.14$ for flying squirrels that were undergoing rapid population growth at low density. Their value was estimated directly from birth and death rates and thus it represented intrinsic growth for a closed population (i.e., in the absence of immigration) well below carrying capacity. We compared their value with r calculated using Eq. (19.4) and assumed that the difference in values represented the contribution to population growth required from immigration to achieve 95% probability of persistence within small OGRs for 25 and 100 years.

To convert the difference in r to an annual number of dispersers, we constructed a simple logistic population growth model:

$$N_{t+1} = N_t + N_t r_i \left(1 - \frac{N_t}{K}\right) + d \qquad (19.5)$$

where N_t is the population at time t, r_i is intrinsic r in the absence of immigration, K is carrying capacity of the reserve, and d is the number of dispersers per year. We assumed that the sex ratio of dispersers was 50:50. We began the simulation with a single breeding pair of flying squirrels and calculated N_t for three time steps that represented population growth at low density relative to K. For each simulation, we added a constant number of dispersers (d) and estimated average per capita rate of growth ($\check{r}$) over the three time steps. We continued to increase d until the estimate of $\check{r}$ equaled r predicted from Eq. (19.4). We used the estimate of $r_i = 0.14$ from Smith and Person (2007) and repeated simulations for $K = 454$, 910, and 1,820 flying squirrels. Our procedure enabled us to predict the number of immigrant dispersers needed annually to achieve our stated goals for persistence of flying squirrel populations in small OGRs, given three different carrying capacities.

We modeled juvenile dispersal within hypothetical landscapes in which a small OGR was positioned between two medium OGRs (~4,000 ha of ≥50% primary habitat; USDA Forest Service 1997). That configuration was consistent with the intended purpose of small OGRs to function as stepping-stones between larger reserves or as components in a source–sink or metapopulation structure. We selected medium OGRs because they represented the smallest reserve with a relatively high (0.85) probability of persisting in isolation for 100 years (Smith and Person 2007). To simulate dispersal from two medium OGRs to a connecting small OGR, we estimated the number of available dispersers by multiplying the average proportion of juveniles occupying primary habitat (0.193, SE = 0.395, $n = 254$; Smith and Nichols 2003) within the medium reserve multiplied by the average autumn population size (i.e., average population density × habitat acreage) in primary habitat (Smith and Person 2007). The number of available juveniles in the medium OGRs was multiplied by the proportion of radio-collared juveniles that emigrated (Pyare and Smith 2005) and was doubled to simulate dispersers emigrating from two source populations (i.e., one on each side). We used this information and output from the logistic growth model to estimate the maximum effective distance that could exist between small OGRs and medium OGRs and still meet demographic requirements in six scenarios that varied in habitat composition and time horizon. To calculate effective distance for each scenario, we substituted the proportion of juveniles needed into Eq. (19.1) (Pyare and Smith 2005).

To estimate the number of dispersers required to recolonize a vacant small OGR, we estimated the number of juvenile emigrants needed to establish a breeding pair, regardless of patch size, habitat composition, or time horizon. The probability that n dispersers are all one sex depends on the sex ratio of the source population and is estimated as $p^n + q^n$ for females (p) or males (q), respectively (Zar 1999: 517). We assumed that the source population did not depart from a 50:50 sex ratio (Smith and Nichols 2003) and calculated the number of juvenile dispersers (d) for a 0.95 probability of at least one immigrant being the opposite sex. As above, we used Eq. (19.1) to estimate the maximum effective distance between small and medium OGRs (Pyare and Smith 2005).

Results

The values of per capita growth rate needed for low populations to persist with a 0.95 probability ranged from 0.19 to 0.39 for populations to persist for 25 years, and from 0.97 to 1.85 for populations to persist for 100 years (Table 19.2). The number of annual dispersers needed to prevent flying squirrels from becoming extirpated for ≥25 years averaged one every 5 years for populations in small OGRs with 100% primary habitat and one per year for small OGRs with a "minimum" (25% primary habitat) prescription. An estimated 162 dispersers per year were needed to sustain populations for 100 years in small OGRs comprising 25% primary habitat (Table 19.3). For small OGRs in which flying squirrel populations are extirpated, ≥6 juvenile dispersers per year were needed in order to have a 0.95 probability that a breeding pair would reach the patch.

TABLE 19.2. *Number of annual dispersers (d) needed to achieve an average intrinsic rate of population growth (r) sufficient to achieve time to extinction (T_e) for time horizons of 25 and 100 years (P_t=0.95; Smith and Person 2007) in reserves with three different carrying capacities (K) and populations that reach a low population density (at least one breeding pair) relative to* K.

Years	K	T_e	Average *r*	Ψ[a]	d
25	454	490	0.39	2.8	1
25	910	490	0.27	1.9	0.5
25	1,820	490	0.19	1.4	0.2
100	454	1,950	1.85	13.2	162
100	910	1,950	1.32	9.4	30
100	1,820	1,950	0.97	6.9	10

[a] The ratio of average *r* and intrinsic rate of growth of the study population (0.14), with a variance in r_i=0.39 (Smith and Person 2007).

TABLE 19.3. *For scenarios that vary in habitat composition* (K) *and time horizons (Years), maximum effective distances between a small OGR serving as a functional stepping-stone and two medium OGRs for flying squirrels* (Glaucomys sabrinus) *to persist with 0.95 probability. Maximum effective distance (m – equivalents) delimits landscape connectivity that facilitates a minimum level of immigration for flying squirrel populations to persist with 0.95 probability (distance$_r$) or colonize a vacant reserve (distance$_c$).*

Scenario	Years	K	Maximum effective distance$_r$ (m)	Maximum effective distance$_c$ (m)
1	25	454	1,172	1,161
2	25	910	1,174	1,161
3	25	1,820	1,174	1,161
4	100	454	844	1,161
5	100	910	1,105	1,161
6	100	1,820	1,151	1,161

Maximum effective distances for 0.95 probability persistence over 100 years ranged from 844 m for small OGRs with 25% primary habitat to 1,151 m for small OGRs comprised of 100% primary habitat (Table 19.3). Corresponding values for persistence in small OGRs over a 25-year time horizon were 1,172 m and 1,174 m. Maximum effective distance to establish a breeding pair in vacant small OGRs was 0.1% less than in circumstances when dispersal supplemented

r_i during 25-year time horizons (Table 19.3). For a 100-year horizon, the maximum effective distance to establish a breeding pair in a vacant small OGR was 38% greater than in circumstances when dispersal augmented populations in small OGRs composed of 25% primary habitat; the difference was <0.1% for small OGRs composed of 100% primary habitat (Table 19.3).

Discussion

Our study makes at least three contributions to the science and application of conservation. First, it demonstrates the value of linking applied ecological research to a sound theoretical foundation. A thorough understanding of source–sink dynamics is fundamental to evaluating the effectiveness of habitat reserves in sustaining viable metapopulations across fragmented landscapes. Also, our study illustrates that achieving a greater understanding of the spatially explicit dynamics of source and sink populations in fragmented landscapes can guide conservation planning toward sustainability. Lastly, our study illustrates the need, feasibility, and value of evaluating the assumptions of conservation plans. The 1997 TLMP reserve network has a sound theoretical framework. Without empirical evidence to support the underlying assumptions, its effectiveness in sustaining well-distributed and viable wildlife populations and the utility of habitat reserve networks remain uncertain.

Probability of dispersal and functional connectivity

We explicitly defined and quantified landscape resistance with empirical data, used those empirical data to parameterize models of functional connectivity, and generated model output that is meaningful for land use planning and conservation of biological diversity (Schumaker 1996; Ruckelshaus *et al.* 1997; Tischendorf and Fahrig 2000; Richards *et al.* 2002). Specifically, we combined results of population modeling (Smith and Person 2007) with logistic growth modeling and simulations of natal dispersal in hypothetical landscapes in order to determine the number of immigrants required for northern flying squirrel populations in small OGRs to function as sinks sustained through immigration from nearby source populations (short term), or to be components of a metapopulation structure that sustains viable populations (long term). We found that the number of dispersers needed per year to achieve an intrinsic rate of population growth to avoid extinction over the short term was well within the range of what source populations can produce annually (Smith and Nichols 2003). The maximum effective distance between small OGRs and source populations required to achieve persistence was not much more than a kilometer. For small OGRs to contribute to metapopulation viability, the number of required dispersers ranged 1–2 orders of magnitude greater than that needed for small OGRs to be sustained in the short term. Moreover, maximum effective distances for the long-term horizon were shorter than short-term scenarios, especially for small OGRs that meet the Tongass Land Management Plan (TLMP) minimum habitat prescription.

For population enhancement, the range of maximum effective distances was relatively narrow and the distribution was skewed toward the maximum value, especially among scenarios with a 25-year horizon. Even the maximum value of 1,174 m falls well within the distance that juveniles can move through intact landscapes (~7 km) over relatively short time periods (48 hours; W. P. Smith, unpublished data). Unfortunately, fewer than half the small OGRs prescribed in the 1997 TLMP for northern Prince of Wales Island were functionally connected to a source population (Pyare and Smith 2005). Since the initial drafting of this manuscript, some revisions have been made to the configuration of small OGRs across the landscapes we analyzed in this chapter (USDA Forest Service 2008). However, revisions have been relatively minor and further analysis revealed that functional connectivity may have been reduced rather than improved (S. Pyare and W. P. Smith, unpublished data). For 51 small OGRs that on average comprised 64% primary habitat, effective distances to large tracts (≥4,000 ha) of protected old-growth forest (i.e., medium or large OGRs) were such that only 20 reserves were located in landscapes where juvenile dispersal would be likely to occur, and these were essentially contiguous (S. Pyare and W. P. Smith, unpublished data). Therefore, the probability of juvenile dispersal between small OGRs and large tracts of old-growth in our study area was either 1 or approached zero.

Because flying squirrel populations are unlikely to persist over the long term in isolated small OGRs (Smith and Person 2007), the role of small reserves within the TLMP conservation framework is to function as stepping-stones connecting larger reserves, i.e., sinks that support ephemeral populations that are frequently replaced and enhance the distribution of flying squirrel populations (i.e., functional components within a metapopulation). Our findings underscore the essential role of immigration in sustaining sinks or facilitating metapopulation viability among unsustainable fragmented populations (i.e., sinks) and the extent to which permeability of landscape elements can influence the probability of dispersal and functional connectivity of subpopulations in a managed matrix (With and Crist 1995; Schumaker 1996; Richards *et al.* 2002; Verbeylen *et al.* 2003; Baum *et al.* 2004; Pyare and Smith 2005; Smith and Person 2007).

Assumptions and limitations

We considered only juvenile emigrants because our goal was to assess the potential for small OGRs to facilitate demographic rescue. Adult males often undertake long forays (4–5 km) during spring in search of estrous females, but it is juvenile squirrels vacating their natal areas that provide a pool of emigrants to supplement existing populations or colonize vacant patches (Wilson 2003, 2010). Also, we limited our simulations to scenarios in which dispersal occurred between two medium OGRs and a single small OGR because the assumption in the 1997 TLMP was that small reserves would serve as "stepping-stones" between larger reserves. For the purpose of this chapter, the consequences of limiting our analysis to those scenarios are probably negligible. When we varied the probability of emigrants selecting a target small OGR (i.e., a situation with multiple small OGRs within the dispersal distance of medium OGRs) the maximum effective distance changed little (<1%). Exceptions did occur for scenarios with the lowest *K* over a 100-year time horizon, which are unlikely to persist and produce dispersers (Smith and Person 2007).

Most of our modeling assumptions resulted in more optimistic scenarios of dispersal and persistence. For example, our estimates of per capita rate of growth (0.14) and available dispersers (juvenile proportion × population size) are generous given the mean recruitment rate of juveniles into the breeding population (i.e., the proportion of juveniles in the autumn population) incorporated no overwinter mortality (Smith and Person 2007). Furthermore, we assumed that all natal dispersers successfully established home ranges and contributed to the breeding population of reserves; a liberal assumption given expected energetic constraints while dispersing in managed landscapes (Flaherty *et al.* 2008) and added risk of predation. Also, our predictions of persistence in small OGRs and our assumption that medium OGRs function as source populations are probably optimistic (Smith and Person 2007).

We did not consider environmental or genetic stochasticity (Foley 1994; Reed 2004), deleterious effects of inbreeding (Frankham 1998; O'Grady *et al.* 2006), or potential Allee effects (Allee *et al.* 1949). Also, we assumed that habitat available to flying squirrels remained unchanged over the time horizon, and there was no reduced fitness from inbreeding depression or genetic drift. However, Southeast Alaska periodically experiences severe catastrophic disturbances from windstorms, avalanches and landslides that dramatically alter hundreds of hectares of forest during a single event (Mitchell 1995; Nowacki and Kramer 1998; Kramer *et al.* 2001). Furthermore, clearcut logging increases the chance of blowdown from windstorms in remaining old-growth forest and reduces the likelihood that patches in managed landscapes remain intact (Concannon 1995; Mitchell 1995). In addition, northern flying squirrel populations on Prince of Wales Island show severely reduced genetic variation, a likely consequence of descent from a single founder population (Demboski *et al.* 1998) that was isolated from mainland or nearshore island populations (Bidlack and Cook 2001). The sensitivity of predictions of T_e and P_t to changes in v (Smith and Person 2007) underscores the potential for additional environmental and genetic variability to increase predicted probabilities of extinction in reserves.

Implications for conservation planning

Land management plans such as TLMP rely on systems of habitat reserves that are sufficient in size and quality, or are adequately distributed and connected, to ensure the viability of wildlife populations (Smith and Zollner 2005). Habitat reserves must either sustain individual insular populations, or the managed matrix between reserves must allow dispersal among reserves to maintain wildlife populations within a metapopulation structure. In Southeast Alaska, the primary habitat component of large old-growth patches may not need to be contiguous because interspersed habitats (e.g., fens, peatland mixed-conifer forests) can clearly support flying squirrels for a short time or facilitate dispersal (Smith and Nichols 2003; Pyare and Smith 2005). However, the contribution of small OGRs to the TLMP conservation strategy for northern flying squirrels will depend on the degree of their isolation and functional connectivity with larger or other small reserves. For this reason, conservation planning that explicitly considers the quality and spatial configuration of habitat across highly modified landscapes (Odom *et al.* 2001; Fuller *et al.* 2006) will have a greater likelihood of sustaining viable metapopulations of flying squirrels (Hanski *et al.* 2000; Selonen and Hanski 2003, 2004).

We estimated the maximum effective distance between larger reserves and a small OGR that would achieve an immigration rate sufficient to persist for 25 or 100 years or colonization by a breeding pair. Maximum effective distance, which accounts for the landscape resistance of different habitats, is dependent upon forest structure (Pyare and Smith 2005). In our study, maximum effective distance for supplementing populations and establishing a breeding pair were essentially the same for $P_t = 25$ (1.1 km). Thus, spacing small OGRs at an effective distance of one kilometer across old-growth habitat would probably facilitate both recolonization of vacant reserves and supplementation of existing populations. For $P_t = 100$ years, the maximum effective distance for supplementing populations was less than the effective distance for recolonization. Small OGRs that meet the TLMP minimum prescription of 25% productive old-growth (POG) should probably be <850 m from other reserves; small OGRs comprised of ≥50% primary habitat (i.e., POG) can probably be spaced at 1-km intervals in landscapes composed of old-growth forests. We suggest the smaller of those distances be used as a guideline for the spacing of reserves, conditional on the time horizon for persistence and the carrying capacity of the reserve.

Clearly, our study does not represent an evaluation of the overall conservation strategy implemented in TLMP with respect to northern flying squirrels. Still, we conclude that most small OGRs on Prince of Wales Island (as described in the 1997 TLMP; USDA Forest Service 1997) were not functionally connected. Maximum effective dispersal distance for northern flying squirrels was ≤1 km, whereas the mean expected distance between small OGRs to achieve well-distributed populations (USDA Forest Service 1997) was ≥5 km (i.e., the radius of circular watersheds, which averaged 7,700 ha). Consequently, the extent of functional connectivity among reserves or between small OGRs and larger patches of primary habitat require fundamental changes in reserve design if metapopulations of flying squirrels are expected to have a high probability (i.e., ≥0.95) of persisting in managed landscapes. Even then, a system of functionally connected reserves does not ensure a viable metapopulation in managed landscapes. The Siberian flying squirrel (*Pteromys volans*) was extirpated from an entire region because industrial-scale logging fragmented continuous populations into isolated patches (Hokkanen *et al.* 1982). That occurred despite the ability of young flying squirrels to disperse long distances through a managed matrix that included residual trees and corridors connecting habitat patches (Selonen and Hanski 2003). Furthermore, as landscapes become increasingly fragmented, a threshold point is reached where metapopulations are limited solely by production in the source habitats (Davis and Howe 1992).

Sustaining viable populations of the northern flying squirrel (and the ecological communities it represents) in Southeast Alaska may require adaptive management, i.e., an iterative process of revising conservation plans and further study. Admittedly, testing the adequacy of a system of reserves for conservation of wildlife can be a daunting task, requiring long-term (decades) population monitoring and studies of animal behavior and demography. However, we demonstrated that short-term evaluations of a plan are possible using empirical data to parameterize simple simulation models designed to evaluate the underlying assumptions.

Nevertheless, we believe the system of small OGRs implemented in the 1997 TLMP would have a substantially higher probability of sustaining metapopulations of flying squirrels if the composition of primary habitat in small OGRs was increased to levels that will provide a relatively high probability of sustaining reserves without immigration (Smith and Person 2007), or if the permeability of habitat comprising the matrix was improved through active management to resemble old-growth. For example, pre-commercial and commercial thinning of second-growth stands, which are less permeable to northern flying squirrels (E. A. Flaherty and M. Ben-David, unpublished data), will increase canopy height and create more open space in the midstory, both of which facilitate efficient gliding (Vernes 2001; Scheibe *et al.* 2006). In any case, active management of second-growth stands will likely hasten succession toward achieving stand conditions that will support additional breeding populations of northern flying squirrels, potentially adding stepping-stones and migrants to a network of old-growth reserves.

Acknowledgments

Earthwatch, USDI Fish and Wildlife Service, and Denver Zoological Foundation provided financial and logistic support for earlier studies upon which this study was dependent for empirical data and analyses. We thank Elizabeth Flaherty and Merav Ben-David, who shared live-trapping data from Kosciusko Island so that we could examine whether the presence of flying squirrels in small, isolated habitat patches conformed to predictions of the probability of persistence. We obtained imagery for our study from the Alaska Mapped website (www.alskamapped.org). We thank Elizabeth Flaherty, Gillian Holloway, and James Wilson for valuable comments on an early draft of this chapter. We thank three anonymous referees for thorough reviews and comments that ultimately improved the quality of this manuscript.

References

Alaback, P. B. (1982). Dynamics of understory biomass in Sitka spruce-western hemlock forests of Southeast Alaska. *Ecology* **63:** 1932–1948.

Allee, W. C., A. E. Emerson, O. Park, T. Park and K. P. Schmidt (1949). *Principles of Animal Ecology*. Saunders, Philadelphia, PA.

Baum, K. A., K. J. Haynes, F. P. Dillemuth and J. T. Cronin (2004). The matrix enhances the effectiveness of corridors and stepping stones. *Ecology* **85:** 2671–2676.

Beier, P. (1993). Determining minimum habitat areas and corridors for cougars. *Conservation Biology* **7:** 94–108.

Bender, D. J., T. A. Contreras and L. Fahrig (1998). Habitat loss and population decline: a meta-analysis of the patch size effect. *Ecology* **79:** 517–533.

Bidlack, A. L. and J. A. Cook (2001). Reduced genetic variation in insular northern flying squirrels (*Glaucomys sabrinus*) along the North Pacific Coast. *Animal Conservation* **4:** 283–290.

Burkey, T. V. (1989). Extinction in nature reserves: the effect of fragmentation and the importance of migration between reserve fragments. *Oikos* **55:** 75–81.

Carey, A. B. (2000). Ecology of northern flying squirrels: implications for ecosystem management in the Pacific Northwest, USA. In *Biology of Gliding Mammals* (R. L. Goldingay and J. S. Scheibe, eds.). Filander Verlag, Fürth, Germany: 45–66.

Carroll, C., R. F. Noss, P. C. Paquet and N. H. Schumaker (2003). Use of population viability analysis and reserve selection algorithms in regional conservation plans. *Ecological Applications* **13:** 1773–1789.

Concannon, J. A. (1995). Characterizing structure, microclimate, and decomposition of peatland, beachfront, and newly-logged forest edges in southeastern Alaska. Dissertation, University of Washington, Seattle, WA.
Davis, G. L. and R. W. Howe (1992). Juvenile dispersal, limited breeding sites, and the dynamics of metapopulations. *Theoretical Population Biology* **41**: 184–207.
Deal, R. L. and W. A. Farr (1994). Composition and development of conifer regeneration in thinned and unthinned natural stands of western hemlock and Sitka spruce in Southeast Alaska. *Canadian Journal of Forestry Research* **24**: 976–984.
DeMars, D. J. (2000). *Stand-Density Study of Spruce–Hemlock Stands in Southeastern Alaska*. General Technical Report PNW-496, USDA Forest Service, Portland, OR.
Demboski, J. R., B. K. Jacobsen and J. A. Cook (1998). Implications of cytochrome b sequence variation for biogeography and conservation of the northern flying squirrels (*Glaucomys sabrinus*) of the Alexander Archipelago, Alaska. *Canadian Journal of Zoology* **76**: 1771–1777.
DeMeo, T., J. Martin and R. A. West (1992). *Forest Plant Association Guide: Ketchikan Area, Tongass National Forest*. USDA Forest Service, Alaska Region R10-MB-210, USDA Forest Service, Juneau, AK.
Everest, F. H., D. N. Swanston, C. G. Shaw III, W. P. Smith, K. R. Julin and S. D. Allen (1997). *Evaluation of the Use of Scientific Information in Developing the 1997 Forest Plan for the Tongass National Forest*. General Technical Report PNW-GTR-415, USDA Forest Service, Portland, OR.
Fahrig, L. and G. Merriam (1985). Habitat patch connectivity and population survival. *Ecology* **66**: 1762–1768.
Fahrig, L. and G. Merriam (1994). Conservation of fragmented populations. *Conservation Biology* **8**: 50–59.
FEMAT (Forest Ecosystem Management Assessment Team) (1993). *Forest Ecosystem Management: An Ecological, Economic, and Social Assessment*. US Department of Agriculture, US Department of the Interior (and others), Portland, OR.
Ferreras, P. (2001). Landscape structure and asymmetrical inter-patch connectivity in a metapopulation of the endangered Iberian lynx. *Biological Conservation* **100**: 125–136.
Flaherty, E. A., W. P. Smith, S. Pyare and M. Ben-David (2008). Experimental trials of the northern flying squirrel (*Glaucomys sabrinus*) traversing managed rainforest landscapes: perceptual range and fine-scale movements. *Canadian Journal of Zoology* **86**: 1050–1058.
Flynn, R. W., T. V. Schumacher and M. Ben-David (2004). *Abundance, Prey Availability, and Diets of American Martens: Implications for the Design of Old-Growth Reserves in Southeast Alaska*. Final Report, US Fish and Wildlife Service Grant DCN 70181-1-G133. Alaska Department of Fish and Game, Douglas, AK.
Foley, P. (1994). Predicting extinction times from environmental stochasticity and carrying capacity. *Conservation Biology* **8**: 124–137.
Frankham, R. (1998). Inbreeding and extinction: island populations. *Conservation Biology* **12**: 665–675.
Fryxell, J. M., J. B. Falls, E. A. Falls and R. J. Brooks (1998). Long-term dynamics of small-mammal populations in Ontario. *Ecology* **79**: 213–225.
Fuller, T., M. Munguia, M. Mayfield, V. Sánchez-Cordero and S. Sarkar (2006). Incorporating connectivity into conservation planning: a multi-criteria case study from Central Mexico. *Biological Conservation* **133**: 131–142.
Hanski, I. (1994). Patch-occupancy dynamics in fragmented landscapes. *Trends in Evolution and Ecology* **9**: 131–135.
Hanski, I. and M. Gilpin (1991). Metapopulation dynamics: brief history and conceptual domain. *Biological Journal of the Linnean Society* **42**: 17–38.
Hanski, I. K., P. C. Stevens, P. Ihalempiä and V. Selonen (2000). Home-range size, movements, and nest-site use in the Siberian flying squirrel, *Pteromys volans*. *Journal of Mammalogy* **81**: 798–809.
Harris, A. S. and W. A. Farr (1974). *The Forest Ecosystem of Southeast Alaska. 7. Forest Ecology and Timber Management*. General Technical Report PNW-25, USDA Forest Service, Portland, OR.

Hokkanen, H., T. Tőrmälä and H. Vuorinen (1982). Decline of the flying squirrel (*Pteromys volans* L.) populations in Finland. *Biological Conservation* **23**: 273–284.
Julin, K. R. and J. P. Caouette (1997). Options for defining old-growth timber volume strata: a resource assessment. In *Assessments of Wildlife Viability, Old-Growth Timber Volume Estimates, Forested Wetlands, and Slope Stability* (C. G. Shaw III, technical coordinator). General Technical Report PNW-GTR-392, USDA Forest Service, Portland OR: 24–37.
Kramer, M. G., A. J. Hansen, M. L. Taper and E. J. Kissinger (2001). Abiotic controls on long-term windthrow disturbance and temperate rain forest dynamics in Southeast Alaska. *Ecology* **82**: 2749–2768.
Leroux, S., F. K. A. Schmiegelow, S. G. Cumming, R. B. Lessard, and J. Nagy (2007). Accounting for system dynamics in reserve design. *Ecological Applications* **17**: 1954–1966.
MacDonald, S. O. and J. A. Cook (1996). The land mammal fauna of southeast Alaska. *Canadian Field Naturalist* **110**: 571–598.
Mitchell, S. J. (1995). A synopsis of windthrow in British Columbia: occurrence, implications, assessment and management. In *Wind and Trees* (M. P. Coutts and J. Grace, eds.). Cambridge University Press, Cambridge, UK: 448–459.
Nowacki, G. J. and M. G. Kramer (1998). *The Effects of Wind Disturbance on Temperate Rain Forest Structure and Dynamics of Southeast Alaska*. General Technical Report PNW-GTR-421, USDA Forest Service, Portland, OR.
Odom, R. H., W. M. Ford, J. W. Edwards, C. W. Stihler and J. M. Menzel (2001). Developing a habitat model for the endangered Virginia northern flying squirrel (*Glaucomys sabrinus fuscus*) in the Allegheny Mountains of West Virginia. *Biological Conservation* **99**: 245–252.
O'Grady, J. J., B. W. Brook, D. H. Reed, J. D. Ballou, D. W. Tonkyn and R. Frankham (2006). Realistic levels of inbreeding depression strongly affect extinction risk in wild populations. *Biological Conservation* **133**: 42–51.
Opdam, P. (1991). Metapopulation theory and habitat fragmentation: a review of holarctic breeding bird studies. *Landscape Ecology* **5**: 93–106.
Opdam, P., G. Rijsdijk and F. Hustings (1985). Bird communities in small woods in an agricultural landscape: effects of area and isolation. *Biological Conservation* **34**: 333–352.
Pulliam, H. R. (1988). Sources, sinks, and population regulation. *American Naturalist* **132**: 652–661.
Pyare, S. and W. P. Smith (2005). *Functional Connectivity of Tongass Old-Growth Reserves: An Assessment Based on Flying Squirrel Movement Capability.* Project Report, Juneau Fish and Wildlife Field Office, Ecological Services, US Fish and Wildlife Service, Juneau, AK [available at http://alaska.fws.gov/fisheries/fieldoffice/juneau/index.htm].
Pyare, S., W. P. Smith and C. Shanley (2010). Den selection by northern flying squirrels in fragmented landscapes. *Journal of Mammalogy* **91**: 886–896.
Reed, D. H. (2004). Extinction risk in fragmented habitats. *Animal Conservation* **7**: 181–191.
Richards, W. H., D. O. Wallin and N. H. Schumaker (2002). An analysis of late-seral forest connectivity in western Oregon. *Conservation Biology* **16**: 1409–1421.
Rolstad, J. (1991). Consequences of forest fragmentation for the dynamics of bird populations: conceptual issues and the evidence. *Biological Journal of the Linnean Society* **42**: 149–163.
Ruckelshaus, M., C. Hartway and P. Kareiva (1997). Assessing the data requirements of spatially explicit dispersal models. *Conservation Biology* **11**: 1298–1306.
Scheibe, J. S., W. P. Smith, J. Basham and D. Magness (2006). Cost of transport in the northern flying squirrel, *Glaucomys sabrinus*. *Acta Theriologica* **51**: 169–178.
Schumaker, N. H. (1996). Using landscape indices to predict habitat connectivity. *Ecology* **77**: 1210–1225.
Searby, H. W. (1968). *Climate of Alaska*. Climatography of the United States No. 60-49. Climates of the States, Alaska. US Department of Commerce, Environmental Science Service Administration, Environmental Data Service.

Selonen, V. and I. K. Hanski (2003). Movements of the flying squirrel *Pteromys volans* in corridors and in matrix habitat. *Ecography* **26**: 641–651.
Selonen, V. and I. K. Hanski (2004). Young flying squirrels (*Pteromys volans*) dispersing in fragmented forests. *Behavioral Ecology* **15**: 564–571.
Shaw, C. G., III (1999). *Use of Risk Assessment Panels During Revision of the Tongass Land and Resource Management Plan*. General Technical Report PNW-460, USDA Forest Service, Portland, OR.
Smith, W. P. (2007). Ecology of *Glaucomys sabrinus*: habitat, demography, and community relations. *Journal of Mammalogy* **88**: 862–881.
Smith, W. P. and J. V. Nichols (2003). Demography of the Prince of Wales flying squirrel: an endemic of Southeastern Alaska temperate rainforest. *Journal of Mammalogy* **84**: 1044–1058.
Smith, W. P. and D. K. Person (2007). Estimated persistence of northern flying squirrel populations in temperate rain forest fragments of Southeast Alaska. *Biological Conservation* **137**: 626–636.
Smith, W. P. and P. A. Zollner (2005). Sustainable management of wildlife habitat and risk to extinction. *Biological Conservation* **125**: 287–295.
Smith, W. P., R. G. Anthony, J. R. Waters, N. L. Dodd and C. J. Zabel (2003). Ecology and conservation of arboreal rodents of the Pacific Northwest. In *Mammal Community Dynamics in Western Coniferous Forests: Management and Conservation* (C. J. Zabel and R. G. Anthony, eds.). Cambridge University Press, Cambridge, UK: 157–206.
Smith, W. P., S. M. Gende and J. V. Nichols (2004). Ecological correlates of flying squirrel microhabitat use and density in temperate rain forest of southeastern Alaska. *Journal of Mammalogy* **85**: 663–674.
Smith, W. P., J. V. Nichols and S. M. Gende (2005). The northern flying squirrel as a management indicator species of north temperate rainforest: test of a hypothesis. *Ecological Applications* **15**: 689–700.
Soulé, M. E., J. A. Estes, J. Berger and C. Martinez Del Rio (2003). Ecological effectiveness: conservation goals for interactive species. *Conservation Biology* **17**: 1238–1250.
Suring, L. A. (compiler) (1993). *Habitat Capability Models for Wildlife in Southeast Alaska*. USDA Forest Service, Alaska Region, Juneau, AK.
Taylor, P. D., L. Fahrig, K. Henein and G. Merriam (1993). Connectivity is a vital element of landscape structure. *Oikos* **68**: 571–573.
Tischendorf, L. and L. Fahrig (2000). On the usage and measurement of landscape connectivity. *Oikos* **90**: 7–19.
USDA Forest Service (1997). *Land and Resource Management Plan: Tongass National Forest*. R10-MB-338dd, USDA Forest Service, Juneau, AK.
USDA Forest Service (2003). *Tongass Land and Resource Management Plan Revision: Final Supplemental Environmental Impact Statement (FSEIS)*. R10-MB-481a, Volume I: Final SEIS, Appendix A, B, D, E. USDA Forest Service, Juneau, AK.
USDA Forest Service (2008). *Land and Resource Management Plan: Tongass National Forest*. R10-MB-603b, USDA Forest Service, Juneau, AK.
Verbeylen, G., L. De Bruyn, F. Adriaensen and E. Matthysen (2003). Does matrix resistance influence red squirrel (*Sciuris vulgaris*) distribution in an urban landscape? *Landscape Ecology* **18**: 791–805.
Vernes, K. (2001). Gliding performance of the northern flying squirrel (*Glaucomys sabrinus*) in mature mixed forests of eastern Canada. *Journal of Mammalogy* **82**: 1026–1033.
Wiens, J. A. (1995). Habitat fragmentation: island vs. landscape perspectives on bird conservation. *Ibis* **137**: S97–S104.
Wiens, J. A., N. C. Stenseth, B. Van Horne and R. A. Ims (1993). Ecological mechanisms and landscape ecology. *Oikos* **66**: 369–380.
Wilcove, D. S., C. H. McLellan and A. P. Dobson (1986). Habitat fragmentation in the temperate zone. In *Conservation Biology: The Science of Scarcity and Diversity* (M. E. Soulé, ed.). Sinauer Associates, Sunderland, MA: 237–256.

Wilson, T. M. (2003). *Sex and the Single Squirrel: A Genetic View of Forest Management in the Pacific Northwest*. USDA Forest Service, Pacific Northwest Research Station, Portland, OR. *Science Findings* **51**: 1–5.

Wilson, T. M. (2010). Limiting factors for northern flying squirrels (*Glaucomys sabrinus*) in the Pacific Northwest: a spatio-temporal analysis. PhD dissertation, Union Institute and University, Cincinnati, OH.

With, K. A. and T. O. Crist (1995). Critical thresholds in species' responses to landscape structure. *Ecology* **76**: 2446–2459.

Zar, J. H. (1999). *Biostatistical Analysis*, 4th edition. Prentice-Hall, Upper Saddle River, NJ.

SCOTT K. ROBINSON AND JEFFREY P. HOOVER

20

Does forest fragmentation and loss generate sources, sinks, and ecological traps in migratory songbirds?

Summary

Forest fragmentation and habitat loss on the breeding grounds of migratory songbirds have long been hypothesized to create source–sink dynamics. Forest fragments have increased the populations of many nest predators that are subsidized by food in the surrounding landscape and are themselves released from top predators. Migratory songbirds are also susceptible to brood parasitism by the brown-headed cowbird (*Molothrus ater*). As a result of reduced nesting success, reproductive output may be driven below the source–sink threshold. Source–sink dynamics occur within forest tracts, among patches in a landscape, and have been hypothesized to occur over entire regions in which sink populations within small tracts in mostly agricultural regions are rescued by distant source populations in mostly forested regions.

In the absence of ongoing habitat loss, habitat fragmentation will only generate continuing population declines if birds lack behavioral mechanisms that enable them to concentrate in the source habitat and avoid population sinks. If sink habitat is attractive, then it can become an ecological trap. There is growing evidence that many birds use decision rules based on nesting success that might enable them to avoid ecological traps created by high rates of nest predation. Most forest birds, however, do not have decision rules to avoid areas with heavy cowbird parasitism.

There are, however, a number of limitations with the fragmentation–source–sink paradigm. First, there is a critical need for improved data on the survival of adult and young birds. Second, we know very

Sources, Sinks and Sustainability, ed. Jianguo Liu, Vanessa Hull, Anita T. Morzillo and John A. Wiens. Published by Cambridge University Press.

little about the spatial scale of dispersal in most species. Third, fragmentation-associated reductions in nesting success appear to be much weaker in some regions, especially in western North America. And fourth, it is far from clear that large habitat patches always act as source populations. Nevertheless, a conservation strategy that maintains or restores large forest tracts in mostly forested regions seems likely to maximize the ratio of source to sink habitat and to lead to sustainable populations.

Introduction

The conservation of long-distance migratory songbirds, those that breed in the temperate zone and winter largely in the tropics, has been a major focus of conservation since a series of influential papers and books linked declining songbird populations in North America with deforestation in the Neotropics (Robbins *et al.* 1989a; Terborgh 1989; Hagan and Johnson 1992). The possibility that many familiar birds of North America may be disappearing because of tropical deforestation made the North American public aware that tropical habitat destruction had global consequences and might even affect wildlife in their own backyards. Forest fragmentation and loss on the breeding grounds, however, also had long been linked to population declines in migratory songbirds, both locally and regionally (Whitcomb *et al.* 1981; Lynch and Whigham 1984; Robbins *et al.* 1989a; Faaborg *et al.* 1995). Most of the early studies of the effects of fragmentation on migratory songbirds focused on long-term declines of migratory songbirds within small forest patches and on the absence of migrants from small forest tracts (area-sensitivity: Whitcomb *et al.* 1981; Lynch and Whigham 1984; Robbins *et al.* 1989b; reviewed in Faaborg *et al.* 1995) and from the edges of forest tracts (edge avoidance: Whitcomb *et al.* 1981; Faaborg *et al.* 1995; Dunford *et al.* 2002). Given these thoroughly documented patterns, declines of migratory songbirds in increasingly fragmented landscapes were readily attributable to both outright habitat loss and to increasing fragmentation of remnant habitat patches (higher edge density and smaller patch size) (Fahrig 2003; Lindenmayer and Fischer 2006).

In practice, it is difficult to separate the effects of habitat amount (a variable related to landscape composition) and habitat fragmentation (a variable related to landscape pattern) (Fahrig 2003). Outright habitat loss inevitably leads to higher edge densities and smaller patch sizes. Therefore, in this chapter we will examine the effects of habitat amount and fragmentation together, as there are no demographic studies we know of that separate the two. It will be important in future studies, however, to control for the two different variables (see below).

Habitat fragmentation can be especially threatening because it potentially leads to the continuing loss of "forest-interior species" from entire landscapes in which only small habitat patches remain (Whitcomb *et al.* 1981; With and King 2001). Fragmented landscapes tend to become dominated by habitat generalists and those that can tolerate fragmentation (see also Pearson and Fraterrigo, Chapter 6, this volume). If land managers try to maximize the biodiversity of each preserve by diversifying habitats within them (increasing beta diversity), this may actually decrease regional (gamma) diversity by creating conditions unfavorable to species that require larger habitat patches (Faaborg 1980; Robinson 1988).

Causes of area-sensitivity and edge avoidance

The link between habitat fragmentation and source–sink dynamics came when researchers began to seek mechanisms to explain edge avoidance and area-sensitivity. Forest tracts are strongly influenced by processes in the matrix surrounding them (Wiens *et al.* 1993; Wiens 1995; Fahrig 1997, 2003; Rodewald 2003; Kupfer *et al.* 2006). Edges of habitat patches are readily invaded by animals that live in the matrix, some of which can create problems for animals that depend upon the habitat patches (Andrèn 1992, 1994; Marzluff and Restani 1999; Dijak and Thompson 2000; Chalfoun *et al.* 2002a, 2002b; Stake *et al.* 2005). Studies of nesting success of songbirds in small tracts and near edges provided the first evidence that reduced nesting success might be causing area-sensitivity and avoidance (Brittingham and Temple 1983; Wilcove 1985).

Brood parasitism

Many migratory songbirds breeding in North America are susceptible to brood parasitism by the brown-headed cowbird (*Molothrus ater*), a species that lays its eggs in the nests of other species (hosts), which raise cowbird young, usually at the expense of many or most of their own young (reviewed in Robinson *et al.* 1995a; Rothstein and Robinson 1998; Chace *et al.* 2005). This species has greatly expanded both its geographic and its host range in the past century, largely attributed to landscape change. Because they are host generalists with more than 200 known hosts, cowbirds can also drive many of their host species to local extinction without necessarily being driven to local extinction themselves (May and Robinson 1985). Nests containing cowbird nestlings require a higher rate of prey delivery to nestlings, which may attract predators and reduce adult survival (Hoover and Reetz 2006). Cowbirds are also known to depredate nonparasitized nests (Arcese *et al.* 1996; Hoover and Robinson

2007) and may even engage in "mafia-like" behavior by depredating nests from which cowbird eggs have been removed (Hoover and Robinson 2007). Cowbirds therefore have the potential to cause severe conservation problems for many of their hosts, some of which have been driven to the brink of extinction by the combined effects of habitat loss and excessive cowbird parasitization of their nests (May and Robinson 1985; Robinson *et al.* 1995a; Trine 2000; Chace *et al.* 2005; reviewed in Rothstein and Robinson 1998). Indeed, some of the earliest source–sink calculations were made to determine whether cowbird parasitism was driving host reproductive success below the levels necessary for population maintenance (Nolan 1978; May and Robinson 1985; Temple and Cary 1988). Cowbirds have been shown to reduce levels of productivity below the source–sink threshold in several species (Trine 1998; Ward and Smith 2000) and have been implicated as a cause of regional declines (Gustafson *et al.* 2002). Compared with their coevolved grassland hosts, cowbirds pose a significant threat to forest hosts because most forest songbirds have no apparent defenses against cowbird parasitism (e.g., they do not reject cowbird eggs or abandon parasitized nests: Rothstein and Robinson 1998; Hosoi and Rothstein 2000; Hoover 2003a, 2003b).

The hypothesized link between cowbird parasitism and habitat fragmentation and landscape composition is based on another unusual life-history trait of cowbirds. Unlike most songbirds, cowbirds generally feed far from the areas where they lay their eggs (Rothstein *et al.* 1984; Thompson 1994; Goguen and Mathews 2000, 2001; Raim 2000). Cowbirds show strong preferences for searching for nests in forests and successional habitats (Hahn and Hatfield 1995; Winfree *et al.* 2006) but feed only in areas where there is short grass or bare ground (reviewed in Robinson *et al.* 1995a; Tewksbury *et al.* 1999; Chace *et al.* 2005). They routinely commute up to 15 km (Goguen and Mathews 2001) between breeding and feeding habitats (pastures, row-crop fields, areas with mowed grass, cattle feedlots, all of which are prevalent in the landscape matrix surrounding most remnant forest patches in human-dominated landscapes). Landscapes with extensive pastures and cattle ranching should therefore have high levels of parasitism because the remaining forest patches are usually surrounded by suitable cowbird foraging habitat and are essentially saturated with cowbirds. In landscapes with limited cowbird feeding habitat compared with forest cover, cowbird parasitism should be higher closer to edges of cowbird feeding sites (Goguen and Matthews 2000).

Empirical studies linking habitat fragmentation to levels of cowbird brood parasitism, starting with the pioneering study of Brittingham and Temple (1983), have yielded somewhat mixed results (Robinson and Wilcove 1994; Bielefeldt and Rosenfield 1997; Tewksbury *et al.* 2002; Knutson *et al.* 2004; Cottam *et al.* 2009). Only a few studies (e.g., Morse and Robinson 1999; Hoover *et al.* 2006) have shown edge effects on levels of cowbird parasitism. Not surprisingly, given the commuting range of the cowbird, these edge effects extend a very long way (more than 1 km) into the interior of the forest. In both of these studies, edge effects were related only to the edges of specific cowbird feeding sites (pastures and feedlots) rather than to edges in general. In agricultural landscapes where cowbird feeding habitat is abundant and forest habitats are limited, parasitism levels can be astonishingly high throughout even the interior of the remaining tracts (Robinson 1992; Donovan *et al.* 1995a, 1995b; Bollinger and Linder 1999; Burke and Nol 2000; Fauth *et al.* 2000; Robinson *et al.* 2000a; Austen *et al.* 2001; Fauth 2001; Tewksbury *et al.* 2002; Knutson *et al.* 2004; Hoover and Hauber 2007). In mostly forested landscapes where cowbird feeding habitat is scarce, parasitism levels are low regardless of distance from edges (Robinson *et al.* 1995b; Donovan *et al.* 1997; Hochachka *et al.* 1999). Cowbird parasitism may therefore be more affected by landscape composition than by effects of habitat configuration *per se*.

Because cowbird parasitism reduces nesting success, it has the potential to create sink habitat in fragmented landscapes. Temple and Cary (1988), for example, used estimates of adult and juvenile survival combined with data on nesting success of parasitized and nonparasitized nests and the frequency of parasitism to show that cowbirds could create host population sinks near edges. They argued that populations near edges may need to be rescued by immigrants from areas with higher nesting success in the forest interior to compensate for losses near edges (see also May and Robinson 1985). Studies from forest tracts in the agricultural Midwest showed such high levels of brood parasitism (90% of all wood thrush [*Hylocichla mustelina*] nests were parasitized, averaging more than 2.5 cowbird eggs per nest: Robinson 1992; Brawn and Robinson 1996; Fauth *et al.* 2000; Fauth 2001; Knutson *et al.* 2004) that there was little doubt that these forest patches were population sinks. The link between cowbird parasitism and declining populations of neotropical migrants was further strengthened by analyses of the North American Breeding Bird Survey, which showed steeper declines in cowbird hosts than in non-host species (Bohning-Gaese *et al.* 1993).

Nest predation

Forest fragmentation has also been linked with increasing levels of nest predation ever since Wilcove's (1985) pioneering study showing dramatic increases in predation rates on artificial nests with decreasing tract sizes and distances from edges. In at least some landscapes, nest predators thrive in forest fragments (Andrèn 1992, 1994; Dijak and Thompson 2000; Chalfoun *et al.* 2002a, 2002b). Just as cowbirds depend upon the matrix surrounding a forest patch for food, many facultative nest predators such as crows (*Corvus* spp.), raccoons (*Procyon lotor*), and opossums (*Didelphis marsupialis*) regularly feed in the matrix surrounding forest patches (reviewed in Chalfoun *et al.* 2002a). Other nest predators such as snakes are more abundant, or at least more active, near habitat edges (Blouin-Demers and Weatherhead 2001; Weatherhead and Blouin-Demers 2004; Stake *et al.* 2005; Carfagno and Weatherhead 2006). The loss of top predators in fragmented landscapes may also lead to "mesopredator release" (Crooks and Soule 1999; Schmidt 2003), in which medium-sized species, many of which are facultative nest predators, reach high population densities.

Although empirical evidence for fragmentation effects on nest predation using studies of real nests has been mixed (Paton 1994; Robinson and Wilcove 1994; Heske 1995; Oehler and Litvaitis 1996; Keyser *et al.* 1998; Lahti 2001; Chalfoun *et al.* 2002a, 2002b; Batáry and Báldi 2004; Peak 2007), many studies have shown increasing nest predation rates near edges, in small tracts, and in mostly agricultural landscapes, especially in the Midwest (Hoover *et al.* 1995; Robinson *et al.* 1995b; Hoover *et al.* 2006). Donovan *et al.* (1997) provided empirical evidence that some of the variation in edge effects may be related to the landscape composition surrounding a patch; edge effects were only present at intermediate levels of forest cover. In contrast, nest predators may saturate habitat patches in mostly agricultural landscapes (Robinson and Wilcove 1994; Heske 1995; Marini *et al.* 1995; Knutson *et al.* 2004; reviewed in Thompson *et al.* 2002).

Does reduced nesting success in fragmented forests lead to source–sink dynamics?

Source–sink models combining the effects of brood parasitism and nest predation strongly indicate that fragmentation creates source–sink dynamics at multiple spatial scales, from local to continental (Gibbs and Faaborg 1990; Donovan *et al.* 1995b; Rosenberg *et al.* 1999; With and King 2001; Donovan and Flather 2002; Gustafson *et al.* 2002; Stephens *et al.* 2003; Driscoll and Donovan 2004; Larson *et al.* 2004; Lampila *et al.* 2005; Lloyd *et al.* 2005). Robinson *et al.* (1995b) documented especially strong gradients in nesting success with fragmentation and landscape composition variables in the midwestern USA. In the largest forest tracts in regions with very high forest cover, brood parasitism was essentially absent and nest predation rates dropped to 30–50%. In contrast, in small fragments in mostly agricultural landscapes, nest predation rates averaged 70–95% (see also Robinson 1992) and parasitism levels averaged 33–95% per species. Using a variety of simple source–sink models, Donovan *et al.* (1995b) and Brawn and Robinson (1996) determined that woodlots in the agricultural landscapes in the Midwest were population sinks for most migratory songbirds, whereas the largest tracts in mostly forested landscapes were likely source habitats. The continued presence of songbirds nesting in sink habitat was used as evidence of the rescue effect (Brown and Kodric-Brown 1977) operating at very large spatial scales, because many small woodlots were hundreds of kilometers from the nearest possible source habitat (Brawn and Robinson 1996). Indeed, Robinson *et al.* (1995b) and Donovan *et al.* (1995b) argued that the presence of songbirds nesting in small woodlots, which are the only forest habitat available in most of the agricultural Midwest, depended upon the presence of very large forest tracts in mostly forested regions such as the Missouri Ozarks. These models were strongly supported by intensive empirical studies measuring the season-long productivity of marked birds in fragmented and unfragmented tracts (Trine 1998; Porneluzi and Faaborg 1999) and by evidence that small tracts were more likely to contain diverse communities of migratory songbirds if they were close to larger forest tracts (Nol *et al.* 2005).

Studies from elsewhere in the eastern USA also showed similar fragmentation effects (Porneluzi *et al.* 1993; Hoover *et al.* 1995), but in many cases, the effects were not as extreme as those documented in the Midwest (e.g., Roth and Johnson 1993; Weinberg and Roth 1998; Friesen *et al.* 1999; Fauth 2001). Fauth (2001), for example, showed generally low nesting success of wood thrushes in most agricultural woodlots, but argued that some small woodlots contained what may be occasional source populations (Fauth *et al.* 2000; see also Roth and Johnson 1993; Friesen *et al.* 1999), which raised the possibility that there may be some local source–sink dynamics within agricultural landscapes, at least for some species (Burke and Nol 2001). Trine (1998), using measures of season-long nesting success of marked pairs of wood thrushes, found that even relatively large tracts (up to 1,000 ha) could act as sinks if these tracts were located in landscapes in which they were surrounded by extensive agricultural habitats. Trine concluded that local source–sink dynamics may be operating in this landscape, but that most source habitats were weak ones, which means that their status as sources was highly uncertain given the problems with estimating adult and juvenile survival discussed below. Porneluzi and Faaborg (1999) and Flaspohler *et al.* (2001) also found evidence for source–sink dynamics within larger midwestern forest tracts in more forested regions in relation to distances from edges. Studies of nesting success in the largest forest tracts in eastern North America (Gale *et al.* 1997; Simons *et al.* 2000; Gram *et al.* 2003) also showed that these tracts do not necessarily serve as refugia from nest predation. Although large tracts in the Smoky Mountains were likely sources, especially in the absence of cowbird parasitism, they do not appear to be strong sources that could sustain regional populations of wood thrushes nesting in fragmented forests throughout the southeastern region (Simons *et al.* 2000).

In western North America, fragmentation effects may operate very differently (Tewksbury *et al.* 1998; Cavitt and Martin 2002; reviewed in Chace *et al.* 2005). Therefore, there is a need to modify source–sink conservation strategies so as to reflect regional differences in fragmentation effects.

Management implications of source–sink dynamics

Tract size and shape

The link between source–sink dynamics and habitat fragmentation and loss has proven to be a powerful tool in developing management recommendations for improving the nesting success of migratory songbirds and, possibly, reversing their decline (Llewellyn *et al.* 1996; Zuidema *et al.* 1996; Bonney *et al.* 1999; Beissinger *et al.* 2000; Carter *et al.* 2000; Rich *et al.* 2004; Betts and Forbes 2005). Most of these management guidelines first identify the largest tracts in a region and assign them the highest priority, assuming that they are source habitat for many forest-interior species (Bonney *et al.* 1999; Carter *et al.* 2000; Rich *et al.* 2004). Next, other areas that are only moderately fragmented have been targeted for reforestation. The Shawnee National Forest (SNF) in Illinois, for example, has been the site of some of the earliest efforts to manage the forest for creating source habitats. This relatively small national forest (*c.* 104,000 ha) in a region of approximately 50% forest cover has extremely fragmented ownership with no tracts of more than 2,500 ha, and therefore was considered to be in need of a plan to reduce this fragmentation. The SNF Management Plan identified Forest Interior Management Units (FIMUs): areas with the potential for at least 1,200 acres (*c.* 450–500 ha) of core area at least 100 m from large openings such as pastures and row-crop fields. Because edge effects in the SNF extend deep into the forest (Fig. 20.1; see also Morse and Robinson 1999; Hoover *et al.* 2006), additional efforts were made to target the largest blocks of forest in the SNF region. The Hutchins Creek area (Fig. 20.2) was identified because there was one series of private inholdings that fragmented a forest block that had the potential to be over 8,000 ha with the reforestation of only about 205 ha of farmland and hayfields. The agricultural inholdings along Hutchins Creek had an additive negative effect on the nesting success of birds in the adjacent forest (Hoover *et al.* 2006). Higher rates of nest predation and brood parasitism were consistently found closer to the forest/agriculture interface and the cumulative negative effects extended at least 600 m into the forest (Fig. 20.1), creating >1,300 ha of sink habitat (Fig. 20.2). Most researchers have found that edge effects penetrate <150 m into forest fragments (Laurence 2000), and literature reviews have suggested that increased nest predation occurs primarily within 50 m of an edge (Paton 1994; Batáry and Báldi 2004). Few of these

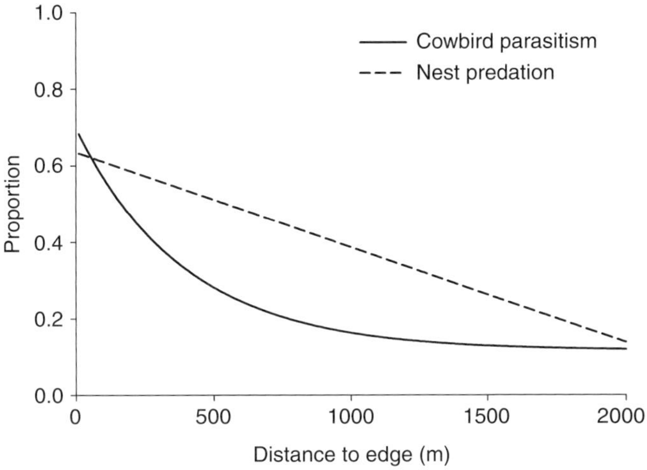

FIGURE 20.1.
Rates of nest predation and cowbird parasitism decrease with increasing distance from the agriculture/forest interface in the Hutchins Creek Valley in Illinois (adapted from Hoover *et al.* 2006).

studies, however, have considered rates of nest predation beyond 500 m from an edge (Laurence 2000; Batáry and Báldi 2004; but see Angelstam 1986). The results of the Hutchins Creek study (Hoover *et al.* 2006) led a consortium of private organizations including The Nature Conservancy to begin purchasing the 205 ha of agricultural land and reforesting it. In this case, the amount of source habitat for forest-nesting birds created by removing a relatively small amount of agricultural land from production should be substantial, with 1,350 ha of source habitat created from the acquisition of just 205 ha (Fig. 20.2).

The growing evidence for the regional scale of source–sink dynamics also has implications for deciding where to invest conservation resources. Smaller tracts closer to large tracts have greater diversity and abundance of breeding migratory songbirds, presumably because of their proximity to source populations (Nol *et al.* 2005). There have even been suggestions that small, isolated tracts be managed solely for their use by birds during migration rather than during the breeding season, when they act as ecological traps (Brawn and Robinson 1996; Robinson *et al.* 1997, 1999).

Cowbird control

In areas where large-scale management to increase or maintain source populations is not possible, cowbird control has been used as a means of turning population sinks into sources. Many endangered species occur in small, isolated patches of habitat surrounded by areas where intense human use creates unlimited cowbird feeding habitat (reviewed in Robinson *et al.* 1995a;

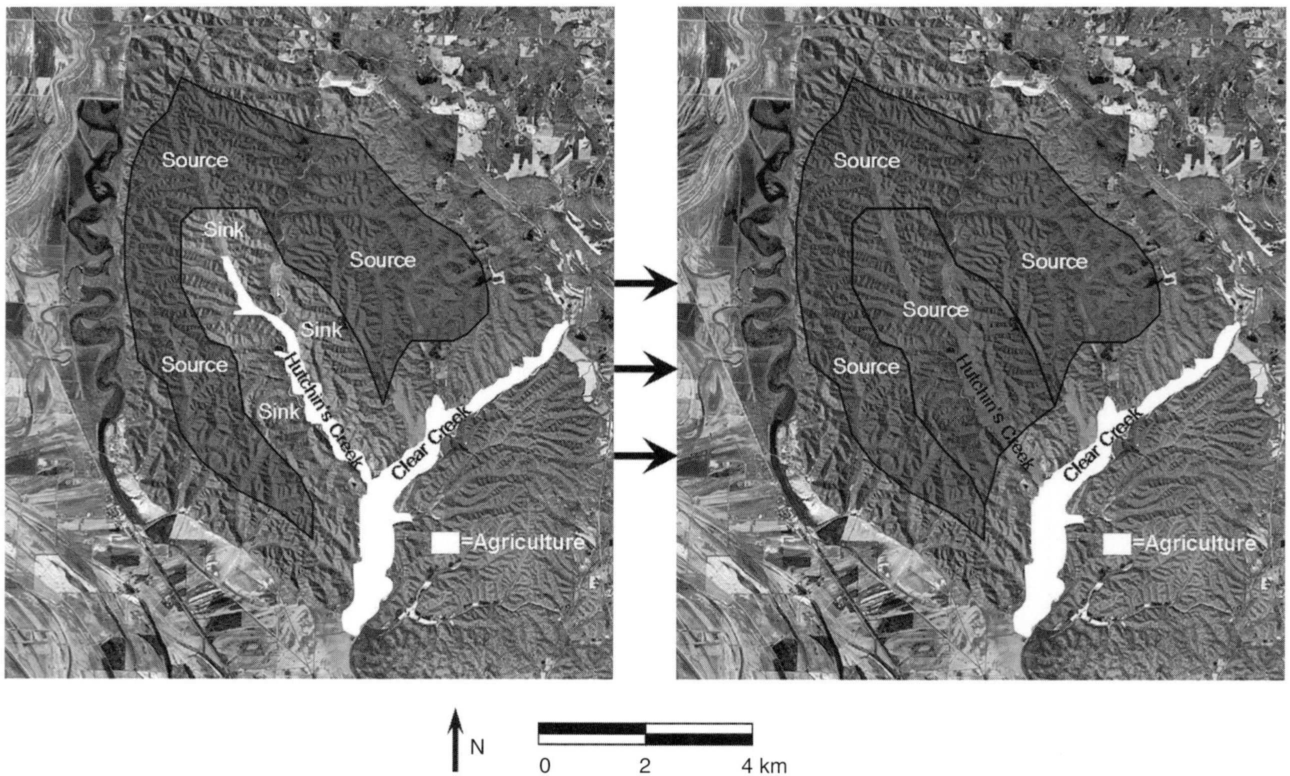

FIGURE 20.2.
Agricultural land use in the Hutchins Creek Valley within the Shawnee National Forest in Illinois resulted in sink habitat in the adjacent forest (left panel). Acquisition and reforestation of 205 ha along the valley should add 1,350 ha of source habitat (right panel).

Rothstein and Robinson 1998). Most of these endangered species depend critically upon frequent disturbances to create their habitat (reviewed in Robinson *et al.* 1995a). Trapping and killing large numbers of cowbirds in these relatively small tracts has proven to be a very effective means of stopping, and even reversing, population declines when combined with intensive habitat management and restoration (e.g., Smith *et al.* 2000; Ward and Schlossberg 2004; Kus and Whitfield 2005; reviewed in Robinson *et al.* 1995a). In some cases, detailed demographic studies have shown that cowbird control projects have turned sink habitats into sources (Smith *et al.* 2002). If cowbirds routinely engage in farming and mafia tactics (e.g., Hoover and Robinson 2007), then we would expect reduced rates of nest predation when adult cowbirds are removed. The presence of a cowbird chick in a nest tends to increase the rate at which the adults provision the brood (Hoover and Reetz 2006) and may facilitate other nest predators in finding the nest (Dearborn 1999; McLaren and Sealy 2000). The removal of cowbirds and subsequent reduction in rates of cowbird parasitism, therefore, may further reduce overall predation rates.

The utility of cowbird removal on a larger scale, however, is much less clear (Rothstein and Robinson 1998). Most of the success stories in cowbird control occur in species that have very small geographic ranges and that are found primarily in disturbance-dependent habitat in landscapes in which cowbird feeding habitat is abundant (Rothstein and Robinson 1998; reviewed in Robinson *et al.* 1995a). It is not clear that cowbird control on a much larger scale would benefit forest species with large geographic ranges. Some removal experiments (Kosciuch and Sandercock 2008), for example, have shown that removals increase the productivity of host species, but also increase the productivity of the cowbirds themselves, which are freed of density-dependent pressures that may actually cause some habitats to be sinks for cowbirds (see also Trine 2000; Winfree *et al.* 2006). In the long run, therefore, cowbird removals may have little impact on parasitism. Another recent removal experiment (Sandercock *et al.* 2008) showed that reducing cowbird parasitism did not increase nesting success because rates of nest predation in these mostly agricultural landscapes were so high that they more than compensated for the reduced losses to cowbirds. This latter experiment illustrates the more general problem with using cowbird removals to mitigate habitat fragmentation. Habitat fragmentation and loss creates many problems in addition to cowbird parasitism, which is a problem unique to birds (Fahrig 1997, 2003; Smith and Hellmann 2002; Lindenmayer and Fischer 2006). Landscape management that focuses on reducing fragmentation and increasing forest cover, on the other hand, does not require continued intervention such as cowbird trapping and should also work for non-avian wildlife that is adversely affected by fragmentation.

Decision rules and ecological traps

Sink habitat will only cause long-term declines in songbirds if they are also ecological traps, i.e., they are equally or more attractive than source habitats (Donovan and Thompson 2001; Schlaepfer *et al.* 2002; Schmidt 2003). If songbirds have mechanisms to avoid sink habitats, then declines will stop as long as source habitats remain (Donovan and Thompson 2001). If animals avoid sink habitats except when populations are high, then sink habitat may actually be useful for long-term population maintenance by providing areas where surplus animals can survive while producing at least some young (Howe *et al.* 1991). In this context, the emerging literature on the use by animals of "public" and "private" information (Doligez *et al.* 1999, 2002; Hoover 2003c; Ward and Schlossberg 2004; Fletcher 2006, 2007; see also Greenwood and Harvey 1982) is relevant to ecological traps. Birds have long been known to vary their site fidelity in response to their own nesting success ("private information") (Greenwood and Harvey 1982; Robinson 1985; Jackson *et al.* 1989). Birds that are more successful are generally more likely to return to a nesting territory than those that are not. In the first experimental study of this phenomenon, Hoover (2003c) determined that these high return rates were not simply the result of higher survival and nesting success of a small subset of birds in superior conditions, which would likely also be those that nested most successfully. He assigned nesting success randomly to individual prothonotary warblers (*Protonotaria citrea*) in study populations and showed that nearly 80% of birds that produced two broods in a season returned, whereas unsuccessful birds avoided returning to sites where they had produced no broods. Such decision rules have undoubtedly evolved under the constant pressure of nest predation, which has been a strong selective agent throughout the evolutionary history of all birds (Ricklefs 1969; Martin 1995). In fragmented landscapes in which nest predation rates are often deterministic and related to distances from edges, landscape composition, and tract size, such decision rules could lead to the abandonment of areas with chronically high nest predation rates and would tend to concentrate birds in source habitats (Donovan and Thompson 2001; Schmidt 2003). Indeed, area-sensitivity (Villard *et al.* 1993; Van Horn *et al.* 1995) in forest songbirds may result from an adaptive avoidance of edges (Schmidt 2003).

The use of "public information," i.e., the distribution and nesting success of other individuals in an area (Doligez *et al.* 2002), could further enhance this tendency of birds to return to source habitat and avoid sink habitat. Conspecific attraction, the tendency of territorial birds to cluster their territories near other conspecifics, was originally thought of as a conservation problem in a metapopulation context (Reed and Dobson 1993) because it would slow down the rate of recolonization of smaller patches once the local population had become extinct. If, however, these patches were sink habitats, then conspecific attraction could represent an adaptive mechanism that could cause birds to avoid these sites. Conversely, if source habitats have consistently high nesting success as a result of low nest predation rates, then conspecific attraction would increase the tendency of birds to concentrate in source habitats, where nesting success is consistently higher (Hoover 2009). Indeed, high concentrations of nesting birds could provide a reliable clue about habitat quality to a dispersing adult that had failed elsewhere and to young birds breeding for the first time (Betts *et al.* 2008).

The lack of decision rules used to avoid cowbird parasitism makes it more likely that some sink habitats could become an ecological trap (Hoover 2003b, 2003c). Unlike nest predation, which has long been an intensive selective agent on nesting birds (Ricklefs 1969; Martin 1995), brood parasitism in large parts of North America is a more recent phenomenon and is much more spatially variable (Mayfield 1977; Rothstein and Robinson 1994, 1998; Robinson *et al.* 1995a; reviewed in Chace *et al.* 2005). The lack of defenses against cowbird parasitism in forest birds has long been noted and has been assumed to reflect the evolutionarily recent contact between most forest birds and cowbirds (Hosoi and Rothstein 2000), which have only recently expanded into forest habitats with the increasing fragmentation of forested habitats (Rothstein and Robinson 1998). Indeed, the lack of parasitism in the largest forest tracts, which act as source populations for many migratory songbirds, is likely to slow the rate of evolution of host defenses against brood parasitism (Barabas *et al.* 2004).

Areas with low rates of nest predation and high rates of parasitism therefore have the potential to act as ecological traps (Gates and Gysel 1978; Donovan and Thompson 2001). This ecological trap scenario exists in prothonotary warblers breeding in Illinois. One particular site has been a habitat sink for many consecutive years (1994–2007; J. P. Hoover, unpublished data). Warblers nesting there suffer high rates of cowbird parasitism (89% of nests parasitized, average of 2.4 cowbird eggs per nest, $n = 550$ nests across all years) but relatively low rates of nest predation (<30%). Female warblers therefore successfully raise an average of 2.6 cowbird chicks but only 2.2 warbler chicks during their first breeding season on the site. These females return to the sink habitat at a relatively high rate (63%), only to produce more cowbird chicks (2.1 on average) and even fewer warbler chicks (1.5 on average) in subsequent breeding seasons. In this context, it may be fortunate that nest predation and brood parasitism rates tend to be positively correlated across large regions (Robinson *et al.* 1995b); the same decision rules that cause birds to concentrate in large forest tracts and in mostly forested landscapes, where rates of nest predation are low, should also tend to keep them away from brood parasitism, both locally and regionally (Robinson *et al.* 1995b; Hochachka *et al.* 1999). Nevertheless, the lack of defenses against cowbirds may explain why some of the steepest declines of North American songbirds have been recorded in species subjected to frequent brood parasitism (Bohning-Gaese *et al.* 1993).

Do all species base their decision rules on nesting success?

There are some species that have not been shown to use private information based upon nesting success when deciding where to nest. Indigo buntings (*Passerina cyanea*), for example, show no tendency to vary their site fidelity based on their previous nesting success (Payne and Payne 1993). Interestingly, indigo buntings have also been identified as being susceptible to ecological traps (Suarez *et al.* 1997; Weldon and Haddad 2005), in which their population densities are just as high in sink habitat as they are in source habitat. Kentucky warblers also show no evidence of varying their site fidelity with previous nesting success; their population densities are just as high in areas with low nesting success as they are in putative source habitats (Morse and Robinson 1999; S. F. Morse, unpublished data). Ovenbirds (*Seiurus auricapillus*) also show no tendency to avoid areas with high nest predation rates (Burke and Nol 2001). Such species may be particularly vulnerable to human modifications of landscapes that create predictable source and sink habitats.

Problems with identifying sources and sinks

With so many studies of avian nesting success in relation to landscape cover and fragmentation, it is not surprising that the methodology has become highly standardized (reviewed in Anders and Marshall 2005). Field estimates of nesting success generally followed the protocols outlined in Martin and Geupel (1993) and Martin *et al.* (1997). Estimators of nesting success now use information-theoretic approaches (White and Burnham 1999; Dinsmore *et al.* 2002; Shaffer 2004; Rotella *et al.* 2007; Shaffer and Thompson 2007). Measuring annual reproductive success is rather complicated when season-long productivity of marked individuals is not available, and requires complex models in populations subjected to both parasitism and nest predation (Pease and Grzybowski 1995; Powell *et al.* 1999; Schmidt and Whelan 1999). Even with marked populations, however, estimating season-long productivity can be extraordinarily complex, especially if there are within-season movements (Nolan 1978; Jackson *et al.* 1989). Survival estimates rely on mark–recapture estimates (Anders and Marshall 2005). Accurate assessments of survival rates and productivity are vital for developing demographic models for conservation and management (Ruth *et al.* 2003), but current source–sink models for migratory songbirds are usually based on assumptions about female survival rates and sometimes poor estimates of fecundity (Podolsky *et al.* 2007). Specifically, the models tend to ignore the potential influence of variations in rates of double-brooding and re-nesting on estimates of fecundity (Nagy and Holmes 2005). Recent modeling efforts have demonstrated that variation in estimates of annual female survival and in rates of re-nesting and double-brooding can all have a large effect on whether habitats are classified as sources or sinks (Podolsky *et al.* 2007).

Problems with survival estimates

All source–sink models depend critically upon accurate estimates of adult and juvenile survival (Pulliam 1988; Pulliam and Danielson 1991; Anders and Marshall 2005; Dunning *et al.*, Chapter 11, this volume). In many respects this has been the Achilles' heel of such studies (Anders and Marshall 2005). Adult survival has been the subject of some of the most intensive modeling efforts in all of population ecology, mostly based on mark–recapture studies. Yet adult dispersal occurs frequently in most species and is very difficult to model accurately because we have so little data on where birds go after they disperse from an area (Cilimburg *et al.* 2002; Marshall *et al.* 2004; Nagy and Holmes 2004). Models that incorporate decision rules into adult survival estimates can solve these problems analytically, but then gathering the relevant data on nest predation rates and decision rules becomes a problem for all but the most intensively studied species (Marshall *et al.* 2004; Anders and Marshall 2005). This may seem like a minor problem in many species, but it can lead to serious underestimates of adult survival and may overestimate the amount of productivity necessary to be at the source–sink threshold (Marshall *et al.* 2004) and overestimate mortality during the migration period, which is thought to be very high in some species (Sillett and Holmes 2002). Analyses of mark–recapture data of prothonotary warblers using widely accepted models that do not control for nesting success, for example, suggest annual survival rates of 49–55% (J. P. Hoover, unpublished data). When nesting success is randomly assigned, however, the warblers with the highest nesting success return at a rate of nearly 80% (Hoover 2003a), which is a minimum estimate of annual survival. Therefore, prothonotary warblers may be living more than twice as long as suggested by mark–recapture models that do not incorporate decision rules.

Juvenile survival, on the other hand, may be overestimated for many species. Recent telemetry studies of fledglings, for example, show very low survival rates during the first two weeks out of the nest (Anders *et al.* 1997; Sillett and Holmes 2002; Gardali *et al.* 2003; Suedkamp Wells *et al.* 2007; Schmidt *et al.* 2008). Recent evidence from stable-isotope studies suggests that juvenile dispersal may occur on very large scales (Studds *et al.* 2008), which makes estimates of juvenile survival based on mark–recapture studies especially problematical. Telemetry studies of post-fledging young have been especially startling because they suggest that one of the key elements that may determine the source–sink status of a population is the presence of safe habitat where juveniles can find food and cover, where they are not vulnerable to predators, and where there are no adults competing for the same space (Anders *et al.* 1998). Managers may therefore need to incorporate the habitat requirements of fledglings into their plans.

Do the interiors of large forest tracts always act as refugia from predation?

One of the central tenets of fragmentation theory is that nest predators are more abundant in fragmented landscapes, largely because of the release of mesopredators from top-down control by predators that formerly regulated their populations, and also through food subsidies from the matrix surrounding the forest patches (reviewed in Lindenmayer and Fischer 2006). Although there is some evidence for this hypothesis, there is also growing evidence that large unfragmented forest tracts do not always act as a refuge from predation. Simons *et al.* (2000), for example, showed that nest predation rates remained high even in the interior of some of the largest forest tracts remaining in North America (see also Gale *et al.* 1997; Gram *et al.* 2003). Nest predation rates in the largest forest remaining in Europe, the Bialowieza National Park, were also extremely high and may be a prime factor in keeping the population densities of most forest-nesting species extremely low (Tomialojc *et al.* 1984). Studies from the American West have shown that fragmented areas may actually serve as refugia from nest predation (Hannon and Cotterill 1998; Tewksbury *et al.* 1998; Bayne and Hobson 2002; Willson *et al.* 2003). In these environments, most nest predators such as red squirrels are adversely affected by habitat loss and fragmentation. Even in the Midwest and East, there are many area-sensitive nest predators, some of which have also experienced major population declines as a result of fragmentation (Heske 1995; Oehler and Litvaitis 1996; Farnsworth and Simons 2000; Heske *et al.* 2001). For some species, landscapes with intermediate levels of forest cover may represent a worst-case scenario, as the forest tracts are large enough to contain forest-interior nest predators but not so large that they can provide a refuge from predators that are subsidized by food from the matrix surrounding the forest patches. Studies from the

Shawnee National Forest in Illinois, for example, generally show few effects of proximity to edge, tract size, or landscape composition, perhaps because of the compensating effect of forest-interior nest predators (e.g., the eastern chipmunk, *Tamias striatus*; and the broad-winged hawk, *Buteo platypterus*) and edge-enhanced nest predators (e.g., raccoons; common grackles, *Quiscalus quiscula*; rat snakes, *Elaphe* spp.) (Heske 1995; Marini *et al.* 1995, Chapa-Vargas and Robinson 2006, 2007; S. K. Robinson, unpublished data). In many tropical forests where there has not been heavy hunting pressure, large troops of monkeys move through the forest and depredate many nests of songbirds, even those in cavities (Robinson *et al.* 2000b). Very high rates of nest predation in these forests may even enhance diversity by preventing the dominance of a few species that would otherwise be able to exploit many of the resources available (Tomialojc *et al.* 1984).

All of these studies suggest that we actually know very little about the role of nest predation in large contiguous tracts. Our viewpoint is historically constrained to a period when almost all of the forests we study are young, and wildlife populations are recovering from the almost complete deforestation that occurred during the late 1800s and early 1900s. The community composition of birds and their nest predators may bear little resemblance to those that prevailed throughout most of the history of the continent, either in North America or Europe.

Does dispersal occur on a large enough spatial scale to recolonize distant patches?

Usually, we assume that migratory songbirds will have few problems recolonizing habitat patches and therefore do not need corridors connecting patches, as has been shown for several non-migratory species (Cooper and Walters 2002; Castellón and Sieving 2007). Nevertheless, there are some indications that the proximity of patches may increase recolonization through its effects on prospecting young birds and adults seeking new territories within a season (Desrochers and Hannon 1997; Norris and Stutchbury 2001). There are also indications that a decline in the recruitment of new individuals to patches that are reproducing at levels above the source–sink threshold can cause long-term declines in populations, even in the absence of further habitat loss and reproductive failure (Ward 2005).

At this stage, we are left with many fundamental questions and gaps in our data that make it very difficult to determine the source–sink status of habitats and the scale on which populations of migratory songbirds are regulated. Data from the North American Breeding Bird Survey showed 1-year time-lagged correlations in abundance between paired wood thrush populations that were 60–80 km apart, suggesting that dispersal and source–sink dynamics may occur over these distances (Tittler *et al.* 2006). Robinson *et al.* (1995b) hypothesized that dispersal occurs on a very large scale, at least hundreds of kilometers, in most forest songbirds, but we lack any studies that have empirically shown that this occurs. One of the few studies of dispersal in a neotropical migrant, the prothonotary warbler, in fact shows that even with extensive searching out to 40 km away from study populations, most natal dispersal occurs within just 10 km of the natal site (J. P. Hoover, unpublished data). Similarly, we know very little about the scale of adult dispersal following breeding failure. Finally, we also know very little about where all but a few local populations spend their winters, which severely hampers our understanding of how winter habitat affects the survival and condition of birds during migration and on the breeding grounds. Answering these questions will require new generations of technology that will facilitate the tracking of individuals year-round. New technologies are becoming available for satellite tracking, and GPS-based technologies may enable to us download the entire movement history of birds when we recapture them. Only by studying migratory songbirds throughout their complex annual cycle will we be able to understand population regulation and design effective conservation strategies.

What are the relative contributions of habitat fragmentation and landscape composition to source–sink dynamics?

As we emphasized at the beginning of this chapter, it is very difficult to separate the effects of landscape composition from habitat fragmentation variables. Fragmentation variables may have a relatively minor influence on nesting success compared with landscape composition (Cottam *et al.* 2009). Controlled experimental studies are rarely possible with songbirds because source–sink dynamics appear to operate on a very large scale (see above), which makes experimental studies logistically unfeasible in most cases. Large-scale restoration projects provide some opportunities, but most restoration projects both increase forest cover and reduce fragmentation at the same time (Fahrig 2003; Lindenmayer and Fischer 2006). In practice, it is important to understand the relative contributions of fragmentation and forest cover; if habitat fragmentation is the most critical variable, then it should be possible to mitigate the effects of loss of forest by designing landscapes to minimize edges (see Fig. 20.2). If, on the other hand, forest loss is most important, then extensive reforestation projects may be the only way to increase the source habitat. In landscapes in which such habitat restoration is unlikely (e.g., in cities and in areas with rich agricultural soils), management to improve avian nesting success is unlikely to succeed and should not be a high priority for conservation planning.

Conclusions

Even given the uncertainties discussed above, it is clear that habitat loss and fragmentation increase sink habitat within and between regions. From a conservation point of view, the preservation and restoration of large tracts and regions with high forest cover and minimal edge effects will give songbird populations the greatest chances of being sustainable over the long run. Birds that have decision rules that enable them to concentrate in source habitat will be far less vulnerable to habitat loss and fragmentation than those that do not. Future studies should perhaps focus on disentangling the effects of habitat fragmentation from habitat loss and on identifying species that do not have effective decision rules that enable them to avoid potential ecological traps.

References

Anders, A. D. and M. R. Marshall (2005). Increasing the accuracy of productivity and survival estimates in assessing landbird population status. *Conservation Biology* **19**: 66–74.

Anders, A. D., D. C. Dearborn, J. Faaborg and F. R. Thompson III (1997). Juvenile survival in a population of neotropical migrant birds. *Conservation Biology* **11**: 698–707.

Anders, A. D., J. Faaborg and F. R. Thompson III (1998). Postfledging dispersal, habitat use, and home range size of juvenile wood thrushes. *Auk* **115**: 349–358.

Andrèn, H. (1992). Corvid density and nest predators in relation to forest fragmentation: a landscape perspective. *Ecology* **73**: 794–804.

Andrèn, H. (1994). Effects of habitat fragmentation on birds and mammals in landscapes with different proportions of suitable habitat: a review. *Oikos* **71**: 355–366.

Angelstam, P. (1986). Predation of ground-nesting birds' nests in relation to predator densities and habitat edge. *Oikos* **47**: 365–373.

Arcese, P., J. N. M. Smith and M. I. Hatch (1996). Nest predation by cowbirds and its consequences for passerine demography. *Proceedings of the National Academy of Sciences* **93**: 4608–4611.

Austen, M. J. W., C. M. Francis, D. M. Burke and M. S. W. Broadstreet (2001). Landscape context and forest fragmentation effects on forest birds in southern Ontario. *Condor* **103**: 701–714.

Barabas, L., B. Gilicze, F. Takasu and C. Moskat (2004). Survival and anti-parasite defense in a host metapoplation under heavy brood parasitism: a source–sink dynamic model. *Journal of Ethology* **22**: 143–151.

Batáry, P. and A. Báldi (2004). Evidence of an edge effect on avian nest success. *Conservation Biology* **18**: 389–400.

Bayne, E. M. and K. L. Hobson (2002). Effects of red squirrel (*Tamiasciurus hudsonicus*) removal on survival of artificial songbird nests in boreal forest fragments. *American Midlands Naturalist* **147**: 72–79.

Beissinger, S. R., J. M. Reed, J. M. Wunderle Jr., S. K. Robinson and D. M. Finch (2000). Report of the AOU conservation committee on the Partners in Flight species prioritization plan. *Auk* **117**: 549–561.

Betts, M. G. and G. J. Forbes (eds.) (2005). *Forest Management Guidelines to Protect Native Biodiversity in the Greater Fundy Ecosystem*. University of New Brunswick Cooperative Fish and Wildlife Research Unit, Fredericton, NB, Canada.

Betts, M. G., A. S. Hadley, N. Rodenhouse and J. J. Nocera (2008). Social information trumps vegetation structure in breeding-site selection by a migrant songbird. *Proceedings of the Royal Society of London Series B – Biological Sciences* **275**: 2257–2263.

Bielefeldt, J. and R. N. Rosenfield (1997). Reexamination of cowbird parasitism and edge effects in Wisconsin forests. *Journal of Wildlife Management* **61**: 1222–1226.

Blouin-Demers, G. and P. J. Weatherhead (2001). Habitat use by black ratsnakes (*Elaphe obsoleta obsoleta*) in fragmented forests. *Ecology* **82**: 2882–2896.

Bohning-Gaese, M. L. Taper and J. H. Brown (1993). Are declines in North American insectivorous songbirds due to causes on the breeding range? *Conservation Biology* **7**: 76–86.

Bollinger, E. K. and E. T. Linder (1999). Reproductive success of neotropical migrants in a fragmented Illinois forest. *Wilson Bulletin* **106**: 46–54.

Bonney, R., D. N. Pashley, R. J. Cooper and L. Miles (eds.) (1999). *Strategies for Bird Conservation: The Partners in Flight Planning Process*. Cornell Laboratory of Ornithology, Ithaca, NY.

Brawn, J. D. and S. K. Robinson (1996). Source–sink dynamics may complicate the interpretation of long-term census data. *Ecology* **77**: 3–12.

Brittingham, M. C. and S. A. Temple (1983). Have cowbirds caused forest songbirds to decline? *BioScience* **33**: 31–35.

Brown, J. H. and A. Kodric-Brown (1977). Turnover rates in insular biogeography: effects of immigration on extinction. *Ecology* **58**: 445–449.

Burke, D. M. and E. Nol (2000). Landscape and fragment size effects on reproductive success of forest-breeding birds in Ontario. *Ecological Applications* **10**: 1749–1761.

Burke, D. M and E. Nol (2001). Age ratios and return rates of ovenbirds in contiguous and fragmented forests. *Journal of Field Ornithology* **72**: 433–438.

Carfagno, G. L. F. and P. J. Weatherhead (2006). Intraspecific and interspecific variation in use of forest edge habitat by snakes. *Canadian Journal of Zoology* **84**: 1440–1452.

Carter, M. F., W. C. Hunter. D. N. Pashley and K. V. Rosenberg (2000). Setting conservation priorities for landbirds in the United States: the Partners in Flight approach. *Auk* **117**: 541–548.

Castellón, T. D. and K. E. Sieving (2007). Sustainable patch networks and landscape connections for an endemic understory bird with limited dispersal ability. *Ecological Applications* **17**: 2152–2163.

Cavitt, J. F. and T. E. Martin (2002). Effects of forest fragmentation on brood parasitism and nest predation in eastern and western landscapes. *Studies in Avian Biology* **25**: 73–80.

Chace, J. F., C. Farmer, R. Winfree, D. R. Curson, W. E. Jensen, C. B. Goguen and S. K. Robinson (2005). Cowbird ecology: a review of factors influencing distribution and abundance of cowbirds across spatial scales. *Ornithological Monographs* **57**: 45–70.

Chalfoun, A. D., M. J. Ratnaswamy and F. R. Thompson III (2002a). Songbird nest predation in a fragmented landscape. *Ecological Applications* **12**: 858–867.

Chalfoun, A. D., F. R. Thompson III and M. J. Ratnaswamy (2002b). Nest predators and fragmentation: a review and meta-analysis. *Conservation Biology* **16**: 306–318.

Chapa-Vargas, L. and S. K. Robinson (2006). Nesting success of a songbird in a complex floodplain forest landscape in Illinois, USA: local fragmentation vs. vegetation structure. *Landscape Ecology* **21**: 525–537.

Chapa-Vargas, L. and S. K. Robinson (2007). Nesting success of Acadian flycatchers (*Empidonax virescens*) in floodplain forest corridors. *Auk* **124**: 1267–1280.

Cilimburg, A. B., M. S. Lindberg, J. J. Tewksbury and S. J. Hejl (2002). Effects of dispersal on survival probability of adult yellow warblers (*Dendroica petechia*). *Auk* **119**: 778–789.

Cooper, C. B. and J. R. Walters (2002). Experimental evidence of disrupted dispersal causing declines of an Australian passerine in fragmented habitat. *Conservation Biology* **16**: 471–478.

Cottam, M. R., S. K. Robinson, E. J. Heske, J. D. Brawn and K. C. Rowe (2009). Use of landscape metrics to predict avian nest survival in a fragmented Midwestern forest landscape. *Biological Conservation* **142**: 2464–2475.

Crooks, K. R. and M. E. Soule (1999). Mesopredator release and avifaunal extinctions in a fragmented system. *Nature* **400**: 563–566.

Dearborn, D. C. (1999). Brown-headed cowbird nestling vocalizations and the risk of nest predation. *Auk* **116**: 448–457.
Desrochers, A. and S. J. Hannon (1997). Gap crossing decisions by forest songbirds during the postfledging period. *Conservation Biology* **11**: 1204–1210.
Dijak, W. and F. R. Thompson III (2000). Landscape and edge effects on the distribution of mammalian predators in Missouri. *Journal of Wildlife Management* **64**: 209–216.
Dinsmore, S. J., G. C. White and F. L. Knopf (2002). Advanced techniques for modeling avian nest survival. *Ecology* **83**: 3476–3488.
Doligez, B., E. Danchin, J. Clobert and L. Gustafsson (1999). The use of conspecific reproductive success for breeding habitat selection in a non-colonial hole-nesting species, the collared flycatcher. *Journal of Animal Ecology* **68**: 1193–1206.
Doligez, B., E. Danchin and J. Clobert (2002). Public information and breeding habitat selection in a wild bird population. *Science* **297**: 1169–1170.
Donovan, T. M. and C. H. Flather (2002). Relationships among North American songbird trends, habitat fragmentation, and landscape occupancy. *Ecological Applications* **12**: 364–374.
Donovan, T. M. and F. R. Thompson III (2001). Modeling the ecological trap hypothesis: a habitat and demographic for migrant songbirds. *Ecological Applications* **11**: 871–882.
Donovan, T. M., F. R. Thompson III, J. Faaborg and J. R. Probst (1995a). Reproductive success of migratory birds in habitat sources and sinks. *Conservation Biology* **9**: 1380–1395.
Donovan, T. M., R. H. Lamberson, A. Kimber, F. R. Thompson III and J. Faaborg (1995b). Modeling the effects of habitat fragmentation on source and sink demography of neotropical migrant birds. *Conservation Biology* **9**: 1396–1407.
Donovan, T. M., P. W. Jones, E. M. Annand and F. R. Thompson III (1997). Variation in local scale edge effects: mechanisms and landscape context. *Ecology* **78**: 2064–2075.
Driscoll, M. J. L. and T. M. Donovan (2004). Landscape context moderates edge effects: nesting success of wood thrushes in central New York. *Conservation Biology* **18**: 1330–1338.
Dunford, W., D. M. Burke and E. Nol (2002). Assessing edge avoidance and area-sensitivity of red-eyed vireos in southcentral Ontario. *Wilson Bulletin* **114**: 79–86.
Faaborg, J. (1980). Potential uses and abuses of diversity concepts in wildlife management. *Transactions of the Missouri Academy of Sciences* **14**: 41–49.
Faaborg, J., M. Brittingham, T. Donovan and J. Blake (1995). Habitat fragmentation in the temperate zone. In *Ecology and Management of Neotropical Migratory Birds* (T. E. Martin and D. M. Finch, eds.). Oxford University Press, New York: 357–380.
Fahrig, L. (1997). Relative effects of habitat loss and fragmentation on population extinction. *Journal of Wildlife Management* **61**: 603–610.
Fahrig, L. (2003). Effects of habitat fragmentation on biodiversity. *Annual Review of Ecology, Evolution, and Systematics* **34**: 487–515.
Farnsworth, G. L. and T. R. Simons (2000). Observation of wood thrush nest predators in a large contiguous forest. *Wilson Bulletin* **112**: 82–87.
Fauth, P. T. (2001). Wood thrush populations are not all sinks in the agricultural Midwestern United States. *Conservation Biology* **15**: 523–527.
Fauth, P. T., E. J. Gustafson and K. N. Rabenold (2000). Using landscape metrics to model source habitat for neotropical migrants in the Midwestern US *Landscape Ecology* **15**: 621–631.
Flaspohler, D. J., S. A. Temple and R. N. Rosenfield (2001). Effects of forest edges on ovenbird demography in a managed forest landscape. *Conservation Biology* **15**: 173–183.
Fletcher, R. J. Jr. (2006). Emergent properties of conspecific attraction in fragmented landscapes. *American Naturalist* **168**: 207–219.
Fletcher, R. J. Jr. (2007). Species interactions and population densities mediate the use of social cues for habitat selection. *Journal of Animal Ecology* **76**: 598–606.
Friesen, L., M. D. Cadman and R. J. Mackay (1999). Nesting success of neotropical migrant songbirds in a highly fragmented landscape. *Conservation Biology* **13**: 338–346.
Gale, G. A., L. A. Hanners and S. R. Patton (1997). Nesting success of worm-eating warblers in a forested landscape. *Conservation Biology* **11**: 246–250.

Gardali, T., D. C. Barton, J. D. White and G. R. Geupel (2003). Juvenile and adult survival of Swainson's thrushes (*Catharus ustulatus*) in coastal California: annual estimates using capture–recapture analysis. *Auk* **120**: 1188–1194.
Gates, J. E. and L. E. Gysel (1978). Avian nest dispersion and fledging success in field-forest ecotones. *Ecology* **59**: 871–883.
Gibbs, J. P. and J. Faaborg (1990). Estimating the viability of ovenbird and Kentucky warbler populations in forest fragments. *Conservation Biology* **4**: 193–196.
Goguen, C. B. and N. E. Mathews (2000). Local gradients of cowbird abundance and parasitism relative to livestock grazing in a western landscape. *Conservation Biology* **14**: 1862–1869.
Goguen, C. B. and N. E. Mathews (2001). Brown-headed cowbird behavior and movements in relation to livestock grazing. *Ecological Applications* **11**: 1533–1544.
Gram, W. K., P. A. Porneluzi, R. L. Clawson, J. Faaborg and S. C. Richter (2003). Effects of experimental forest management on densities and nesting success of bird species in Missouri Ozark forests. *Conservation Biology* **17**: 1324–1337.
Greenwood, P. J. and P. H. Harvey (1982). The natal and breeding dispersal of birds. *Annual Review of Ecology and Systematics* **13**: 1–21.
Gustafson, E. J., M. G. Knutson, G. J. Niemi and M. Friberg (2002). Evaluation of spatial models to predict vulnerability of forest birds to brood parasitism by cowbirds. *Ecological Applications* **12**: 412–426.
Hagan, J. M. III and D. W. Johnson (eds.) (1992). *Ecology and Conservation of Neotropical Migrant Landbirds*. Smithsonian Institution Press, Washington, DC.
Hahn, D. C. and J. S. Hatfield (1995). Parasitism at the landscape scale: cowbirds prefer forests. *Conservation Biology* **9**: 1415–1424.
Hannon, S. J. and S. E. Cotterill (1998). Nest predation in aspen woodlots in an agricultural area in Alberta: the enemy from within. *Auk* **115**: 16–25.
Heske, E. J. (1995). Mammalian abundances on forest-farm edges versus forest interiors in southern Illinois: is there an edge effect? *Journal of Mammalogy* **76**: 562–568.
Heske, E. J., S. K. Robinson and J. D. Brawn (2001). Nest predation and neotropical migrant songbirds: piecing together the fragments. *Wildlife Society Bulletin* **29**: 52–61.
Hochachka, W. M., T. E. Martin, V. Artman, C. R. Smith, S. J. Hejl, D. E. Andersen, D. Curson, L. Petit, N. Mathews, T. Donovan, E. E. Klaas, P. B. Wood, J. C. Manolis, K. P. McFarland, J. V. Nichols, J. C. Bednarz, D. M. Evans, J. P. Duguay, S. Garner, J. Tewksbury, K. L. Purcell, J. Faaborg, C. B. Goguen, C. Rimmer, R. Dettmers, M. Knutson, J. A. Collazo, L. Garner, D. Whitehead and G. Geupel (1999). Scale dependence in the effects of forest coverage on parasitization by brown-headed cowbirds. *Studies in Avian Biology* **18**: 80–88.
Hoover, J. P. (2003a). Multiple effects of brood parasitism reduce the reproductive success of prothonotary warblers *Protonotaria citrea*. *Animal Behaviour* **65**: 923–937.
Hoover, J. P. (2003b). Experiments and observations of prothonotary warblers indicate a lack of adaptive responses to brood parasitism. *Animal Behaviour* **65**: 935–944.
Hoover, J. P. (2003c). Decision rules for site fidelity in a migratory bird, the prothonotary warbler. *Ecology* **84**: 416–430.
Hoover, J. P. (2006). Water depth influences nest predation for a wetland-dependent bird in fragmented bottomland forests. *Biological Conservation* **127**: 37–45.
Hoover, J. P. (2009). Effects of hydrologic restoration on birds breeding in forested wetlands. *Wetlands* **29**: 563–573.
Hoover, J. P. and M. E. Hauber (2007). Individual patterns of habitat and nest-site use by hosts promote transgenerational transmission of avian brood parasitism status. *Journal of Animal Ecology* **76**: 1208–1214.
Hoover, J. P. and M. J. Reetz (2006). Brood parasitism increases provisioning rate and reduces offspring recruitment and adult return rates in a cowbird host. *Oecologia* **149**: 165–173.
Hoover, J. P. and S. K. Robinson (2007). Retaliatory mafia behavior by a parasitic cowbird favors host acceptance of parasitic eggs. *Proceedings of the National Academy of Sciences* **104**: 4479–4483.

Hoover, J. P., M. C. Brittingham and L. J. Goodrich (1995). Effects of forest patch size on nesting success of wood thrushes. *Auk* **112**: 146–155.
Hoover, J. P., T. H. Tear and M. E. Baltz (2006). Edge effects reduce the nesting success of Acadian flycatchers in a moderately fragmented forest. *Journal of Field Ornithology* **77**: 425–436.
Hosoi, S. A. and S. I. Rothstein (2000). Nest desertion and cowbird parasitism: evidence for evolved responses and evolutionary lag. *Animal Behaviour* **59**: 823–840.
Howe, R. W., G. J. Davis and V. Mosca (1991). The demographic significance of "sink" populations. *Biological Conservation* **57**: 239–255.
Jackson, W. M., S. Rohwer and V. Nolan Jr. (1989). Within-season breeding dispersal in prairie warblers and other passerines. *Condor* **91**: 233–241.
Keyser, A. J., G. E. Hill and E. C. Soehren (1998). Effects of forest fragment size, nest density, and proximity to edge on the risk of predation to ground-nesting passerine birds. *Conservation Biology* **12**: 986–994.
Knutson, M. G., G. J. Niemi, W. E. Newton and M. A. Friburg (2004). Avian nest success in Midwestern forests fragmented by agriculture. *Condor* **106**: 116–130.
Kosciuch, K. L. and B. K. Sandercock (2008). Cowbird removals unexpectedly increase productivity of a brood parasite and the songbird host. *Ecological Applications* **18**: 537–548.
Kupfer, J. A., G. P. Malanson and S. B. Franklin (2006). Not seeing the ocean for the islands: the mediating influence of matrix-based processes on forest fragmentation effects. *Global Ecology and Biogeography* **15**: 8–20.
Kus, B. E. and M. J. Whitfield (2005). Parasitism, productivity and population growth: responses of Least Bell's vireos and southwestern willow flycatchers to cowbird control. *Ornithological Monographs* **57**: 16–27.
Lahti, D. (2001). The "edge effect" on nest predation hypothesis after 20 years. *Biological Conservation* **99**: 365–374.
Lampila, P., M. Monkkonen and A. Desrochers (2005). Demographic responses by forest birds to forest fragmentation. *Conservation Biology* **19**: 1537–1546.
Larson, M. A., F. R. Thompson III, J. J. Millspaugh, W. D. Dijak and S. R. Shifley (2004). Linking population viability, habitat suitability, and landscape simulation models for conservation planning. *Ecological Modeling* **180**: 103–118.
Laurence, W. F. (2000). Do edge effects occur over large spatial scales? *Trends in Ecology and Evolution* **15**: 134–135.
Lindenmayer, D. B. and J. Fischer (2006). *Habitat Fragmentation and Landscape Change: An Ecological and Conservation Synthesis*. Island Press, Washington, DC.
Llewellyn, D. W., G. P. Shaffer, N. J. Craig, L. Creasman, D. N. Pashley, M. Swan and C. Brown (1996). A decision support system for prioritizing restoration sites on the Mississippi River Alluvial Plain. *Conservation Biology* **10**: 1446–1455.
Lloyd, P., T. E. Martin, R. L. Redmond, U. Langner and M. M. Hart (2005). Linking demographic effects of habitat fragmentation across landscapes to continental source–sink dynamics. *Ecological Applications* **15**: 1504–1514.
Lynch, J. F. and D. F. Whigham (1984). Effects of forest fragmentation on breeding bird communities in Maryland, USA. *Biological Conservation* **28**: 287–324.
Marini, M. A., S. K. Robinson and E. J. Heske (1995). Edge effects on nest predation in the Shawnee National Forest, southern Illinois. *Biological Conservation* **74**: 203–213.
Marshall, M. R., D. R. Diefenbach, L. A. Wood and R. J. Cooper (2004). Annual survival estimation of migratory songbirds confounded by incomplete breeding site-fidelity: study designs that may help. *Animal Biodiversity and Conservation* **27**: 59–72.
Martin, T. E. (1995). Avian life-history evolution in relation to nest sites, nest predation, and food. *Ecological Monographs* **65**: 101–127.
Martin, T. E. and G. R. Geupel (1993). Nest-monitoring plots: methods for locating nests and monitoring success. *Journal of Field Ornithology* **64**: 507–519.

Martin, T. E., C. R. Paine, C. J. Conway, W. M. Hochachka, P. Allen and W. Jenkins (1997). *BBIRD Field Protocol*. Montana Cooperative Fish and Wildlife Research Unit, University of Montana, Missoula, MT.

Marzluff, J. M. and M. Restani (1999). The effects of forest fragmentation on avian nest predation. In *Forest Fragmentation: Wildlife Management Implications* (J. A. Rochelle, L. A. Lehmann and J. Wisniewski, eds.). Brill, Boston, MA: 155–169.

May, R. M. and S. K. Robinson (1985). Population dynamics of avian brood parasitism. *American Naturalist* **126**: 475–494.

Mayfield, H. (1977). Brown-headed cowbird: agent of extermination? *American Birds* **31**: 107–113.

McLaren, C. M. and S. G. Sealy (2000). Are nest predation and brood parasitism correlated in yellow warblers? A test of the cowbird predation hypothesis. *Auk* **117**: 1056–1060.

Morse, S. F. and S. K. Robinson (1999). Nesting success of a neotropical migrant in a multiple-use forested landscape. *Conservation Biology* **13**: 327–337.

Nagy, L. R. and R. T. Holmes (2004). Factors influencing fecundity in migratory songbirds: is nest predation the most important? *Journal of Avian Biology* **35**: 487–491.

Nagy, L. R. and R. T. Holmes (2005). To double-brood or not? Individual variation in the reproductive effort in black-throated blue warblers (*Dendroica caerulescens*). *Auk* **122**: 902–914.

Nol, E., C. M. Francis and D. M. Burke (2005). Using distance from putative source woodlots to predict occurrence of forest birds in putative sinks. *Conservation Biology* **19**: 836–844.

Nolan, V. Jr. (1978). *The Ecology and Behavior of the Prairie Warbler (Dendroica discolor)*. Ornithological Monographs 26. American Ornithologists' Union, Washington, DC.

Norris, D. R. and B. J. M. Stutchbury (2001). Extraterritorial movements of a forest songbird in a fragmented landscape. *Conservation Biology* **15**: 729–736.

Oehler, J. D. and J. A. Litvaitis (1996). The role of spatial scale in understanding responses of medium-sized carnivores to forest fragmentation. *Canadian Journal of Zoology* **74**: 2070–2079.

Paton, P. W. C. (1994). The edge effect on avian nesting success: how strong is the evidence? *Conservation Biology* **8**: 17–26.

Payne, R. B. and L. L. Payne (1993). Breeding dispersal in indigo buntings: circumstances and consequences for breeding success. *Condor* **95**: 1–24.

Peak, R. G. (2007). Forest edges negatively affect golden-cheeked warbler nest survival. *Condor* **109**: 628–637.

Pease, C. M. and J. A. Grzybowski Jr. (1995). Assessing the consequences of brood parasitism and nest predation on seasonal fecundity in passerine birds. *Auk* **112**: 343–363.

Podolsky, A. L., T. R. Simons and J. A. Collazo (2007). Modeling population growth of the ovenbird (*Seiurus aurocapilla*) in the southern Appalachians. *Auk* **124**: 1359–1372.

Porneluzi, P. A. and J. Faaborg (1999). Season-long fecundity, survival, and viability of ovenbirds in fragmented and unfragmented landscapes. *Conservation Biology* **13**: 1151–1161.

Porneluzi, P., J. C. Bednarz, L. J. Goodrich, N. Zawada and J. Hoover (1993). Reproductive performance of territorial ovenbirds occupying forest fragments and a contiguous forest in Pennsylvania. *Conservation Biology* **7**: 618–622.

Powell, L. A., M. J. Conroy, D. G. Krementz and J. D. Lang (1999). A model to predict breeding season productivity for multibrooded songbirds. *Auk* **116**: 1001–1008.

Pulliam, H. R. (1988). Sources, sinks, and population regulation. *American Naturalist* **132**: 652–661.

Pulliam, H. R. and B. J. Danielson (1991). Sources, sinks, and habitat selection: a landscape perspective on population dynamics. *American Naturalist* **137**(Suppl.): S50–S66.

Raim, A. (2000). Spatial patterns of breeding female brown-headed cowbirds on an Illinois site. In *Ecology and Management of Cowbirds and Their Hosts* (J. N. M. Smith, T. Cook, S. I. Rothstein, S. K. Robinson and S. G. Sealy, eds.). University of Texas Press, Austin, TX: 87–99.

Reed, J. M. and A. P. Dobson (1993). Behavioural constraints and conservation biology: conspecific attraction and recruitment. *Trends in Ecology and Evolution* **9**: 108–110.

Rich, T. D., C. J. Beardmore, H. Berlanga, P. J. Blancher, G. S. Butcher, D. W. Demarest, E. H. Dunn, W. C. Hunter, F. E. Inigo-Elias, J. A. Kennedy, A. M. Martell, K. V. Rosenberg, C. W. Rustay, J. S.

Wendt and T. C. Will (2004). *Partners in Flight North American Landbird Conservation Plan*. Cornell Laboratory of Ornithology, Ithaca, NY.
Ricklefs, R. E. (1969). An analysis of nesting mortality in birds. *Contributions in Zoology* **9**: 1–48.
Robbins, C. S., J. R. Sauer, R. S. Greenberg and S. Droege (1989a). Population declines in North American birds that migrate to the Neotropics. *Proceedings of the National Academy of Sciences USA* **86**: 7658–7662.
Robbins, C. S., D. K. Dawson and B. A. Dowell (1989b). Habitat area requirements of breeding forest birds of the Middle Atlantic States. *Wildlife Monographs* **103**: 1–34.
Robinson, S. K. (1985). Coloniality as a defense against nest predators of the yellow-rumped cacique. *Auk* **102**: 509–519.
Robinson, S. K. (1988). Reappraisal of the costs and benefits of habitat heterogeneity for nongame wildlife. *Transactions of the North American Natural Resources Conference* **53**: 145–155.
Robinson, S. K. (1992). Population dynamics of breeding neotropical migrants in a fragmented Illinois landscape. In *Ecology and Conservation of Neotropical Migrant Land Birds* (J. M. Hagan III and D. W. Johnston, eds.). Smithsonian Institution Press, Washington, DC: 408–418.
Robinson, S. K. and D. S. Wilcove (1994). Impacts of forest fragmentation on breeding populations of neotropical migrants. *Bird Conservation International* **4**: 233–249.
Robinson, S. K., S. I. Rothstein, M. C. Brittingham, L. J. Petit and J. A. Grzybowski Jr. (1995a). Ecology and behavior of cowbirds and their impact on host populations. In *Ecology and Management of Neotropical Migratory Landbirds* (T. E. Martin and D. M. Finch, eds.). Oxford University Press, New York: 428–460.
Robinson, S. K., F. R. Thompson III, T. M. Donovan, D. R. Whitehead and J. Faaborg (1995b). Regional fragmentation and the nesting success of migratory birds. *Science* **267**: 1987–1990.
Robinson, S. K., J. D. Brawn and J. P. Hoover (1997). Effectiveness of small nature preserves for breeding birds. In *Conservation in Highly Fragmented Landscapes* (M. W. Schwartz, ed.). Chapman and Hall, New York: 154–188.
Robinson, S. K., J. D. Brawn, J. R. Herkert and S. A. Morse (1999). Use of different habitats by breeding brown-headed cowbirds in fragmented midwestern landscapes. *Studies in Avian Biology* **18**: 52–61.
Robinson, S. K., J. P. Hoover and J. R. Herkert (2000a). Cowbird parasitism in a fragmented landscape: effects of tract size, habitat, and abundance of cowbirds and hosts. In *Ecology and Management of Cowbirds and Their Hosts* (J. N. M. Smith, T. Cook, S. I. Rothstein, S. K. Robinson and S. G. Sealy, eds.). University of Texas Press, Austin, TX: 280–297.
Robinson, W. D., J. D. Brawn, T. R. Robinson and S. K. Robinson (2000b). Nesting success of understory forest birds in Panama. *Journal of Avian Biology* **31**: 151–164.
Rodewald, A. D. (2003). The importance of land uses within the landscape matrix. *Wildlife Society Bulletin* **31**: 586–592.
Rosenberg, K. V., J. D. Lowe and A. A. Dhondt (1999). Effects of forest fragmentation on breeding tanagers: a continental perspective. *Conservation Biology* **13**: 568–583.
Rotella, J., M. Taper, S. Stephens and M. Lindberg (2007). Extending methods for modeling heterogeneity in nest survival data using generalized mixed models. *Studies in Avian Biology* **34**: 34–44.
Roth, R. R. and R. K. Johnson (1993). Long-term dynamics of a wood thrush population breeding in a forest fragment. *Auk* **110**: 37–48.
Rothstein, S. I. and S. K. Robinson (1994). Conservation and coevolutionary implications of brood parasitism by cowbirds. *Trends in Ecology and Evolution* **9**: 162–164.
Rothstein, S. I. and S. K. Robinson (1998). Avian brood parasitism: an overview. In *Avian Brood Parasitism: Studies in Coevolution* (S. I. Rothstein and S. K. Robinson, eds.). Oxford University Press, Oxford, UK: 3–56.
Rothstein, S. I., J. Verner and E. Steven (1984). Radio-tracking confirms a unique diurnal pattern of spatial occurrence in the brown-headed cowbird. *Ecology* **65**: 77–88.
Ruth, J. M., D. R. Petit, J. R. Sauer, M. D. Samuel, F. A. Johnson, M. D. Fornwall, C. E. Korschgen and J. P. Bennett (2003). Science for avian conservation: priorities for the new millennium. *Auk* **120**: 204–211.

Sandercock, B. K., E. N. Hewett and K. L. Kosciuch (2008). Effects of experimental cowbird removals on brood parasitism and nest predation in a grassland songbird. *Auk* **125**: 820–830.
Schlaepfer, M. A., M. C. Runge and P. W. Sherman (2002). Ecological and evolutionary traps. *Trends in Ecology and Evolution* **117**: 474–480.
Schmidt, K. A. (2003). Nest predation and population declines in Illinois songbirds: a case for mesopredator effects. *Conservation Biology* **17**: 1141–1150.
Schmidt, K. A. and C. J. Whelan (1999). The relative impacts of nest predation and brood parasitism on seasonal fecundity in songbirds. *Conservation Biology* **13**: 46–57.
Schmidt, K. A., S. A. Rush and R. S. Ostfeld (2008). Wood thrush nest success and post-fledging survival across a temporal pulse of small mammal abundance in an oak forest. *Journal of Animal Ecology* **77**: 830–837.
Shaffer, T. L. (2004). A unified approach to analyzing nesting success. *Auk* **121**: 526–540.
Shaffer, T. L. and F. R. Thompson III (2007). Making meaningful estimates of nest survival with model-based methods. *Studies in Avian Biology* **34**: 84–95.
Sillett, T. S. and R. T. Holmes (2002). Variation in survivorship of a migratory songbird throughout its annual cycle. *Journal of Animal Ecology* **71**: 296–308.
Simons, T. R., G. L. Farnsworth and S. A. Shriner (2000). Evaluating Great Smoky Mountains National Park as a population source for the wood thrush. *Conservation Biology* **14**: 1133–1144.
Smith, J. N. M. and J. T. Hellmann (2002). Population persistence in fragmented landscapes. *Trends in Ecology and Evolution* **17**: 397–399.
Smith, J. N. M., T. L. Cook, S. I. Rothstein, S. K. Robinson and S. G. Sealy (eds.) (2000). *Ecology and Management of Cowbirds and their Hosts*. University of Texas Press, Austin, TX.
Smith, J. N. M., M. J. Taitt and L. Zanette (2002). Removing brown-headed cowbirds increases seasonal fecundity and population growth in song sparrows. *Ecology* **83**: 3037–3047.
Stake, M. M., F. R. Thompson III, J. Faaborg and D. E. Burhans (2005). Patterns of snake predation at songbird nests in Missouri and Texas. *Journal of Herpetology* **39**: 215–222.
Stephens, S. E., D. N. Koons, J. J. Rotella and D. N. Willey (2003). Effects of habitat fragmentation on avian nesting success: a review of evidence at multiple spatial scales. *Biological Conservation* **115**: 101–110.
Studds, C. E., T. K. Kyser and P. P. Marra (2008). Natal dispersal driven by environmental conditions interacting across the annual cycle of a migratory songbird. *Proceedings of the National Academy of Sciences* **105**: 2924–2933.
Suarez, A. V., K. S. Pfennig and S. K. Robinson (1997). Nesting success of a disturbance-dependent songbird on different kinds of edges. *Conservation Biology* **11**: 928–935.
Suedkamp Wells, K. M., M. R. Ryan, J. J. Millspaugh, F. R. Thompson III and M. W. Hubbard (2007). Survival of postfledging grassland birds in Missouri. *Condor* **109**: 781–794.
Temple, S. A. and J. R. Cary (1988). Modeling dynamics of habitat-interior birds in fragmented forest landscapes. *Conservation Biology* **2**: 340–347.
Terborgh, J. (1989). *Where Have All the Birds Gone?* Princeton University Press, Princeton, NJ.
Tewksbury, J. J., S. J. Hejl and T. E. Martin (1998). Breeding productivity does not decline with increasing fragmentation in a western landscape. *Ecology* **79**: 2890–2903.
Tewksbury, J. J., T. E. Martin, S. J. Hejl, T. S. Redman and F. J. Wheeler (1999). Cowbirds in a western valley: effects of landscape structure, vegetation, and host density. *Studies in Avian Biology* **18**: 23–33.
Tewksbury, J. J., A. E. Black, N. Nur, V. A. Saab, B. D. Logan and D. S. Dobkin (2002). Effects of anthropogenic fragmentation and livestock grazing on western riparian bird communities. *Studies in Avian Biology* **25**: 158–202.
Thompson, F. R. III (1994). Temporal and spatial patterns of breeding brown-headed cowbirds in the Midwestern US. *Auk* **111**: 979–990.
Thompson, F. R. III, T. M. Donovan, R. M. DeGraaf, J. Faaborg and S. K. Robinson (2002). A multi-scale perspective of the effects of forest fragmentation on birds in eastern forests. *Studies in Avian Biology* **25**: 8–19.

Tittler, R., L. Fahrig and M. A. Villard (2006). Evidence of large-scale source–sink dynamics and long-distance dispersal among wood thrush populations. *Ecology* **87**: 3029–3036.
Tomialojc, L., T. Wesolowski and W. Walankiewicz (1984). Breeding bird community of a primeval temperate forest (Bialoweza National Park, Poland). *Acta Ornithologica* **20**: 241–306.
Trine, C. L. (1998). Wood thrush population sinks and implications for the scale of regional conservation strategies. *Conservation Biology* **12**: 576–585.
Trine, C. L. (2000). Effects of multiple parasitism on cowbird and wood thrush nesting success. In *Ecology and Management of Cowbirds and Their Hosts* (J. N. M. Smith, T. L. Cook, S. I. Rothstein, S. K. Robinson and S. G. Sealy, eds.). University of Texas Press, Austin, TX: 135–144.
Van Horn, M. A., R. M. Gentry and J. Faaborg (1995). Patterns of ovenbird (*Seiurus aurocapillus*) pairing success in Missouri forest tracts. *Auk* **112**: 98–106.
Villard, M.-A., P. R. Mankin, and C. G. Drummond (1993). Habitat fragmentation and pairing success in the ovenbird (*Seiurus aurocapillus*). *Auk* **110**: 759–768.
Ward, D. and J. N. M. Smith (2000). Brown-headed cowbird parasitism results in a sink population in warbling vireos. *Auk* **117**: 337–344.
Ward, M. P. (2005). The role of immigration in the decline of an isolated migratory bird population. *Conservation Biology* **19**: 1528–1536.
Ward, M. P. and S. Schlossberg (2004). Conspecific attraction and the conservation of territorial songbirds. *Conservation Biology* **18**: 519–524.
Weatherhead, P. J. and G. Blouin-Demers (2004). Understanding avian nest predation: why ornithologists should study snakes. *Journal of Avian Biology* **35**: 185–190.
Weinberg, H. J. and R. R. Roth (1998). Forest area and habitat quality for nesting wood thrushes. *Auk* **115**: 879–889.
Weldon, A. J. and M. M. Haddad (2005). The effects of patch shape on indigo buntings: evidence for an ecological trap. *Ecology* **86**: 1422–1431.
Whitcomb, R. F., C. S. Robbins, J. F. Lynch, B. L. Whitcomb, M. K. Klimkiewicz and D. Bystrak (1981). Effects of forest fragmentation on avifauna of the eastern deciduous forest. In *Forest Island Dynamics in Man-Dominated Landscapes* (R. L. Burgess and D. M. Sharpe, eds.). Springer-Verlag, New York: 125–205.
White, G. L. and K. P. Burnham (1999). Program MARK: survival estimation from populations of marked animals. *Bird Study* 46: S120–S139.
Wiens, J. A. (1995). Habitat fragmentation: island versus landscape perspectives on bird conservation. *Ibis* **137**: S97–S104.
Wiens, J. A., N. C. Stenseth, B. Van Horne and R. A. Ims (1993). Ecological mechanisms and landscape ecology. *Oikos* **66**: 369–380.
Wilcove, D. S. (1985). Nest predation in forest tracts and the decline of migratory songbirds. *Ecology* **66**: 1211–1214.
Willson, M. F., T. L. De Santo and K. E. Sieving (2003). Red squirrels and predation risk to bird nests in northern forests. *Canadian Journal of Zoology* **81**: 1202–1208.
Winfree, R., J. Dushoff, S. K. Robinson and D. Bengali (2006). A Monte Carlo model for estimating the productivity of a generalist brood parasite across multiple host species. *Evolutionary Ecology Research* **8**: 213–236.
With, K. A. and A. W. King (2001). Analysis of landscape sources and sinks: the effects of spatial pattern on avian demography. *Biological Conservation* **100**: 75–88.
Zuidema, P. A., J. A. Sayer and W. Dijkman (1996). Forest fragmentation and biodiversity: the case for medium-sized conservation areas. *Environmental Conservation* **23**: 290–297.

ERIC J. BERRY, DAVID L. GORCHOV AND BRYAN A. ENDRESS

21

Source–sink population dynamics and sustainable leaf harvesting of the understory palm *Chamaedorea radicalis*

Summary

In this study we assessed the sustainability of leaf harvesting of the palm *Chamaedorea radicalis* by modeling the dynamics of harvested populations using stage-structured transition matrices. Within the study site, El Cielo Biosphere Reserve, palm demography and population growth is dependent on substrate type; a relationship that is due to the role of rock outcrops as a refuge from herbivory by free-ranging livestock. We accounted for this environmental heterogeneity by using a source–sink model in which non-browsed palms on rock outcrops act as a source population for browsed palms on the forest floor (sink). To evaluate the impact of leaf harvesting on these populations we incorporated the demographic effects of local harvesting practices into population models using data from leaf harvesting experiments. Results showed that when the demographic effects of leaf harvesting were combined with the effects of livestock browsing, population growth dropped significantly below the replacement rate, indicating that the combination of the two was not sustainable. This result is explicable in the context of the source–sink dynamic described above, where browsed palms on the forest floor are dependent on the migration of seeds from protected palms on rock outcrops. Incorporating leaf harvesting into the model reduces the survival and fecundity of all non-browsed palms, including important "source" palms on rock outcrops, with the result that non-browsed rock outcrops are no longer a sufficient source of recruitment for the entire population. The source–sink model was critical in projecting the consequences of this interaction between browsing and harvesting.

Sources, Sinks and Sustainability, ed. Jianguo Liu, Vanessa Hull, Anita T. Morzillo and John A. Wiens. Published by Cambridge University Press.

Background

The harvesting of non-timber forest products (NTFP) from wild plant populations provides an important source of income for millions of people throughout the world. Palms (Arecaceae) constitute one of the most widely used plant families of NTFPs (Uhl and Dransfield 1987; Balick and Beck 1990), with products derived from nearly every plant part, including leaves, stems, apical meristems and fruits. The growing demand for these resources has led to increased harvesting, creating a concern about the long-term sustainability of many economically important species (Flores and Ashton 2000; Svenning and Macía 2002; Ticktin 2004).

Assessing the impact of leaf harvesting is of particular importance to the well-established trade in palm leaves from species of *Chamaedorea*, a genus of understory palms that are harvested throughout Mexico, Central America, and Colombia. Leaves from more than 20 different species of this genus are sold to international floral and horticultural markets for use in floral displays, with peak demand during Palm Sunday and the Easter holiday (CEC 2002). Due to this large and increasing demand for *Chamaedorea* leaves, some species have been overexploited in their native habitats and are currently threatened. *Chamaedorea radicalis* is one of several species of *Chamaedorea* that are considered vulnerable in Mexico or Central America (FAO 1997), due – at least in part – to leaf harvesting (Hodel 1992). Leaves from *C. radicalis* are harvested from wild populations in the mountains of northeastern Mexico and provide the main source of income for most families within the El Cielo Biosphere Reserve (hereafter El Cielo) in Tamaulipas, Mexico (Peterson 2001).

To better evaluate the effects of NTFP extraction on the population viability of economically valuable species such as *C. radicalis*, recent studies have incorporated local harvesting and management practices into experimental designs (Joyal 1996; Kainer *et al.* 1998; Velásquez Runk 1998; Svenning and Macía 2002; Ticktin *et al.* 2002; Endress *et al.* 2004b). This approach provides a more accurate assessment of harvest sustainability because it reflects management and conservation strategies that are currently practiced by local communities. For *C. radicalis*, Endress *et al.* (2004a) combined leaf harvest experiments and matrix models to conclude that population growth was reduced by leaf harvesting, but these effects were not enough to reduce finite rates of population growth (λ) significantly below the replacement rate of 1, suggesting that current leaf harvest practices are ecologically sustainable. However, similar models of *C. radicalis* that included the effect of burro browsing on demographic rates did result in projected population declines (Endress *et al.* 2004a).

This finding underscores the reality that, for NTFPs, harvesting is only one of many environmental factors, both natural and anthropogenic, that affect population growth. While ecologists recognize the potentially important effects of environmental heterogeneity on plant demography and population dynamics, most studies fail to account for this variability when assessing NTFP sustainability (Ticktin 2004). For *C. radicalis* populations, substrate type is an important environmental variable, as rock outcrops have higher palm density (Jones and Gorchov 2000), proportionally more large adults (Endress *et al.* 2004a), and greater fruit production (Berry and Gorchov 2006) than the surrounding substrates. However, such differences are not due to natural variation in microsite conditions between the substrates, but instead are due to differences in intensity of browsing by the livestock that range freely throughout the El Cielo forests (Berry *et al.* 2008). Non-browsed palms on rock outcrops have higher rates of survival, growth and reproduction than browsed palms on the forest floor. Berry *et al.* (2008) incorporated this environmental heterogeneity into a modified source–sink model (*sensu* Pulliam 1988) in which non-browsed palms on rock outcrops act as a source population for browsed palms on the forest floor (sink). These models showed that for *C. radicalis* populations exposed to livestock browsing, seed migration from rock outcrops was both necessary and sufficient to sustain palm populations on the forest floor.

In this study, we incorporated the demographic effects of local leaf harvesting practices into the source–sink model in order to evaluate the impact of leaf harvesting on *C. radicalis* populations that are exposed to livestock browsing. By including source–sink dynamics between substrates as a source of natural environmental variation in population models of livestock browsing and leaf harvesting, we were able to examine how *C. radicalis* populations respond to these anthropogenic disturbances in conjunction with the underlying natural environmental heterogeneity within El Cielo. This combined approach has been advocated recently (Endress *et al.* 2004a; Ticktin 2004, 2005) but few studies have incorporated landscape heterogeneity into population models for NTFP sustainability.

Objectives

We had two primary objectives for this study:

1. to examine the sustainability of leaf harvesting by modeling population growth in *C. radicalis* populations exposed to leaf harvesting and livestock browsing;

2. to apply the findings from model projections in order to provide recommendations for leaf harvesting and livestock management strategies that facilitate sustainable leaf harvesting.

To meet these objectives, we used population projection matrices, which have become the favored model for population studies in plant conservation (Silvertown *et al.* 1996). These models are well suited to conservation applications because, in addition to estimating the finite rate of population growth (λ), they can be used for elasticity analysis. Elasticity measures the relative sensitivity of λ to small changes in each stage transition of a plant's life cycle, providing a straightforward and quantitative way to rank the importance of different demographic rates (see, e.g., van Groenendael *et al.* 1988; Benton and Grant 1999; Heppell *et al.* 2000; Wisdom *et al.* 2000). This analysis is particularly useful in conservation biology (Silvertown *et al.* 1996), where such information can direct conservation efforts toward a manageable subset of individuals within a population that have the most influence on population growth. We made use of elasticity analysis to identify key life-history stages for *C. radicalis* population growth in harvested populations within El Cielo, which provided a basis for specific management recommendations.

Research methods

Study site

We conducted this study in the forests near the communities of San José and Alta Cimas within El Cielo, Tamaulipas, Mexico (22°55′–23°30′ N and 99°02′–99°30′ W). El Cielo straddles the Sierra de Guatemala mountain range on the eastern slopes of the Sierra Madre Oriental in northeast Mexico. Although El Cielo contains a wide range of vegetation types (González-Medrano 2005), this study was conducted primarily in montane mesophyll (tropical cloud) forests at elevations of 950–1,500 m. The forest near San José is in a transition zone between montane mesophyll forest and pine–oak forest (Puig and Bracho 1987).

The communities of Alta Cimas and San José are both legally recognized *ejidos*, which is an arrangement where each community collectively holds and manages a portion of the land. Within El Cielo, *C. radicalis* leaves are the only natural product that these communities are authorized to harvest, and leaf harvesting represents the main source of income for most families (Peterson 2001). In addition to leaf harvesting, families supplement their income by managing small herds of livestock that include cattle, mules, burros, horses, goats and sheep. These animals at times range freely in El Cielo forests and appear to account for more herbivory on *C. radicalis* than native herbivores do (E. J. Berry, personal observation).

Study species

Chamaedorea radicalis Mart. (Arecaceae) is a dioecious understory palm occurring in wild populations in the mountains of northeast Mexico. Villagers within the El Cielo area harvest leaves from adult *C. radicalis* to sell to international cut-foliage markets. Marketable leaves are typically ≥ 40 cm in length, with minimal damage from insects or pathogens (Endress *et al.* 2004b). Adult palms have approximately 4–8 pinnately compound leaves (Hodel 1992) and most appear stemless because their stem typically forms a "heel" that grows into the substrate. However, individuals with an erect, above-ground stem can reach 2–4 m in height (Gorchov and Endress 2005). For this study we recognized the five life-history stages for *C. radicalis* as described in Endress *et al.* (2004a): seeds, seedlings (bifid leaves), juveniles (3–9 leaflets on youngest fully expanded leaf; YFL), small adults (10–25 leaflets on YFL), and large adults (>25 leaflets on YFL). Within El Cielo, *C. radicalis* distribution ranges from 200 to 1,500 m elevation (Mora–Olivo *et al.* 1997), and the highest densities are associated with forest stands having low tree basal area and a high percentage of rock substrate (Jones and Gorchov 2000).

Population matrix models

Population models in this study were constructed as stage-structured transition (Lefkovitch) matrices (Fig. 21.1). Each population model was based either on a Basic model for non-browsed palms (Table 21.1) or on a modified source–sink model for palms exposed to livestock browsing (Table 21.2). In the Basic model, projection matrix **A** incorporates the transition probabilities for each of the palm's five life-history stages. Demographic parameters for this model were calculated using pooled data from palms both on rock outcrops and on the forest floor. Pooling the data to construct a one-substrate model is justified for non-browsed palms because, in the absence of animal browsing, *C. radicalis* survival, growth and fecundity are similar on the two substrates (Berry *et al.* 2008).

However, for a browsed population, Berry *et al.* (2008) found herbivory to be limited to the forest floor, creating demographic differences between non-browsed palms on rock outcrops and browsed individuals on the forest floor. Therefore, population models that incorporated livestock browsing were modeled using the two-substrate source–sink model described in Berry *et al.* (2008). In this model (Fig. 21.1), projection matrix **A** is a 10 × 10 transition matrix that is partitioned as submatrices, where $\mathbf{B}_r$ is the demography of non-browsed palms on rock outcrops (source), $\mathbf{B}_f$ is the demography of palms on the forest floor (sink), $\mathbf{M}_{r\rightarrow f}$ is seed dispersal from rock outcrops to the forest floor, and $\mathbf{M}_{f\rightarrow r}$ is

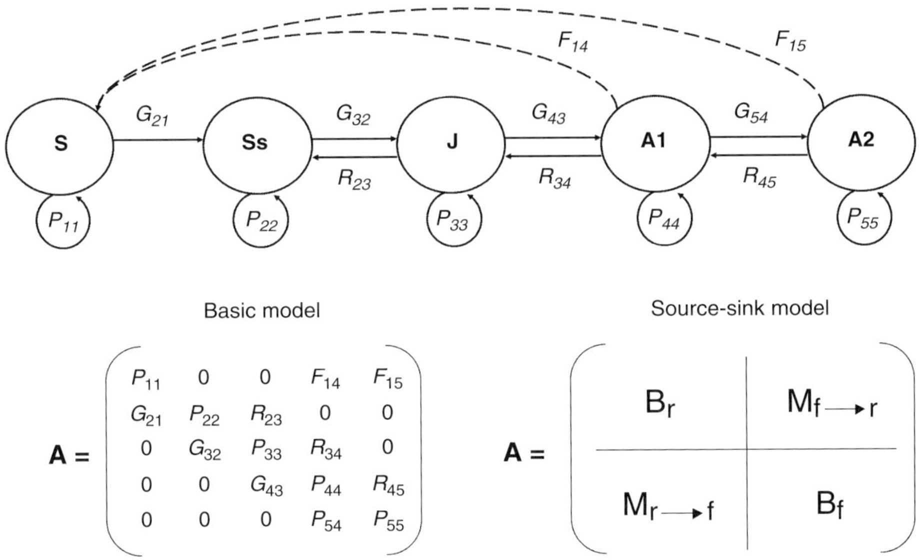

FIGURE 21.1.
Stage-structured life-cycle diagram of *C. radicalis* and its associated population projection matrix (Basic model; bottom left). Solid arrows represent annual transitions between stages and dashed arrows indicate fecundity. The letters above or below each arrow show how transitions in the life-cycle diagram correspond with each entry in the projection matrix. To project populations exposed to livestock browsing we used a modified source–sink model (bottom right; Berry *et al.* 2008), which links non-browsed palms on rock outcrops (submatrix B_r) with palms on the forest floor (submatrix B_f) via seed migration (submatrices $M_{r \rightarrow f}$ and $M_{f \rightarrow r}$). Submatrices B_r and B_f follow the same 5 × 5 transition matrix structure as in the Basic model. Abbreviations: S = seed, Ss = seedling, J = juvenile, A1 = small adult, A2 = large adult; *R* = regressing to a smaller stage, *P* = remaining at same stage, *G* = growing to a larger stage, *F* = per capita seed production.

seed dispersal from the forest floor to rock outcrops. Because not all palms on the forest floor are browsed, $\mathbf{B_f}$ was parameterized using weighted averages of demographic rates of browsed and non-browsed palms that reflected the proportion of palms that were browsed in each life-history stage.

To quantify the impact of leaf harvesting and/or livestock browsing on population growth, we parameterized different models based on the demography of palms that were:

1. protected from browsing and leaf harvesting (Control)
2. protected from browsing but exposed to two different intensities of leaf harvesting (4m Harvest and 4x Harvest – see below)
3. protected from leaf harvesting but exposed to burro browsing (Browse)
4. exposed to both leaf harvesting and livestock browsing (4m Harvest+Browse, 4x Harvest+Browse).

TABLE 21.1. *Stage transition matrices for non-browsed* C. radicalis *(Basic model, Fig. 21.1). Transitions for these matrices were calculated using pooled demographic data from palms on rock outcrops and the forest floor. Vital rates were based on annual stage transitions during 2003/4 for all matrices except Control year 2, which was parameterized from 2004/5 data. Life-history stage abbreviations as in Figure 21.1.*

Treatment	Stage	S	Ss	J	A1	A2
Control	S	0.028	0	0	0.243	7.360
	Ss	0.143	0.716	0.050	0	0
	J	0	0.222	0.433	0.044	0
	A1	0	0	0.467	0.679	0.073
	A2	0	0	0	0.248	0.894
Control year 2	S	0.015	0	0	0.429	8.773
	Ss	0.235	0.707	0.042	0	0
	J	0	0.207	0.458	0.032	0
	A1	0	0	0.438	0.722	0.053
	A2	0	0	0	0.206	0.924
4m Harvest	S	0.028	0	0	0.000	0.056
	Ss	0.143	0.716	0.050	0	0
	J	0	0.222	0.433	0.029	0
	A1	0	0	0.467	0.441	0.111
	A2	0	0	0	0.412	0.889
4x Harvest	S	0.028	0	0	0.000	0.111
	Ss	0.143	0.716	0.050	0	0
	J	0	0.222	0.433	0.000	0
	A1	0	0	0.467	0.581	0.081
	A2	0	0	0	0.258	0.892

Since only adult palms produce leaves that are harvestable, the demographic effects of leaf harvesting were incorporated into models by using the demographic rates of small and large adults from their respective harvesting treatments. In the more intense 4x treatment ($n = 68$ adults), all harvestable leaves were extracted four times per year (in February, May, August and November). The 4m treatment ($n = 70$ adults) was a modification of 4x Harvest, in which – for each harvest – at most only one leaf per palm was harvested and, for a leaf to be harvested, the palm had to have at least two leaves (i.e., palms were never entirely defoliated).

We also modeled the effects of livestock browsing *frequency* on population growth by constructing simulations based on different browsing frequencies of every 1, 2, and 5 years. These models followed the linear population projection matrix model $\mathbf{n}(t+1) = \mathbf{A} * \mathbf{n}(t)$ (Caswell 2001), where $\mathbf{n}(t)$ represents a

TABLE 21.2. *Stage transition matrices for source–sink models (Fig. 21.1, bottom right). Each projection matrix was parameterized using demographic data from non-browsed palms on rock outcrops and browsed palms on the forest floor. Vital rates were calculated from 2003/4 demographic data for all matrices, except Browse recovery, which was parameterized from 2004/5 data. Life-history stage abbreviations as in Figure 21.1.*

			Rock outcrop					Forest floor				
Treatment	Substrate	Stage	S	Ss	J	A1	A2	S	Ss	J	A1	A2
Browse	Rock outcrop	S	0.028	0	0	0.020	0.609	0	0	0	0.001	0.008
		Ss	0.143	0.716	0.050	0	0	0	0	0	0	0
		J	0	0.222	0.433	0.044	0	0	0	0	0	0
		A1	0	0	0.467	0.679	0.073	0	0	0	0	0
		A2	0	0	0	0.248	0.894	0	0	0	0	0
	Forest floor	S	0	0	0	0.223	6.751	0.028	0	0	0.091	1.361
		Ss	0	0	0	0	0	0.143	0.634	0.084	0	0
		J	0	0	0	0	0	0	0.160	0.435	0.040	0
		A1	0	0	0	0	0	0	0	0.198	0.544	0.097
		A2	0	0	0	0	0	0	0	0	0.261	0.855
Browse recovery	Rock outcrop	S	0.015	0	0	0.073	1.500	0	0	0	0.007	0.146
		Ss	0.235	0.707	0.042	0	0	0	0	0	0	0
		J	0	0.207	0.458	0.032	0	0	0	0	0	0
		A1	0	0	0.438	0.722	0.053	0	0	0	0	0
		A2	0	0	0	0.206	0.924	0	0	0	0	0
	Forest floor	S	0	0	0	0.355	7.273	0.015	0	0	0.155	3.259
		Ss	0	0	0	0	0	0.235	0.769	0.134	0	0
		J	0	0	0	0	0	0	0.133	0.366	0.101	0
		A1	0	0	0	0	0	0	0	0.298	0.569	0.064
		A2	0	0	0	0	0	0	0	0	0.256	0.914

TABLE 21.2. (*cont.*)

			Rock outcrop					Forest floor				
Treatment	Substrate	Stage	S	Ss	J	A1	A2	S	Ss	J	A1	A2
4m Harvest	Rock	S	0.028	0	0	0.000	0.005	0	0	0	0.000	0.000
+Browse	outcrop	Ss	0.143	0.716	0.050	0	0	0	0	0	0	0
		J	0	0.222	0.433	0.029	0	0	0	0	0	0
		A1	0	0	0.467	0.441	0.111	0	0	0	0	0
		A2	0	0	0	0.412	0.889	0	0	0	0	0
	Forest	S	0	0	0	0.000	0.051	0.028	0	0	0.000	0.010
	floor	Ss	0	0	0	0	0	0.143	0.634	0.084	0	0
		J	0	0	0	0	0	0	0.160	0.435	0.035	0
		A1	0	0	0	0	0	0	0	0.198	0.454	0.104
		A2	0	0	0	0	0	0	0	0	0.323	0.854
4x Harvest	Rock	S	0.028	0	0	0.000	0.009	0	0	0	0.000	0.000
+Browse	outcrop	Ss	0.143	0.716	0.050	0	0	0	0	0	0	0
		J	0	0.222	0.433	0.000	0	0	0	0	0	0
		A1	0	0	0.467	0.581	0.081	0	0	0	0	0
		A2	0	0	0	0.258	0.892	0	0	0	0	0
	Forest	S	0	0	0	0.000	0.102	0.028	0	0	0.000	0.021
	floor	Ss	0	0	0	0	0	0.143	0.634	0.084	0	0
		J	0	0	0	0	0	0	0.160	0.435	0.024	0
		A1	0	0	0	0	0	0	0	0.198	0.507	0.099
		A2	0	0	0	0	0	0	0	0	0.265	0.855

column-vector whose elements are the population's stage structure **n** at time *t*. **A** is the transition matrix for the population, which was parameterized as the demography of palms that were either exposed to browsing (**B**), were recovering from browsing (**B** + 1), or were non-browsed (**C**, control). The different browsing frequencies were modeled by projecting the initial population structure ($\mathbf{N}_0$) using an appropriate series of transition matrices (e.g., Hoffmann 1999; Zuidema and Werger 2000). For example, the series of matrices simulating browsing every 5 years was (*note*: sequence read right to left): $\mathbf{N}_{25} = \ldots$ (**C**) (**C**) (**C**) (**B** + 1) (**B**) (**C**) (**C**) (**C**) (**B** + 1) (**B**) ($\mathbf{N}_0$). $\mathbf{N}_0$ was based on the observed stage distribution of *C. radicalis* in El Cielo reported in Endress *et al.* (2004a). These data were collected from 15 belt transects along five hillsides within the valley Cañón del Diablo near Alta Cimas during July and August of 2000 (n = 922 palms). Unlike the other model simulations, which utilized eigenanalysis to calculate λ as the finite rate of increase, for the browsing frequency models we used transient analyses to calculate λ as the geometric mean of the annual population growth rates for years 21–25.

Model analyses

Population projections and eigenanalysis were conducted using Matlab version 7.0.1 (MathWorks 1989). To project long-term population dynamics, eigenanalysis was used to estimate the finite rate of population increase (λ). λ is the dominant eigenvalue (largest real root) of the matrix (Caswell 2001) and is a measure of population fitness, as populations with $\lambda > 1$ grow, $\lambda < 1$ decline, and $\lambda = 1$ remain at size n. The 95% confidence intervals for λ were obtained using bootstrap analysis by resampling the original dataset in order to create 1,000 resampled matrices for each treatment and calculating λ for each resampled matrix. We then calculated 95% confidence intervals using the "percentile method," a nonparametric approach (Scheiner and Gurevitch 1993).

Eigenanalysis was also used to calculate the population's stable stage distribution and elasticity values. The stable stage distribution was obtained from the right eigenvector associated with λ, and is the percentage of individuals in each stage when the population is growing at λ. Elasticity values indicate the relative sensitivity of λ to changes in a particular life-history transition (Caswell 2001). Because these values are additive and sum to unity, they can be interpreted as the relative contribution of each life-history stage to λ (Heppell *et al.* 2000). Therefore, it is possible to compare elasticity values between life-history stages within an elasticity matrix (e.g., seedlings vs. small adults), as well as to compare between elasticity matrices of different treatments (e.g.,

Browse vs. Harvest). To compare between specific life-history stages in different treatments, composite elasticities were constructed for each life-history stage by summing within columns for each elasticity matrix.

Transient population dynamics for harvested *C. radicalis* were projected for 25 years using the same linear population projection matrix model and initial population structure described above for browse frequency, but were based on transition matrices parameterized with demographic data from leaf harvest experiments. Based on the transient population dynamics, we estimated leaf yield under the different leaf harvesting and livestock browsing regimes by multiplying the number of adult palms in the population for a given year by the average annual leaf yield from 1999 to 2004 for palms exposed to the corresponding leaf harvest treatments (Endress *et al.* 2006).

Although *C. radicalis* is dioecious, we used a one-sex female-dominant model (Caswell 2001) for our analyses. One-sex models are common for dioecious species and have previously been applied to model *C. radicalis* populations (Endress *et al.* 2004a; Berry *et al.* 2008). These models are based on two assumptions, which have been supported for *C. radicalis*: that there is no apparent sexual dimorphism in vital rates (Endress 2002) and that female fecundity is not dependent upon male distribution or abundance (Berry and Gorchov 2004, 2006).

Model parameters

Based on the reproductive phenology of *C. radicalis* (Endress *et al.* 2004a), we used a birth-pulse model based on a post-breeding census that was conducted each August. Demographic data used to construct the transition matrix models in Tables 21.1 and 21.2 were obtained from the results of a leaf harvest experiment and data from burro browsing and seed dispersal experiments reported in Berry *et al.* (2008). For each transition matrix model we calculated the following demographic parameters for each life-history stage: R = regressing to a smaller stage, P = remaining at same stage, G = growing to a larger stage, F = per capita seed production (Fig. 21.1). Although demographic data were collected for two consecutive years (2003/4 and 2004/5), most results were reported for model projections from the year of the burro browsing experiment (2003/4) only, in order to emphasize comparisons between treatments rather than between years.

Demographic rates for control palms in forest plots that were protected from harvesting and browsing were calculated using pooled data from palms both on rock outcrops and on the forest floor (Berry *et al.* 2008). These data were collected annually from August 2003 to August 2005 on 429 tagged palms within ten permanent plots in Cañón del Diablo near the *ejido* of Alta Cimas. These plots had been free from leaf harvesting and livestock browsing since they were established in January 1999 (Endress 2002). During each annual census, data were collected on the number of leaves, number of leaflets on the youngest fully expanded leaf (YFL), leaf length (of YFL), the phenological stage and sex of each inflorescence, and fruit number. In addition to the annual census, adults were also examined quarterly (February, May, August and November) to quantify flowering and fruiting.

Demographic rates for harvested palms were collected from adults in each of two harvest treatments within the same ten permanent plots described above ($n = 70$ for 4m Harvest and 68 for 4x Harvest). Harvest treatments were initiated in May 1999 by Endress *et al.* (2004b) and were continued by E. Padrón-Serrano, a local palm collector and collaborator on this project. Demographic data were obtained from censuses in the same manner as for control plants.

Demographic rates for browsed palms were based on results from a burro browsing experiment (Berry *et al.* 2008). In this experiment, data were collected on the proportion of palms browsed on each substrate that was browsed, and on the demographic rates of browsed palms during the year of the browsing episode (August 2003–August 2004; Browse model) and one year following (August 2004–August 2005; Browse recovery model). Since rock outcrops are not accessible to burros or other livestock, the burros only browsed palms on the forest floor during the experiment. Therefore, the overall demography for populations exposed to livestock browsing is a function of both non-browsed palms on rock outcrops and browsed and non-browsed palms on the forest floor. To incorporate these demographic differences into transition matrices, the demographic rates for palms on rock outcrops were calculated using data from non-browsed palms in the control plots. The demographic rates for palms on the forest floor were calculated using data from both browsed and non-browsed palms, weighted according to the proportion of palms browsed for each life-history stage.

To estimate seed migration for the two-substrate source–sink models (i.e., populations exposed to burro browsing), we conducted a fruit trap experiment in the main study site – Cañón del Diablo – from 2003 to 2005, with ten fruiting females on each substrate. For rock outcrop palms, seed migration to the forest floor was quantified as total mature fruit production of females on rock outcrops minus the number of fruits remaining on the rock outcrop after dispersal. Dispersed fruit remaining on each rock outcrop was determined with a fruit trap composed of fiberglass mesh under each fruiting female. To minimize seed migration onto these sites, which would have confounded calculations, rock outcrops within 15 m of other fruiting palms were avoided. Seed migration in the opposite direction (i.e., from females on the forest floor to rock outcrops) was quantified as the number of fruits captured in fruit traps on rock outcrops where there were currently no fruiting palms ($n = 10$ outcrops). To minimize seed migration from other rock outcrops, these sites were within 1–3 m of fruiting palms on the forest floor, but at least 15 m from other rock outcrops with fruiting palms.

Rates of seed germination and dormancy on each substrate were based on results from paired seed plots placed near the ten permanent plots, with one plot of each pair on a rock outcrop and the other on the forest floor. See Berry *et al.* (2008) for details of these experiments.

Results

In the absence of leaf harvesting and livestock browsing there were no significant differences in *C. radicalis* population growth (λ) between 2003/4 (mean = 1.14, 95% CI = 1.09–1.18) and 2004/5 (mean = 1.18, 95% CI = 1.12–1.23). Therefore, we reported only the model projections from the year of the burro browsing experiment (2003/4) in order to highlight comparisons between treatments rather than between years in Figures 21.2–21.7. During this year, the model for the control palms projected population growth at a rate significantly above the replacement rate of $\lambda = 1$, and significantly greater than model projections for browsed palms and for palms in either leaf harvest treatment (4m or 4x Harvest) (Fig. 21.2).

Assuming no seed flow between the substrates in the Browse model, the forest floor fits the definition of a sink ($\lambda = 0.95$, CI = 0.92–0.98) and rock outcrops a source ($\lambda = 1.14$, CI = 1.10–1.18). Including seed flow in the two-substrate Browse model resulted in overall population growth slightly above the replacement rate (Fig. 21.2). Assuming no seed flow, the forest floor was a sink for both of the combined models: the 4m Harvest+Browse ($\lambda = 0.95$, CI = 0.88–1.00) and the 4x Harvest+Browse ($\lambda = 0.92$, CI = 0.86–0.97). However, in these models, rock outcrops were not a source, having projected population growth rates of $\lambda = 0.96$ (CI = 0.92–1.00) for the 4m Harvest+Browse and $\lambda = 0.92$ (CI = 0.86–0.97) for the 4x Harvest+Browse. Including seed flow in these models resulted in overall population growth for each model that was significantly below the replacement rate (Fig. 21.2).

Transient analyses revealed that palm populations subject to either leaf harvest treatment were expected to decline very slightly over the next 25 years (Fig. 21.3). Estimated leaf yield was projected to increase for about 5 years and then decrease gradually (Fig. 21.3). The stable stage distribution for the Browse model was similar to that of the control. However, populations exposed to leaf harvesting were projected to have few seedlings and juveniles, and more large adults than the controls: an effect that was slightly more pronounced in the combined Harvest+Browse populations (Fig. 21.4).

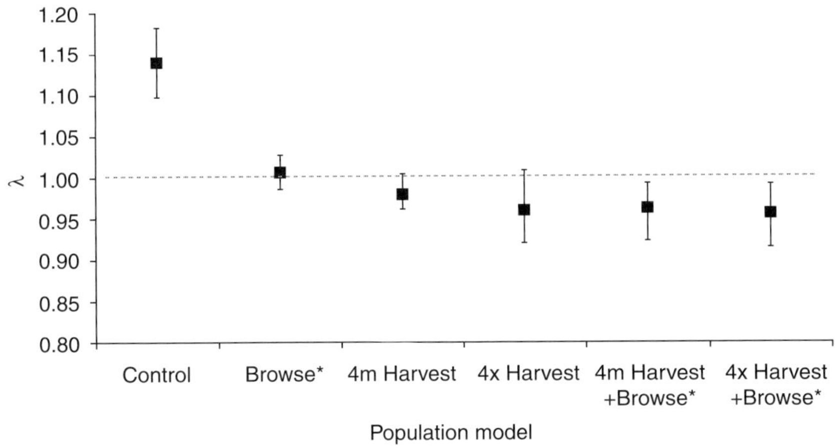

FIGURE 21.2.
The effect of leaf harvesting and livestock browsing on finite rate of population growth (λ) in *C. radicalis*. Reported are mean values and 95% confidence intervals (CI) from bootstrap analysis ($n = 1{,}000$ iterations). Population projection matrices for non-browsed palms were modeled using the one-substrate Basic model (Fig. 21.1, Table 21.1). Projection matrices for browsed palms (indicated by asterisks) were modeled as two-substrate source–sink models with seed dispersal between non-browsed palms on rock outcrops and browsed palms on the forest floor (Fig. 21.1, Table 21.2).

Elasticity analysis revealed that, for all models, population growth was most sensitive to changes in the vital rates of large adult palms (Figs. 21.5 and 21.6). For models that included livestock browsing, which partition the contribution of each life-history stage by substrate, the impact on λ of large adults on rock outcrops was much greater than for large adults on the forest floor (Fig. 21.6). For all models incorporating browsing, except the model for recovery from browsing, the impact on λ of forest floor palms was projected to be negligible. For each model, the greater elasticity values for large adults was mostly due to the sensitivity of λ to changes in large adult stasis rather than regression, growth or fecundity. This trend was apparent in all models including the control, but was most pronounced in models that simulated leaf harvesting (Fig. 21.5) or leaf harvesting and livestock browsing (Fig. 21.6).

Model simulations based on different frequencies of livestock browsing showed that greater time intervals between browsing events had a positive effect on projected population growth (Fig. 21.7). While this effect was apparent in population projections for models with and without harvesting, the effect of browsing frequency on λ was greatest for non-harvested palms. For both the 4m and 4x Harvest models, mean growth rate remained below the

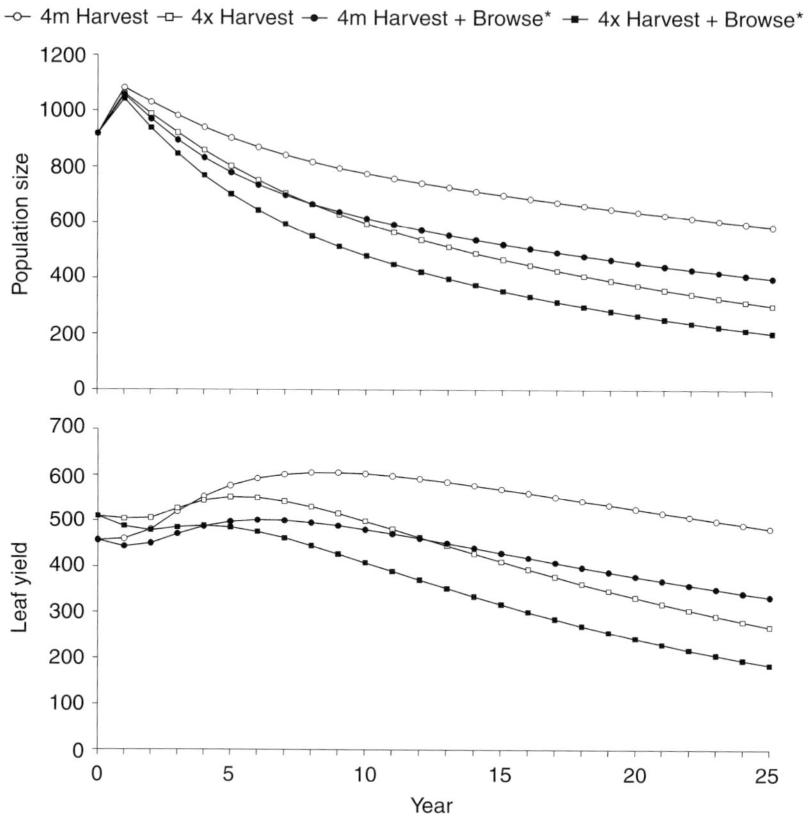

FIGURE 21.3.
Projected transient dynamics of *C. radicalis* populations (excluding seeds, top) and estimated leaf yield (bottom) under two leaf harvest treatments and in the presence and absence of browsing. Leaf yield was estimated by multiplying the number of adult palms in the population for a given year by the average leaf yield of palms exposed to the two leaf harvest treatments (Endress *et al.* 2006). The initial population structure ($n = 922$) was based on the observed stage distribution of palms in El Cielo reported in Endress *et al.* (2004a). Population projection matrices for the harvested but non-browsed palms were modeled using the one-substrate Basic model (Fig. 21.1, Table 21.1), and projection matrices for browsed palms (indicated by asterisks) were modeled using the two-substrate source–sink model (Fig. 21.1, Table 21.2).

replacement rate of 1, even in the scenario of lowest browsing frequency (once every 5 years).

Conclusions

Impact of harvesting and browse on source–sink dynamics and sustainability

Population models for *C. radicalis* revealed that the negative effects of either leaf harvesting or livestock browsing alone was sufficient to reduce

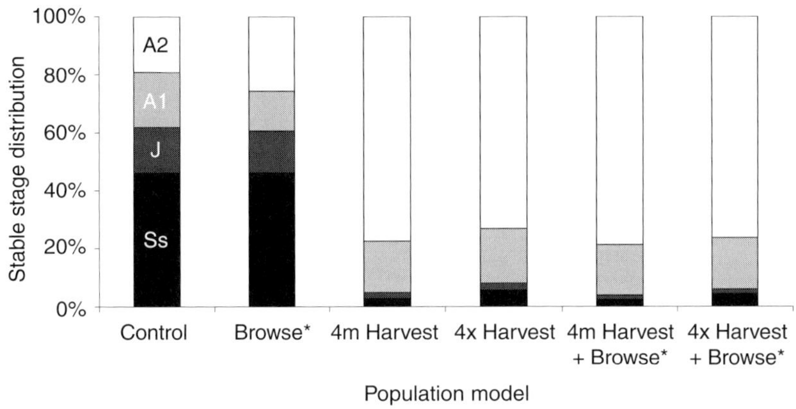

FIGURE 21.4.
Stable stage distributions (excluding seeds) for *C. radicalis* under different harvest and browse regimes. For models that include browsing (marked by an asterisk) and are based on the two-substrate source–sink model (Fig. 21.1, Table 21.2), stage distributions represent the combined proportion of palms on rock outcrops and the forest floor. Life-history stages are as defined in Fig. 21.1.

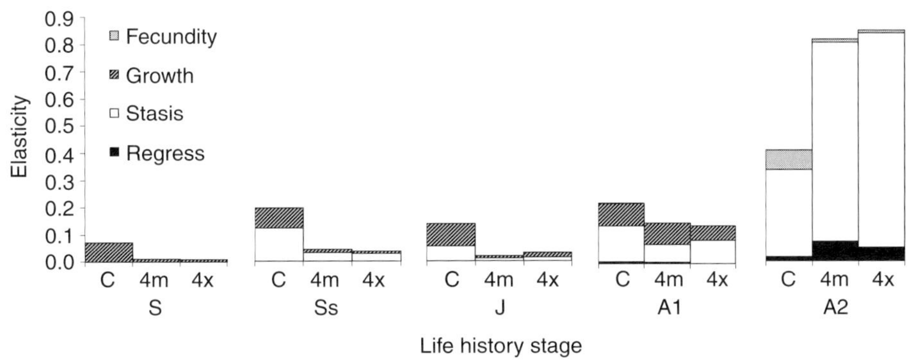

FIGURE 21.5.
Elasticities for non-browsed *C. radicalis* in control and leaf harvest models. Bars represent composite elasticities for each life-history stage, and are shaded to show the relative contribution from each possible life-history transition: regression, stasis, growth and fecundity. Model abbreviations: C = control, 4m = 4m Harvest, 4x = 4x Harvest. Life-history stages are as defined in Fig. 21.1.

the rate of population growth significantly below that of undisturbed palms (Control model). The finite rates of increase from our models of each year of demographic data from protected palms in control plots (λ = 1.14 and 1.18, respectively) were very similar to the geometric mean from the four previous years (1999–2003: λ = 1.14, CI = 1.09–1.21; Endress *et al.* 2006). Model projections for browsed populations had slightly higher rates of growth than either of the two leaf harvesting treatments; a finding that appears to be inconsistent with previous results for *C. radicalis* that reported

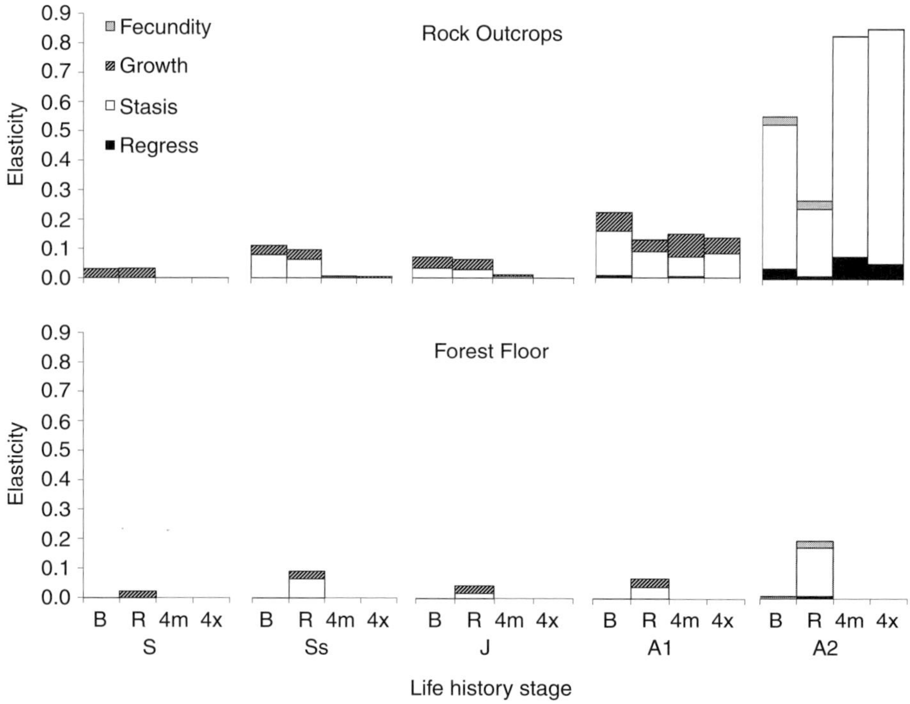

FIGURE 21.6.
Substrate-specific elasticities for populations of browsed *C. radicalis*. Bars represent composite elasticities for each stage on a given substrate, and are shaded to show the relative contribution from each possible life-history transition: regression, stasis, growth and fecundity. For each model the demography of forest floor palms was calculated as a weighted average between browsed and non-browsed individuals. For Browse models that also incorporated leaf harvest (4m and 4x models), the transition matrix elements for non-browsed adults were those of the corresponding Harvest model. Model abbreviations: B = Browse, R = Recovery from browsing (one year after browse), 4m = 4m Harvest+Browse, 4x = 4x Harvest+Browse. Life-history stages are as defined in Fig. 21.1.

livestock browsing had a more negative effect on palm demography than leaf harvesting (Endress *et al.* 2004a). This difference may be due to other studies overestimating the impact of browsing on palm populations by not accounting for the fact that livestock browsing is limited to accessible individuals only (those on the forest floor), unlike leaf harvesting, which affects all adults. Therefore, in browsed populations, a source–sink relationship (*sensu* Pulliam 1988, 1996) between non-browsed palms on rock outcrops (source) and browsed palms on the forest floor (sink) creates an overall population dynamic where the enhanced survival and fruit production of rock outcrop palms is sufficient to sustain the entire population (Berry *et al.* 2008).

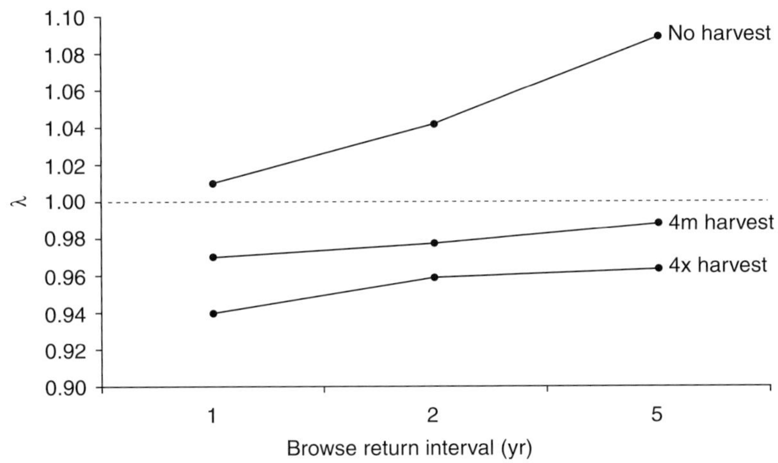

FIGURE 21.7.
The effect of livestock browsing frequency on the finite rate of population growth (λ) as projected by transition matrix models. Each model was parameterized by demographic data from browsed palms that were either protected from leaf harvest (control) or subject to two different levels of leaf harvest intensity (4m or 4x Harvest).

Because our source–sink model was spatially implicit, rather than spatially explicit, we are not able to explore how seed migration rates and population dynamics would be influenced by the spatial pattern of the two substrates, e.g., the size and spacing of rock outcrops (Berry *et al.* 2008). However, evidence that our calculated seed migration rates are qualitatively valid comes from the close match between the relative proportions of palms on outcrops vs. forest floor predicted from our source–sink model and that observed in the study area (Berry *et al.* 2008). Nevertheless, it would be interesting to investigate how seed migration is affected by topography, and to parameterize a spatially explicit model.

Model projections for each of the two leaf harvesting models had growth rates that were not significantly less than the replacement rate of $\lambda = 1$, suggesting that current levels of leaf harvesting in El Cielo are ecologically sustainable. Our estimates of the impact of leaf harvesting on population growth were very similar to values reported in Endress *et al.* (2006), which monitored palms under similar leaf harvesting treatments from 1999 to 2003 (4-year geometric mean for 4m Harvest $\lambda = 1.02$, and 4x Harvest $\lambda = 0.99$). However, when the demographic effects of leaf harvesting were combined with browsed palms (Harvest+Browse models), population growth dropped significantly below the replacement rate, indicating that the combination of the two is not sustainable. This result is explicable in the context of the source–sink dynamic described above, where browsed populations are dependent on the fecundity of inaccessible large adults on rock outcrops. Incorporating leaf harvesting into the model reduces the survival and fecundity of all non-browsed palms, including important "source" palms on rock outcrops, with the result that non-browsed rock outcrops are no longer a sufficient source of recruitment for the entire population.

Elasticity values confirm that in these source–sink models it is the life-history transitions of large adults, specifically on rock outcrops, that have the greatest effect on overall population growth. Indeed, all population models in our study, both Basic and source–sink models, identified large adults as the most crucial stage for population growth. This finding is consistent with the elasticities previously reported for *C. radicalis* (Endress *et al.* 2004a), as well as for other harvested palms, such as *Coccothrinax readii*, *Euterpe precatoria* (Olmsted and Alvarez-Buylla 1995), *Neodypsis decaryi* (Ratsirarson *et al.* 1996), and *Geonoma orbignyana* (Rodríguez-Buriticá *et al.* 2005). However, this trend does not extend to all harvested palms, as population growth rates for *Phytelephas seemanii* (Bernal 1998), *Astrocaryum mexicanum* and *Podococcus barteri* (Piñero *et al.* 1984) were found to be less sensitive to adult growth and fecundity, and much more sensitive to demographic variation in the smaller, mostly non-reproductive, life-history stages.

The relative importance of large adults in populations of *C. radicalis* is perhaps not surprising, given that this life-history stage has the highest survival rate and accounts for nearly all the reproduction in the population (Endress *et al.* 2004b). Survival of these large adults has a greater impact on population growth than their reproduction, as indicated by the greater elasticities associated with stasis versus fecundity. This pattern is strongest in populations exposed to leaf harvesting where reproduction is limited. In these populations, stable stage distributions revealed that small and large adults account for over 90% of the total population versus fewer than 40% in undisturbed populations. This pattern is explained by the lack of recruitment of new individuals to the population, producing a population whose persistence is overwhelmingly dependent on adult palm survival. This greater sensitivity of *C. radicalis* to changes in survival, rather than to growth or fecundity, is similar to patterns reported for other long-lived woody species, but is quite different from that of short-lived herbs that depend much more on reproduction for population persistence (Silvertown *et al.* 1996).

Interpreting elasticities for conservation

Population models from this study suggest that harvesting *C. radicalis* leaves in forest areas that are protected from free-range livestock is ecologically sustainable ($\lambda \approx 1$). However, given the concerns expressed by *palmilleros* about insufficient numbers of marketable leaves in El Cielo (Endress *et al.* 2004b), economic sustainability may require more than sustaining current levels, but rather an increase in the size of harvested populations. At first glance, our findings suggest that management efforts to increase population growth should focus on adult survival, which had the largest elasticities. While an increase in adult survival would help to ensure population persistence, prolonging the life of harvestable palms would only increase the rate of population growth closer to the replacement rate, but not above it. To achieve positive growth ($\lambda > 1$) requires substantial recruitment, as well as low mortality. Therefore, current harvesting regimes, which do not harm adult survival but do greatly reduce fecundity, may produce populations that are persistent but are incapable of significant growth.

The importance of additional adult fecundity for population growth is evident in our model comparisons between harvested and non-harvested palms. Population growth was significantly lower in harvested populations, and the main demographic difference between the two models was adult fecundity, not growth or survival. This important contribution of adult fecundity to population growth was not reflected in our elasticity results, which were heavily weighted toward adult survival, as described above. This discrepancy illustrates one limitation of interpreting elasticities for conservation applications. As discussed in Silvertown *et al.* (1996), elasticities are useful for indicating the relative sensitivity of population growth to small changes in a particular life-history transition, but they are not useful for indicating how sensitive any of these life-history transitions are to environmental perturbation. Therefore, a life-history transition that has a small elasticity, but has a large response to disturbance, may have a greater effect on population growth than a life-history transition with a large elasticity, but a relatively small response to the same disturbance. This is the case with *C. radicalis*, where the larger magnitude of response to harvesting by adult fecundity had a bigger impact on population growth than adult survival, which had a much larger elasticity.

Management recommendations

Estimates of leaf yield from our population models project a slight increase in leaf production during the next 5–10 years, despite the fact that the population is projected to decline slightly. This pattern can be explained by an increase in the number of adults during the first 10 years in harvested populations and the first 5 years in harvested + browsed populations, as a large cohort of seeds and seedlings from the initial population grew into adults. This short-term increase in leaf yield may mislead harvesters into believing that the population is healthy and growing. Our results indicate that this is not the case, and that intense leaf harvesting practices are producing populations that are becoming more biased toward adults, which will ultimately lead to population decline due to lack of reproduction and seedling recruitment.

Given that *C. radicalis* fecundity responds dramatically to protection from leaf harvesting, one way to achieve an increase in fruit production in harvested populations would be to restrict leaf harvesting from fruiting palms. Although this would require an agreement by the community to restrict the harvesting of these palms, this approach is practical for *C. radicalis* because relatively few large females account for nearly all fruit production in a population (Endress 2002). Protecting these few palms would have a minimal effect on leaf harvesting, but would ensure recruitment in harvested populations. For example, if we simulated the effect of protecting fruiting palms by incorporating the fecundity values of non-harvested palms in model projections of leaf harvesting treatments (Endress *et al.* 2006), population growth for palms exposed to leaf harvesting would increase to significantly above the replacement rate (4m $\lambda = 1.14$, 4x $\lambda = 1.11$). Population growth for palms exposed to both livestock browsing and leaf harvesting would increase from significantly below the replacement rate to $\lambda \approx 1$ (4m+browse $\lambda = 1.00$, 4x+browse $\lambda = 0.99$). An alternative approach to increase recruitment in harvested populations is through seed and seedling enrichment plantings. Enrichment planting, especially of seedlings, is more labor-intensive, as it requires a small nursery to germinate and grow seedlings, as well as the labor to plant the seedlings in the forest (Kilroy and Gorchov 2011).

Female fecundity is also negatively affected by herbivory from free-range livestock. For palms on the forest floor that are accessible to livestock, we documented that browsing has a greater impact on *C. radicalis* than leaf harvesting, because in addition to reducing adult fecundity, it also negatively affects the survival and growth of smaller palms. Therefore, removing livestock from harvested populations would have a positive effect on both fecundity and recruitment, which would increase the rate of population growth. This management approach has been undertaken by the *ejido* of Alta Cimas, where free-range livestock have been banned and fencing is being constructed to protect El Cielo forests, which are not only habitat for *C. radicalis* but are also home to a number of other rare or endangered plants and animals (Sánchez-Ramos *et al.* 2005). This and other efforts by Alta Cimas to protect their natural forest resources recently earned the community federal recognition and the official status of Campesino Reserve.

Another way to increase the economic viability of leaf harvesting in El Cielo is to increase the value of each leaf sold. Palm harvesters (*palmilleros*) within El Cielo receive little income for their work (~\$0.01/leaf; Endress 2002), which indirectly leads to the overharvesting of palm populations as *palmilleros* seek to increase their income. To address this, the Commission for Environmental

Cooperation of North America (CEC 2002) and others advocate "ecological certification" of sustainably harvested *Chamaedorea* leaves as a means to promote their conservation. Such certification initiatives provide a mechanism to identify natural products that meet minimum ecological and social standards, and reward them with a higher market price (Kiker and Putz 1997). This approach rewards only sustainably harvested leaves and is therefore preferable to simply raising the market price for everyone, which might induce overexploitation – as more people would be attracted to the higher price.

Pilot certification programs are now being pursued for *C. radicalis* leaf harvesting within El Cielo, as well as for other *Chamaedorea* species throughout Central America and Mexico. However, to accurately assess the sustainability of harvesting requires an understanding of the biology and population dynamics of harvested palms. While there is currently a small and growing body of literature on the ecological impacts of defoliation on *Chamaedorea* palms (Oyama and Mendoza 1990; Anten *et al.* 2003; Endress *et al.* 2004a, 2004b), there is a need to establish a set of ecological criteria by which to judge the sustainability of harvest and award certification. Establishing a set of agreed-upon ecological measures would maintain minimum resource management standards and help promote "best practices" among leaf harvesters; an outcome that benefits both those interested in conservation and the collectors who depend on *Chamaedorea* for their livelihood (Wilsey and Current 2004).

Acknowledgments

We would like to thank the villagers of Alta Cima and San José, where this research was conducted, for their hospitality and cooperation. We express our gratitude to Eduardo Padrón Serrano, Don Pedro Gonzaléz, and Pedro Gonzaléz for help with field research. Permission to study in El Cielo Biosphere Reserve was granted by the Secretary for Urban Development and Ecology (SEDUE), Cd. Victoria, Tamaulipas, Mexico. Funding was provided by the Ohio Academic Challenge Grant and Summer Workshop in Field Research through the Department of Botany, Miami University. This manuscript was much improved thanks to the thoughtful and thorough reviews provided by T. Crist, H. Stevens, and three anonymous reviewers.

References

Anten, N. P. R., M. Martiínez-Ramos and D. D. Ackerly (2003). Defoliation and growth in an understory palm: quantifying the contributions of compensatory responses. *Ecology* **84**: 2905–2918.

Balick, M. J. and H. S. Beck (1990). *Useful Palms of the World: A Synoptic Bibliography*. Columbia University Press, New York.

Benton, T. G. and A. Grant (1999). Elasticity analysis as an important tool in evolutionary and population ecology. *Trends in Ecology and Evolution* **14**: 467–471.

Bernal, R. (1998). Demography of the vegetable ivory palm *Phtelephas seemannii* in Colombia, and the impact of seed harvesting. *Journal of Applied Ecology* **35**: 64–74.

Berry, E. J. and D. L. Gorchov (2004). Reproductive biology of the dioecious understory palm *Chamaedorea radicalis* in a Mexican cloud forest: pollination vector, flowering phenology, and female fecundity. *Journal of Tropical Ecology* **20**: 369–376.

Berry, E. J. and D. L. Gorchov (2006). Female fecundity is dependent on substrate, rather than male abundance, in the wind-pollinated, dioecious understory palm *Chamaedorea radicalis*. *Biotropica* **39**: 186–194.

Berry, E. J., D. L. Gorchov, B. A. Endress and M. H. H. Stevens (2008). Source–sink dynamics within a plant population: the impact of substrate and herbivory on palm demography. *Population Ecology* **50**: 63–77.

Caswell, H. (2001). *Matrix Population Models*, 2nd edition. Sinauer Associates, Sunderland, MA.

CEC (Commission for Environmental Cooperation of North America) (2002). In search of a sustainable palm market in America [available at www.cec.org/pubs—docs/documents/index.cfm?varlan=english&id=1028].

Endress, B. A. (2002). Population dynamics, conservation and management of the palm *Chamaedorea radicalis* Mart. in the El Cielo Biosphere Reserve, Tamaulipas, Mexico. Doctoral Dissertation, Miami University, Oxford, OH.

Endress, B. A., D. L. Gorchov and R. B. Noble (2004a). Nontimber forest product extraction: effects of harvest and browsing on an understory palm. *Ecological Applications* **14**: 1139–1153.

Endress, B. A., D. L. Gorchov, M. B. Peterson and E. Padrón-Serrano (2004b). Harvest of the palm *Chamaedorea radicalis*, its effect on leaf production, and implications for sustainable management. *Conservation Biology* **18**: 822–830.

Endress, B. A., D. L. Gorchov and E. J. Berry (2006). Sustainability of a non-timber forest product: effects of alternative leaf harvest practices over 6 years on yield and demography of the palm *Chamaedorea radicalis*. *Forest Ecology and Management* **234**: 181–191.

FAO (Food and Agriculture Organization of the United Nations) (1997). *Non-wood Forest Products. 10. Tropical Palms*. FAO, Bangkok.

Flores, C. F. and P. M. S. Ashton (2000). Harvesting impact and economic value of *Geonoma deversa*, Arecaceae, an understory palm used for roof thatching in the Peruvian Amazon. *Economic Botany* **54**: 267–277.

González-Medrano, F. (2005). La vegetación. In *Historia Natural de la Reserva de la Biósfera El Cielo* (G. Sánchez-Ramos, P. Reyes-Castillo and R. Dirzo, eds.). Universidad Autónoma de Tamaulipas, Tamaulipas, Mexico: 38–50.

Gorchov, D. L. and B. A. Endress (2005). Historia natural de *Chamaedorea radicalis*. In *Historia Natural de la Reserva de la Biósfera El Cielo* (G. Sánchez-Ramos, P. Reyes-Castillo and R. Dirzo, eds.). Universidad Autónoma de Tamaulipas, Instituto de Ecología AC, and Universidad Nacional Autónoma de Mexico.

Heppell, S. S., C. Pfister and H. de Kroon (2000). Elasticity analysis in population biology: methods and applications. *Ecology* **81**: 605–606.

Hodel, D. R. (1992). *Chamaedorea Palms: The Species and Their Cultivation*. Allen Press, Lawrence, KS.

Hoffman, W. A. (1999). Fire and population dynamics of woody plants in a neotropical savanna: matrix model projections. *Ecology* **80**: 1354–1369.

Jones, F. A. and D. L. Gorchov (2000). Patterns of abundance and human use of the vulnerable understory palm, *Chamaedorea radicalis* (Arecaceae), in a montane cloud forest, Tamaulipas, Mexico. *Southwestern Naturalist* **45**: 421–430.

Joyal, E. (1996). The palm has its time: an ethnoecology of *Sabal uresana* in Sonora, Mexico. *Economic Botany* **50**: 446–462.

Kainer, K. A., M. L. Duryea, N. C. Macêdo and K. Williams (1998). Brazil nut seedling establishment and autecology in extractive reserves of Acre, Brazil. *Ecological Applications* **8**: 397–410.
Kilroy, H. and D. L. Gorchov (2010). Enrichment planting of an understory palm: Effect of micro-environmental factors on seedling establishment, growth and survival. *International Journal of Biodiversity and Conservation* **2**: 105–113.
Kiker, C. F. and Putz, F. E. (1997). Ecological certification of forest products: economic challenges. *Ecological Economics* **20**: 37–51.
Mora-Olivo, A., J. L. Mora Lopez, J. P. Jiménez Pérez and J. Sifuentes Silva (1997). Vegetación y flora asociada a la palmilla (*Chamaedorea radicalis* Mart.) en la Reserva de la biosfera "El Cielo". *Biotam* **8**: 1–10.
Olmsted, I. and E. R. Alvarez-Buylla (1995). Sustainable harvesting of tropical trees: demography and matrix models of two palm species in Mexico. *Ecological Applications* **5**: 484–500.
Oyama, K. and A. Mendoza (1990). Effects of defoliation on growth, reproduction, and survival of a neotropical dioecious palm, *Chamaedorea tepejilote*. *Biotropica* **22**: 86–93.
Peterson, M. B. (2001). Resource use and livelihood strategies of two communities in El Cielo Biosphere Reserve, Tamaulipas, Mexico. MS thesis, Miami University, Oxford, OH.
Piñero, D., M. Martinez-Ramos and J. Sarukhán (1984). A population model of *Astrocaryum mexicanum* and a sensitivity analysis of its finite rate of increase. *Journal of Ecology* **72**: 977–991.
Puig, H. and R. Bracho (eds.) (1987). *El bosque mesófilo de montaña de Tamaulipas*. Instituto de Ecología, México DF.
Pulliam, H. R. (1988). Sources, sinks, and population regulation. *American Naturalist* **132**: 652–661.
Pulliam, H. R. (1996). Sources and sinks: empirical evidence and population consequences. In *Population Dynamics in Ecological Space and Time* (J. O. E. Rhodes, R. K. Chesser and M. H. Smith, eds.). University of Chicago Press, Chicago.
Ratsirarson, J., J. A. Silander and A. F. Richard (1996). Conservation and management of a threatened Madagascar palm species, *Neodypsis decaryi*, Jumelle. *Conservation Biology* **10**: 40–52.
Rodríguez-Buriticá, S., M. A. Orjuela and G. Galeano (2005). Demography and life history of *Geonoma orbignyana*: an understory palm used as foliage in Colombia. *Forest Ecology and Management* **211**: 329–340.
Sánchez-Ramos, G., P. Reyes-Castillo and R. Dirzo (2005). *Historia Natural de la Reserva de la Biósfera El Cielo*. Universidad Autónoma de Tamaulipas, Tamaulipas, Mexico.
Scheiner, S. M. and J. C. Gurevitch (1993). *Design and Analysis of Ecological Experiments*. Chapman and Hall, New York.
Silvertown, J., M. Franco and E. Menges (1996). Interpretation of elasticity matrices as an aid to the management of plant populations for conservation. *Conservation Biology* **10**: 591–597.
Svenning, J. C. and M. J. Macía (2002). Harvesting of *Geonoma macrostachys* Mart. leaves for thatch: an exploration of sustainability. *Forest Ecology and Management* **167**: 251–262.
Ticktin, T. (2004). The ecological implications of harvesting non-timber forest products. *Journal of Applied Ecology* **41**: 11–21.
Ticktin, T. (2005). Applying a metapopulation framework to the management and conservation of a non-timber forest species. *Forest Ecology and Management* **206**: 249–261.
Ticktin, T., P. Nantel, F. Ramírez and T. Johns (2002). Effects of variation on harvest limits for non-timber forest species in Mexico. *Conservation Biology* **16**: 691–705.
Uhl, N. W. and J. Dransfield (1987). *Genera Palmarum: A Classification of Palms Based on the Work of H. E. Moore, Jr.* Allen Press, Lawrence, KS.
van Groenendael, J., H. de Kroon and H. Caswell (1988). Projection matrices in population biology. *Trends in Ecology and Evolution* **3**: 264–269.
Velásquez Runk, J. (1998). Productivity and sustainability of a vegetable ivory palm (*Phytelephas aequatoralis*, Arecaceae) under three management regimes in northwest Ecuador. *Economic Botany* **52**: 168–182.

Wilsey, D. and D. Current (2004). *In Search of a Sustainable Palm: Progress and Work Plan, 2004–2005* [available at www.cinram.umn.edu/ecopalms/palm/Palm_Planning_Revised.pdf].

Wisdom, M. J., L. S. Mills and D. F. Doak (2000). Life-stage simulation analysis: estimating vital rate effects on population growth for conservation. *Ecology* **81**: 628–641.

Zuidema, P. A. and M. J. A. Werger (2000). Impact of artificial defoliation on ramet and genet demography in a neotropical understory palm. In *Demography of Exploited Tree Species in the Bolivian Amazon* (P. A. Zuidema, ed.). Programa Manejo de Bosques de la Amazonía Boliviana (PROMAB), Riberalta: 108–131.

NICK M. HADDAD, BRIAN HUDGENS, ELLEN I. DAMSCHEN, DOUGLAS J. LEVEY, JOHN L. ORROCK, JOSHUA J. TEWKSBURY AND AIMEE J. WELDON

22

Assessing positive and negative ecological effects of corridors

Summary

The most popular landscape-level strategy to conserve biodiversity is to link reserves with corridors. Despite much theoretical and empirical support for their benefits in creating or maintaining population sources, corridors may have negative effects and create sinks by altering the dynamics of competitors and natural enemies. In this chapter, we synthesize results from the largest and longest-running experiment to test the effects of corridors, the Savannah River Site Corridor Experiment, and assess their positive and negative ecological effects. In addition to reviewing previously published studies from this experiment, we present new findings about corridor effects on seed mass and number, bird-dispersed seed rain, and bird nest predation and density. Taken together, these empirical studies broadly affirm the positive effects of corridors, particularly on dispersal and diversity. Where there are negative impacts of corridors, the underlying processes are nearly always linked to edge effects, a side-effect of creating corridors. These negative edge effects have the potential to change source patches into sink patches. To further explore the balance of positive and negative corridor effects, we conducted a modeling study, and found that corridors can benefit populations despite edge effects, as long as the edge effects associated with corridors are not too large. Our synthesis serves to highlight areas for future research, particularly on the effects of corridors on population persistence and how corridor characteristics (e.g., width, length) and matrix permeability alter corridor efficacy. As long as efforts are taken to reduce the negative effects of edges, our findings generally support efforts to reconnect landscapes for biodiversity conservation.

Sources, Sinks and Sustainability, ed. Jianguo Liu, Vanessa Hull, Anita T. Morzillo and John A. Wiens. Published by Cambridge University Press.

Background

Habitat loss and fragmentation have created many small sink populations. The most popular landscape-level strategy to overcome the negative effects of habitat loss is to maintain or restore landscape connectivity through habitat corridors, with the goal of restoring source populations. Since 2006, there have been three books and one major review devoted to the rationale, design, implementation, and effectiveness of corridors (Anderson and Jenkins 2006; Chetkiewicz *et al.* 2006; Crooks and Sanjayan 2006; Hilty *et al.* 2006); all describe many examples of corridors being implemented to reconnect small and large patches of habitat. With the increasing implementation of corridors come more questions about their effectiveness. In this chapter, we evaluate the positive and negative ecological effects of corridors on plants and animals. To do so, we specifically evaluate the influence of corridors on the key population parameters (birth, immigration, death, and emigration; BIDE) that are used in the original model of source–sink dynamics (Pulliam 1988) and that provide the theoretical underpinning for corridor function. Unlike most studies that focus on populations through local dynamics (birth and death), our corridor studies place the emphasis squarely on immigration and emigration – that is, the mechanisms that can determine whether a population becomes a source or sink when birth and death rates are relatively constant from one population to the next. We illustrate how corridors, through both positive and negative impacts on birth, death, emigration and immigration, may have the capacity to change populations of conservation concern from sinks into sources, or from sources into sinks.

Theory supporting the positive effects of corridors is clear: corridors reduce isolation between habitat fragments, thereby promoting dispersal and gene flow, and increasing population viability and species diversity. Empirical studies have confirmed most of these key benefits of corridors (e.g., Beier and Noss 1998; Gonzalez *et al.* 1998; Haddad *et al.* 2003; Damschen *et al.* 2006). In a source–sink context, a positive effect of corridors would be to create or maintain sources, i.e., population growth rates are positive (Pulliam 1988). This could happen if corridors increase birth or decrease death rates, increase dispersal to maintain sink populations, or direct dispersal away from sink habitat.

What is striking about the literature on corridors is that the empirical evidence is overwhelmingly positive, with virtually no evidence of negative ecological effects. Intuitively, this should not be the case. Because corridors connect physical places used by many organisms, their construction likely facilitates the movement of both target organisms (typically of conservation concern) and non-target or potentially detrimental organisms (Simberloff and Cox 1987; Simberloff *et al.* 1992). In particular, corridors could increase dispersal of antagonistic species, including predators, diseases and invasives (Hess 1994; McCallum and Dobson 2002). Corridors could also modify the landscape in unintended ways, for example by increasing edge effects (Orrock *et al.* 2003; Weldon 2006). Furthermore, corridors could synchronize population dynamics and cause the simultaneous collapse of many different subpopulations (Earn *et al.* 2000; Hudgens and Haddad 2003), increase the fixation of deleterious mutations (Orrock 2005), or both. Under the worst-case scenario, corridors could turn networks of patches from sources into sinks, reducing population viability and species diversity. Yet, despite the potential for such negative effects, authors typically do not go much beyond Simberloff and colleagues' original critique (Simberloff and Cox 1987; Simberloff *et al.* 1992), and empirical demonstrations of negative impacts are rare.

The generally positive view of corridors is encouraging from a management perspective. However, because corridors have the potential to affect the dynamics of many species and their interactions, it is premature to embrace corridors as uniformly beneficial. What are needed are long-term, community-oriented studies of corridor function that examine direct and indirect effects of corridors across a wide range of taxa and ecosystems.

We draw on 15 years of research from the largest and longest-running experimental study devoted to understanding corridors, the Savannah River Site Corridor Experiment, to assess the positive and negative effects of corridors. Within our experimental landscapes, we have studied how corridors affect animal behavior, plant and animal dispersal, population distribution and dynamics, species interactions, and biodiversity. These studies cover a wide range of taxa and interactions, and our experimental design provides for explicit consideration of potential positive and negative effects of corridors. In this chapter, we synthesize this diverse literature from our experimental landscapes, focusing on elements of BIDE, particularly immigration and emigration, and their potential for source and sink creation. We then build a population model to ask how analyzing tradeoffs between one positive effect of corridors – dispersal – and one negative effect – edge creation – can be used to inform conservation decisions about corridor implementation.

We recognize that weighing the positive and negative effects of corridors involves analysis of their economic and ecological impacts. Economic impacts are predicated upon ecological impacts because we must first assess the potential for corridors to increase the conservation of biodiversity and the function of ecosystems beyond what could be achieved by increasing habitat area alone. As such, ecological impacts are a logical place to begin, and they are the focus of this chapter.

We also recognize that weighing the ecological costs and benefits of corridors involves value judgments requiring, for example, hard decisions about

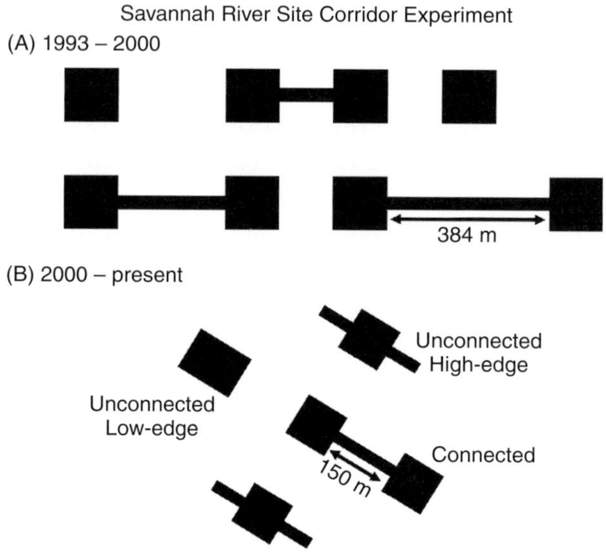

FIGURE 22.1.
The Savannah River Site Corridor Experiment. Black areas are openings created by clearcutting and, in (B), being restored to longleaf pine woodland. Openings are surrounded by plantation pine forest planted at high density and extending at least 150 m beyond patch boundaries in all directions.

which species are "detrimental." In the case of invasive exotics, the decision is easy. But what about predators and pathogens? What is beneficial for them is likely detrimental to their prey or host species, and vice versa. Which side does one take? In this chapter, we focus on the role of corridors in maintaining biodiversity, and we use focal species as indicators of changes in important processes. Determining which species are "most important" will always depend on the species under consideration and the nature of their interactions.

The Savannah River Site Corridor Experiment

The Savannah River Site Corridor Experiment in South Carolina, USA uses a series of replicated experimental landscapes created and maintained in cooperation with the US Forest Service–Savannah River. Patches consist of cleared forest within a matrix of mature pine plantation. There have been two separate but related experiments. The first, lasting from 1993 to 2000, manipulated connectivity by creating patches with and without corridors of varying lengths (Fig. 22.1A). A total of 27 patches, each 1.64 ha in size, were separated from the nearest patch by 64, 128, 256 or 384 m. Some patches were connected by a 32 m wide corridor and others were not. Details and rationale are provided in Haddad (1999a). Effects of corridors on connectivity could be assessed by comparing movement between pairs of connected or unconnected patches, or by comparing density within connected or unconnected patches. In a few instances, distance between patches was included as an additional covariate. Although responses such as density could be measured along a gradient with respect to the forest edge, the effects of corridors on patch shape were not controlled.

The second experiment, which began in 2000 and is ongoing, focuses explicitly on the separation of two corridor functions: their impact on connectivity and their impact on patch shape. Changes in patch shape and, in particular, increases in the prevalence of patch edge relative to the size of the patch are unavoidable consequences of corridor creation. Yet, very few studies include sufficient controls to separate these effects (but see Gonzalez *et al.* 1998). This is problematic because responses to corridors and edges can be confounded (e.g., corridors or edges could increase population density) and because corridors and edges may have opposing effects on populations in a source–sink context (e.g., corridors may increase dispersal, but edges created by corridors may increase predation during dispersal). To separate connectivity and edge effects, we manipulated connectivity by creating corridors while controlling edge-to-area ratios and patch size, and we manipulated edge-to-area ratios while controlling for connectivity and patch size (Fig. 22.1B).

The experiment consists of eight blocks with five patches each. One center patch (1 ha) is surrounded by four peripheral patches, each 150 m from the center patch. One peripheral patch is 1 ha and connected to the center patch by a 25-m-wide corridor. The three remaining peripheral patches are unconnected, and are equal in area to a patch plus a corridor (1.375 ha), and of two shapes. Low-edge unconnected patches are rectangular in shape, and control for the effects of corridors on increasing the area of patches. High-edge unconnected patches have two blind-end 75 × 25 m corridors extending from two sides parallel to the edge facing the center patch. They control for the effects of corridors on patch shape, including increased edge-to-area ratios. Details of the experimental design are described in Tewksbury *et al.* (2002). To test for effects of connectivity while controlling for edge effects, we can compare responses in connected and high-edge unconnected patches. To test for edge effects, we can compare responses in high-edge and low-edge unconnected patches. We can also measure responses to edges by comparing processes within patches along a gradient of distance from the matrix.

Methods

To assess the effectiveness of corridors, including their positive and negative effects, we assembled all published and relevant unpublished data from the experiment (Table 22.1). For each taxon and response variable (e.g., dispersal, density, predation, diversity), we asked two questions:

TABLE 22.1. *Synthesis of studies across taxa and response variables in the Savannah River Site (SRS) Corridor Experiment. The relation to BIDE (birth-immigration-death-emigration) relates each study to its relevance for source–sink dynamics.*

Species	Taxon	Study design	Response	Relation to BIDE	Corridor effect?	Edge effect?	Reference	Notes
Peromyscus polionotus	Mammal	Experiments with predator cues and seeds	Foraging behavior	BD	Positive	No effect	Brinkerhoff *et al.* (2005)	Presence of predators increased foraging in connected patches
Sialia sialis	Bird	Observational	Movement behavior	IE	Positive	Positive and Negative	Levey *et al.* (2005)	No effect on movement distance, negative effect on perch time, and positive effect on movement direction along edges
Eurema nicippe, Papilio troilus, Phoebis sennae	Insect	Observational	Movement behavior	IE	Positive and No effect	Positive	Haddad (1999a)	More turning nearer edge and faster movement rates directed along corridors for some species
Sigmodon hispidus	Mammal	Radiotelemetry	Emigration	E	Positive	NA	Bowne *et al.* (1999), Peles *et al.* (1999)	

Junonia coenia	Insect	Experimental release	Colonization	E	Positive	NA	Haddad (2000)	Interaction between corridor and distance, negative distance effect in matrix
Musca domestica	Insect	Experimental release	Dispersal	IE	Positive	Positive	Fried *et al.* (2005)	Evidence for a drift-fence effect of edges
Morella cerifera	Plant	Mark–recapture	Dispersal	IE	Positive	No effect	Levey *et al.* (2005)	
Euptoieta claudia, *Junonia coenia*	Insect	Mark–recapture	Dispersal	IE	Positive	No effect	Tewksbury *et al.* (2002)	
Ilex opaca, *Morella cerifera*, *Phytolacca americana*, *Rhus copallina*	Plant	Experiment removed target plants in peripheral patches	Dispersal	IE	Positive	NA	Haddad *et al.* (2003)	Nonsignificant trend in other species
Euptoieta claudia, *Junonia coenia*	Insect	Mark–recapture	Dispersal	IE	Positive	NA	Haddad (1999b)	Negative distance effect
Peromyscus polionotus	Mammal	Experimental introduction of nests	Dispersal	IE	Positive	NA	Haddad *et al.* (2003)	
Ilex vomitoria, *Morella cerifera*	Plant	Experiment and mark–recapture	Dispersal	IE	Positive	No effect	Tewksbury *et al.* (2002)	
Ilex verticillata	Plant	Experiment and mark–recapture	Pollination	B	Positive	No effect	Tewksbury *et al.* (2002)	

TABLE 22.1. (*cont.*)

Species	Taxon	Study design	Response	Relation to BIDE	Corridor effect?	Edge effect?	Reference	Notes
Lantana camara, Rudbeckia hirta	Plant	Tracked dye powder movement	Pollination	B	Positive	No effect	Townsend and Levey (2005)	Negative distance effect tested outside experimental plots
Ilex verticillata	Plant	Experimental placement of plants	Seed number and size	B	Positive	No effect	This chapter	
Euptoieta claudia, Junonia coenia, Papilio troilus, Phoebis sennae	Insect	Survey	Density	BD	Positive	Negative	Haddad and Baum (1999)	Could not separate corridor and edge effects
Natives of longleaf pine woodlands	Plant	Survey	Diversity		Positive	No effect	Damschen *et al.* (2006)	
Peromyscus polionotus	Mammal	Experimental introduction of nests	Emigration	E	No effect	NA	Danielson and Hubbard (2000)	
Peromyscus polionotus	Mammal	Experimental distribution of seeds	Foraging behavior	BD	No effect	Negative	Orrock and Danielson (2005)	More activity in centers of edgy patches
Xylocopa virginica	Insect	Mark–recapture	Dispersal	IE	No effect	NA	Haddad *et al.* (2003)	Nonsignificant trend
Sigmodon hispidus	Mammal	Radiotelemetry	Dispersal	IE	No effect	NA	Bowne *et al.* (1999), Peles *et al.* (1999)	

Peromyscus gossypinus, *Peromyscus polionotus*, *Sigmodon hispidus*	Mammal	Survey and radiotelemetry	Home range size, dispersal	BIDE	No effect	NA	Mabry and Barrett (2002)	
Passerina cyanea	Bird	Survey	Nest predation	D	No effect	Positive	Weldon and Haddad (2005), Weldon (2006)	
Phytolacca americana	Plant	Experiment controlling seed predators	Seed predation	D	No effect	Positive and Negative	Orrock *et al.* (2003)	Depends on predator, ants (+) or mammals (−)
Euptoieta claudia, *Junonia coenia*	Insect	Survey	Density	BD	No effect	Negative	Haddad and Tewksbury (2005)	Less common near edges and in corridors
Asclepias tuberosa, *Carduus repandus*, *Cassia* sp., *Crotolaria* sp., *Gelsemium sempervirens*, *Gerardia purpurea*, *Linaria canadensis*, *Rubus* sp., *Sassafras albidum*	Plant	Survey	Density	BD	No effect	NA	Haddad and Baum (1999)	

TABLE 22.1. (*cont.*)

Species	Taxon	Study design	Response	Relation to BIDE	Corridor effect?	Edge effect?	Reference	Notes
Peromyscus gossypinus, *Peromyscus polionotus*, *Sigmodon hispidus*	Mammal	Survey	Density	BD	No effect	NA	Danielson and Hubbard (2000)	
Passerina cyanea	Bird	Survey	Density	BD	No effect	Positive	Weldon and Haddad (2005), Weldon (2006)	Equal nest densities in different shaped patches, but nests placed near edges
Bird-dispersed seed rain	Plant	Survey	Seed rain	I	No effect	Positive	This chapter	
Exotics	Plant	Survey	Diversity		No effect	No effect	Damschen *et al.* (2006)	
Peromyscus gossypinus, *Peromyscus polionotus*, *Sigmodon hispidus*	Mammal	Survey	Density	BD	Negative and No effect	Negative	Mabry *et al.* (2003)	Negative corridor effect on 1 of 3 species in 1 of 3 years
Prunus serotina, *Rubus allegheniensis*	Plant	Experiment controlling seed predators	Seed predation	D	Negative	No effect	Orrock and Damschen (2005)	More seed predation in connected patches for larger seeded species

Asclepias tuberosa, Linaria canadensis, Passiflora incarnata	Plant	Survey	Density	BD	NA	No effect	Haddad and Tewksbury (2005)	
Gueraca caerulea, Dendroica discolor	Bird	Survey	Density	BD	NA	No effect	This chapter	
Cardinalis cardinalis, Pipilo erythrophthalmus, Toxostoma rufum	Bird	Survey	Density	BD	NA	Trend toward positive	This chapter	More nests in edge patches
Gueraca caerulea,Dendroica discolor	Bird	Survey	Nest predation	D	NA	No effect	This chapter	
Cardinalis cardinalis, Pipilo maculatus,Toxostoma rufum	Bird	Survey	Nest predation	D	NA	No effect	This chapter	

1. Did corridors affect the response variable through increased connectivity?
2. Did edges affect the response?

For a limited set of studies from the first phase of the experiment, we also asked how the distance between patches or corridor length affected movement rates. Most of the results presented here come from published studies, and we refer readers to the relevant papers for detailed methods (Table 22.1). Results from three studies are previously unpublished. We now elaborate the methods for these studies.

Tewksbury *et al.* (2002) tested the effects of corridors and patch shape on pollination and dispersal in holly, *Ilex verticillata*, a dioecious shrub. Males were planted in the center patch and three females were planted in each peripheral patch. In addition to measuring the proportion of flowers that set fruit on two randomly sampled branches on each plant (fruit set; Tewksbury *et al.* 2002), previously unpublished data were collected on the seed number and seed mass of fruit produced on these branches. Here we evaluate the relationships between fruit set and the number and mass of seeds, both important variables for plant fitness (Howe *et al.* 1985; Alcantara and Rey 2003). We restricted our analyses to plants in which at least ten fruits were measured, so that the relationship between fruit set and fruit characteristics could be accurately determined. This restriction, while necessary, resulted in data on a total of only 29 plants spread unevenly across 15 patches in five blocks. In these circumstances, our traditional blocked analysis is not possible. We thus used linear regression to examine the effect of fruit set on seed number per fruit and seed mass. Because we have rigorous estimates of the effects of connectivity and patch shape on fruit set, we used these data, along with the overall regression results, to predict seed mass and seed number differences between patch shapes (connected, unconnected high-edge, unconnected low-edge) based solely on differential pollen limitation.

Levey *et al.* (2005) tested for effects of corridors and patch shape on seed dispersal of a woody shrub, *Morella cerifera* (wax myrtle). Seed traps used in that study also collected seeds of other bird-dispersed plants. These traps were erected on a grid of polyvinyl chloride poles, each 3 m tall, placed 25 m apart and at least 12.5 m from the nearest edge. From the top of all the poles, we suspended seed traps made from 25 cm diameter flowerpots. A hole in the bottom of each pot allowed us to pass the pole through the pot and center the pot below the top of the pole. Seeds deposited in the traps were collected in all peripheral patches during the winters (December–March) of 2000/1 and 2001/2, when many plants at our site were in fruit and many birds were frugivorous (McCarty *et al.* 2002). Because we sampled during the non-breeding season, when most birds were not territorial, movements by bluebirds (the most common frugivore using our perches) were attributable to landscape features and not confounded with territorial boundaries.

To analyze the abundance of seed rain, seed trap data were used from the four center poles in each patch from both field seasons. *Phytolacca americana* (American pokeweed) and *Rhus copallina* (winged sumac) seeds were excluded from the analysis. These species had strongly aggregated distributions and produced huge quantities of fruits in nearly every patch, creating biases in seed rain, as the seed traps near individuals of these species received mostly seeds from within the same patch, regardless of patch shape. When *P. americana* and *R. copallina* were included, results were qualitatively similar to those we report below. The dependent variable was the sum of all seeds across years in each patch. Differences in species richness were standardized to the same sample size ($n = 10$ seeds/trap) with rarefaction. To test for the effects of connectivity and patch shape on the diversity and abundance of seed rain, we used a linear mixed effects model. Experimental block was included as a random effect and patch shape (connected, unconnected high-edge, unconnected low-edge) as a fixed effect.

Weldon and colleagues (Weldon and Haddad 2005; Weldon 2006) tested the effects of corridors and patch shape on nest success of indigo buntings (*Passerina cyanea*), the most abundant breeding bird in our experimental patches. Nests of other bird species located during indigo bunting nest searches were also monitored for success, including other migrants such as blue grosbeaks (*Passerina caerulea*) and prairie warblers (*Dendroica discolor*) and residents such as brown thrashers (*Toxostoma rufum*), northern cardinals (*Cardinalis cardinalis*) and eastern towhees (*Pipilo erythrophthalmus*). Regular nest searches were concentrated in unconnected high-edge and unconnected low-edge patches (Weldon and Haddad 2005). Nest success was determined using Mayfield statistics for all nests pooled across patches of similar type, and a binomial test was used to compare differences in nest success between unconnected high-edge and unconnected low-edge patches (Williams *et al.* 2002). We analyzed nest density using the statistical model described above for seed rain.

Results

We summarize results from 22 published papers and from the three unpublished studies described above. Although studies varied in their response variables, level of control, and statistical power, we included all the studies in our initial analysis. We then discuss the implications of our findings based on response variables, connection to the BIDE model, level of control, and statistical power. Some papers reported more than one taxon or response variable, and we treated those results separately. A few papers reported results that were also reported in another paper, and we treated those results together. In all, we divided the studies into 37 groups that included detailed studies of 15 different animal species and 18 different plant species (Table 22.1). For each study, we evaluated the positive or negative ecological effect of corridors or edges from the perspective of the focal taxon. Thus, for studies of seed predation, our focus was on the seeds rather than on their predators.

Corridor effects

Corridors had positive effects in 17 of 32 studies in which corridor effects could be assessed (Table 22.1). Positive effects were typically on aspects of immigration and/or emigration, including aspects of behavior, dispersal and pollination. Corridors also increased butterfly density and plant diversity. Of the remaining 15 studies, 13 showed no statistically significant evidence of corridor effects on aspects of behavior, dispersal, demography, density or diversity (Table 22.1). Negative effects of corridors were found in two studies, and were concentrated on the birth and death elements of BIDE. Orrock and Damschen (2005) found a negative effect of corridors on seed predation by small mammals (i.e., seed predation rates were higher in connected patches). Mabry *et al.* (2003) found a negative effect of corridors on the population densities of one of three small mammal species in one of three years.

Both experiments were established to test for effects of corridors on movement, and rigorous studies of movement (e.g., behavioral tracking, mark–recapture, and radiotelemetry) were most likely to show positive effects of corridors (Fig. 22.2; Table 22.1). Of the studies that showed no effect of corridors, half were surveys of natural populations occurring in the experimental landscapes but otherwise unmanipulated. Insects and plants were more likely to respond positively to corridors, whereas small mammals were likely to show no or negative responses to corridors (Table 22.1).

Results from studies of holly seed number and mass illustrate one way in which corridors may have positive effects on populations by increasing recruitment, increasing births in the BIDE framework. Holly fruits had an average of 5.9 seeds (SE = 0.05, n = 1,233 fruits on 29 plants) and the average seed mass was 4.5 mg (SE = 0.16). Both seed number per fruit and average seed size increased significantly as fruit set increased (Fig. 22.3; seed number: $F_{1,27} = 8.7$, $P = 0.0067$; seed mass: $F_{1,27} = 4.3$, $P = 0.047$), and these relationships were not dominated by differences between blocks, as the slopes for the individual sites were also positive and similar to the overall relationship, suggesting general relationship between fruit set and seed number (slopes ranging from 0.3 to 1.4) and fruit set and seed mass (slopes ranging from 0.3 to 1.5). When combined with data on fruit set, which was on average 20% higher in connected

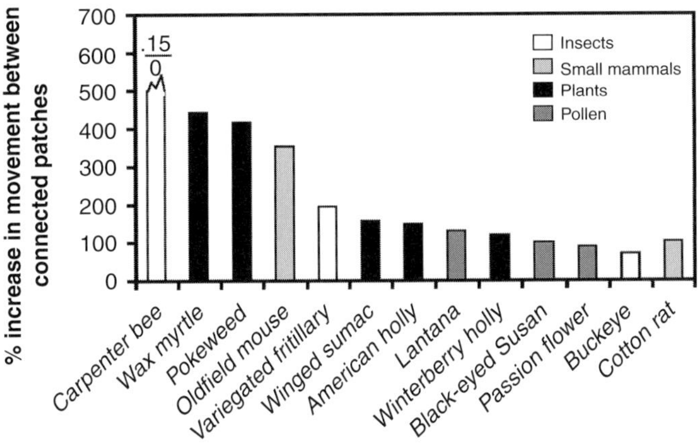

FIGURE 22.2.
Percentage increase in movement rate of animals, plants and pollen between connected patches over movement between unconnected patches. All detailed studies of movement of individuals or pollen between patches are included. Results are expanded from Haddad *et al.* (2003). For carpenter bees, the increase is undefined (0.15 of bees moved between connected patches, zero moved between unconnected patches). Only for cotton rats was movement not significantly different between connected and unconnected patches.

than unconnected patches (Tewksbury *et al.* 2002), we can predict the relationship between patch type and seed number and mass: plants in connected patches should have about 2% more seeds per fruit and 6% heavier fruit than plants in unconnected patches (Fig. 22.3). We note that there is no indication that the relationships between fruit set and seed number or seed mass vary by patch type (Ps > 0.4).

Edge effects

Of 26 studies that examined edge effects, either by comparing the effects of patch shape in high-edge vs. low-edge unconnected patches or by measuring responses at different distances from the edge, eight found an increase and six found a decrease in the response variable near the edge (Table 22.1).

Results of studies of the diversity and abundance of bird-dispersed seeds and of nest predation illustrate the importance of controlling for edge effects in testing for connectivity effects, as we were able to do in our second experiment. Connected and high-edge unconnected patches received similar numbers of seeds, with low-edge unconnected patches receiving 31% fewer seeds than connected patches ($t_{54} = 1.96$, $P = 0.05$) and 37% fewer than high-edge patches (Fig. 22.4A; $t_{54} = 2.21$, $P = 0.01$). Connected and high-edge unconnected patches received approximately 20% more species than low-edge unconnected patches (Fig. 22.4B; $F_{2,54} = 4.4$, $P = 0.016$). This difference in species richness

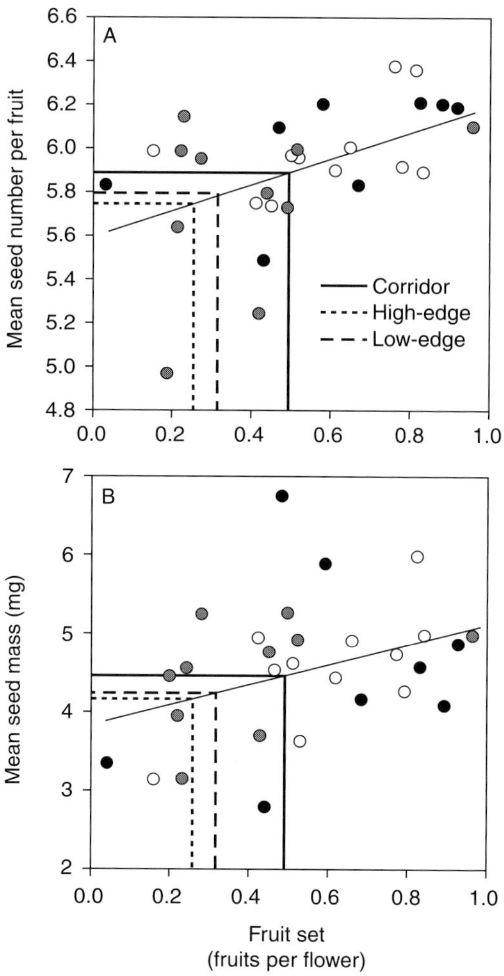

FIGURE 22.3.
Effects of pollen limitation (measured as the proportion of flowers that set fruit) on (A) the number of seeds per fruit, and (B) seed mass. Solid black regression lines indicate the overall relationship including all patches. Thick drop lines indicate mean fruit set for connected (solid lines, black circles), low-edge (long dash, grey circles), and high-edge (short dash, open circles) patches.

between patches, however, was entirely due to differences in overall seed numbers; when we controlled for total seed rain, patch connectivity and shape had no effect on rarefied species richness (Fig. 22.4C; $F_{2,54} = 1.9, P = 0.15$).

Studies of nest success also show how corridor effects can be confounded with edge effects. Weldon (2006) and Weldon and Haddad (2005) showed how indigo buntings experienced higher nest predation in high-edge patches, whether connected or not. Data collected on other migrant (in addition to the indigo bunting) and non-migrant birds in this study illustrate how low statistical power was an issue in some of our studies (Haddad *et al.* 2003). Nest success

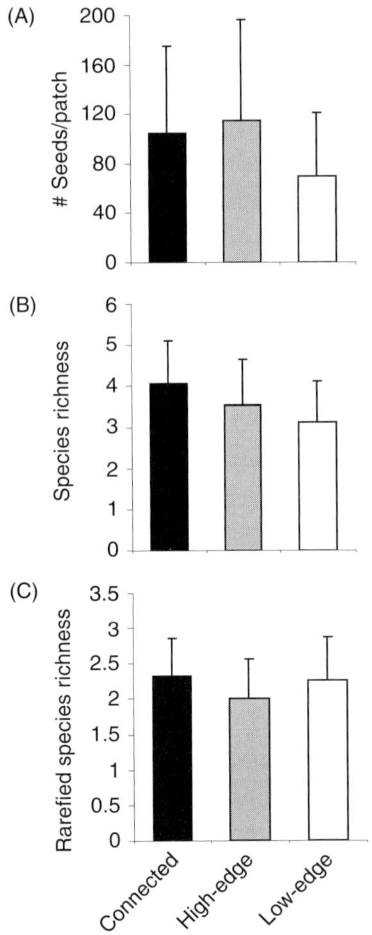

FIGURE 22.4.
Effects of patch connectivity and shape (black columns = connected, gray columns = high-edge unconnected, white columns = low-edge unconnected) on bird-dispersed seed rain. (A) Mean (+1 SD) number of plant species dispersed into the center four seed traps in each patch type. Species richness is higher in connected patches than in unconnected low-edge patches. (B) Mean number of seeds (+1 SD) collected from the center four poles in each patch type. Controlling for year and location of experimental landscape (see text), low-edge patches received significantly fewer seeds than high-edge and connected patches. (C) Mean rarefied species richness (+1 SD). Differences in rarefied richness between patch types are not significant. Therefore, the higher species richness in connected patches (A) was primarily caused by higher seed deposition in connected patches (B).

was calculated for the first brood, which included most nests. No differences in total nest success were found among different patch types for migrant (high-edge: 0.198; 95% CI 0.05, 0.76; low-edge: 0.250; 95% CI 0.06, 0.95) or resident (high-edge: 0.152; 95% CI 0.03, 0.78; low-edge: 0.156; 95% CI 0.09, 0.27) birds. Differences for indigo buntings are consistent with, although larger than, those for the other migrant birds, and the wide range of the confidence intervals is caused by low nest densities for all species. Migrant birds showed no difference in nest density between patch types (high-edge: 1.19 (SD 1.13) nests/patch; low-edge: mean 1.31 (SD 0.70) nests/patch; $F_{1,7} = 0.26$; $P = 0.63$), while resident birds showed a trend toward higher abundances in high-edge (mean 4.06 (SD 2.16) nests/patch) than low-edge patches (mean 2.62 (SD 1.63) nests/patch; $F_{1,7} = 4.33$; $P = 0.08$).

Distance effects

The effects of corridor length were analyzed in three studies (Haddad 1999a, 2000; Townsend and Levey 2005). All studies found that fewer individuals moved long distances than short distances. Studies with butterflies found an interaction between connectivity and distance. Butterflies were equally likely to disperse between nearby patches, whether connected or unconnected. As distances between patches increased, so did dispersal rates between connected patches relative to dispersal rates between unconnected patches (Haddad 1999a). When butterflies were released up to 50 m outside patches, they were more likely to colonize from matrix habitat than from corridors. But as distances outside patches increased, butterflies were more likely to colonize patches from corridors than from the matrix (Haddad 2000). In a study of plants, pollen transfer measured in powerline corridors near our experimental landscapes decreased rapidly with distance, but was still observed up to 1.5 km (Townsend and Levey 2005).

Modeling tradeoffs between positive and negative effects of corridors

Our empirical results raise questions such as: When corridors increase both connectivity and edge effects, under what conditions do corridors still benefit populations? Can corridors change sinks into sources, or vice versa? We cannot address these questions empirically. Instead, we used a previously developed model (Hudgens and Haddad 2003) in which we could vary the strength of connectivity and edge effects. We modeled two logistically growing populations that were connected by dispersal. Intrinsic growth rates, emigration rates, and survival during dispersal were modeled as stochastic processes. We considered two kinds of populations, slow- and fast-growing. Slow-growing populations had intrinsic growth rates low enough for them to be susceptible to extinction following sequences of relatively bad years, and represented the typical case for species of management concern. Fast-growing populations represented weedy species that had intrinsic growth rates high enough to drive cyclical dynamics, making them susceptible to exaggerated boom–bust cycles following sequences of relatively good years. We evaluated the effects of corridors by comparing paired simulations that differed only in the presence or absence of a corridor. Corridors influenced populations by increasing emigration rates, changing the survival of dispersers, and increasing edge effects. Edge effects were modeled as changes to the intrinsic growth rate caused by the creation of a corridor.

We describe the impacts of our modeled corridors by a composite measure of the intrinsic growth rate within patches and dispersal-related mortality (the *contribution rate*). By subtracting the mortality of dispersing individuals from the intrinsic growth rate, we can describe the contribution of a population to the numbers of individuals in the next generation within the patch network. Including edge effects of corridors did not change the primary results reported in Hudgens and Haddad (2003). For slow-growing populations, corridors that increased the contribution rate increased persistence through the rescue effect. When there was very little dispersal through the matrix, corridors could also increase persistence via the rescue effect simply by increasing dispersal rates. This was true even when corridors reduced the contribution rate by promoting risky dispersal behaviors. In contrast, for fast-growing populations, corridors that increased the contribution rate decreased persistence by enhancing and synchronizing boom–bust dynamics.

Edges created by corridors increased or decreased population growth rate and extinction risk. For slow-growing populations, if edges had negative effects on growth rates, corridors tended to increase extinction risk. Conversely, when edges had positive effects on growth rates, corridors tended to reduce extinction risk (Fig. 22.5A). When there was relatively low dispersal through the matrix, corridors that enhanced negative edge effects could still benefit connected populations by increasing dispersal and strengthening rescue effects (Fig. 22.5A). For fast-growing populations, if edges created by the addition of corridors increased growth rates, corridors tended to increase the risk of extinction and, conversely, if edges decreased growth rates, corridors tended to decrease extinction risk. The magnitude of corridor effects on fast-growing populations was greater if dispersal through the matrix was relatively common than if it was relatively rare (Fig. 22.5B). These results demonstrate that strong, negative edge effects can reduce boom–bust population cycles and actually reduce extinction risk in some species.

Discussion

Across many studies and taxa in our long-term experiment, there were two consistent results. First, corridors increased movement rates for most taxa that were tested, supporting the most fundamental rationale for corridor

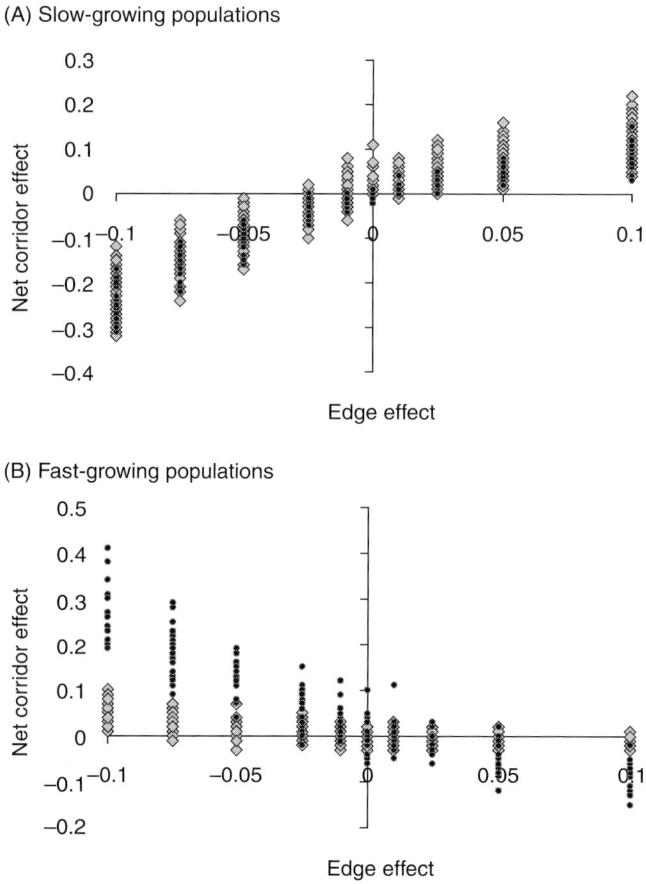

FIGURE 22.5.
Influence of corridors on edge effects and extinction probability for (A) slow-growing populations ($r = 1.5$) and (B) fast-growing populations ($r = 3.5$). Populations are stationary when $r = 1.0$. Edge effects are measured as a proportional change to the population growth rate within patches. The net corridor effect is measured as the difference in the extinction risk between pairs of simulated populations that differ only in the presence or absence of a corridor. Light-colored diamonds indicate paired runs with low dispersal rates (<5%) between unconnected patches, dark circles indicate paired runs with relatively high dispersal rates (>15%) between unconnected patches.

creation. Whether increased movement is positive or negative in conservation depends on the context. Regardless, increased dispersal via corridors is likely to increase population size in a source–sink context. Second, when we observed negative effects of corridors, they were exerted mainly through edge effects. Because corridors are relatively long and narrow, they are inherently "edgy." Although edge effects can be positive or negative, their role in reducing birth rates and increasing death rates must be considered when corridors are created, as these local effects of edges on intrinsic population growth rates could change the status of a population from source to sink and vice versa (Battin 2004; Robinson and Hoover, Chapter 20, this volume). In the following two sections, we generalize our results first with respect to corridor effects, then with respect to edge effects.

Corridor effects

In nearly all cases, corridors increased or had no effect on the processes we studied. Corridor effects were most positive in detailed experiments and capture–recapture studies involving movement (Fig. 22.2). For example, in nine studies of dispersal, corridors increased dispersal in seven, had no effect in one, and had a nonsignificant but positive trend in another. In no case did corridors decrease movement rates. Our ability to detect corridor effects depended in part on which experimental landscape was used; studies that used the most recent set of patches were more likely to detect effects because the design controlled for effects of corridors on patch shape. Our ability to detect corridor effects also depended on methodology: results were strongest when we conducted rigorous studies of movement in which marked individuals were tracked between connected or unconnected patches (Fig. 22.2). Although not as straightforward to interpret as movement studies, surveys of populations and diversity suggest the impact of increased movement. Most significantly, corridors caused ~20% higher plant species diversity relative to unconnected patches after just 6 years (Damschen *et al.* 2006). All evidence to date suggests that these increases are due primarily to the effects of corridors on plant dispersal (Damschen *et al.* 2008).

Whether corridors have positive or negative effects depends on the perspective of the organism under study, and how it interacts with other species that may be of conservation interest. For example, increasing movement by small mammals may be positive if the small mammals are of conservation concern, but negative if they are competing with other rare small mammals or are consuming the seeds of plants of conservation concern. The strongest evidence for negative effects of corridors comes from studies of seed predation by small mammals, because small mammals consume more seeds in connected patches (Orrock and Damschen 2005). The mechanism by which corridors increase seed predation has not been determined; it is not through increased density of small mammals, in connected patches. The best-supported mechanism, even in this case, is related to edge effects, as small mammals appear to change their activity patterns within connected patches that have high edge-to-area ratios (Orrock and Danielson 2005). It is also possible that small mammals move more frequently between connected patches while foraging, as we have observed for one species, the oldfield mouse, *Peromyscus polionotus* (Haddad *et al.* 2003). The other study to show negative effects of corridors observed lower densities of one of three species (the cotton mouse, *P. gossypinus*) in one of three years (Mabry *et al.* 2003). In this case, it was impossible to separate the effects of corridors on connectivity from their effects on edge creation.

Corridors affect many different species in different ways, but after 15 years of study in these landscapes, we are now able to make some broad generalizations about the types of species for which corridors are most effective. Corridor effects on dispersal were most commonly observed in insects and in bird-dispersed plants, and were least likely to be observed for small mammals. Species that did not use corridors for movement, including the spicebush swallowtail butterfly (*Papilio troilus*) and small mammals such as the cotton rat (*Sigmodon hispidus*) and cotton mouse, were larger and/or more generalist in habitat use than other species. In a community context, we would not expect all species to benefit from corridors, and our results provide some guidance as to which species will and will not use them to disperse through fragmented landscapes (see also Damschen *et al.* 2008).

In evaluating the positive and negative effects of corridors, the information that is most noticeably missing is how movement affects population persistence or patch occupancy. Population persistence is a key metric in evaluating corridor impact, as corridors should work within a spatial context to overcome local extinction through dispersal. Yet persistence is notoriously difficult to detect, requiring long-term monitoring and demographic data (as detailed in Dunning *et al.*, Chapter 11, this volume). Although our experiments have been running for 15 years, our longest continuous measure of any taxon (the presence or absence of 300+ plant species found within our patches) is 8 years. Many species native to the ecosystem we work in – longleaf pine woodland – are perennials. Community dynamics are strongly influenced by fire, and we have instituted a fire return interval of ~3 years, consistent with pre-settlement rates. Given these dynamics, longer time series will be needed to test the population consequences of corridors.

Some of our results allow us to assess the population consequences of corridors. The best direct measure of corridors affecting population demography comes from our study of plant pollination: not only do corridors increase pollen transfer but, at least for holly, they also increase the number of seeds produced per fruit and seed mass. These data hint at ways in which corridors can increase plant fitness and, ultimately, the persistence of populations. More indirectly, from our results that corridors increase plant diversity we can infer that corridors do affect persistence. Another metric that we have used to assess population impacts of corridors is population density, and various taxa have shown positive and negative responses to corridors. However, population density responses were only found in the first phase of our experiment, where we were unable to separate effects of corridor-created edges from effects of corridors as conduits for dispersal.

Our first experiment specifically manipulated the distances between patches. Although few studies were able to incorporate this variable, those that did focused on effects of corridors and distance on movement and have shown that (1) dispersal decreases with increasing distances between patches, and (2) there is an interaction, whereby dispersal decreases less rapidly between connected than unconnected patches. Thus, the effects of corridors become stronger as the distances between patches increase.

Edge effects

Our narrow corridors created strong edge effects on density and animal behavior. These effects could be viewed as either positive or negative depending on the conservation context: increased seed rain and nest densities near edges would be positive for plants and birds, but increased seed and nest predation in edgy patches would be negative for plants and birds. Positive effects of edges were on birds or species affected by them. Some bird species nested near edges, and others perched near edges, depositing seeds in edgy patches. Negative effects of edges on density and behavior were concentrated on insects and small mammals. Perhaps because of thermal needs, insects were often concentrated in the centers of our patches (Haddad and Baum 1999). Although their total densities typically did not vary between connected and unconnected patches, small mammals foraged more intensely near the centers of edgier patches. Even though these animals avoided edges, they still preferentially used corridors to move between patches (Haddad and Tewksbury 2005).

Whether or not animals preferred habitats near or far from edges, the movement behavior of many species was strongly influenced by edges. Boundaries restricted or guided animal movements, causing some animals like birds (Levey *et al.* 2005), butterflies (Haddad 1999b), and small mammals to preferentially disperse through or along corridors. The effects of edges for birds were sometimes complex: indigo buntings preferred to nest near edges and selected high-edge patches, but these locations had higher rates of nest predation than low-edge patches (Weldon and Haddad 2005).

Our results demonstrate the need to separate the effects of edges in studies of corridors (see also Gonzalez *et al.* 1998). In our first experiment, there were several cases in which we could not determine whether corridors or edges affected responses, particularly density. In our second experiment, there were several studies in which, if we had not controlled for edges created by corridors, we might have wrongly concluded that corridors were affecting responses. For example, if we had not had high-edge unconnected patches in our second experiment, we would have concluded that seed rain and nest predation were higher in connected patches due to the presence of corridors. In fact, seed rain and nest predation increase in high-edge patches, regardless of the presence of the corridor.

The effects of edges on populations and biodiversity can extend long distances into patches (Laurance 2000; Ewers and Didham 2008), to the extent that nearly any corridor will increase edge effects. The negative effects of edges may in some cases make corridors a poor conservation investment, causing results counter to conservation goals. However, ecologists know a lot about edge effects – they have been studied intensively since Leopold (1933), and hundreds of studies demonstrate how and when they should be important (Ries *et al.* 2004). Furthermore, because we know a lot about when edges should have negative effects, we have a good foundation on which to build when attempting to mitigate them, whether it is through increasing corridor width, or by reducing the abruptness of edges between corridors and matrix habitat.

Putting in perspective the positive and negative effects of corridors

Although our synthesis focuses on studies within one experiment, our results are generally consistent with those found in the literature. Corridors often have positive effects, but in many cases have no effect on target taxa (figure 6 in Beier and Noss 1998; Chetkiewicz *et al.* 2006). Our studies occurred in a model system of open, suitable habitat surrounded by unsuitable forest; a key aspect of the design that makes it similar to most other landscapes where corridors are considered. As in our studies, habitat specialization emerges as a key predictor of corridor effectiveness (Gillies and St. Clair 2008). The spatial extent of corridors is yet another key consideration – our study is well suited to studies of plants and insects that disperse at scales that match our experiment (Kinlan and Gaines 2003; Haddad and Tewksbury 2005), whereas species that disperse over longer distances would need larger corridors. As we found in our experiment, there are no reported studies in which corridors create barriers to dispersal.

Less attention has been given to corridor effects on populations and communities, and this is a key area of future study. Yet, in the few studies that do exist, there is some evidence that corridors increase population persistence and diversity (Haddad and Tewksbury 2006). As we found in our experiment, negative effects of corridors generally come through edge effects (Schmiegelow *et al.* 1997), and studies that have not properly controlled for edge effects should reevaluate their results in light of how edge effects are confounded with corridor effects. Furthermore, there are no published studies documenting that corridors created for habitat conservation promote dispersal of invasive species.

Indeed, with regard to negative effects of corridors, the empirical literature fails to support most of Simberloff *et al.*'s (1992) concerns. Taken together, our results should generalize to a wide variety of taxa and landscapes.

Just because we did not observe negative effects does not mean they do not exist. Corridors increase the dispersal of antagonists, and all of our focal species of plants and animals are likely competitors or consumers of other species. Other negative effects, such as corridors causing population synchronization and simultaneous collapse, seem less likely for populations of conservation concern with slow population growth rates. In trying to evaluate the negative effects of corridors, one thing is clear: we, and the conservation community more generally, have struggled to assess the consequences for populations. Such an assessment is particularly challenging for slow-growing populations when there is little dispersal through the matrix. The net effect of corridors on these populations may be beneficial, even when corridors also have a negative impact on population growth through edge effects or by promoting risky dispersal behaviors. Our synthesis points to an urgent need to test the population consequences of corridors; long-term studies will be needed to do so.

The next step will be the integration of corridor science into the landscape planning arena (Beier *et al.* 2006, 2008). There is a lot left to learn about the function of corridors in conservation: how does corridor width affect function? How fast does corridor function degrade with increasing corridor length? How much does this depend on matrix quality? We are unlikely to secure landscapes large enough to answer these questions definitively unless we integrate studies of corridor function into landscape planning that is already taking place at the local, state, national and international level. By using ongoing efforts to preserve landscape connectivity as an opportunity to study the effectiveness of large-scale corridors, we can move the science of landscape conservation forward and increase the effectiveness of working corridors, whether they are wildlife bridges over highways or 100-mile-long linkages between reserves.

In the meantime, our experimental landscapes have gone a long way toward validating corridors as an important tool for landscape managers. The ultimate function of corridors is to create, link or sustain sources. Our focus on key aspects of the BIDE model, particularly corridor effects on dispersal, provides some assurance that corridors will maintain or rescue source–sink populations, and should increase population persistence. If there has been any surprise from our experiments, it is the consistency with which our results support theory (corridor effects on dispersal and diversity) or well-studied responses to landscape patterns (edges; Table 22.2). When we focus on species in which detailed studies of movement were conducted, the proportion of species that moved more frequently between connected than unconnected patches was remarkable. Also remarkable was the speed with which plant diversity diverged in

TABLE 22.2. *Summary of positive and negative ecological effects of corridors found at the Savannah River Site Corridor Experiment. Key citations are not an exhaustive list.*

Response	Summary of results	Key citations
Positive effects		
Increased dispersal	Corridors increase dispersal for most taxa	Haddad *et al.* (2003); Tewksbury *et al.* (2002); Levey *et al.* (2005)
Increased pollination	Corridors increase pollination rates for several plants	Tewksbury *et al.* (2002); Townsend and Levey (2005)
Increased population persistence	No evidence	NA
Increased diversity	Corridors increase plant diversity	Damschen *et al.* (2006)
Negative effects		
Create edges	Increase or decrease abundances, and predation intensity	Orrock and Damschen (2005); Weldon and Haddad (2005); Haddad and Baum (1999)
Increased dispersal of antagonists:		
- predators	Corridors increase dispersal of animals that are predators or herbivores of other species	Haddad *et al.* (2003); Brinkerhoff *et al.* (2005)
- invasives	Corridors do not spread invasives	Damschen *et al.* (2006)
- disease	No evidence	NA
Synchronize population fluctuations	Evidence from models only	Hudgens and Haddad 2003)
Reduce population persistence	No evidence	NA
Reduce biodiversity	No evidence	NA

connected and unconnected patches (~20% higher in connected patches within 6 years; Damschen *et al.* 2006). We did not observe surprising dynamics, such as negative effects caused by disease, population synchronization that caused extinction, or spread of invasive species (Table 22.2). What we have learned over the last 15 years in our experiment and from other studies is that we should not be distracted by unlikely contingencies; rather, we can concentrate our efforts in mitigating the negative effects of habitat edges while restoring connectivity to fragmented landscapes.

Acknowledgments

This chapter is dedicated to Ron Pulliam, whose mentorship of Nick Haddad led to the creation of this experiment. We have benefited from collaborations with many faculty, postdocs, and students, and would like to extend our particular thanks to Robert Cheney, Brent Danielson, and Sarah Sargent. Our experiment has been made possible by an ongoing partnership with the US Forest Service–Savannah River, especially John Blake, Chris Hobson, Ed Olson, Jim Segar, Kim Wright, and many others. The experiment has been supported by grants from the Department of Agriculture Forest Service–Savannah River, under Interagency Agreement DE-AI09-00SR22188 with the Department of Energy, Aiken, SC, and the National Science Foundation (DEB-0613701, DEB-0613975, DEB-0733746, and DEB-9907365). We thank the graduate class in Connectivity Conservation at NC State University and Chris Ives for comments.

References

Alcantara, J. M. and P. J. Rey (2003). Conflicting selection pressures on seed size: evolutionary ecology of fruit size in a bird-dispersed tree, *Olea europaea*. *Journal of Evolutionary Biology* **16**: 1168–1176.

Anderson, A. B. and C. N. Jenkins (2006). *Applying Nature's Design: Corridors as a Strategy for Biodiversity Conservation*. Columbia University Press, New York.

Battin, J. (2004). When good animals love bad habitats: ecological traps and the conservation of animal populations. *Conservation Biology* **18**: 1482–1491.

Beier, P. and R. F. Noss (1998). Do habitat corridors really provide connectivity? *Conservation Biology* **12**: 1241–1252.

Beier, P., D. R. Majka and W. D. Spencer (2008). Forks in the road: choices in procedures for designing wildland linkages. *Conservation Biology* **22**: 836–851.

Beier, P., K. L. Penrod, C. Luke, W. D. Spencer and C. Cabañero (2006). South coast missing linkages: restoring connectivity to wildlands in the largest metropolitan area in the USA. In *Connectivity Conservation* (K. R. Crooks and M. A. Sanjayan, eds.). Cambridge University Press, Cambridge, UK: 555–586.

Bowne, D. R., J. D. Peles and G. W. Barrett (1999). Effects of landscape spatial structure on movement patterns of the hispid cotton rat (*Sigmodon hispidus*). *Landscape Ecology* **14**: 53–65.

Brinkerhoff, R. J., N. M. Haddad and J. L. Orrock (2005). Corridors and olfactory predator cues affect small mammal behavior. *Journal of Mammalogy* **86**: 662–669.

Chetkiewicz, C. L. B., C. C. S. Clair and M. S. Boyce (2006). Corridors for conservation: integrating pattern and process. *Annual Review of Ecology and Systematics* **37**: 317–342.

Crooks, K. R. and M. A. Sanjayan (2006). *Connectivity Conservation*. Cambridge University Press, Cambridge, UK.

Damschen, E. I., N. M. Haddad, J. L. Orrock, J. J. Tewksbury and D. J. Levey (2006). Corridors increase plant species richness at large scales. *Science* **313**: 1284–1286.

Damschen, E. I., L. A. Brudvig, N. M. Haddad, D. J. Levey, J. L. Orrock and J. J. Tewksbury (2008). The movement ecology and dynamics of plant communities in fragmented landscapes. *Proceedings of the National Academy of Sciences* **105**: 19078–19083.

Danielson, B. J. and M. W. Hubbard (2000). The influence of corridors on the movement behavior of individual *Peromyscus polionotus* in experimental landscapes. *Landscape Ecology* **15**: 323–331.

Earn, D. J. D., S. A. Levin and P. Rohani (2000). Coherence and conservation. *Science* **290**: 1360–1364.

Ewers, R. M. and R. K. Didham (2008). Pervasive impact of large-scale edge effects on a beetle community. *Proceedings of the National Academy of Sciences* **105**: 5426–5429.

Fried, J. H., D. J. Levey and J. A. Hogsette (2005). Habitat corridors function as both drift fences and movement conduits for dispersing flies. *Oecologia* **143**: 645–651.

Gillies, C. S. and C. C. St. Clair (2008). Riparian corridors enhance movement of a forest specialist bird in fragmented tropical forest. *Proceedings of the National Academy of Sciences* **105**: 19774–19779.

Gonzalez, A., J. H. Lawton, F. S. Gilbert, T. M. Blackburn and I. Evans-Freke (1998). Metapopulation dynamics, abundance, and distribution in a microecosystem. *Science* **281**: 2045–2047.

Haddad, N. M. (1999a). Corridor and distance effects on interpatch movements: a landscape experiment with butterflies. *Ecological Applications* **9**: 612–622.

Haddad, N. M. (1999b). Corridor use predicted from behaviors at habitat boundaries. *American Naturalist* **153**: 215–227.

Haddad, N. M. (2000). Corridor length and patch colonization by a butterfly, *Junonia coenia*. *Conservation Biology* **14**: 738–745.

Haddad, N. M. and K. A. Baum (1999). An experimental test of corridor effects on butterfly densities. *Ecological Applications* **9**: 623–633.

Haddad, N. M. and J. J. Tewksbury (2005). Low-quality habitat corridors as movement conduits for butterflies. *Ecological Applications* **15**: 250–257.

Haddad, N. M. and J. J. Tewksbury (2006). Impacts of corridors on populations and communities. In *Connectivity Conservation* (K. R. Crooks and M. A. Sanjayan, eds.). Cambridge University Press, Cambridge, UK: 390–415.

Haddad, N. M., D. R. Bowne, A. Cunningham, B. J. Danielson, D. J. Levey, S. Sargent and T. Spira (2003). Corridor use by diverse taxa. *Ecology* **84**: 609–615.

Hess, G. R. (1994). Conservation corridors and contagious disease: a cautionary note. *Conservation Biology* **8**: 256–262.

Hilty, J. A., W. Z. Lidicker Jr. and A. M. Merenlender (2006). *Corridor Ecology: The Science and Practice of Linking Landscapes for Biodiversity Conservation*. Island Press, Washington, DC.

Howe, H. F., E. W. Schupp and L. C. Westley (1985). Early consequences of seed dispersal for a neotropical tree (*Virola surinamensis*). *Ecology* **66**: 781–791.

Hudgens, B. R. and N. M. Haddad (2003). Predicting which species will benefit from corridors in fragmented landscapes from population growth models. *American Naturalist* **161**: 808–820.

Kinlan, B. P. and S. D. Gaines (2003). Propagule dispersal in marine and terrestrial environments: a community perspective. *Ecology* **84**: 2007–2020.

Laurance, W. F. (2000). Do edge effects occur over large spatial scales? *Trends in Ecology and Evolution* **15**: 134–135.

Leopold, A. (1933). *Game Management*. Charles Scribner's Sons, New York.

Levey, D. J., B. M. Bolker, J. J. Tewksbury, S. Sargent and N. M. Haddad (2005). Effects of landscape corridors on seed dispersal by birds. *Science* **309**: 146–148.

Mabry, K. E. and G. W. Barrett (2002). Effects of corridors on home range sizes and interpatch movements of three small mammal species. *Landscape Ecology* **17**: 629–636.

Mabry, K. E., E. A. Dreelin and G. W. Barrett (2003). Influence of landscape elements on population densities and habitat use of three small-mammal species. *Journal of Mammalogy* **84**: 20–25.

McCallum, H. and A. Dobson (2002). Disease, habitat fragmentation and conservation. *Proceedings of the Royal Society of London Series B – Biological Sciences* **269**: 2041–2049.
McCarty, J., D. Levey, C. Greenberg and S. Sargent (2002). Spatial and temporal variation in fruit use by wildlife in a forested landscape. *Forest Ecology and Management* **164**: 277–291.
Orrock, J. L. (2005). Conservation corridors affect the fixation of novel alleles. *Conservation Genetics* **6**: 623–630.
Orrock, J. L. and E. I. Damschen (2005). Corridors cause differential seed predation. *Ecological Applications* **15**: 793–798.
Orrock, J. L. and B. J. Danielson (2005). Patch shape, connectivity, and foraging by the oldfield mouse, *Peromyscus polionotus*. *Journal of Mammalogy* **86**: 569–575.
Orrock, J. L., B. J. Danielson, M. J. Burns and D. J. Levey (2003). Spatial ecology of predator–prey interactions: corridors and patch shape influence seed predation. *Ecology* **84**: 2589–2599.
Peles, J. D., D. R. Bowne and G. W. Barrett (1999). Influence of landscape structure on movement patterns of small mammals. In *Landscape Ecology of Small Mammals* (G. W. Barrett and J. D. Peles, eds.). Springer-Verlag, New York: 41–62.
Pulliam, H. R. (1988). Sources, sinks, and population regulation. *American Naturalist* **132**: 652–661.
Ries, L., R. J. Fletcher Jr., J. Battin and T. D. Sisk (2004). Ecological responses to habitat edges: mechanisms, models, and variability explained. *Annual Review of Ecology and Systematics* **35**: 491–522.
Schmiegelow, F. K. A., C. S. Machtans and S. J. Hannon (1997). Are boreal birds resilient to forest fragmentation? An experimental study of short-term community responses. *Ecology* **78**: 1914–1932.
Simberloff, D. and J. Cox (1987). Consequences and costs of conservation corridors. *Conservation Biology* **1**: 63–71.
Simberloff, D., J. A. Farr, J. Cox and D. W. Mehlman (1992). Movement corridors: conservation bargains or poor investments? *Conservation Biology* **6**: 493–504.
Tewksbury, J. J., D. J. Levey, N. M. Haddad, S. Sargent, J. L. Orrock, A. Weldon, B. J. Danielson, J. Brinkerhoff, E. I. Damschen and P. Townsend (2002). Corridors affect plants, animals, and their interactions in fragmented landscapes. *Proceedings of the National Academy of Sciences* **99**: 12923–12926.
Townsend, P. A. and D. J. Levey (2005). Do habitat corridors affect pollen transfer? An experimental test. *Ecology* **86**: 466–475.
Weldon, A. J. (2006). How corridors reduce indigo bunting nest success. *Conservation Biology* **20**: 1300–1305.
Weldon, A. J. and N. M. Haddad (2005). The effects of patch shape on indigo buntings: evidence for an ecological trap. *Ecology* **86**: 1422–1431.
Williams, B. K., J. D. Nichols and M. J. Conroy (2002). *Analysis and Management of Animal Populations*. Academic Press, New York.

Part V

Synthesis

JOHN A. WIENS AND BEATRICE VAN HORNE

23

Sources and sinks: what is the reality?

Summary

We consider four "realities" that may affect the usefulness of source–sink theory and thinking as applied to conservation and resource management. First, documenting source–sink systems requires detailed information on demography, dispersal, and habitat selection – information that is hard to come by. Second, source and sink patches are embedded in heterogeneous landscape mosaics, in which the details of spatial relationships are important. Third, source–sink systems are dynamic: what is a source or a sink can change with variations in regional population abundance or environmental conditions. And fourth, considering source–sink systems at too fine a spatial or temporal scale may artificially truncate their dynamics, while viewing them at too broad a scale may obscure or average away the critical interactions between sources and sinks; source–sink dynamics are scale-dependent. The upshot of these realities is that populations in different parts of a landscape may have quite different dynamics, which may change with time or changes in scale. Conserving or managing such populations requires direct or indirect information about landscape-specific demography and dispersal. It also means that protected areas should be established and managed with regard to current *and future* source–sink dynamics, including adequate landscape heterogeneity at a sufficient scale to allow source–sink dynamics to play out.

Introduction

As landscape ecologists, we should enthusiastically embrace the concept of sources and sinks and all that it implies. After all, landscape ecology deals

Sources, Sinks and Sustainability, ed. Jianguo Liu, Vanessa Hull, Anita T. Morzillo and John A. Wiens. Published by Cambridge University Press.

with the effects of spatial patterns on ecological processes, and what more clearly epitomizes this than an array of habitat patches among which population dynamics differ, while the fates of populations in individual patches are influenced by the dynamics of other patches in the landscape? But we are also conservation scientists, so we look at concepts with an eye toward their applicability to real-world resource management and conservation as well. These differing perspectives on sources and sinks do not always smoothly mesh with one another, thus creating tensions that sometimes border on schizophrenia.

What is the reality (or usefulness) of the concept of sources and sinks? Is it an intellectual plaything, serving as fodder for theory and models and analyses, or is it a valuable tool that can provide insights and guidance about how populations might be managed and conserved more effectively? The concept itself, of course, is not new, nor was it really new when Pulliam's signature paper was published more than two decades ago (Pulliam 1988). Lidicker (1975), Van Horne (1983) and others had expressed much the same idea previously, and one could probably find the notion hiding among the writings of Charles Darwin or R. A. Fisher, or certainly Sewall Wright. The great contribution, and the beauty, of Pulliam's paper was to frame these ideas in the formalisms of population theory and models, providing the soil and nutrients in which elaborations and extensions of the ideas could take root and blossom. And blossom they have, as the varied contributions to this volume exemplify. The citation statistics reviewed by Hull *et al.* (Chapter 1, this volume) merely hint at the impact that Pulliam's crystallization of the source–sink concept has had in, and beyond, population ecology.

But we have not come to praise Pulliam's work, nor (unlike Mark Antony in *Julius Caesar*) to bury it. Rather, we would like to draw attention to four realities that may constrain source–sink theory and then consider its potential applicability to real-world conservation and resource management. We will approach this more from the viewpoint of our personal perspectives than as a traditional synthesis or review richly annotated with literature citations. This is not to imply that our thoughts are new or original; indeed, they probably occurred to Darwin or Fisher or Wright, and they are embedded in many of the contributions to this volume. But these realities are not always recognized in attempts to apply the source–sink concept to conservation and resource management, so there may be value in highlighting them.

Reality 1: It takes a lot of information

The thinking behind the source–sink concept is simple: populations in some habitat patches (i.e., sources) do well, producing offspring in excess of those necessary to maintain a stable population within those patches. Some of these excess individuals emigrate, and some of them end up in other habitat patches (i.e., sinks) in which production is inadequate to maintain population numbers. Populations in the sink patches persist by virtue of replenishment from the source patches in the landscape.

All well and good, in theory. Documenting the extent to which patches act as sources or sinks in a real-world landscape and whether the overall dynamics accord with source–sink models, however, requires information that is not easily obtained, especially for the small or fragmented populations that are often the target of conservation efforts. One must begin with some knowledge of patch-specific demography – identifying which patches or habitat types in a landscape support what levels of net potential recruitment (simply, births minus deaths per time period). Population size or density in a patch or habitat is not by itself the relevant statistic, even though this is what is most easily and most often recorded. It is the relation between net recruitment and population size, not population size *per se*, that determines source/sink value (i.e., whether $\lambda > 0$). This is what underlies Van Horne's (1983) observation that density can be a misleading indicator of habitat quality.

If information about patch-specific demography is difficult to obtain, that about dispersal is even more so. Yet dispersal is the key to source–sink dynamics. If individuals cannot or do not disperse from a patch, then the dynamics envisioned in source–sink models simply do not occur and the concept itself becomes irrelevant. If dispersal is ignored or assumptions about dispersal are incorrect, sources and sinks may be misclassified, leading to incorrect conclusions about source–sink dynamics and their conservation implications (Pulliam *et al.*, Chapter 9, this volume).

Virtually all species include some form of dispersal in their life-history strategy, yet dispersal is rarely observed directly and is correspondingly difficult to measure (although new technologies for tracking individual movements over long distances are changing this). The factors that determine whether individuals disperse or not, which individuals disperse, how far they go, and where they end up are largely unknown for all but a few populations. To keep things simple, metapopulation models may assume that dispersal or settlement probability is a decreasing function of interpatch distance (i.e., a diffusion process; Okubo 1980; Turchin 1998). This may be fine, depending on the objectives of the modeling exercise, but understanding real-world population dynamics will require incorporating landscape heterogeneity into dispersal behavior (see Heinz *et al.* 2006). Source and sink patches are embedded in landscape mosaics composed of elements that may facilitate or hamper the movement of individuals and whose boundaries may be differentially permeable to dispersers (Wiens *et al.* 1993). As a result, the movement of dispersing individuals through a landscape may be anything but linear (Wiens 2001), affecting both the distance traveled by an individual and its probability of encountering another source or sink patch. If the composition of a landscape changes as a result of natural processes such as fire, or human land uses such as agriculture or forestry, dispersal probabilities and the linkages between patches may be altered, even if conditions within those patches remain unchanged.

Conservationists and land managers are increasingly advocating corridors or other forms of habitat connectivity as ways of enhancing the movement of individuals among otherwise isolated habitat patches in a fragmented landscape (see Crooks and Sanjayan 2006; Hilty *et al.* 2006). If properly designed, connectivity networks can facilitate dispersal among source and sink patches, contributing to source–sink dynamics and the persistence of metapopulations, and even turning some sinks into sources. As Haddad *et al.* (Chapter 22, this volume) note, however, the value of corridors as dispersal conduits may be compromised by edge effects, even when corridors are seemingly broad. More to the point, it is not just whether individuals use corridors or even whether dispersal is enhanced that is important, but how population demography and persistence are affected by the greater connectivity. These outcomes are more often assumed than known.

Dispersal is driven by the factors that prompt individuals to leave a patch (emigration) and its consequences are determined by the processes that lead individuals to settle in another patch (immigration). Emigration may be prompted by internal factors such as physiology, sex, or age status ("innate dispersal") or environmental factors such as food supply or competition ("environmental dispersal"; Howard 1960; Ims and Hjermann 2001). The former is generally density independent and the latter density dependent; the difference may have important effects on whether, how, and how many individuals of what status disperse from a source patch to serve as potential replenishment for sinks. Whether or not a dispersing individual settles in a particular patch may be determined by the density of individuals already present in a patch (if patch choice is density dependent) and by the factors underlying the behavioral process of habitat selection (Pulliam and Danielson 1991; Morris, Chapter 3, this volume). Ideally, individuals should select habitats on the basis of their individual fitness prospects there (even in sink habitats in which such prospects are low); this is the foundation of ideal free distribution theories (Fretwell and Lucas 1969). Other factors, such as the attraction of dispersers to patches already occupied by other individuals (conspecific attraction) or habitat associations formed through early experience (habitat imprinting; Davis and Stamps 2004), can distort such theoretical expectations, leading individuals to select particular sink patches ("attractive sinks"; Diez and Giladi, Chapter 14, this volume) even when better options are available.

It may seem obvious, but the demographic effects of dispersal depend largely on the demographic contribution of successful dispersers. If, for instance, female numbers drive population reproductive rates, then obtaining detailed information on male dispersal might not be very useful.

Life history, dispersal, density-dependent patch occupancy, and habitat selection – these are all factors that determine whether particular places may function as sources or sinks and whether source–sink dynamics play out as projected by the theories and models. All are difficult to measure in natural populations, although the contributions to this volume indicate that it is possible to obtain detailed information on some of these things for some populations under some circumstances. We will return later to consider whether the difficulty in obtaining such information hopelessly compromises the applicability of source–sink concepts to conservation and management.

Reality 2: Sources and sinks are parts of heterogeneous landscapes

Although it is easy (and perhaps appropriate) in a model to portray sources and sinks as featureless blobs floating in a dimensionless space, the reality is very different. Patches are usually embedded in a complex landscape mosaic, and if there is any single message that emerges from the discipline of landscape ecology it is that the spatial arrangement of landscape structure and composition matters. We noted above how landscapes may affect dispersal pathways and probabilities, but there are other ways in which landscape features become important when considering source–sink dynamics. What goes on within a patch, for example, may be influenced by what is in the landscape immediately surrounding the patch. Benkman and Siepielski (Chapter 4, this volume) note how the presence or absence of a predator, competitor, or mutualist in a patch may affect the population dynamics of a target species within that patch, and thus the potential of the patch to act as a source or a sink. The same may hold for areas adjacent to a patch; within-patch dynamics of a species, for example, may be affected by the presence in adjoining patches of predators that make foraging forays into the patch of interest. This is why populations in protected areas may be vulnerable to forces external to the area (Janzen 1983; Hansen, Chapter 16, this volume). Likewise, the effectiveness of corridors involves more than strips of habitat connecting otherwise isolated patches. Network analysis and graph theory suggest that *how* patches in a landscape are interconnected – the configuration or topology of connectivity – may affect transmission rates (i.e., dispersal probabilities) over the landscape as a whole (Jordán, Chapter 12, this volume). And clearly, the addition or loss of source or sink patches from a landscape can affect dispersal patterns, immigration rates to patches, and whether a given patch remains a source or a sink. Conversely, wide-ranging species such as raptors or mountain lions may forage in areas outside the habitat patches where they nest or den, in which case reproductive success may depend on access to different types of patches.

All of this is to say that, to be useful, source–sink models and studies should be spatially explicit. The locations of particular source and sink patches, and the characteristics of other places in a landscape, may affect how or whether potential source–sink systems function as expected. Landscape ecology offers a variety of tools and metrics (e.g., GIS, remote sensing, spatial statistics, FRAGSTATS; Fortin and Dale 2005; Wiens and Moss 2005) for describing and analyzing landscapes; these should be incorporated into the framework of source–sink research and applications.

Reality 3: Source–sink systems are dynamic

When Pulliam (1988) formulated his model of source–sink dynamics, he did not consider temporal variation. Sources and sinks were assumed to be fixed: once a source (or sink), always a source (or sink). Indeed, one of the primary messages of source–sink theory was that this sort of population structure can enhance overall population stability by damping fluctuations and increasing the probability of population persistence. Pulliam was keeping it simple.

Even if the broad conclusion that a source–sink population structure enhances population stability were to hold, however, this does not mean that individual source and sink patches are necessarily unchanging. Recent modeling has begun to incorporate temporal dynamics more explicitly, although perhaps not as fully as might be needed. We noted above that the source/sink characteristics of a patch may be affected by the surrounding landscape. If this is so, then changes in either the composition or the structural configuration of a landscape might be expected to affect population demographics in particular patches, and thus their propensity to export or import dispersing individuals. Changing environmental conditions, such as a drought or an unusual summer frost (Pearson and Fraterrigo, Chapter 6, this volume; Matter and Roland, Chapter 15, this volume), can also alter the relations between sources and sinks in a landscape. Patches may serve as sinks during periods of high metapopulation densities, but remain stable – or even serve as sources – at lower population densities ("pseudo-sinks"; see Watkinson and Sutherland 1995).

One example of these dynamics is particularly instructive. Van Horne and colleagues (Van Horne *et al.* 1997) conducted a 4-year study of Townsend's ground squirrel (*Spermophilus townsendii*) population dynamics in several study plots in an Idaho shrubsteppe mosaic of sagebrush (*Artemisia*) and grassland (*Poa*) patches. Through intensive capture–recapture analysis, they were able to derive habitat-specific reproduction and survival estimates. Conditions were good during the first two years of the study, and population densities, reproductive output, and overwinter survival of adults and young were greater in grassland than in sagebrush-dominated habitats. This was followed by an extraordinarily severe 1½-year drought that had disproportionately large effects on populations in grassland habitats. Reproduction was nil, overwinter survival of adults was low, few – if any – juveniles survived, and densities plummeted. Populations in sagebrush habitats were also affected, but reproduction and overwinter survival were greater than in grasslands, and populations recovered from the effects of the drought more rapidly. In short, the potential to act as a population source shifted from grasslands to sagebrush in association with the change in weather conditions. As Holt *et al.* (2003) have shown theoretically and experimentally, temporal synchronization (autocorrelation) of population dynamics over large areas can lead to situations in which abundance in "sink" populations may be temporarily inflated during environmentally benign times, potentially altering competitive relationships and creating conditions in which the "sinks" act as sources.

Obtaining this critical population and demographic information required a massive effort, yet Van Horne and her colleagues were unable to continue their studies long enough to determine whether population recruitment in the sagebrush areas provided immigrants to replenish populations in grassland patches, in a true source–sink fashion. The study nonetheless provided the demographic information needed to determine the shifts in the potential of the habitats to act as sources or sinks in response to environmental variations. It was also clear that densities in the habitat types were not satisfactory indicators of potential habitat quality (i.e., reproduction and survival) under changing conditions.

Empirical studies and models both suggest that habitats normally regarded as population sinks may play an important role in the overall dynamics of populations in variable environments, by serving as refugia that may "rescue" populations in source habitats during stress periods such as droughts (Foppen *et al.* 2000; Frouz and Kindlmann 2001; Falcy and Danielson, Chapter 7, this volume). This suggestion bears on the somewhat arcane debate about whether or not sink habitats should be included in the determination of a species' ecological niche. Pulliam (1988) originally suggested that because, by definition, $\lambda < 1$ in sink habitats, such conditions do not fall within the fundamental niche of a species. Including sink habitat conditions in the definition of a species' niche would therefore make the realized niche occupy a greater niche space than the fundamental niche. According to Hutchinson's (1957) original niche formulation, this is not supposed to happen. On the other hand, if sinks can at times function as sources, or if populations in either source or sink patches (and their habitats) suffer local extinctions and subsequent recolonizations (as envisioned in metapopulation theory), then definition of niche envelopes based only on the environmental correlates of occupied areas at a given time will provide an incomplete picture of the actual niche space of a species (Pulliam 2000). Such considerations become relevant in the context of modeling how species' distributions may change in response to climate change (e.g., Elith and Leathwick 2009; Wiens *et al.* 2009).

Reality 4: Source–sink dynamics are scale-dependent

The definition of what is a source and what is a sink, and an understanding of the consequences of interactions between them, depends on the spatial and temporal scales at which they are viewed. For example, if sources and sinks are dynamic, changing in status over time in response to environmental changes such as drought, then a study conducted at a short temporal scale would provide only a snapshot of these dynamics. A patch or habitat would be only source or sink. If Van Horne and her colleagues had confined their studies to a single year, they would have concluded that grasslands were always a source or always a sink, depending on the conditions at the time of their study. The broader temporal scale allowed them to see the shifts in the potential source or sink status of habitats at play.

Scale and its effects have become frequent topics of discussion among ecologists, conservationists, and resource managers. The emerging message is that virtually everything in ecology is scale-dependent. This means that the processes that drive population dynamics change with the spatial and temporal scale of resolution. As we just noted, whether one sees a population persisting, disappearing, or winking in and out of existence (or acting as a source or a sink) depends on the scale of temporal resolution. It also depends on the spatial scale of resolution. What is a "habitat patch," and therefore a separate unit for demographic analysis as a source or a sink, may change with the grain (fineness of resolution) or extent (breadth of area) considered in a study. (Most models, being spatially dimensionless, do not suffer from this problem.) How we perceive or measure the structure and composition of a landscape that contains putative source and sink patches depends on the scale of resolution of landscape pattern. This, in turn, will influence how we factor such things as landscape connectivity or permeability and their effects on dispersal into considerations of source–sink dynamics.

The need to define bounded populations in order to calculate population parameters may cause us to misunderstand populations that are essentially continuously distributed across habitats of varying quality. Individuals in such populations may move among these habitats depending on environmental conditions. Defining genetic neighborhoods in such populations (e.g., Antolin *et al.* 2001) can be an initial step in understanding this flexibility.

Rather than ignoring such potential complications of scale or silently putting up with them, ecologists should develop ways of dealing with them. The starting point is to recognize that how *we* view a system imposes *our* anthropogenic scales on the system. It is perhaps trite to observe that a beetle, a bird, and a bison all view and respond to the same prairie habitat at vastly different scales, so when an investigator decides to study the system at some arbitrary scale (square-meter samples, for example), the system and its dynamics may be artificially truncated or submerged.

So how should one determine the appropriate scale(s) for assessing source–sink dynamics? A simple rule of thumb is to let the organisms and landscapes make the decision. This is much easier said than done. Diez and Giladi (Chapter 14, this volume) suggest that the choice of scale(s) for a particular study system should be based on the grain of habitat heterogeneity and the dispersal distances of the target organisms (which, as noted above, are interrelated). The finest grain at which heterogeneity is detected (by the organisms) sets a lower limit to the relevant scale, while the upper limit may be set by species mobility and dispersal. The former may determine what is a "patch," whereas the latter determines the distances over which populations may be demographically connected. Setting aside the bothersome detail that dispersal distances and their sensitivity to landscape patterns are rarely known, the latter measure makes good sense: habitat patches that lie beyond the reach of dispersing individuals cannot, by definition, be functional parts of a source–sink system. Using the grain of heterogeneity to determine the lower scale boundary, however, is more problematic. As Pearson and Fraterrigo (Chapter 6, this volume) note, there may be considerable heterogeneity within habitat patches, especially for organisms such as plants that respond strongly to edaphic variations and have limited mobility. If the broadly defined habitat patches are larger than the dispersal distance of individuals, such within-patch heterogeneity may create source–sink dynamics *within* individual patches.

Such propositions relate to the question of "what is a patch?" The habitat patches that are relevant to source–sink considerations should be defined demographically: population parameters within the patch differ from those in the surrounding landscape or other habitat patches. Thus, to determine the relevant scale(s) for investigating source–sink systems in the real world, information on dispersal and within-patch demography seems necessary, bringing us full circle to the need for lots of information in order to assess source–sink dynamics.

What are the implications for conservation and resource management?

Most studies of source–sink systems have relied on models. Few have gathered empirical information from real populations, and still fewer have addressed the conservation or resource management implications. Given the spatial heterogeneity and increasing fragmentation of most environments, the models lead us to expect that source–sink relations may be common in nature. Models tell us what *could occur*, given certain information and assumptions; however, they do not tell us what *is occurring*. That is the information needed to guide conservation and management efforts. The overarching implication of source–sink theory is that populations in different patches or habitat types in a landscape may have quite different dynamics, some sustainable in isolation and others not. Managing populations of a species within habitat patches (e.g., natural reserves or protected areas) or across broader landscapes and predicting how they will respond to management actions or habitat alterations requires that demographic rates be linked to specific landscape features or environmental conditions (Etterson *et al.*, Chapter 13, this volume).

Obtaining such information is difficult, but not impossible. Management plans for northern spotted owls (*Strix occidentalis caurina*) in the Pacific Northwest or northern goshawks (*Accipiter gentilis*) in northern Arizona were based on detailed demographic information gathered over multiple years (Franklin *et al.* 1996; Wiens *et al.* 2006). The effort required to obtain this information was massive, however, and it was possible only because these species were of particular conservation concern and conservation needs clashed with forestry practices, generating both legal action and funding to support the research. Even so, it was not possible in either study to unambiguously define stable source or sink habitats or populations on which management efforts could be focused.

Most species do not benefit from such attention. Is source–sink thinking therefore irrelevant to most situations? Perhaps. But some of the insights emerging from source–sink studies may be useful in framing broad approaches to conservation and management, even if the information needed to address specific situations and species is lacking. One clear message is that conservation and management efforts should be focused on source populations and habitats, as these represent (in theory) the most stable elements of the broader population and are critical to ensuring its long-term persistence. The flip-side of this proposition is the assumption (usually implicit) that areas set aside for protection do in fact harbor source elements of a population. Protected areas in the USA are located disproportionately in environmentally harsh areas (e.g., high elevations, deserts, low-productivity soils; Scott *et al.* 2001), however, and one might expect these to be sinks more often than sources (Hansen, Chapter 16, this volume). They may also be more susceptible to the effects of climate change.

As modeling and thinking about sources and sinks has developed from the foundation established by Pulliam (1988), it has become increasingly apparent that sinks may have important conservation value, especially in variable environments. We have already noted the potential role of sinks in replenishing populations in source habitats and how sinks may become sources if conditions change. The modeling analyses of Falcy and Danielson (Chapter 7, this volume) suggest that under conditions of frequent disturbance, management aimed at reducing the rate of population decline may have a greater influence on the persistence of the population as a whole than efforts to increase the rate of habitat recovery in source areas. In addition, if populations intermittently disappear from sink habitats, managers may regard the unoccupied habitat as unsuitable and divert management resources from such areas or abandon them altogether. This is especially likely to occur if the time scale of management is short or the spatial scale is mismatched with the dynamics of the source–sink system.

So we are faced with a conundrum: distinguishing source and sink habitats (or the different dynamics of populations in such habitats) may be important to effective conservation, especially in variable environments, yet making this distinction may require information far in excess of what can be obtained with available resources. Short of forgetting about such distinctions and ignoring source–sink dynamics, the alternative may be to search for measures that can serve as surrogates for the demographic and dispersal information that is so difficult to obtain. For some time, ecologists and managers believed that population density within a habitat could serve as such a surrogate measure, but this notion has been rather thoroughly discounted. Dunning *et al.* (Chapter 11, this volume) have suggested that (for birds, at least) measures such as territory occupancy, age-class ratios of new breeders, and production of emigrants could serve as surrogates for the demographic and dispersal parameters of source–sink models. The efficacy of such surrogate measures should be validated in populations in which the actual demography and dispersal functions are known (no easy task!). Increased use of non-invasive techniques of genetic analysis may allow us to obtain better estimates of effective dispersal. New developments in technologies for tracking individual movements can also contribute. If the relevance of source–sink thinking and theory to on-the-ground conservation and management is to be maintained, good surrogate measures must be developed and validated. We should bear in mind, however, the admonition of Dunning *et al.* (Chapter 11, this volume): just because something is easy to measure does not make it the right thing to measure.

The applicability of source–sink theory to conservation and management of actual populations in real habitats remains to be shown. But the ideas and the conclusions from modeling are too important to be ignored. "If we do not understand how organisms use the landscape … how they move, how they interact in space and behave in new community contexts, we probably have no chance of achieving the right conservation practice" (Jordán, Chapter 12, this volume).

Acknowledgments

We appreciate thoughtful comments on an earlier version of this chapter from Ron Pulliam and Bob Holt.

References

Antolin, M. F., B. Van Horne, M. D. Berger Jr., A. K. Holloway, J. L. Roach and R. D. Weeks Jr. (2001). Effective size and genetic structure of a Piute ground squirrel population (*Spermophilus mollis*). *Canadian Journal of Zoology* **79**: 26–34.
Crooks, K. R. and M. Sanjayan (2006). *Connectivity Conservation*. Cambridge University Press, Cambridge, UK.
Davis, J. M. and J. A. Stamps (2004). The effect of natal experience on habitat preferences. *Trends in Ecology and Evolution* **19**: 411–416.
Elith, J. and J. Leathwick (2009). Species distribution models: ecological explanation and prediction across space and time. *Annual Review of Ecology, Evolution and Systematics* **40**: 677–697.
Foppen, R. P. B., J. P. Chardon and W. Liefveld (2000). Understanding the role of sink patches in source–sink metapopulations: reed warbler in an agricultural landscape. *Conservation Biology* **14**: 1881–1892.
Fortin, M.-J. and M. R. T. Dale (2005). *Spatial Analysis: A Guide for Ecologists*. Cambridge University Press, Cambrige, UK.
Franklin, A. B., R. J. Gutierrez, B. R. Noon and J. P. Ward Jr. (1996). Demographic characteristics and trends of northern spotted owl populations in northwestern California. In *Demography of the Northern Spotted Owl* (E. D. Forsman, S. DeStefano, M. G. Raphael and R. J. Gutierrez, eds.). Studies in Avian Biology No. 17, Allen Press, Lawrence, KS: 83–91.
Fretwell, S. D. and H. L. Lucas (1969). On territorial behavior and other factors influencing habitat distribution in birds. I. Theoretical development. *Acta Biotheoretica* **19**: 16–36.
Frouz, J. and P. Kindlmann (2001). The role of sink to source re-colonisation in the population dynamics of insects living in unstable habitats: an example of terrestrial chironomids. *Oikos* **93**: 50–58.
Heinz, S. K., C. Wissel, L. Conradt and K. Frank (2006). Integrating individual movement behavior into dispersal functions. *Journal of Theoretical Biology* **245**: 601–609.
Hilty, J. A., W. Lidicker Jr. and A. Merenlender (2006). *Corridor Ecology: The Science and Practice of Linking Landscapes for Biodiversity Conservation*. Island Press, Washington, DC.
Holt, R. D., M. Barfield and A. Gonzalez (2003). Impacts of environmental variability in open populations and communities: "inflation" in sink environments. *Theoretical Population Biology* **64**: 315–330.
Howard, W. E. (1960). Innate and environmental dispersal of individual vertebrates. *American Midland Naturalist* **63**: 152–161.
Hutchinson, G. E. (1957). Concluding remarks. *Cold Spring Harbor Symposia on Quantitative Biology* **22**: 415–427.
Ims, R. A. and D. Ø. Hjermann (2001). Condition-dependent dispersal. In *Dispersal* (J. Clobert, E. Dachin, A. A. Dhondt and J. D. Nichols, eds.). Oxford University Press, Oxford, UK: 203–216.
Janzen, D. H. (1983). No park is an island: increase in interference from outside as park size decreases. *Oikos* **41**: 402–410.
Lidicker, W. Z., Jr. (1975). The role of dispersal in the demography of small mammals. In *Small Mammals: Their Productivity and Population Dynamics* (F. B. Golley, K. Petrusewicz and L. Ryszkowski, eds.). Cambridge University Press, Cambridge, UK: 103–128.
Okubo, A. (1980). *Diffusion and Ecological Problems: Mathematical Models*. Springer-Verlag, Berlin.
Pulliam, H. R. (1988). Sources, sinks, and population regulation. *American Naturalist* **132**: 652–661.

Pulliam, H. R. (2000). On the relationship between niche and distribution. *Ecology Letters* **3**: 349–361.
Pulliam, H. R. and B. J. Danielson (1991). Sources, sinks, and habitat selection: a landscape perspective on population dynamics. *American Naturalist* **137**(Suppl.): S50–S66.
Scott, J. M., R. J. F. Abbitt and C. R. Groves (2001). What are we protecting? The United States Conservation Portfolio. *Conservation Biology in Practice* **2**: 18–19.
Turchin, P. (1998). *Quantitative Analysis of Movement: Measuring and Modeling Population Redistribution in Animals and Plants*. Sinauer Associates, Sunderland, MA.
Van Horne, B. (1983). Density as a misleading indicator of habitat quality. *Journal of Wildlife Management* **47**: 893–901.
Van Horne, B., G. S. Olson, R. L. Schooley, J. G. Corn and K. P. Burnham (1997). Effects of drought and prolonged winter on Townsend's ground squirrel demography in shrubsteppe habitats. *Ecological Monographs* **67**: 295–315.
Watkinson, A. R. and W. J. Sutherland (1995). Sources, sinks and pseudo-sinks. *Journal of Animal Ecology* **64**: 126–130.
Wiens, J. A. (2001). The landscape context of dispersal. In *Dispersal: Individual, Population, and Community* (J. Clobert, E. Danchin, A. A. Dhondt and J. D. Nichols, eds.). Oxford University Press, Oxford, UK: 96–109.
Wiens, J. and M. Moss (eds.) (2005). *Issues and Perspectives in Landscape Ecology*. Cambridge University Press, Cambridge, UK.
Wiens, J. A., N. C. Stenseth, B. Van Horne and R. A. Ims (1993). Ecological mechanisms and landscape ecology. *Oikos* **66**: 369–380.
Wiens, J. A., D. Stralberg, D. Jongsomjit, C. A. Howell and M. A. Snyder (2009). Niches, models, and climate change: assessing the assumptions and uncertainties. *Proceedings of the National Academy of Sciences* **106**(Suppl. 2): 19729–19736.
Wiens, J. D., B. R. Noon and R. T. Reynolds (2006). Post-fledging survival of northern goshawks: the importance of prey abundance, weather, and dispersal. *Ecological Applications* **16**: 406–418.

Index